Review of Neuroscience

BEN PANSKY, Ph.D., M.D.
Professor of Anatomy,
Medical College of Ohio,
Toledo, Ohio

DELMAS J. ALLEN, Ph.D.
Professor and Associate Dean,
College of Health Sciences,
Georgia State University,
Atlanta, Georgia

G. COLIN BUDD, Ph.D.
Professor of Physiology,
Medical College of Ohio,
Toledo, Ohio

Illustrations by Ben Pansky, Ph.D., M.D.

Review of Neuroscience

Second Edition

Macmillan Publishing Company
New York

Collier Macmillan Canada, Inc.
Toronto

Collier Macmillan Publishers
London

Macmillan Publishing Company
866 Third Avenue, New York, New York 10022

Collier Macmillan Canada, Inc.
Collier Macmillan Publishers • London

Library of Congress Cataloging-in-Publication Data

Pansky, Ben.
  Review of neuroscience.

  Bibliography: p.
  Includes index.
  1. Neurology. I. Allen, Delmas J. II. Budd, G.
Colin. III. Title. [DNLM: 1. Nervous System—
anatomy & histology. 2. Nervous System—physiology.
WL 101 P196r]
QP355.2.P36   1988   612'.8   87-34772
ISBN 0-02-390611-1

Printing: 1 2 3 4 5 6 7 8   Year: 8 9 0 1 2 3 4 5 6

Preface to the Second Edition

This book has been written for students and practitioners in the health sciences as well as others who require a visual comprehension, important facts, and an overview of the basic concepts of modern neuroscience.

The overwhelming amount of material necessary for a fundamental understanding of the nervous system becomes evident to the student approaching the neurosciences for the first time. The rapid progress in the field, research, continual evaluation and reevaluation of new data and information, and the mass of material available make one's ability to "keep up" exceedingly difficult and review almost impossible. Basic anatomic and functional aspects of the "neurologic body" become buried as time and clinical concepts push basic knowledge into the background, i.e., as one proceeds from studies as a beginning student to busy schedules as a clinician, graduate student, or teacher. Although the student of neuroscience continues to expand and reinforce his information as an essential part of his work, others in related fields are often overwhelmed and left behind, even in areas that deal with subject matter that relies on information relating to the nervous system.

Review of Neuroscience is presented in a concise, simplified, and modified outline form but is, nevertheless, complete. With only a few exceptions, illustrations appear on right-hand pages and corresponding text on opposite left-hand pages. The terminology used throughout is an anglicized version of the standard nomenclature (*Nomina Anatomica,* 5th ed.) approved at the Eleventh International Congress of Anatomists, held at Mexico City in 1980. The text is functionally oriented and clinically informative, and emphasizes fundamentals of structure as well as function. In this second edition, the authors have made a concerted effort to bring together a comprehensive review of the major neuroscience areas—namely, neuroembryology, microscopic and gross neuroanatomy, neurophysiology, and neuropathology. The basic aspects of each area have been carefully reviewed, reorganized, and reevaluated to present a clear and concise format for learning, an opportunity for review, and incentive for further study. Greater emphasis has been placed on understanding the physiologic and functional aspects of neuroscience, since full appreciation of the beauty of structure is dependent upon knowledge of both normal and abnormal function.

The text is divided into three parts. Part I, "Structural and Functional Basis of Neuroscience," includes developmental neuroanatomy, gross and microscopic neuroanatomy, and material usually found in the standard textbooks of neuroanatomy, i.e., cranial nerves, diencephalon, basal ganglia, autonomic nervous system, cerebellum, reticular formation, limbic and olfactory systems, and organs of special sensation. To further enhance one's understanding of the anatomic structure of the central nervous system as it relates to clinical studies, special sections on computerized tomography (CT) and magnetic resonance imaging (MRI) of the brain have been included since these procedures have become standard tools in most hospitals and clinical practice. In addition, there is an expanded unit dealing with the functional cerebral cortex, in which there are relevant discussions of cortical physiologic and integrated functional concepts dealing with the aphasias, cortical syndromes, language, learning and memory, consciousness, sleep, and migraine.

Part II, "Physiologic Mechanisms of Neuroscience," has been changed from the first edition and now deals with neurophysiology and physiologic activities of the nervous system as well as the fundamental basis of sensory and motor pathways, including skeletal muscle regulation. Once the reader has acquired a basic understanding of structure and function he/she can then appreciate alterations in the nervous system that result from developmental changes, trauma, and disease processes. Part III, "Clinical Neuroscience," deals with the essential clinical aspects of neurology and neuropathology. As each part is subdivided into essential units that comprise the specific subject matter of that part, an automatic outline is created that makes study easier, comprehension more reasonable, and learning more organized. Of importance is the fact that although the parts and units of the text can be studied in sequence, they may also be studied independently depending on the particular course one is involved with.

Appendix I, "Atlas of the Brainstem," should enhance the student's overall comprehension and emphasize the need for an understanding of structure as the basis for appreciating function. Most neuroscience courses require study of comparable sections as a necessary background for understanding neurologic pathways—their location, interruption, and clinical significance. Appendix II is an expanded glossary, and Appendix III is an up-to-date selection of essential books, monographs, and papers that can be used as a "starting-point" in one's study of the nervous system.

Students are usually overwhelmed by excess explanation, references, and minute details and thereby lose sight of the real essentials of the subject. Thus, we have used a modified outline format that includes only the necessary words needed to convey basic thoughts and ideas. In addition, since illustrations play such an important role in the visualization and understanding of the biomedical sciences, over 1,000 original line-cut drawings, as well as numerous photomicrographs and photographs, were created for this text.

The authors would like to express their appreciation to the students at the Medical College of Ohio, whose comments and suggestions have helped to make this book possible. We are grateful to Dr. Frank Redmond, who provided a collection of gross and microscopic specimens that were used in Part III; to Dr. A. John Christoforidas, Chairman and Professor of Radiology at Ohio State University, and Dr. Edward Savolaine, Department of Radiology, Medical College of Ohio, for their help in the selection and interpretation of the CT scans and MRIs; and to Ms. Faye Keen, Library Media Technician Assistant in the Department of Radiology, who assisted with the photographic work necessary to present the CT scans and x-rays offered by the department.

The authors also wish to express their appreciation and obligations to the publisher and particularly to Ms. Joan C. Zulch, Vice President, for her infinite patience, continuous enthusiasm, courtesy, and support in all matters related to this textbook.

B.P
D.J.A.
G.C.B.

Contents

Unit Three. CRANIAL NERVES

Unit Four. THE DIENCEPHALON

Unit Five. THE BASAL GANGLIA

Unit Six. FUNCTIONAL CEREBRAL CORTEX

Unit Seven. AUTONOMIC NERVOUS SYSTEM

Unit Eight. THE CEREBELLUM

Unit Nine. THE RETICULAR FORMATION; LIMBIC AND OLFACTORY SYSTEMS

Unit Ten. ORGANS OF SPECIAL SENSATION

PART II. PHYSIOLOGIC MECHANISMS OF NEUROSCIENCE

Unit Eleven. NEUROPHYSIOLOGY

Unit Twelve. SENSORY AND MOTOR PATHWAYS, SKELETAL MUSCLE REGULATION, AND SPINAL CORD LESIONS

PART III. CLINICAL NEUROSCIENCE

Unit Thirteen. INTRACRANIAL VASCULAR ACCIDENTS, TRAUMA, AND EPILEPSY

Unit Fourteen. CONGENITAL MALFORMATIONS AND PERINATAL PATHOLOGY

Unit Fifteen. DEMYELINATING, HEREDITARY, AND FAMILIAL DISEASES

Unit Sixteen. NEUROLOGIC DISTURBANCES WITH SKELETAL ABNORMALITIES; DEGENERATIVE DISEASES

PART I

Structural and Functional Basis of Neuroscience

DEVELOPMENT OF
THE NERVOUS
SYSTEM

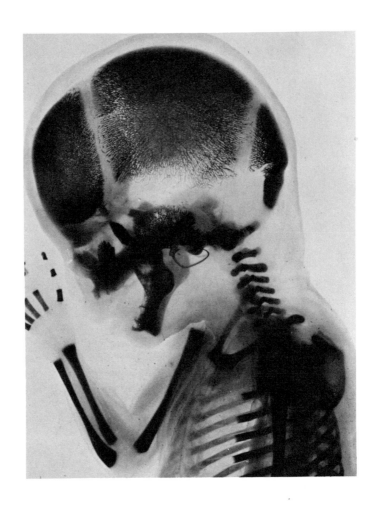

1. EMBRYOLOGY OF THE CENTRAL NERVOUS SYSTEM: GENERAL DEVELOPMENT

I. **Central nervous system (CNS)** appears at the beginning of the 3rd week of development as an elongated plate of thickened ectoderm, the neural or medullary plate

A. THE PLATE is found in the middorsal region in front of the primitive pit occupying the anterior portion of the embryonic disk

   1. The lateral edges of the plate soon elevate to form the neural folds; the depressed longitudinal furrow between the folds is the neural or medullary groove

B. THE NEURAL FOLDS elevate with development, approach each other in the midline, fuse, and form the neural tube, which then loses connection to overlying ectoderm

   1. Alongside the developing neural tube are strips of mesoderm which begin to show segmentation into somites near the end of the 3rd week of development

      a. The 1st somite appears behind the cephalic tip of the notochord

      b. Somites appear in a craniocaudal sequence, and 42–44 pairs are present by day 31

      c. Most of the axial skeleton and musculature develop from the somites

   2. Closure of the neural tube begins near the 4th somite (future cervical level) and proceeds both cephalically and caudally

      a. At the cranial and caudal ends of the embryo, the fusion is delayed and the anterior and posterior neuropores temporarily form open connections between the lumen of the neural tube and the amniotic cavity

         i. The anterior neuropore indicates the lamina terminalis (most cephalic limit of developing brain) and closes at the 25th day (18–20 somite stage); the posterior neuropore closes in the 25 somite stage

   3. The CNS then forms a closed tubular structure with a long caudal portion, the future spinal cord, and a broader cephalic portion, the future brain

C. DIVISIONS OF THE NEURAL TUBE: even before the neural tube is closed cephalically, the cephalic end shows 3 distinct dilatations, the primary brain vesicles:

   1. The prosencephalon or forebrain is the most anterior vesicle

   2. The mesencephalon or midbrain is the central vesicle

   3. The rhombencephalon or hindbrain is the most posterior vesicle

D. UNEQUAL GROWTH RATES and cell migration result in flexures, constrictions, thickening, invaginations, and evaginations

   1. Coincidental with the appearance of the vesicles (5-mm embryo), the neural tube bends ventrally to form 2 flexures: a cervical flexure at the junction of spinal cord and hindbrain; and a cephalic flexure, seen in the midbrain

E. AT 5 WEEKS OF AGE, 5 components are distinguished in the developing brain:

   1. The prosencephalon which consists of 2 portions:

      a. An anterior telencephalon endbrain whose midposterior portion, the diencephalon, is characterized by the outgrowth of the optic vesicles

         i. Dorsal and ventral evaginations from the diencephalon become the primordia of the pineal gland and posterior hypophysis, respectively

      b. Two anterolateral expansions called the primitive cerebral hemispheres form

   2. The mesencephalon undergoes little change

   3. The rhombencephalon consists of 2 parts:

      a. Metencephalon (anterior) which later forms the pons and cerebellum

      b. Myelencephalon (posterior) which later forms the medulla oblongata

      c. The boundary between the metencephalon and myelencephalon is marked by a 3rd flexure, the pontine flexure, compensatory to the other two

F. THE LUMEN OF THE SPINAL CORD, the central canal, is continuous with the brain vesicles permitting CSF to circulate freely between brain and cord

   1. The cavity of the rhombencephalon is the 4th ventricle, of the diencephalon the 3rd ventricle, and of the cerebral hemispheres the lateral ventricles

   2. The lumen between the 3rd and 4th ventricles is the narrow cerebral aqueduct

   3. The interventricular foramina of Monro connect the 3rd and lateral ventricles

   4. 4th ventricle opens into subarachnoid space via foramina of Luschka and Magendie

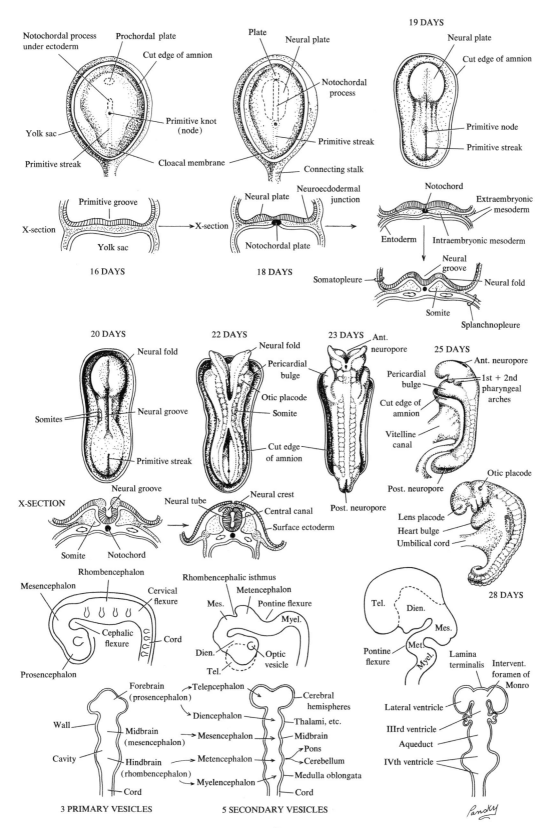

Notochordal process under ectoderm
Prochordal plate
Cut edge of amnion
Yolk sac
Primitive streak
Primitive knot (node)
Cloacal membrane

Plate
Neural plate
Notochordal process
Primitive streak
Connecting stalk

19 DAYS
Neural plate
Cut edge of amnion
Primitive node
Primitive streak

X-section
Primitive groove
Yolk sac

X-section
Neural plate
Neuroecdodermal junction
Notochordal plate

Notochord
Extraembryonic mesoderm
Entoderm
Intraembryonic mesoderm

16 DAYS

18 DAYS

Neural groove
Somatopleure
Neural fold
Somite
Splanchnopleure

20 DAYS
Neural fold
Somites
Neural groove
Primitive streak

22 DAYS
Neural fold
Pericardial bulge
Otic placode
Somite
Cut edge of amnion

23 DAYS
Ant. neuropore
Pericardial bulge
Cut edge of amnion
Vitelline canal
Post. neuropore

25 DAYS
Ant. neuropore
1st + 2nd pharyngeal arches

X-SECTION
Neural groove
Somite
Notochord

Neural tube
Neural crest
Central canal
Surface ectoderm

Post. neuropore

Otic placode
Lens placode
Heart bulge
Umbilical cord

Rhombencephalon
Mesencephalon
Cervical flexure
Cephalic flexure
Cord
Prosencephalon

Rhombencephalic isthmus
Metencephalon
Mes.
Pontine flexure
Myel.
Dien.
Optic vesicle
Tel.

28 DAYS
Tel.
Dien.
Mes.
Met.
Myel.
Pontine flexure
Lamina terminalis
Intervent. foramen of Monro

Wall
Cavity

Forebrain (prosencephalon)
Telencephalon
Diencephalon
Midbrain (mesencephalon)
Mesencephalon
Hindbrain (rhombencephalon)
Metencephalon
Myelencephalon
Cord

Cerebral hemispheres
Thalami, etc.
Midbrain
Pons
Cerebellum
Medulla oblongata
Cord

Lateral ventricle
IIIrd ventricle
Aqueduct
IVth ventricle

3 PRIMARY VESICLES

5 SECONDARY VESICLES

Pansky

-5-

2. SPINAL CORD: NORMAL DEVELOPMENT

I. Neuroepithelial, mantle, and marginal layers

A. THE WALL of a recently closed neural tube consists of only 1 cell type, the neuroepithelial cell, which extends over the entire thickness of the wall and forms a thick pseudostratified epithelium
   1. The cells are connected to each other by terminal bars at the lumen
   2. During interphase, when DNA synthesis takes place, the cells are wedge-shaped with the broader portion containing the nucleus in the outer zone of the wall and a slender cytoplasmic part extending toward the lumen
   3. Just after DNA synthesis, the nucleus begins to move toward the lumen, while the cell contracts toward the terminal bars
   4. During metaphase, the cells are round and in broad contact with the lumen, squeezing the thin cytoplasmic processes of neighboring nondividing cells
B. DURING THE NEURAL GROOVE STAGE and just after tube closure, the neuroepithelial cells divide rapidly, resulting in the production of more cells, and the thickened epithelium in the recently closed neural tube is referred to as the neuroepithelial layer or neuroepithelium
C. ONCE THE TUBE IS CLOSED, the neuroepithelial cells give rise to another cell type characterized by a round nucleus with pale cytoplasm and a dark-staining nucleolus, the primitive nerve cells or neuroblasts
   1. The neuroblasts form a zone that surrounds the neuroepithelial layer, called the mantle layer (the future gray matter of the spinal cord)
   2. The outermost layer of the cord contains the nerve fibers emerging from the neuroblasts in the mantle layer and is called the marginal layer
      a. As a result of myelination of the nerve fibers, the marginal layer takes on a "white" appearance and is called the white matter of the cord

II. Basal, alar, roof, and floor plates

A. WITH THE CONTINUAL ADDITION OF NEUROBLASTS to the mantle layer, each side of the neural tube shows a ventral and dorsal thickening
   1. The ventral thickenings, the basal or motor plates, contain the anterior motor horn cells and form the motor areas of the spinal cord
   2. The dorsal thickenings, the alar or sensory plates, form the sensory areas
   3. A longitudinal groove, the sulcus limitans, is found bilaterally on the inner surface of the tube. It marks the boundary between the anterior motor and posterior sensory areas and ends in the region of the mamillary recess in the ventral portion of the diencephalon
      a. One should therefore not expect any truly motor nerves emerging from the brain rostral to the mamillary recess
      b. Since all neural tissue rostral to this point is an extension of the alar plate, its function is regarded as sensory or associational
B. THE THIN DORSAL PORTION and ventral midline parts of the tube are the roof and floor plates, respectively. They contain no neuroblasts and serve primarily as pathways for nerve fibers crossing from one side of the cord to the other
C. THE BASAL PLATES bulge ventrally on each side of the midline as a result of the continuous enlargement of the neuroblasts, creating a deep longitudinal groove called the ventral fissure, which will later contain the anterior spinal artery
D. THE ALAR PLATES expand predominantly in a medial direction compressing the dorsal portion of the lumen of the neural tube
   1. The posterior median septum is formed where the 2 alar plates fuse in the midline
E. ACCUMULATION OF NEURONS between alar and basal plates causes the formation of the intermediate horn which contains motor neurons of the autonomic nervous system. The spinal cord acquires its definitive form: motor horns anteriorly, sensory horns posteriorly, intermediate horns laterally, and small lumen, the central canal

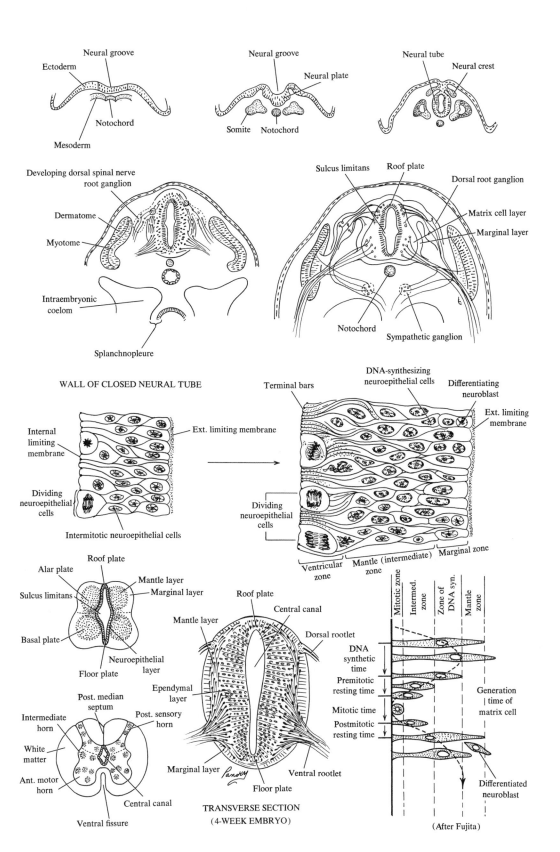

Ectoderm — Neural groove — Neural plate — Neural tube — Neural crest

Notochord — Mesoderm — Somite — Notochord

Developing dorsal spinal nerve root ganglion — Sulcus limitans — Roof plate — Dorsal root ganglion

Dermatome — Matrix cell layer

Myotome — Marginal layer

Intraembryonic coelom

Splanchnopleure — Notochord — Sympathetic ganglion

WALL OF CLOSED NEURAL TUBE — Terminal bars — DNA-synthesizing neuroepithelial cells — Differentiating neuroblast

Internal limiting membrane — Ext. limiting membrane — Ext. limiting membrane

Dividing neuroepithelial cells — Dividing neuroepithelial cells

Intermitotic neuroepithelial cells — Ventricular zone — Mantle (intermediate) zone — Marginal zone

Roof plate — Mantle layer — Marginal layer

Alar plate — Roof plate — Central canal — Mitotic zone — Intermed. zone — Zone of DNA syn. — Mantle zone

Sulcus limitans — Mantle layer — Dorsal rootlet

Basal plate — DNA synthetic time

Neuroepithelial layer — Premitotic resting time

Floor plate — Ependymal layer — Generation time of matrix cell

Post. median septum — Mitotic time

Intermediate horn — Post. sensory horn — Postmitotic resting time

White matter — Ant. motor horn — Central canal — Marginal layer — Ventral rootlet — Differentiated neuroblast

Ventral fissure — Floor plate

TRANSVERSE SECTION
(4-WEEK EMBRYO)

(After Fujita)

3. SPINAL CORD: NERVE AND GLIAL CELL DIFFERENTIATION

I. **Nerve cells:** the neuroblasts or primitive nerve cells arise exclusively by division of neuroepithelial cells. Once the neuroblasts are formed, they lose their ability to divide

A. THE NEUROBLASTS OF THE ANTERIOR HORN are formed first, and only when most of these have migrated to the mantle layer does formation of nerve cells for the alar plate begin

B. THE NEUROBLASTS initially have a central process extending to the lumen, the transient dendrite, but it disappears when the cells migrate to the mantle zone. The neuroblasts are temporarily round, the so-called apolar neuroblasts

C. THERE IS FURTHER DIFFERENTIATION and 2 new cytoplasmic processes appear on opposite sides of the cell body. Thus, the bipolar neuroblasts are formed

D. THE PROCESS at one end elongates rapidly to form the primitive axon, whereas that at the other end develops a number of cytoplasmic arborizations which are known as the primitive dendrites

E. THE CELL is now a multipolar neuroblast and with further development becomes the adult nerve cell or neuron

F. THE AXONS of neurons in the posterior sensory horn behave differently than those in the anterior motor horn:
   1. Those in the posterior horn penetrate the marginal layer of the cord and then ascend or descend to a higher or lower level (association neurons)
   2. Those in the anterior horn break through the marginal zone and are seen on the ventral aspect of the cord where they form the anterior motor root of the spinal nerve (they conduct motor impulses from cord to muscles)

II. **Glia cells:** majority of primitive supporting cells are called glioblasts, formed by the neuroepithelial cells after the production of neuroblasts has ceased

A. THE GLIOBLASTS migrate from the neuroepithelial layer to the mantle layer (some even to the marginal layer) where they differentiate into the fibrous and protoplasmic astrocytes
   1. The fibrous astrocytes are abundant in the white matter, where they provide both support and binding for the tracts of nerve fibers via long, slender, smooth processes
   2. The protoplasmic astrocytes are present in great numbers in the gray matter, where they occupy the spaces between axons and dendrites, and their processes form close contacts (bridges) between neuronal cell bodies, blood capillaries, and the pia mater. They constitute water-ion compartments for transport of metabolites as well as a highly selective blood-CSF barrier

B. ANOTHER TYPE OF SUPPORTING CELL, possibly of glioblast origin, is the oligodendroglia cell
   1. This cell is mainly found in the marginal layer and forms the myelin sheath around the ascending and descending axons in that layer
   2. They may be derived from mesenchyme cells which have penetrated into the CNS, thus their origin from neuroepithelial cells is in doubt

C. IN THE 2ND HALF OF DEVELOPMENT, a 3rd type of supporting cell, the microglia cell, appears in the CNS. It is believed that its origin is from mesoderm which surrounds the neural tube.

D. WHEN THE NEUROEPITHELIAL CELLS cease to produce neuroblasts and glioblasts, they finally differentiate into the ependymal cells (as seen in the adult)

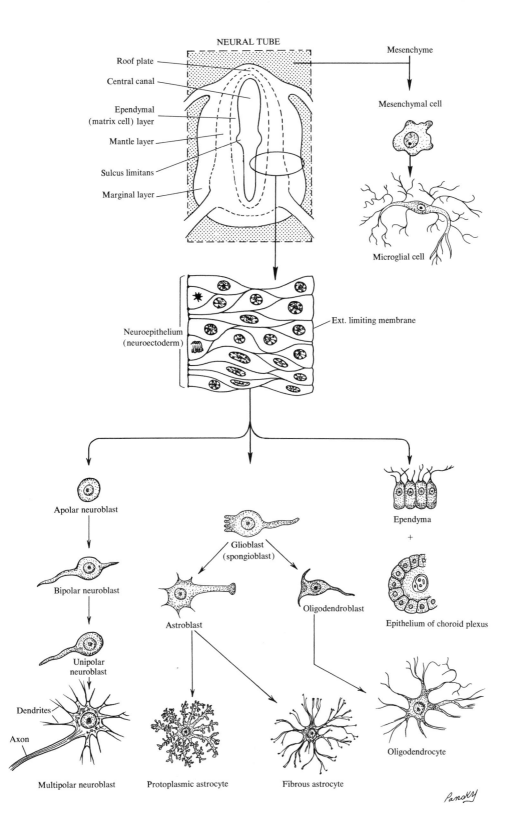

NEURAL TUBE

Roof plate

Central canal

Ependymal
(matrix cell) layer

Mantle layer

Sulcus limitans

Marginal layer

Mesenchyme

Mesenchymal cell

Microglial cell

Neuroepithelium
(neuroectoderm)

Ext. limiting membrane

Apolar neuroblast

Bipolar neuroblast

Unipolar
neuroblast

Dendrites

Axon

Multipolar neuroblast

Glioblast
(spongioblast)

Astroblast

Protoplasmic astrocyte

Oligodendroblast

Fibrous astrocyte

Ependyma

+

Epithelium of choroid plexus

Oligodendrocyte

Pansky

4. SPINAL CORD: NEURAL CREST CELLS AND MYELINATION

I. Neural crest cells

A. DURING INVAGINATION of the neural plate, a distinct group of cells appears along each edge of the neural groove, are ectodermal in origin, and are called the neural crest cells

1. The cells form an intermediate zone between the tube and surface ectoderm
2. The zone extends from the mesencephalon to the level of the caudal somite and in time divides into 2 parts, each of which migrates to the dorsolateral aspect of the neural tube
   a. Here the cells of the neural crest form a series of cell clusters that give rise to the sensory or dorsal root ganglia of the spinal and cranial nerves (Vth, VIIth, IXth, and Xth cranial nerves)

B. DURING DEVELOPMENT, the neuroblasts of the sensory ganglia form 2 processes:

1. One grows centrally and penetrates the dorsal portion of the neural tube.
   a. In the spinal cord, they either end in the dorsal horn or ascend through the marginal layer to one of the higher brain centers
   b. Collectively, these processes are called the dorsal sensory root of the spinal nerve
2. The other process grows peripherally and forms fibers of the ventral motor root and participates in the formation of the trunk of the spinal nerve.
   a. These processes eventually terminate in the sensory receptor organs

C. IN ADDITION TO FORMING THE SENSORY GANGLIA, the cells of the neural crest differentiate into sympathetic neuroblasts, Schwann cells, pigment cells, odontoblasts, meninges, and cartilage cells of the branchial arches

a. Removal of neural crest cells of the trigeminal region results in facial abnormalities, including clefts of the primary palate

II. Myelination

A. MYELINATION OF THE PERIPHERAL NERVE is brought about by the neurilemma cells

1. These cells originate from the neural crest, migrate peripherally, and wrap around the axons to form the neurilemma sheath
2. Axons, varying in number from 1 to 20, can be enwrapped by one neurilemma cell

B. AT THE 4TH MONTH OF FETAL LIFE, the nerve fibers gradually obtain a whitish appearance as a result of myelin deposition between the axon and neurilemma

1. This substance is formed by repeated coiling of the membrane around the axon

C. BOTH THE NEURILEMMA AND THE MYELIN SHEATH of the peripheral nerve fibers are formed by the cells of Schwann

D. THE MYELIN SHEATH of nerve fibers within the spinal cord is of different origin, being formed by the oligodendroglia cells

E. THOUGH MYELINATION OF NERVE FIBERS in the cord generally begins at about the 4th month of fetal life, some motor fibers that descend from higher brain centers to the cord do not become myelinated until the 1st year of postnatal life

1. The tracts in the nervous system apparently become myelinated at about the time they begin to function

F. MYELINATION is associated with development of functional neuron capacity. Myelinated neurons fire rapidly and have a long period of activity before they fatigue

G. MYELIN first appears in the cervical part of the cord and then extends progressively to lower levels

1. During fetal month 4, the ventral root fibers are the first to acquire a myelin sheath, followed by the dorsal sensory roots
2. The last spinal tracts to be myelinated are the descending motor tracts (during years 1 and 2 after birth)

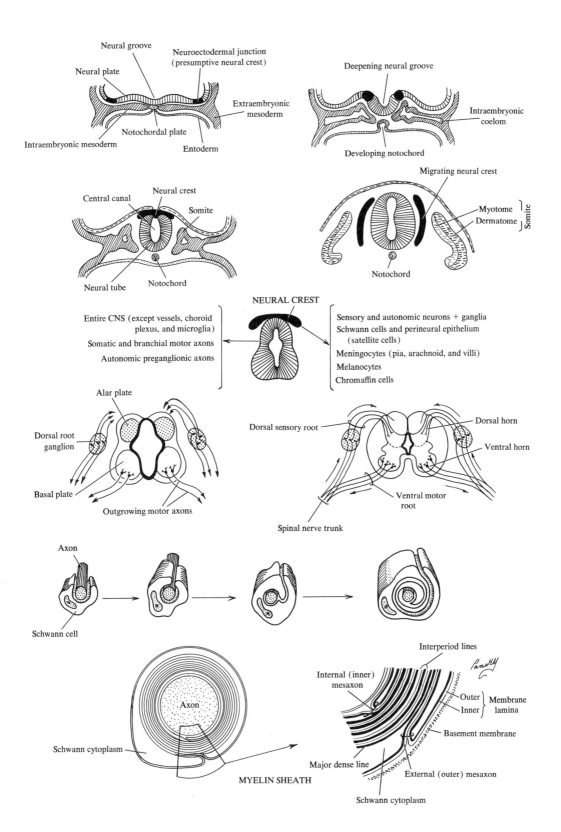

Neural groove

Neural plate

Neuroectodermal junction (presumptive neural crest)

Extraembryonic mesoderm

Notochordal plate

Intraembryonic mesoderm

Entoderm

Deepening neural groove

Intraembryonic coelom

Developing notochord

Central canal

Neural crest

Somite

Neural tube

Notochord

Migrating neural crest

Myotome

Dermatome

Somite

Notochord

NEURAL CREST

Entire CNS (except vessels, choroid plexus, and microglia)

Somatic and branchial motor axons

Autonomic preganglionic axons

Sensory and autonomic neurons + ganglia

Schwann cells and perineural epithelium (satellite cells)

Meningocytes (pia, arachnoid, and villi)

Melanocytes

Chromaffin cells

Alar plate

Dorsal root ganglion

Basal plate

Outgrowing motor axons

Dorsal sensory root

Dorsal horn

Ventral horn

Ventral motor root

Spinal nerve trunk

Axon

Schwann cell

Interperiod lines

Internal (inner) mesaxon

Outer

Inner

Membrane lamina

Axon

Basement membrane

Schwann cytoplasm

Major dense line

External (outer) mesaxon

MYELIN SHEATH

Schwann cytoplasm

5. SPINAL CORD: LENGTH AND CONGENITAL MALFORMATIONS

I. Changes in cord position during development

A. WHEN THE CROWN-RUMP LENGTH of the embryo is about 30 mm (in the 3rd month of development), the spinal cord extends the entire length of the embryo and the spinal nerves pass through the intervertebral foramina at their level of origin

B. WITH INCREASING AGE, the vertebral column and dura mater lengthen more rapidly than the neural tube and the terminal end of the spinal cord gradually shifts to a higher level

C. AT BIRTH, the end of the cord is located at the level of the 3rd lumbar vertebra.
1. Due to disproportionate growth, the spinal nerves run obliquely from their segment of cord origin to the corresponding level of the vertebral column
2. The dura remains attached to the bony column at the coccygeal level

D. IN THE ADULT, the spinal cord terminates at the level of L2
1. Below this point, the CNS is represented only by the filum terminale internum, marking the tract of spinal cord regression
2. Nerve fibers below the terminal end of the cord are called, collectively, the cauda equina

E. WHEN CEREBROSPINAL FLUID is taken in a lumbar puncture, the needle is inserted at a lower lumbar level to avoid injuring the lower end of the spinal cord.

II. Congenital malformations of the cord and column

A. THE CNS forms a closed tubular structure which is detached from the overlying ectoderm by the end of the 4th week
1. Occasionally, however, the neural groove fails to close, either as a result of faulty induction by the underlying notochord, or because of the action of environmental teratogenic factors on the neuroepithelial cells
   a. The neural tissue then remains exposed to the surface
   b. Such a defect may extend the entire length of the embryo or be restricted to a small area only—complete or partial rachischisis
   c. Failure of closure in the cephalic region is called anencephalis
   d. If localized in the cord region, the abnormality is called spina bifida
2. Spina bifida refers to a wide range of defects
   a. In its most simple form, it is seen as a failure of the dorsal portions of the vertebrae to fuse with one another (bifid spine, literally)
      i. Usually localized in the sacrolumbar region, covered by skin and not noticeable on the surface except for the presence of a small tuft of hair over the affected area; called spina bifida occulta
      (a) Cord and nerves are normal; neurologic symptoms are absent
   b. If more than one or two vertebrae are involved in the defect, the meninges of the cord bulge through the opening and a sac covered with skin is seen on the surface; called meningocele
   c. If the sac is so large that it contains not only the meninges but also the spinal cord and its nerves, this abnormality is meningomyelocele
      i. Usually covered by a thin, easily torn membrane
      ii. Neurologic symptoms are usually present
   d. If the neural groove fails to close, the nervous tissue is widely exposed to the surface; called a myelocele or rachischisis
      i. Occasionally the neural tissue shows much overgrowth, but the excess tissue invariably becomes necrotic before or after birth
   e. Myelomeningoceles are usually associated with a caudal displacement of the medulla oblongata and a part of the cerebellum into the spinal canal
      i. The myelomeningocele is frequently combined with hydrocephaly; called Arnold-Chiari malformation

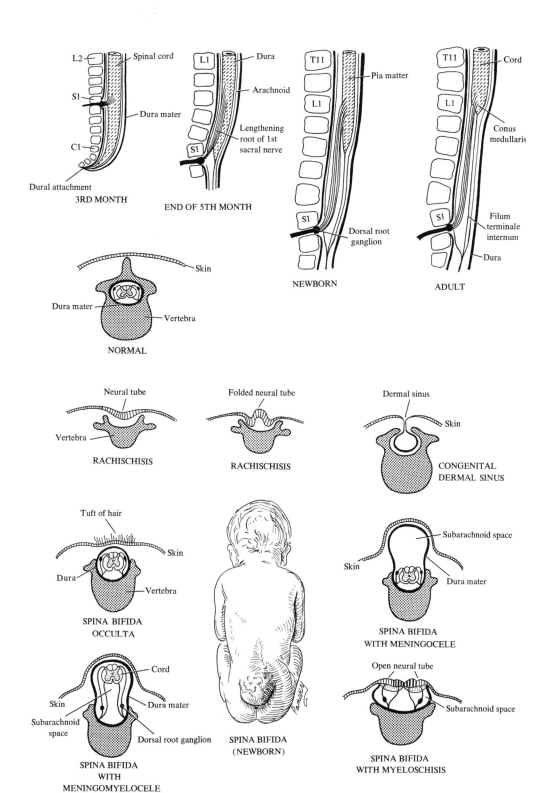

L2
Spinal cord
S1
Dura mater
C1
Dural attachment
3RD MONTH

L1
Dura
Arachnoid
Lengthening root of 1st sacral nerve
S1
END OF 5TH MONTH

T11
Pia matter
L1
S1
Dorsal root ganglion
NEWBORN

T11
Cord
L1
Conus medullaris
S1
Filum terminale internum
Dura
ADULT

Skin
Dura mater
Vertebra
NORMAL

Neural tube
Vertebra
RACHISCHISIS

Folded neural tube
RACHISCHISIS

Dermal sinus
Skin
CONGENITAL DERMAL SINUS

Tuft of hair
Skin
Dura
Vertebra
SPINA BIFIDA OCCULTA

Cord
Skin
Dura mater
Subarachnoid space
Dorsal root ganglion
SPINA BIFIDA WITH MENINGOMYELOCELE

SPINA BIFIDA (NEWBORN)

Subarachnoid space
Skin
Dura mater
SPINA BIFIDA WITH MENINGOCELE

Open neural tube
Subarachnoid space
SPINA BIFIDA WITH MYELOSCHISIS

6. BRAIN: MYELENCEPHALON—BASAL MOTOR PLATE

I. **Introduction:** the external shape of the cephalic part of the neural tube changes markedly with the appearance of brain vesicles and the development of cervical and cephalic flexures. Despite changes, some morphologic characteristics seen in the spinal cord are recognizable in most of the brain vesicles

A. DISTINCT BASAL AND ALAR PLATES, representing the motor and sensory areas, respectively, are seen on each side of the midline in most of the brain vesicles

B. THE SULCUS LIMITANS, which forms the boundary line between alar and basal plates in the cord, is present in the rhombencephalon and mesencephalon where it also forms the divider between sensory and motor areas

II. **The myelencephalon** is the most caudal brain compartment and extends from the 1st spinal nerve to the pontine flexure and gives rise to the medulla oblongata

A. THE MEDULLA differs from the cord in that its lateral walls rotate around an imaginary long axis in the floor plate (like opening a textbook); as a result, the roof plate stretches and becomes a single layer of cells

   1. The lateral wall structure is similar to that of the cords. Alar and basal plates are separated by the sulcus limitans

B. BASAL MOTOR PLATE OF THE MYELENCEPHALON, like the cord, contains the motor nuclei, but these are divided into 3 groups:

   1. A medial somatic efferent group: forms the cephalic continuation of the anterior horn cells containing the motor neurons which innervate striated muscle (derived from myotomes in the cephalic region)

      a. Since the somatic efferent group continues rostrally into the mesencephalon (through the metencephalon), it is referred to as the somatic efferent motor column

      b. It is represented in the myelencephalon by the neurons of the hypoglossal (XII) nerve which supplies the 4 occipital myotomes (tongue musculature)

      c. In the metencephalon and mesencephalon, the column is represented by the neurons of the abducens (VI), trochlear (IV), and oculomotor (III) cranial nerves, respectively, supplying eye muscles thought to be derived from preoptic myotomes

        i. The neurons of nerves III, IV, VI, and XII are located near the midline

   2. An intermediate special visceral efferent group extends into the metencephalon and forms the special visceral efferent motor column which contains motor neurons supplying striated muscles derived from the mesenchyme of the pharyngeal or branchial arches

      a. In the myelencephalon, the column is represented by neurons of the accessory (XI), vagus (X), and glossopharyngeal (IX) cranial nerves

        i. In the adult, motor neurons of the above nerves are formed by the nucleus ambiguus and the bulbar portion of the accessory nerve

   3. A lateral general visceral efferent group contains the neurons whose axons grow out as preganglionic fibers to synapse in the parasympathetic ganglia supplying involuntary muscles of the heart, respiratory tract, and intestinal tract, as well as innervating the salivary glands

      a. In the myelencephalon, this group is represented by the dorsal nucleus of the vagus and the inferior salivatory nucleus (which, by way of cranial nerve IX innervates the parotid gland)

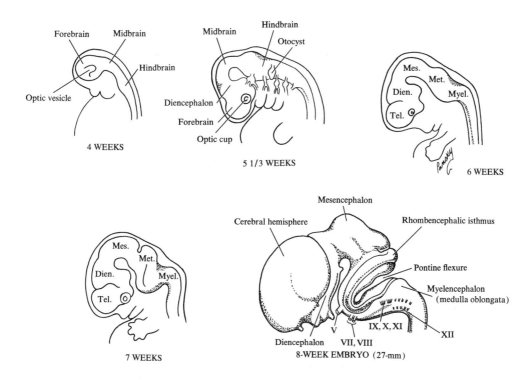

4 WEEKS

Forebrain
Midbrain
Hindbrain
Optic vesicle

5 1/3 WEEKS

Midbrain
Hindbrain
Otocyst
Diencephalon
Forebrain
Optic cup

6 WEEKS

Mes.
Met.
Dien.
Myel.
Tel.

7 WEEKS

Mes.
Met.
Dien.
Myel.
Tel.

8-WEEK EMBRYO (27-mm)

Mesencephalon
Cerebral hemisphere
Rhombencephalic isthmus
Pontine flexure
Myelencephalon (medulla oblongata)
V
VII, VIII
IX, X, XI
XII
Diencephalon

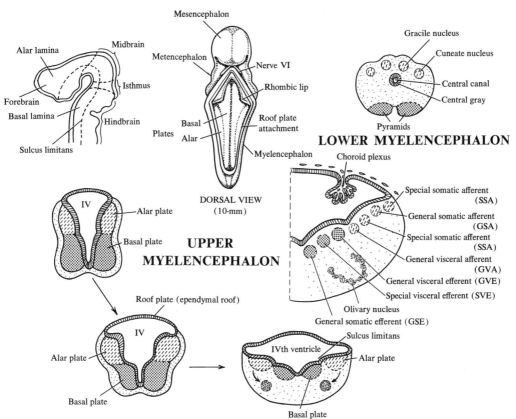

Alar lamina
Midbrain
Isthmus
Forebrain
Basal lamina
Hindbrain
Sulcus limitans

Mesencephalon
Metencephalon
Nerve VI
Rhombic lip
Roof plate attachment
Basal
Alar
Plates
Myelencephalon

DORSAL VIEW
(10-mm)

Gracile nucleus
Cuneate nucleus
Central canal
Central gray
Pyramids

LOWER MYELENCEPHALON

IV
Alar plate
Basal plate

**UPPER
MYELENCEPHALON**

Roof plate (ependymal roof)
IV
Alar plate
Basal plate

IVth ventricle
Sulcus limitans
Alar plate
Basal plate

Choroid plexus
Special somatic afferent (SSA)
General somatic afferent (GSA)
Special somatic afferent (SSA)
General visceral afferent (GVA)
General visceral efferent (GVE)
Special visceral efferent (SVE)
Olivary nucleus
General somatic efferent (GSE)

7. BRAIN: MYELENCEPHALON—ALAR SENSORY AND ROOF PLATE

I. The myelencephalon (cont.)

A. THE ALAR SENSORY PLATE contains the sensory relay nuclei, which like the basal plate are divided into 3 groups:
   1. The most lateral is the somatic afferent group, which receives impulses from the ear and surface of the head via the statoacoustic (VIII) and bulbospinal part of the trigeminal (V) nerves
   2. The intermediate is the special visceral afferent group, which receives impulses from the taste buds of the tongue and from the palate, oropharynx, and epiglottis
      a. These neurons later form the nucleus of the solitary tract
   3. The medial is the general visceral afferent group, represented by the dorsal sensory nucleus of the vagus with its neurons receiving interoceptive information from the heart and GI tract
   4. In addition to the sensory relay nuclei, other cells of the alar plate migrate downward to be ventrolateral to the basal plate and form a part of the olivary nuclear complex

B. ROOF PLATE, CHOROID PLEXUS, AND FORAMINA OF LUSCHKA AND MAGENDIE
   1. The roof plate of the myelencephalon consists of a single layer of ependymal cells which is later covered by vascular mesenchyme, the pia mater. Together, they make up the tela choroidea
      a. As a result of active proliferation of vascular mesenchyme, the tela forms a series of saclike invaginations that project into the underlying ventricular cavity in the region of the pontine flexure, forming the choroid plexus (tuftlike invaginations that produce the CSF of the CNS)
   2. At about 4 months of age, areas in the roof plate of the rhombencephalon thin out, bulge outward, and finally disappear, and the apertures formed are the 2 lateral foramina of Luschka and a median foramen of Magendie which allow the CSF to move freely between the ventricles and the surrounding subarachnoid space
      a. The major site of absorption of CSF into the venous system is by way of the arachnoid villi, which project into the dural sinuses (superior sagittal sinus). The villi are made up of a thin cellular layer derived from the epithelium of the arachnoid and the sinus endothelium

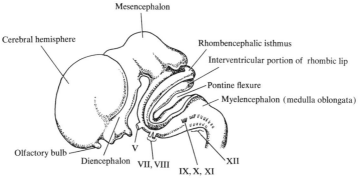

Mesencephalon

Cerebral hemisphere

Rhombencephalic isthmus

Interventricular portion of rhombic lip

Pontine flexure

Myelencephalon (medulla oblongata)

Olfactory bulb

Diencephalon

V

VII, VIII

IX, X, XI

XII

LATERAL VIEW—8-WEEK EMBRYO (27-mm)

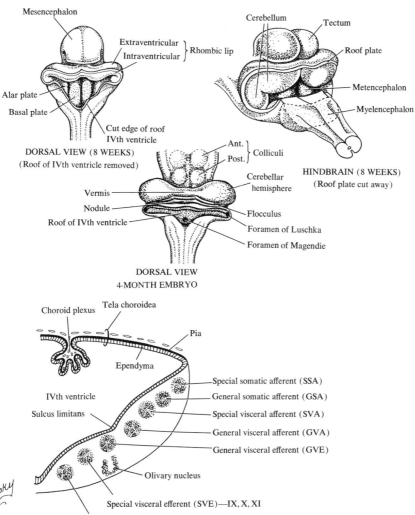

Mesencephalon

Extraventricular

Intraventricular

Rhombic lip

Alar plate

Basal plate

Cut edge of roof
IVth ventricle

DORSAL VIEW (8 WEEKS)
(Roof of IVth ventricle removed)

Cerebellum

Tectum

Roof plate

Metencephalon

Myelencephalon

HINDBRAIN (8 WEEKS)
(Roof plate cut away)

Ant.
Post.
Colliculi

Cerebellar
hemisphere

Vermis

Nodule

Roof of IVth ventricle

Flocculus

Foramen of Luschka

Foramen of Magendie

DORSAL VIEW
4-MONTH EMBRYO

Choroid plexus

Tela choroidea

Pia

Ependyma

IVth ventricle

Sulcus limitans

Special somatic afferent (SSA)

General somatic afferent (GSA)

Special visceral afferent (SVA)

General visceral afferent (GVA)

General visceral efferent (GVE)

Olivary nucleus

Special visceral efferent (SVE)—IX, X, XI

General somatic efferent (GSE)—XII

– 17 –

8. BRAIN: METENCEPHALON

I. **The metencephalon** develops from the anterior part of the rhombencephalon and extends from the pontine flexure to the rhombencephalic isthmus. It differs from the myelencephalon in forming 2 specialized components: the dorsal portion forms the cerebellum which functions as a coordination center for posture and movement and a ventral portion which becomes the pons which serves as the pathway for nerve fibers between the cord and cerebral and cerebellar cortices

A. BASAL PLATE AND PONS: the major morphologic features of the metencephalon do not change even though the lateral walls reapproach each other and the basal motor and sensory alar plates are easily seen. Each basal plate contains 3 groups of motor neurons:

1. The medial somatic efferent group which gives rise to the abducens (VI) nerve
2. The special visceral efferent group, containing the nuclei of the trigeminal (V) and facial (VII) nerves, which innervate the muscles of the 1st and 2nd branchial arches
3. The general visceral efferent group contains the superior salivatory nucleus, axons of which grow out of the facial nerve to supply the submandibular and sublingual glands as well as the nasal and lacrimal glands
4. The marginal layer of the basal plates expands to serve as a bridge for the nerve fibers connecting the cerebral cortex and cerebellar cortex with the spinal cord and is known as the pons
   a. In addition to nerve fibers, the pons contains pontine nuclei which originate in the alar plates of the metencephalon and myelencephalon
   b. The axons of the nuclei grow toward the cerebellum and form the middle cerebellar peduncles

B. ALAR PLATE AND RHOMBIC LIP: development is complicated

1. The ventromedial portion of the plate contains 3 groups of sensory nuclei:
   a. The lateral somatic afferent group contains neurons of the pontine portion of the trigeminal (V) nerve and a small part of the vestibulocochlear complex
   b. The special visceral afferent group is represented by the cranial portion of the nucleus of the solitary tract
   c. The general visceral afferent group is represented by the most cranial part of the dorsal sensory nucleus of the vagus (X) nerve
2. The dorsolateral parts of the alar plates bend medially and form the rhombic lips which project partly into the lumen of the 4th ventricle and partly above the attachment of the roof plate (extraventricular part) to give rise to the cerebellum

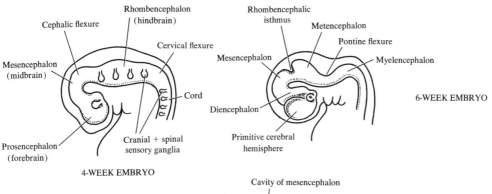

4-WEEK EMBRYO

6-WEEK EMBRYO

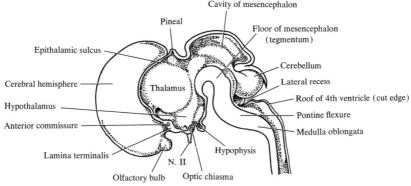

9-WEEK EMBRYO (43-mm)

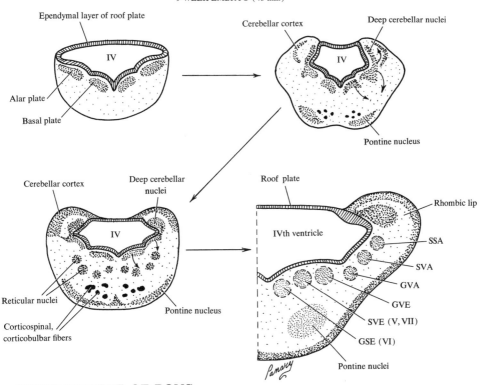

DEVELOPMENT OF PONS

CAUDAL METENCEPHALON

9. BRAIN: METENCEPHALON—CEREBELLUM

I. The metencephalon (cont.)

A. THE CEREBELLUM is derived from the dorsal part of the alar lamina of the metencephalon. This part of the alar lamina, of each side, forms the so-called rhombic lip. The rhombic lips on the caudal portion of the metencephalon are widely separated, but just below the mesencephalon, they approach each other in the midline. With further deepening of the pontine flexure, the rhombic lips become compressed in a cephalocaudal direction to form the cerebellar plate

1. In the 12-week embryo, the plate shows a small midline portion, the vermis, and 2 lateral portions, the hemispheres
2. A transverse fissure soon separates the nodule from the vermis; and the lateral flocculus from the hemispheres to form the flocculonodular lobe (the most primitive part of the cerebellum), which maintains connections with the vestibular septum
3. Many other transverse fissures appear, giving the cerebellum its adult appearance
4. Initially, the cerebellar plate consists of a neuroepithelial, a mantle, and a marginal layer
   a. With development, a number of cells formed by the neuroepithelium migrate through the marginal layer to the cerebellar surface to form the external granular layer
      i. The cells of this layer retain their abilty to divide and form a proliferative zone on the cerebellar surface
5. In the 6th developmental month, the external granular layer begins to release various cell types, which migrate inward toward the differentiating Purkinje cells
   a. Thus, the external granular layer gives rise to the granule cells, the basket cells, and the stellate cells, in sequential order
      i. The production and migration of these small neurons continue for about $1\frac{1}{2}$ years after birth
6. Thus, the cerebellar cortex consists of Purkinje cells, Golgi II neurons, and the neurons produced by the external granular layer and reaches its definitive size a considerable time after birth
7. The deep cerebellar nuclei, such as the dentate, are formed by neuroblasts in the mantle layer and reach their definitive position long before birth
8. The greater portion of the original roof plate of the 4th ventricle forms the pia mater on the cerebellar surface
   a. The part located in front of, and behind, the cerebellum becomes specialized and forms the anterior and posterior medullary velum, respectively
9. The axons of the Purkinje cells pass to the dentate nuclei, and the fibers from the latter grow forward into the mesencephalon to form the major part of the superior cerebellar peduncles
10. Afferent connections come from many sources. The earliest is from the vestibular nerve itself. Next, fibers come from the pontine nucleus of the Vth nerve, and then anterior and posterior spinocerebellar tracts. Development of pontocerebellar and corticopontine tracts brings the cerebellum into functional connection with the cerebral cortex

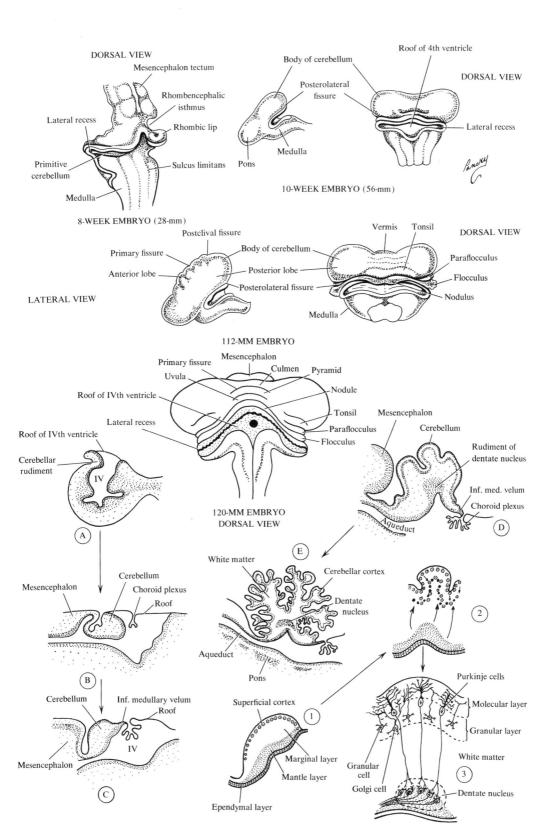

DORSAL VIEW

Mesencephalon tectum

Rhombencephalic isthmus

Rhombic lip

Lateral recess

Primitive cerebellum

Sulcus limitans

Medulla

8-WEEK EMBRYO (28-mm)

Body of cerebellum

Roof of 4th ventricle

DORSAL VIEW

Posterolateral fissure

Pons

Medulla

Lateral recess

10-WEEK EMBRYO (56-mm)

Postclival fissure

Primary fissure

Anterior lobe

Body of cerebellum

Posterior lobe

Posterolateral fissure

LATERAL VIEW

Vermis   Tonsil

DORSAL VIEW

Paraflocculus

Flocculus

Nodulus

Medulla

112-MM EMBRYO

Primary fissure

Mesencephalon

Culmen

Pyramid

Uvula

Nodule

Roof of IVth ventricle

Lateral recess

Tonsil

Paraflocculus

Flocculus

Mesencephalon

Cerebellum

Rudiment of dentate nucleus

Inf. med. velum

Choroid plexus

Aqueduct

D

120-MM EMBRYO DORSAL VIEW

Roof of IVth ventricle

Cerebellar rudiment

IV

A

E

White matter

Cerebellar cortex

Dentate nucleus

Mesencephalon

Cerebellum

Choroid plexus

Roof

Aqueduct

Pons

B

2

Cerebellum

Inf. medullary velum

Roof

Mesencephalon

IV

C

Superficial cortex

Marginal layer

Mantle layer

Ependymal layer

1

Purkinje cells

Molecular layer

Granular layer

White matter

Granular cell

Golgi cell

Dentate nucleus

3

- 21 -

10. BRAIN: MESENCEPHALON

I. **The mesencephalon** undergoes less change than other parts of the brain except for the caudal part of the hindbrain and is morphologically the most primitive of the brain vesicles. Its basal and alar plates, separated by the sulcus limitans (as seen in transverse sections), are easily identified

A. BASAL PLATE (WHICH GIVES RISE TO NEURONS OF THE TEGMENTUM) AND CRUS CEREBRI (BASIS PEDUNCULI)

    1. Each basal plate contains 2 groups of motor nuclei:

        a. A medial somatic efferent group, represented by the oculomotor (III) and trochlear (IV) nerves which innervate the preoptic (eye) muscles

        b. A small general visceral efferent group, represented by the Edinger-Westphal nucleus, which innervates the sphincter pupillary muscle

    2. The marginal layer of each basal plate enlarges and forms the basis pedunculi or crus cerebri

        a. The crura serve as pathways for the nerve fibers descending from the cerebral cortex to the lower centers in the pons and spinal cord

        b. In the adult, the crura contain the corticospinal, the corticobulbar, and the corticopontine tracts

        c. All fibers from the IIIrd nerve nucleus grow ventrally through the red nucleus and substantia nigra and emerge on the medial surface of the peduncle

        d. Fibers of the IVth nerve grow dorsally around the cerebral aqueduct, decussate, and emerge on the dorsal surface through the superior medullary velum

B. ALAR PLATE AND COLLICULI

    1. Neuroblasts migrate from the alar plates into the roof or tectum and aggregate to form four large groups of neurons, the paired superior and inferior colliculi

    2. The alar plates initially appear as 2 longitudinal elevations separated by a shallow midline depression.

        a. With development, a transverse groove divides each longitudinal elevation into an anterior (superior) and a posterior (inferior) colliculus

            i. The nuclei of the posterior colliculus serve as synaptic relay stations for the auditory reflexes

            ii. The nuclei of the anterior colliculus serve as correlation and reflex centers for the visual impulses

            iii. The colliculi are formed by waves of neuroblasts produced by the neuroepithelial cells that migrate into the overlying marginal zone and become arranged in stratified layers

    3. Some feel that the cells of the alar plate also give rise to the nucleus ruber (red nucleus) and the substantia nigra (others feel they develop *in situ*)

C. THE CAVITY OF THE MESENCEPHALON is called the aqueduct of Sylvius (cerebral aqueduct) and connects the 3rd and 4th ventricles and is derived from a narrowed neural canal

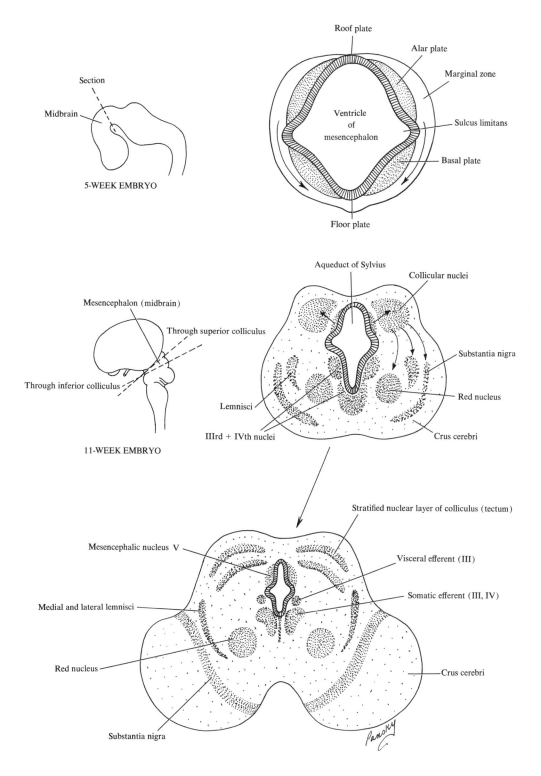

Roof plate

Alar plate

Marginal zone

Ventricle of mesencephalon

Sulcus limitans

Basal plate

Floor plate

Section

Midbrain

5-WEEK EMBRYO

Aqueduct of Sylvius

Collicular nuclei

Mesencephalon (midbrain)

Through superior colliculus

Through inferior colliculus

Substantia nigra

Red nucleus

Lemnisci

IIIrd + IVth nuclei

Crus cerebri

11-WEEK EMBRYO

Stratified nuclear layer of colliculus (tectum)

Mesencephalic nucleus V

Visceral efferent (III)

Somatic efferent (III, IV)

Medial and lateral lemnisci

Red nucleus

Crus cerebri

Substantia nigra

Pansky

DEVELOPMENT OF MESENCEPHALON

11. BRAIN: DIENCEPHALON

I. **The diencephalon** develops from the median portion of the prosencephalon and consists of a roof plate, 2 alar plates, and the 3rd ventricle, but has no floor or basal plates. It is bounded posteriorly by a plane passing behind the pineal gland and mamillary bodies and anteriorly by a plane passing just rostral to the optic chiasma and encircling the interventricular foramen of Monro (the lamina terminalis is considered to be a part of the telencephalon)

A. ROOF PLATE AND PINEAL BODY (EPIPHYSIS)
   1. The roof plate consists of a single layer of ependymal cells covered by vascular mesenchyme
      a. The two later combine to form the choroid plexus of the 3rd ventricle
   2. The most caudal part of the roof plate does not participate in the formation of the choroid plexus but develops into the pineal body or epiphysis
      a. The pineal initially appears as an epithelial thickening in the midline but begins to evaginate by the 7th week and eventually forms a solid organ located on the roof of the mesencephalon
   3. Occasionally the roof plate forms another evagination near the interventricular foramen, called the paraphysis, which sometimes persists into postnatal life and may give rise to a small cyst
   4. The roof plate is also thought to give rise to the epithalamus, a group of nuclei located on each side of the midline close to the pineal gland; however, the epithalamus may arise from the alar plate
      a. The epithalamic region is originally large but regresses to a small area where the habenular nuclei are found
         i. The habenular nuclei form a link in the olfactory conduction path and are connected to each other across the midline by a group of nerve fibers collectively called the habenular commissure (found rostral to the pineal stalk)
   5. Another commissure, the posterior commissure, appears caudal to the pineal stalk and also connects 2 nuclear areas on each side of the midline

B. THALAMUS AND HYPOTHALAMUS
   1. The alar plates form both the lateral walls and floor of the diencephalon
   2. A distinct longitudinal groove is seen on the side facing the lumen, the hypothalamic sulcus, which divides the alar plate into a dorsal and ventral region, the thalamus and hypothalamus, respectively
      a. The sulcus is of a different nature than the sulcus limitans since it does not form a dividing line between motor and sensory areas
   3. Following high proliferative activity, the thalami (right and left) gradually bulge into the lumen of the diencephalon and, frequently, the 2 may fuse in the midline to form the massa intermedia or interthalamic connexus (adhesion)
   4. The thalamic nuclear areas eventually form 2 distinct nuclear groups:
      a. A dorsal thalamic group: important for the reception and transmission of visual and auditory impulses
      b. A ventral thalamic group: serving as a passage and relay station to higher centers
   5. The hypothalamus which forms from the lower portion of the alar plate also differentiates into a number of nuclear groups:
      a. Nuclei that serve as regulation centers of visceral functions, i.e., sleep, digestion, body temperature, emotional behavior, etc.
      b. Two groups, the mamillary bodies, form the rounded elevations on the ventral surface of the hypothalamus on either side of the midline

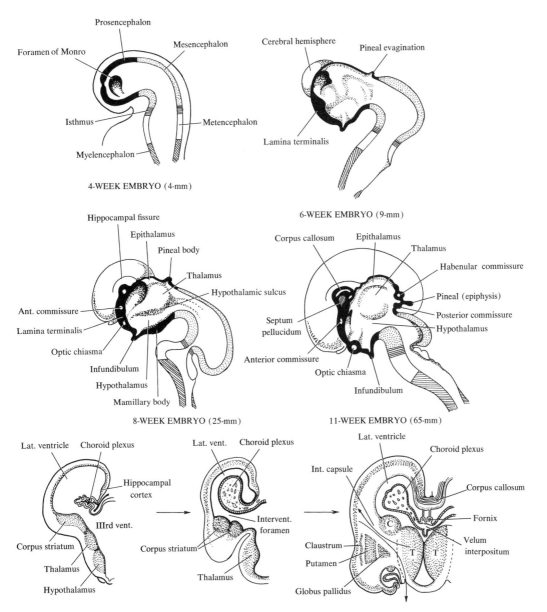

Prosencephalon

Foramen of Monro

Mesencephalon

Isthmus

Metencephalon

Myelencephalon

4-WEEK EMBRYO (4-mm)

Cerebral hemisphere

Pineal evagination

Lamina terminalis

6-WEEK EMBRYO (9-mm)

Hippocampal fissure

Epithalamus

Pineal body

Thalamus

Hypothalamic sulcus

Ant. commissure

Lamina terminalis

Optic chiasma

Infundibulum

Hypothalamus

Mamillary body

8-WEEK EMBRYO (25-mm)

Corpus callosum

Epithalamus

Thalamus

Habenular commissure

Pineal (epiphysis)

Posterior commissure

Septum pellucidum

Hypothalamus

Anterior commissure

Optic chiasma

Infundibulum

11-WEEK EMBRYO (65-mm)

Lat. ventricle

Choroid plexus

Hippocampal cortex

IIIrd vent.

Corpus striatum

Thalamus

Hypothalamus

Lat. vent.

Choroid plexus

Intervent. foramen

Corpus striatum

Thalamus

Lat. ventricle

Choroid plexus

Int. capsule

Corpus callosum

Fornix

Velum interpositum

Claustrum

Putamen

Globus pallidus

C

T   T

SECTIONS AT LEVEL OF INTERVENTRICULAR FORAMEN (Early to late stages)

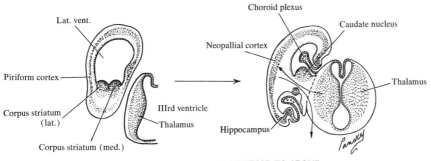

Lat. vent.

Choroid plexus

Caudate nucleus

Neopallial cortex

Piriform cortex

Thalamus

Corpus striatum (lat.)

IIIrd ventricle

Thalamus

Corpus striatum (med.)

Hippocampus

SECTIONS POSTERIOR TO ABOVE

12. BRAIN: DIENCEPHALON—THE HYPOPHYSIS

I. The diencephalon (cont.)

A. THE HYPOPHYSIS (PITUITARY GLAND): develops from 2 completely different parts:

1. An ectodermal outpocketing of the stomodeum just in front of the buccopharyngeal membrane called Rathke's pouch
   a. At 3 weeks of embryonic life, Rathke's pouch is seen as an invagination of the stomodeum and grows dorsally toward the infundibulum
   b. By the end of month 2, it loses its connection with the oral cavity and is in close contact with the infundibulum
   c. Occasionally, a small part of the pouch persists in the pharyngeal wall and is referred to as the pharyngeal hypophysis
   d. With development, the cells of the anterior wall of Rathke's pouch increase rapidly in number and form the anterior lobe of the hypophysis or the adenohypophysis
      i. A small extension of the adenohypophysis, the pars tuberalis, later grows along the stalk of the infundibulum and surrounds it
   e. The posterior wall of the pouch forms the pars intermedia of the hypophysis
   f. In man, the lumen of the pouch is usually obliterated, but may persist as a narrow cleft
   g. A not uncommon remnant of the pouch is the craniopharyngioma or Rathke's pouch tumor which is found intracranially and whose symptoms resemble those of anterior lobe tumors. The onset of symptoms is usually before the age of 15 years

2. A downward neuroectodermal extension of the diencephalon forms the infundibulum
   a. The infundibulum gives rise to the median eminence, the infundibular stem, and the pars nervosa. All three make up the neurohypophysis
      i. It is composed of neuroglia cells which later differentiate to form pituicytes
      ii. It also contains many nerve fibers coming from the hypothalamic area of the diencephalon

DEVELOPMENT OF PITUITARY GLAND

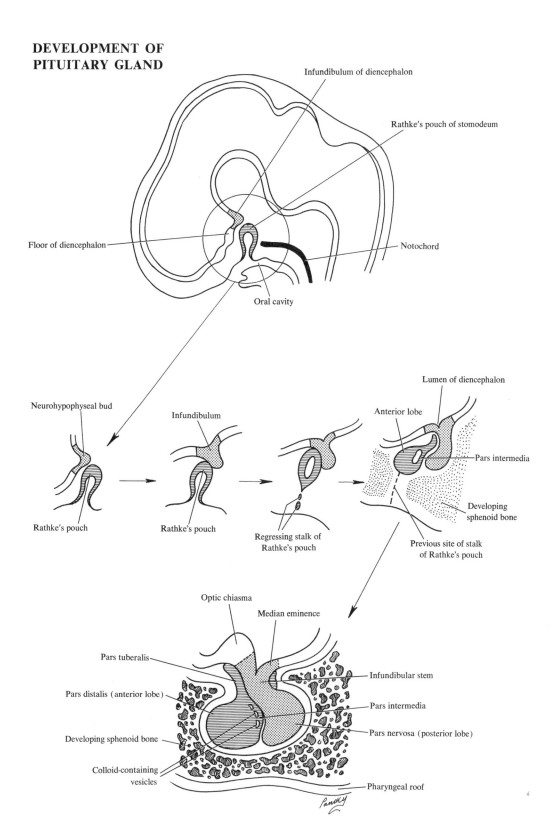

Infundibulum of diencephalon

Rathke's pouch of stomodeum

Floor of diencephalon

Notochord

Oral cavity

Neurohypophyseal bud

Infundibulum

Lumen of diencephalon

Anterior lobe

Rathke's pouch

Rathke's pouch

Regressing stalk of Rathke's pouch

Pars intermedia

Developing sphenoid bone

Previous site of stalk of Rathke's pouch

Optic chiasma

Median eminence

Pars tuberalis

Infundibular stem

Pars distalis (anterior lobe)

Pars intermedia

Developing sphenoid bone

Pars nervosa (posterior lobe)

Colloid-containing vesicles

Pharyngeal roof

13. BRAIN: TELENCEPHALON—CEREBRAL HEMISPHERES

I. **The telencephalon,** the most rostral portion of the brain vesicle, consists of 2 lateral outpocketings, the cerebral hemispheres, and a median portion, the lamina terminalis. The cavities of the hemispheres, the lateral ventricles, communicate with the lumen of the diencephalon via the interventricular foramina of Monro

  A. THE CEREBRAL HEMISPHERES arise at the beginning of week 5 as bilateral evaginations of the lateral wall of the prosencephalon

    1. Cell proliferation occurs in the neuroepithelial layer and large numbers of neuroblasts are produced for the mantle layer

    2. By the middle of month 2, the mantle layer in the basal portions of the hemispheres (the part which initially formed the forward extension of the thalamus) increases in size

      a. This rapidly growing area bulges into the lumen of the lateral ventricle as well as into the floor of the foramen of Monro, has a striated appearance, and is called the corpus striatum

      b. The rest of the hemisphere wall temporarily stays thin and is called the pallium, the primordium of the cerebral cortex

    3. In the region where the wall of the hemisphere is attached to the roof of the diencephalon, it fails to develop neuroblasts and stays very thin

      a. Here the hemisphere wall consists of a single layer of ependymal cells covered by vascular mesenchyme, and together they form the choroid plexus

        i. The position of choroid plexuses on the medial surface of the hemisphere is a result of disproportionate growth of parts of the hemisphere

        ii. The plexus should have formed the roof of the hemisphere, but it protrudes into the lateral ventricle along a line, the choroidal fissure

      b. Just above the choroidal fissure, the wall of the pallium is thickened to form the hippocampus, which has an olfactory function and bulges out into the lateral ventricle

    4. As the hemispheres expand, they gradually cover the lateral aspect of the diencephalon, mesencephalon, and cephalic portion of the metencephalon

    5. The corpus striatum, a part of the hemisphere wall, also expands posteriorly and forms a longitudinal ridge in the floor of the lateral ventricle

      a. During this growth, the corpus striatum is divided into 2 portions (as a result of afferent and efferent axons passing to and from the cortex of the hemisphere and breaking through the nuclear mass of the corpus striatum)

        i. A dorsomedial portion which forms the caudate nucleus

        ii. A ventromedial portion which forms the lentiform nucleus

        iii. The fiber bundles between them known as the internal capsule

        iv. The lentiform nucleus is later divided into a lateral part, the putamen, and a medial, lightly staining part, the globus pallidus

    6. As a result of posterior expansion of the hemispheres, its medial surface approaches the lateral surface of the diencephalon, and they fuse

      a. With fusion, the caudate nucleus and the thalamus come into close contact, while the nerve fibers descending from the cortex of the hemisphere pass through the plane of fusion to enter the peduncles of the mesencephalon

    7. Growth of the hemispheres in an anterior, dorsal, and inferior direction results in the formation of the frontal, temporal, and occipital lobes

      a. A lag in growth in the region overlying the corpus striatum, between the frontal and temporal lobes, results in its depression and is called the insula. (This region is later overgrown by the adjacent lobes and at birth it is fully covered.)

    8. During late fetal life, the surface of the hemispheres grows so rapidly that many convolutions or gyri, separated by sulci and fissures, are seen on its surface

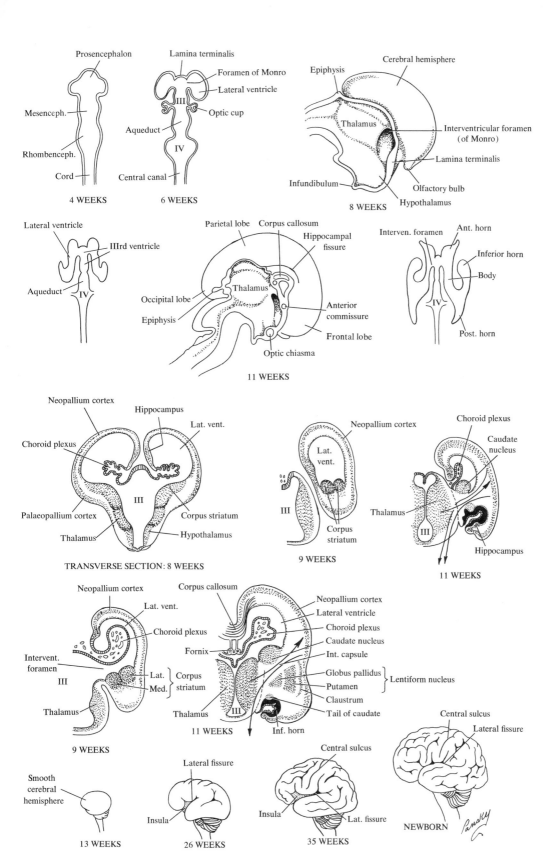

4 WEEKS

Prosenceph.
Mesenceph.
Rhombenceph.
Cord

6 WEEKS

Lamina terminalis
Foramen of Monro
Lateral ventricle
Optic cup
III
Aqueduct
IV
Central canal

8 WEEKS

Epiphysis
Cerebral hemisphere
Thalamus
Interventricular foramen (of Monro)
Lamina terminalis
Infundibulum
Olfactory bulb
Hypothalamus

Lateral ventricle
IIIrd ventricle
Aqueduct
IV

11 WEEKS

Parietal lobe
Corpus callosum
Hippocampal fissure
Thalamus
Occipital lobe
Epiphysis
Anterior commissure
Frontal lobe
Optic chiasma

Interven. foramen
Ant. horn
Inferior horn
Body
IV
Post. horn

TRANSVERSE SECTION: 8 WEEKS

Neopallium cortex
Hippocampus
Lat. vent.
Choroid plexus
III
Palaeopallium cortex
Corpus striatum
Thalamus
Hypothalamus

9 WEEKS

Neopallium cortex
Lat. vent.
III
Corpus striatum

11 WEEKS

Choroid plexus
Caudate nucleus
Thalamus
III
Hippocampus

9 WEEKS

Neopallium cortex
Lat. vent.
Choroid plexus
Intervent. foramen
III
Lat. / Med. Corpus striatum
Thalamus

11 WEEKS

Corpus callosum
Neopallium cortex
Lateral ventricle
Choroid plexus
Caudate nucleus
Int. capsule
Fornix
Globus pallidus
Putamen } Lentiform nucleus
Claustrum
Thalamus
III
Inf. horn
Tail of caudate

13 WEEKS

Smooth cerebral hemisphere

26 WEEKS

Lateral fissure
Insula

35 WEEKS

Central sulcus
Insula
Lat. fissure

NEWBORN

Central sulcus
Lateral fissure

Pansky

14. BRAIN: TELENCEPHALON—CEREBRAL CORTEX AND COMMISSURES

I. The telencephalon (cont.)

A. CEREBRAL CORTEX DEVELOPMENT is from the pallium and is divided into 2 regions:
1. Paleo- or archipallium: an area just lateral to the corpus striatum
   a. Appears in the 7th week of development and is formed by a mixture of cells migrating from the striated mantle layer into the marginal zone
   b. The cells here are established from a thin nuclear layer near the surface that serves as a relay station for olfactory impulses
2. Neopallium occupies the rest of the surface of the hippocampus and paleopallium
   a. The neuroepithelial cells of the neopallium begin to release a series of waves of neuroblasts shortly after the paleopallium appears
   b. When new neuroblasts arrive, they migrate through the earlier formed layer of cells to reach subpial position. Thus, the early formed neuroblasts get a deeper position in the cortex than those formed later
3. At birth, the cortex appears stratified due to layers of neuroblasts and specific cell differentiation. In addition, different cortical areas acquire specific cell types; i.e., the motor cortex has many pyramidal cells, the sensory cortex is characterized by granular cells

B. COMMISSURES are fiber bundles, in the adult, which cross the midline and connect the right and left halves of the hemispheres
1. The most important make use of the lamina terminalis, which extends from the roof plate of the diencephalon to the optic chiasma in the midportion of the telencephalon
2. The 1st commissure to appear is the anterior commissure which is seen by the 3rd month of development and consists of fibers connecting the olfactory bulb and interconnects the left and right temporal lobes
3. The 2nd one to appear is the hippocampal or fornix commissure
   a. Its fibers arise in the hippocampus, converge on the lamina terminalis near the roof plate of the diencephalon, and continue to form an arching system, just outside of the choroid fissure, to the mamillary bodies and hypothalamus
   b. It regresses greatly as the corpus callosum grows
4. The most important commissure is the corpus callosum, which appears by week 10 of development and connects nonolfactory areas of the two hemispheres
   a. Initially it forms a small bundle in the lamina terminalis just rostral to the hippocampal commissure, but with continuous expansion of neopallium, it extends anteriorly and then posteriorly, arching over the thin roof of the diencephalon
   b. Growth of the corpus callosum in an anterior direction pulls the area of the lamina terminalis away from the fornix commissure and thins out the lamina terminalis locally, forming the septum pellucidum (may contain a small cavity which has no relation to the brain vesicles)
      i. It has been suggested that the septum represents the apposed walls of the 2 hemispheres anterior to the lamina terminalis
5. Several other commissures appear in the lamina terminalis:
   a. Two appear just below and rostral to the stalk of the pineal gland, the posterior commissure, and the habenular commissure
   b. The third, the optic chiasma, seen in the rostral wall of the diencephalon contains fibers from the medial halves of the retinae which cross on their way to the lateral geniculate bodies and anterior colliculi

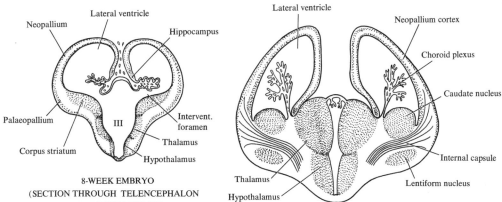

Neopallium
Lateral ventricle
Hippocampus
Palaeopallium
III
Intervent. foramen
Corpus striatum
Thalamus
Hypothalamus

8-WEEK EMBRYO
(SECTION THROUGH TELENCEPHALON
AND DIENCEPHALON)

(After Hochstetter)

Lateral ventricle
Neopallium cortex
Choroid plexus
Caudate nucleus
Internal capsule
Thalamus
Lentiform nucleus
Hypothalamus

10-WEEK EMBRYO
(THROUGH HEMISPHERE
AND DIENCEPHALON)

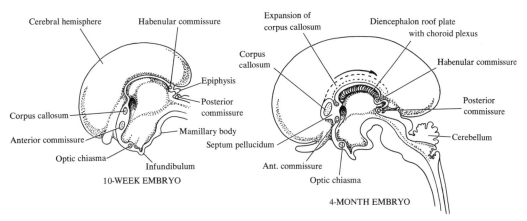

Cerebral hemisphere
Habenular commissure
Corpus callosum
Epiphysis
Posterior commissure
Corpus callosum
Mamillary body
Anterior commissure
Optic chiasma
Infundibulum

10-WEEK EMBRYO

Expansion of corpus callosum
Diencephalon roof plate with choroid plexus
Corpus callosum
Habenular commissure
Posterior commissure
Cerebellum
Septum pellucidum
Ant. commissure
Optic chiasma

4-MONTH EMBRYO

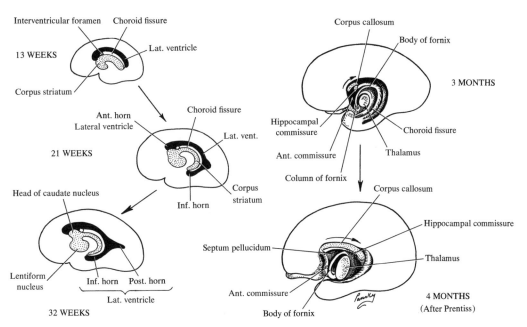

Interventricular foramen
Choroid fissure
Lat. ventricle
13 WEEKS
Corpus striatum

Ant. horn
Lateral ventricle
Choroid fissure
Lat. vent.
21 WEEKS
Corpus striatum
Inf. horn

Head of caudate nucleus
Lentiform nucleus
Inf. horn
Post. horn
Lat. ventricle
32 WEEKS

Corpus callosum
Body of fornix
3 MONTHS
Hippocampal commissure
Choroid fissure
Ant. commissure
Thalamus
Column of fornix

Corpus callosum
Hippocampal commissure
Septum pellucidum
Thalamus
Ant. commissure
Body of fornix
4 MONTHS
(After Prentiss)

Pansky

15. BRAIN: CONGENITAL MALFORMATIONS

I. Primary cause of many malformations: ossification defect in the skull bones

A. THE MOST FREQUENTLY AFFECTED BONE is the squamous portion of the occipital bone, which may be partially or totally missing, and the resultant opening is often confluent with the foramen magnum

   1. If the opening of occipital bone is small, only the meninges bulge through it, creating a meningocele
   2. If the defect is large, part of the brain also bulges out; called meningoencephalocele
   3. If the defect is very large, part of the brain and part of the ventricle may penetrate through the opening into the meningeal sac, creating a meningohydroencephalocele

II. Anencephalus: characterized by failure of cephalic part of neural tube to close

A. AT BIRTH, the brain consists of a mass of degenerated tissue exposed to the surface

   1. The defect is almost always continuous with an open cord in the cervical region
   2. The skull vault is gone, giving the head a characteristic appearance: the eyes bulge forward; neck absent; surfaces of face and chest form a continuous plane
   3. The last 2 months of pregnancy are characterized by hydramnios since the fetus lacks the control mechanism for swallowing
   4. The abnormality is recognized by x-ray since the skull vault is missing

B. THIS IS A COMMON ABNORMALITY (1:1,000): 4 times more frequently in females than males; 4 times more frequently in whites than blacks

III. Hydrocephalus: characterized by an abnormal accumulation of CSF in ventricular system (may be an external hydrocephalus with fluid between brain and dura)

A. HYDROCEPHALUS is thought to be due to an obstruction of the aqueduct which prevents CSF of the 3rd and lateral ventricles from passing into the 4th ventricle and from there into the subarachnoid space

   1. Malformation is often accompanied by a widening of skull sutures and the bones gradually thin out

IV. Other malformations: those above are serious and often incompatible with life. In addition, other defects of the CNS occur without external manifestations:

A. THE CORPUS CALLOSUM may be partially or completely absent without much functional disturbance

B. COMPLETE OR PARTIAL CEREBELLAR ABSENCE may show only a slight coordination disturbance

C. SEVERE IMBECILITY OR IDIOCY may show little morphologic brain abnormalities

V. Chromosomal and environmental phenomena

A. CONGENITAL MALFORMATION may be caused by chromosomal abnormalities, particularly mental retardation

   1. Mongolism is associated with 47 chromosomes, #21 being represented $3\times$ instead of twice; condition called trisomy
   2. Klinefelter's syndrome also shows 47 chromosomes with an extra X chromosome and accounts for about 1% of mentally defective children
   3. Many sex chromosome abnormalities are associated with mental retardation
   4. Congenital metabolic defects, i.e., phenylketonurea, cause or accompany mental retardation

B. ENVIRONMENTAL TERATOGENS cause CNS malformations in offspring of experimental animals

C. CNS ABNORMALITIES can occur from fetal infection by Toxoplasma organism

D. RADIATION during early developmental stages may lead to microcephalus

E. VIRAL DISEASES and their effects are uncertain and variable

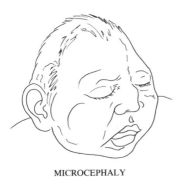

MICROCEPHALY

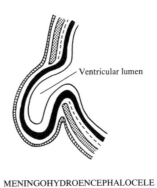

HYDROCEPHALY

ANENCEPHALY

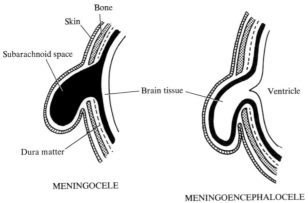

Skin
Bone

Subarachnoid space

Brain tissue

Ventricle

Dura matter

MENINGOCELE

MENINGOENCEPHALOCELE

Ventricular lumen

MENINGOHYDROENCEPHALOCELE

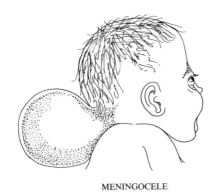

MENINGOCELE

MENINGOHYDROENCEPHALOCELE

16. PHYSIOLOGIC BRAIN DEVELOPMENT: CRITICAL PERIODS

I. **The growth and development of interconnections** between neurons and the development of the behavioral activities of the brain are determined by genetic (phylogenetic and ontogenetic) factors and environmental factors (prenatal and postnatal experience)

A. DURING BRAIN DEVELOPMENT, stages of irreversible differentiation occur (critical periods) in structure (formation of fiber pathways and patterns of synaptic interconnections) and function (development of instinctual behavior, sensory perception, and social behavior)

B. NORMAL ENVIRONMENTAL INTERACTION is essential during the critical periods for normal structural and functional brain development to occur

II. **Outgrowth of axons and development of synaptic connections:** experimental evidence from various vertebrate species indicates that the positions of nerve cell bodies are "chemically marked" and that their axons grow to specific "target sites" along relatively specific pathways, guided by chemical and structural cues and chemical recognition mechanisms. The timing of axon growth has been shown to be critical in some instances to ensure correct linkage between neurons (e.g., frog retinal development)

III. **Critical periods in behavioral development** are also known.

A. IMPRINTING: perceptual or social stimuli are important for early development in many species. Soon after birth, birds become socially attached to a nearby animal, usually the mother, but sometimes imprinting occurs between a newly hatched bird and a human observer. This occurs during a critical period of a few hours

B. NORMAL DEVELOPMENT of human infants is compromised if normal social interactions with other adults (and/or children) is absent or reduced. Severely socially deprived infants exhibit abnormal behavior, often for their entire life. This is in contrast to those who may experience sensory or social deprivation later in life yet retain or recover normal behavior

IV. **Sensory deprivation:** there are critical periods for sensory perception which must not be exceeded if normal sensory functions are to be developed

A. NORMAL VISUAL PERCEPTION develops during the first few months of postnatal life. If one eye of a monkey or kitten is sutured shut during the first 6 months of life, there is permanent loss of vision in that eye after the sutures are removed. The critical period may be as short as 1 week for the development of visual perception

B. BINOCULAR EYELID CLOSURE has a much less severe effect on visual perception development after the eyes open
   1. The development of normal binocular vision depends on there being balanced and synchronous activity in the sensory pathways from the two eyes
   2. If one eye is closed, it is at a competitive disadvantage with respect to the open eye. The action potentials developed in the open eye may be critical for development of connections to the lateral geniculate nucleus and striate cortex

Ocular dominance columns

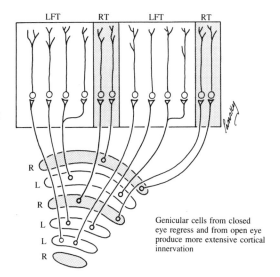

NORMAL

Ocular columns are of equal
size when both eyes
are open during early
postnatal development

Cerebral striate visual
cortex

Lateral geniculate
body

RT. EYE CLOSED

If rt. eye is closed, the
rt. eye columns are
narrower than those for
lft. eye.

Genicular cells from closed
eye regress and from open eye
produce more extensive cortical
innervation

Stages in the Developmental of Ocular Dominance Columns in the Monkey

Stage	Time period
Beginning of afferent fiber segregation	6–3 weeks prenatal
Beginning of critical period	Birth
Height of sensitivity to monocular deprivation	From birth to 6 weeks postnatal
Nearly complete afferent segregation	3–6 weeks
End of layer IVc susceptibility to monocular deprivation	6–8 weeks
End of susceptibility of upper and lower layers of cortex to monocular deprivation	6 months to 1 year postnatal

17. EMBRYOLOGY OF THE AUTONOMIC NERVOUS SYSTEM: GENERAL DEVELOPMENT

I. **Introduction:** functionally the system is divided into 2 parts: a sympathetic part, found in the thoracolumbar region, and a parasympathetic part, found in the cephalic (cranial) and sacral regions. Both divisions are composed of characteristic two-neuron efferent chains

II. The sympathetic nervous system

A. SYMPATHETIC NEUROBLASTS, originating in the neural crest of the thoracic region, in the 5th week of development migrate on each side of the spinal cord to the region just behind the dorsal aorta, where they form a bilateral chain of segmentally arranged sympathetic ganglia interconnected by longitudinal nerve fibers—the sympathetic chains found on each side of the vertebral column

1. From their thoracic position, the neuroblasts migrate toward the cervical and lumbosacral regions extending the sympathetic chains to full length
2. Though the ganglia are initially segmentally arranged, this arrangement is obscured, particularly in the cervical region, by fusion of the ganglia
3. Some neuroblasts migrate in front of the aorta to form preaortic ganglia, i.e., celiac and mesenteric ganglia found at the roots of the main aortic branches
4. Other neuroblasts migrate to the heart, lungs, and GI tract to give rise to sympathetic organ plexuses, i.e., submucosal plexus in the gut wall

B. WITH ESTABLISHMENT OF THE SYMPATHETIC CHAINS, nerve fibers originating in the visceroefferent column of the thoracolumbar segments of the spinal cord penetrate the ganglia of the chains and form synapses around the developing neuroblasts

1. Some fibers either extend to higher or lower levels in the sympathetic chains or to preaortic or collateral ganglia before synapsing; called preganglionic fibers
   a. They have myelin sheaths
   b. They stimulate the sympathetic ganglion cells into action
   c. Passing from the spinal nerves to the ganglia, they form the so-called white rami communicantes (white communicating rami)
   d. Since the visceroefferent column extends only from T1 to L2 segments of the cord, the white rami are found only between these levels
2. The axons of the sympathetic ganglion cells are called postganglionic fibers
   a. They have no myelin
   b. They pass either to other levels of the sympathetic chain or extend to the heart, lungs, and GI tract
   c. Other fibers called gray communicating rami pass from the chain to the spinal nerves and from there to peripheral blood vessels, hair, and sweat glands

III. The parasympathetic nervous system

A. THE ORIGIN of the parasympathetic ganglia found along the oculomotor (III), facial (VII), glossopharyngeal (IX), and vagus (X) nerves is controversial

1. Some believe the cells of these ganglia migrate out of the CNS along the preganglionic fibers of the nerves
2. Some feel the cells arise from neuroblasts originating in the sensory ganglia of the Vth, VIIth, and IXth nerves

B. THE POSTGANGLIONIC FIBERS of the parasympathetic ganglia pass to the branchial arches and to the cardiac, pulmonary, and intestinal plexuses

C. THE ACTION OF THIS SYSTEM is said to be antagonistic to those of the sympathetic system

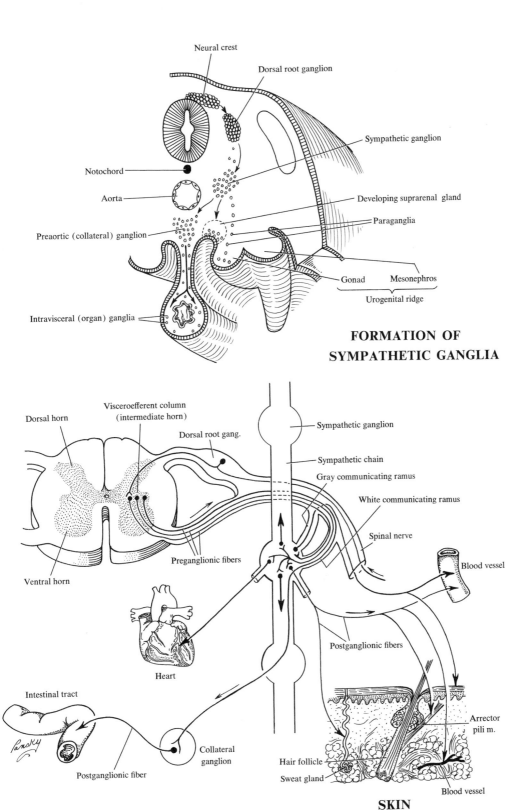

Neural crest

Dorsal root ganglion

Sympathetic ganglion

Notochord

Aorta

Developing suprarenal gland

Paraganglia

Preaortic (collateral) ganglion

Gonad   Mesonephros

Urogenital ridge

Intravisceral (organ) ganglia

FORMATION OF SYMPATHETIC GANGLIA

Dorsal horn

Visceroefferent column (intermediate horn)

Dorsal root gang.

Sympathetic ganglion

Sympathetic chain

Gray communicating ramus

White communicating ramus

Spinal nerve

Blood vessel

Preganglionic fibers

Ventral horn

Postganglionic fibers

Heart

Intestinal tract

Pansky

Collateral ganglion

Arrector pili m.

Hair follicle

Sweat gland

Postganglionic fiber

Blood vessel

SKIN

18. AUTONOMIC NERVOUS SYSTEM: SUPRARENAL GLAND AND CAROTID BODY

I. The suprarenal gland

A. THE GLAND develops from 2 components:
1. A mesodermal portion which forms the cortex
2. An ectodermal portion which forms the medulla

B. THE CORTEX
1. During the 5th week of development, mesothelial cells located between the root of the mesentery and developing gonad on the posterior abdominal wall begin to proliferate and penetrate the underlying mesenchyme
2. They differentiate into large acidophilic organs which form the fetal or primitive suprarenal cortex
3. Shortly after, a 2nd wave of cells from the mesothelium penetrates the mesenchyme and surrounds the original acidophilic cell mass
   a. These cells are smaller than the 1st wave
   b. These smaller cells form the definitive cortex of the gland
4. After birth, the fetal cortex regresses rapidly except for its outer layer which differentiates into the reticular zone
5. The adult structure of the cortex is not achieved until near puberty

C. THE MEDULLA
1. While the fetal cortex is forming, cells originating in the sympathetic system invade its medial part where they become arranged in clusters and cords
   a. These cells give rise to the adrenal medulla, but do not form nerve processes but rather stain yellow-brown with chromic salts and are thus called chromaffin cells
      i. The staining is probably due to epinephrine and norepinephrine in the cells
2. During embryonic life, the chromaffin cells are widely scattered throughout the embryo, but in the adult, the only persisting group is found in the adrenal medulla.

D. THE ZONA GLOMERULOSA AND ZONA FASCICULATA are present at birth, but the zona reticularis is not recognizable until the end of the 3rd year

E. THE FETAL ADRENAL GLAND is about 20 times its relative size in the adult, due to an extensive fetal cortex
1. The medulla remains relatively small until after birth

F. THE GLANDS rapidly decrease in size postnatally as the fetal cortex regresses, and this cortex normally disappears by the end of the first year
1. The glands lose ⅓ of their weight during the first 2 or 3 weeks after birth and regain their original weight by the end of the 2nd year

G. FEMALE PSEUDOHERMAPHRODITISM: hyperplasia of the fetal adrenal cortex during the fetal period

H. CONGENITAL ADRENAL HYPERPLASIA: may lead to adrenogenital syndrome and is manifested in a variety of clinical forms, and is probably due to enzymatic deficiencies of cortisol biosynthesis

II. The carotid bodies

A. THE CAROTID BODIES are formed by a mesodermal condensation around the origin of the internal carotid arteries and are supplied by branches of the glossopharyngeal (IX) nerve (3rd arch nerve)
1. They are invaded by cells from nearby ganglia and then develop into chemoreceptor organs

B. THE CAROTID BODIES serve as reflex systems in the regulation of blood pressure.

SUPRARENAL GLAND

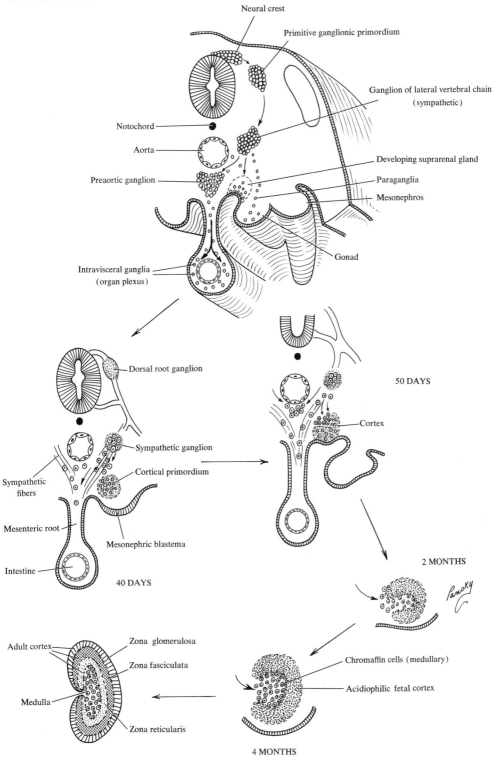

Neural crest

Primitive ganglionic primordium

Ganglion of lateral vertebral chain (sympathetic)

Notochord

Aorta

Developing suprarenal gland

Preaortic ganglion

Paraganglia

Mesonephros

Gonad

Intravisceral ganglia (organ plexus)

Dorsal root ganglion

50 DAYS

Cortex

Sympathetic ganglion

Cortical primordium

Sympathetic fibers

Mesenteric root

Mesonephric blastema

Intestine

40 DAYS

2 MONTHS

Panoky

Adult cortex

Zona glomerulosa

Zona fasciculata

Chromaffin cells (medullary)

Acidiophilic fetal cortex

Medulla

Zona reticularis

4 MONTHS

19. THE EYE: OPTIC CUP, LENS VESICLE, RETINA, IRIS, AND CILIARY BODY

I. The optic cup and lens vesicle
A. THE EARLIEST SIGN of eye development is seen in the 22-day embryo as a pair of shallow grooves on each side of the invaginated forebrain
   1. With closure of the neural tube, the grooves form outpocketings of the forebrain, the optic vesicles, which are in contact with surface ectoderm
   2. In days, the vesicles cause chemical changes in surface ectodermal cells needed for lens formation, the optic vesicles invaginate and form the double-walled optic cup
      a. The inner and outer walls of the cup are initially separated by the intraretinal space, but with development, the lumen disappears and the walls oppose each other
   3. The invagination is not only centrally restricted in the cup but also involves its ventral rim where the choroid fissure is formed which extends along the undersurface of the optic stalk where it tapers off
      a. Fissure formation allows the hyaloid artery to reach the inner eye chamber
      b. During the 7th week, the lips of the choroid fissure fuse, and the mouth of the optic cup becomes a rounded opening, the future pupil
   4. In the meantime, the cells of the surface ectoderm, initially in contact with the optic vesicle, elongate and form the lens placode
      a. The placode subsequently invaginates and develops into the lens vesicle
      b. During week 5, the lens vesicle loses surface contact and is then located in the mouth of the optic cup

II. Retina, iris, and ciliary body
A. DEVELOPMENT OF THE OUTER LAYER of the optic cup is characterized by the appearance of small pigment granules during week 5 to form the pigment layer of the retina
B. THE POSTERIOR $\frac{4}{5}$ OF THE INNER LAYER of the optic cup (pars optica retinae):
   1. This part thickens and undergoes a series of changes similar to those seen in the wall of the brain vesicle.
   2. Bordering the intraretinal space, the ependymal layer differentiates here into the light-receptive elements, the rods and cones
   3. Adjacent to the photoreceptive layer, the mantle layer, as in the brain, gives rise to neurons and supporting cells
      a. In the adult, the outer nuclear layer, the inner nuclear layer, and the ganglion cell layer are distinguished
   4. On the surface of the mantle layer, the marginal zone contains the axons of the nerve cells of the deeper layers; and the nerve fibers in this marginal layer converge toward the optic stalk, which gradually forms the optic nerve
C. THE ANTERIOR $\frac{1}{5}$ of the inner layer of the optic cup (pars caeca retinae):
   1. Changes very little and remains one-cell layer thick but later divides into the pars iridica retinae, forming the inner layer of the iris, and the pars ciliaris retinae, which participates in the formation of the ciliary body
      a. The pars ciliaris retinae is recognized by its marked folding
         i. Externally it is covered by a layer of mesenchyme which forms the ciliary muscle
         ii. Internally, it is connected to the lens by a network of elastic fibers which form the suspensory ligament or zonula
         iii. Contraction of the ciliary muscle changes the tension in the ligament and controls the curvature of the lens
D. IN THE HUMAN, the sphincter and dilator pupillae muscles develop from ectoderm underlying the optic cup in the area between optic cup and overlying surface epithelium
E. THE IRIS in the adult is formed by the pigment-containing internal and external layers of the optic cup and by a layer of vascularized connective tissue which also contains the pupillary muscles

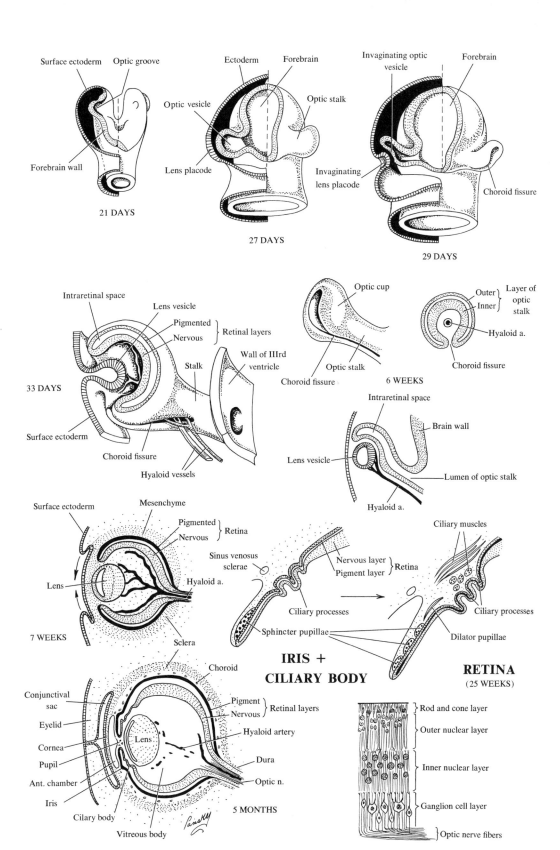

21 DAYS

Surface ectoderm — Optic groove
Forebrain wall

27 DAYS

Ectoderm — Forebrain
Optic vesicle
Optic stalk
Lens placode

29 DAYS

Invaginating optic vesicle — Forebrain
Invaginating lens placode
Choroid fissure

33 DAYS

Intraretinal space — Lens vesicle
Pigmented
Nervous } Retinal layers
Stalk
Wall of IIIrd ventricle
Surface ectoderm
Choroid fissure
Hyaloid vessels

Optic cup
Optic stalk
Choroid fissure

6 WEEKS

Outer
Inner } Layer of optic stalk
Hyaloid a.
Choroid fissure

Intraretinal space
Brain wall
Lens vesicle
Lumen of optic stalk
Hyaloid a.

7 WEEKS

Surface ectoderm — Mesenchyme
Pigmented
Nervous } Retina
Lens
Hyaloid a.
Sclera

Sinus venosus sclerae
Nervous layer
Pigment layer } Retina
Ciliary processes
Sphincter pupillae

IRIS + CILIARY BODY

Ciliary muscles
Ciliary processes
Dilator pupillae

RETINA
(25 WEEKS)

Conjunctival sac
Eyelid
Cornea
Pupil
Ant. chamber
Iris
Cilary body
Vitreous body

Choroid
Pigment
Nervous } Retinal layers
Hyaloid artery
Dura
Optic n.
Lens

5 MONTHS

Rod and cone layer
Outer nuclear layer
Inner nuclear layer
Ganglion cell layer
Optic nerve fibers

-41-

20. THE EYE: LENS, CHOROID, SCLERA, CORNEA, AND OPTIC NERVE

I. The lens
A. SHORTLY AFTER FORMATION of the lens vesicle, the cells of the posterior wall elongate anteriorly and form long fibers which gradually fill the vesicle lumen
B. THE PRIMARY LENS FIBERS reach the epithelium of the anterior wall of the vesicle by the end of the 7th week and form the nucleus of the lens
  1. Lens growth is not finished at this stage, but new, secondary lens fibers are continuously added to the central core
  2. The new fibers arise from cells in the equatorial zone

II. The choroid, sclera, and cornea
A. BY THE END OF THE 5TH WEEK and when the optic cup and lens vesicle have formed, the eye primordium is completely surrounded by loose mesenchyme
  1. The mesenchyme differentiates into an inner layer (comparable to the pia mater of the brain) and an outer layer (comparable to the dura mater)
    a. The inner layer forms a highly vascularized pigmented layer, the choroid
    b. The outer layer develops into the sclera and is continuous with the dura mater around the optic nerve
  2. Mesenchyme layers over the anterior aspect of the eye differentiate differently
    a. The cells arrange themselves so that a space, the anterior chamber, splits the mesenchyme into a thin inner layer just in front of the lens and iris (the iridopupillary membrane) and a thick outer layer which is continuous with the sclera
      i. The anterior chamber itself is lined by flattened mesenchymal cells which form the posterior lining of the cornea as well as the anterior covering of the iridopupillary membrane. The membrane in front of the lens normally disappears completely
B. THE CORNEA, from outside to inside, is formed by:
  1. An epithelial layer derived from surface ectoderm
  2. A layer of dense connective tissue, the substantia propria or stroma
  3. An epithelial layer that borders the anterior chamber
C. THE MESENCHYME, which surrounds the eye primordium, also invades the inside of the optic cup via the choroid fissure:
  1. It participates in formation of the hyaloid vessels (during intrauterine life supply the lens and form the vascular layer found on the retinal inner surface)
  2. Forms a delicate network of fibers between the lens and the retina, the interstitial spaces, which later fill with a transparent gelatinous substance, the vitreous body

III. The optic nerve
A. THE OPTIC CUP is initially connected to the brain by the optic stalk
  1. The choroid fissure is a groove on the stalk's ventral surface
  2. The nerve fibers of the retina returning to the brain are found among the cells of the inner wall of the stalk
B. DURING WEEK 7 the choroid fissure closes and a narrow tunnel is formed inside the optic stalk
C. WITH AN INCREASING NUMBER of nerve fibers growing toward the brain, the inner stalk walls increase in size and fuse with the outer walls
  1. The cells of the inner layer provide a network of neuroglia cells that support the optic nerve fibers
D. THUS, THE OPTIC STALK is transformed into the optic nerve and in its center is the hyaloid artery, later called the central artery of the retina

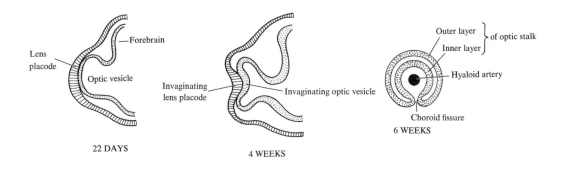

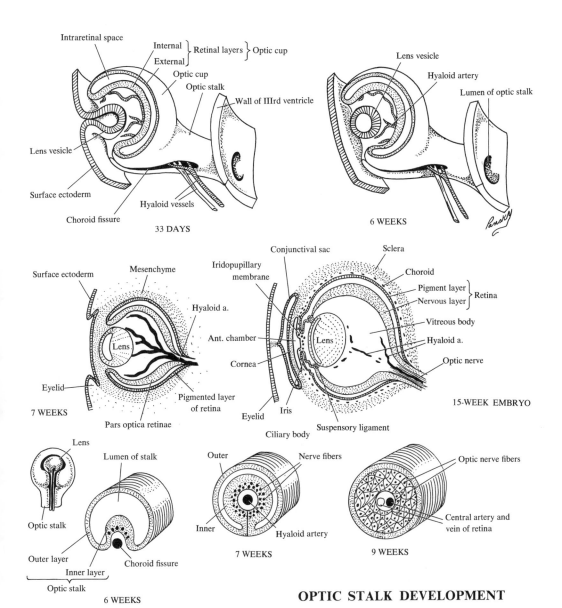

OPTIC STALK DEVELOPMENT

- 43 -

21. THE EYE: CONGENITAL MALFORMATIONS

I. Coloboma iridis
A. NORMALLY, THE CHOROID FISSURE CLOSES during the 7th week of development, but when it fails to close, a cleft persists which is usually located in the iris only and is called coloboma iridis
   1. The cleft may extend into the ciliary body, the retina, the choroid, and the optic nerve
B. THIS MALFORMATION is frequently seen in combination with other abnormalities

II. Persistent iridopupillary membrane
A. THE MEMBRANE usually disappears during intrauterine life, but if reabsorption is incomplete, a network of connective tissue is seen suspended in front of the pupil
   1. This usually does not cause visual disturbance

III. Microphthalmia
A. THE OVERALL SIZE of the eye is too small, and the eyeball may be reduced to $\frac{2}{3}$ of its normal volume
B. THIS IS USUALLY ASSOCIATED with other ocular abnormalities
C. FREQUENTLY RESULTS from intrauterine infections such as cytomegalovirus or toxoplasmosis

IV. Anophthalmia
A. SOMETIMES THE EYE is grossly absent, and there is no trace of the eyeball except as seen by histologic means
B. THIS CONDITION is usually accompanied by other serious craniocerebral abnormalities

V. Cyclopia
A. BOTH ORBITS are joined, and there exists only one median eye
B. THIS IS A RARE MALFORMATION and is often accompanied by a proboscis and other cerebral abnormalities

VI. Buphthalmos or congenital glaucoma
A. THE EYE IS GREATLY ENLARGED as a result of insufficient drainage of its aqueous humor
B. THERE IS HIGH INTRAOCULAR PRESSURE caused by an absence of the sinus venosus sclerae
C. THE CONDITION is frequently of genetic origin, but it is also seen after maternal rubella infection

VII. Congenital cataract
A. THE LENS has become opaque during intrauterine life
B. THE CONDITION is usually genetically determined, but it may be caused by environmental factors
   1. Seen in children of mothers who suffered from German measles between the 4th and 7th week of pregnancy
   2. If the mother is infected after the 7th week of pregnancy, the lens escapes damage, but the child may be deaf as a result of imperfect differentiation of the cochlea

VIII. Retinocele: herniation of the retina into the sclera, caused by failure of the choroid fissure to close. If severe, can cause eye protrusion

IX. Other conditions
A. A VIRUS may produce abnormalities, but the eye is only sensitive during certain developmental states
B. OTHER TERATOGENIC AGENTS related to eye abnormalities are hypoxia, vitamin deficiency, hypervitaminosis A, thyroid deficiency, and x-irradiation

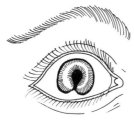

COLOBOMA IRIDIS

PARTIALLY PERSISTENT
IRIDOPUPILLARY MEMBRANE

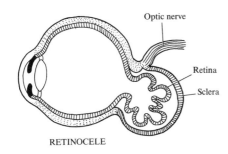

Optic nerve

Retina

Sclera

RETINOCELE

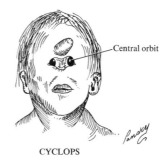

Central orbit

CYCLOPS

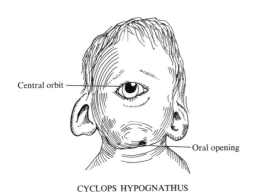

Central orbit

Oral opening

CYCLOPS HYPOGNATHUS
(nose absent; mouth rudimentary)

22. THE INTERNAL EAR

I. **Introduction:** the adult ear forms one unit serving both hearing and equilibrium. In the embryo, the ear develops from 3 different parts: the external ear, the sound-collecting organ, develops from the dorsal portion of the 1st pharyngeal cleft and 6 mesenchymal swellings; the middle ear, a sound conductor between the external and internal ear, arises from the 1st pharyngeal pouch; and the internal ear (converts sound waves into nerve impulses, registers equilibrium changes) forms from the ectodermal otic vesicle

II. **The internal ear**
A. THE OTIC PLACODES are the first indication of ear development seen in embryos of about 22 days as thickening of the surface ectoderm on each side of the rhombencephalon which invaginate rapidly and form the otic (or auditory) vesicles or otocysts
B. DURING LATER DEVELOPMENT, each vesicle divides into:
   1. A ventral component which gives rise to the saccule and the cochlear duct
   2. A dorsal component forms the utricle, semicircular canals, endolymphatic duct
   3. The structures are collectively known as the membranous labyrinth
C. THE MEMBRANOUS LABYRINTH is embedded in mesenchyme, but, with time, the surrounding mesenchyme converts into a cartilaginous shell, which in turn is ossified to form the bony labyrinth
   1. Thus the membranous labyrinth is entirely encased in a bony labyrinth, with only narrow perilymphatic spaces separating the two
D. DURING OTIC VESICLE FORMATION, a small group of cells breaks away from its wall and forms the statoacoustic ganglion (other cells probably come from neural crest)
   1. The ganglion subsequently splits into cochlear and vestibular portions, with saccule, utricle, and semicircular canals (equilibrium)
E. THE SACCULE, COCHLEA, AND ORGAN OF CORTI
   1. In the 6th week of development, the saccular part of the otic vesicle forms a tubular-shaped outpocketing at its lower pole called the cochlear duct
      a. The duct penetrates the surrounding mesenchyme in a spiral fashion until, at the end of the 8th month, it has completed $2\frac{1}{2}$ turns
      b. Its connection with the remaining part of the saccule is confined to a narrow pathway, the ductus reuniens
   2. The mesenchyme around the cochlear duct differentiates into a thin, fibrous basement membrane lining the outside of the duct and a large shell of cartilage
   3. In week 10, the cartilage shell undergoes vacuolization and 2 perilymphatic spaces are formed, the scala vestibuli and the scala tympani
   4. The cochlear duct then is separated from the scala vestibuli by the vestibular membrane and from the scala tympani by the basilar membrane
   5. The lateral wall of the cochlear duct is attached to the surrounding cartilage by the spiral ligament. The duct is also connected and partly supported by a long cartilaginous process, the modiolus, the future axis of the bony cochlea
   6. The epithelial cells of the cochlear duct are initially alike, but with development they form 2 ridges:
      a. An inner ridge, the future spiral limbus, is the larger of the two and is located close to the cochlear center
      b. An outer ridge, which forms one row of inner and 3–4 rows of outer hair cells, the sensory cells of the auditory system
      c. The hair cells are covered by the tectorial membrane, initially a fibrillar gelatinous substance, which is carried by the spiral limbus and rests with its tip on the hair cells.
   7. The neuroepithelial cells and the covering tectorial membrane together are known as the organ of Corti, the true organ of hearing
   8. The impulses received by the organ are transmitted to the spiral ganglion and then to the nervous system via auditory fibers of the VIIIth cranial nerve

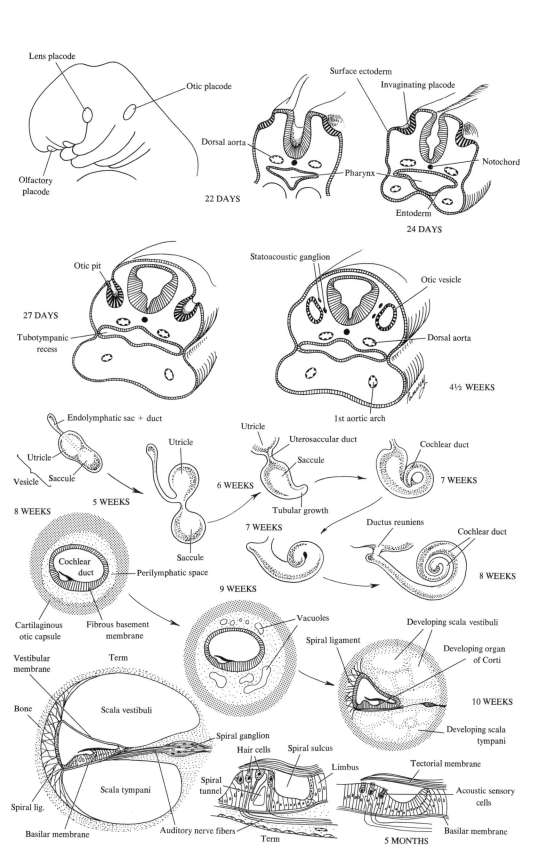

Lens placode

Otic placode

Olfactory placode

Surface ectoderm

Invaginating placode

Dorsal aorta

Pharynx

Notochord

Entoderm

22 DAYS

24 DAYS

Otic pit

Statoacoustic ganglion

Otic vesicle

27 DAYS

Tubotympanic recess

Dorsal aorta

4½ WEEKS

1st aortic arch

Endolymphatic sac + duct

Utricle

Utricle

Uterosaccular duct

Cochlear duct

Utricle

Vesicle

Saccule

Saccule

7 WEEKS

8 WEEKS

5 WEEKS

6 WEEKS

Tubular growth

Saccule

Ductus reuniens

Cochlear duct

Cochlear duct

Perilymphatic space

7 WEEKS

8 WEEKS

9 WEEKS

Cartilaginous otic capsule

Fibrous basement membrane

Vacuoles

Spiral ligament

Developing scala vestibuli

Developing organ of Corti

Vestibular membrane

Term

10 WEEKS

Bone

Scala vestibuli

Developing scala tympani

Spiral ganglion

Spiral sulcus

Hair cells

Limbus

Tectorial membrane

Spiral tunnel

Acoustic sensory cells

Scala tympani

Spiral lig.

Basilar membrane

Auditory nerve fibers

Term

5 MONTHS

Basilar membrane

-47-

23. THE INTERNAL AND EXTERNAL EAR

I. The internal ear (cont.)

A. UTRICLE AND SEMICIRCULAR CANALS
1. During the 6th week of development, the semicircular canals are seen as flattened outpocketings of the utricular portion of the otic vesicle
2. The central parts of the walls of the outpocketings eventually become apposed to each other and disappear, giving origin to the 3 semicircular canals
   a. One end of each canal dilates to form the crus ampullare, while the other end does not widen and is called the crus nonampullare
   b. Two of the latter type fuse, and thus only 5 crura enter the utricle, 3 with an ampulla and 2 without
3. In week 7, the cells in the crus ampullare of each semicircular canal form a crest, the crista ampullaris, containing the sensory cells for maintenance of equilibrium
4. Other sensory areas develop in the walls of the utricle and saccule to form the maculae acousticae
5. Impulses generated in the sensory cells of the cristae and maculae, as a result of position changes of the body, are carried to the brain by the vestibular fibers of the VIIIth cranial nerve

II. The external ear

A. EXTERNAL AUDITORY MEATUS develops from the dorsal portion of the 1st pharyngeal cleft, which grows inward as a funnel-shaped tube to reach the entodermal lining of the tympanic cavity
1. At the beginning of the 3rd month, the epithelial cells deep in the meatus proliferate to form a solid epithelial plate, the meatal plug
2. In the 7th month, the plug dissolves, and the epithelial lining in the floor of the enlarged meatus then participates in the formation of the eardrum (tympanic membrane)
   a. If the meatal plug persists until birth, it may result in congenital deafness

B. THE EARDRUM OR TYMPANIC MEMBRANE
1. The drum is made up of:
   a. The ectodermal epithelial lining at the bottom of the external auditory meatus
   b. The entodermal epithelial lining of the tympanic cavity
   c. An intermediate layer of loose connective tissue
2. The major portion of the eardum is firmly attached to the malleus handle and is formed only after dissolution of the mesenchyme surrounding the ossicles
3. The rest of the eardrum forms the separation between the external meatus and the original tubotympanic recess

C. THE AURICLE OR PINNA develops from a number of mesenchymal proliferations found at the dorsal ends of the 1st and 2nd pharyngeal arches and surrounding the 1st pharyngeal cleft
1. These swellings, 3 on each side of the external meatus, are seen during the 6th week of development
2. The hillocks fuse and are gradually formed into a definitive auricle
3. Developmental abnormalities are not uncommon

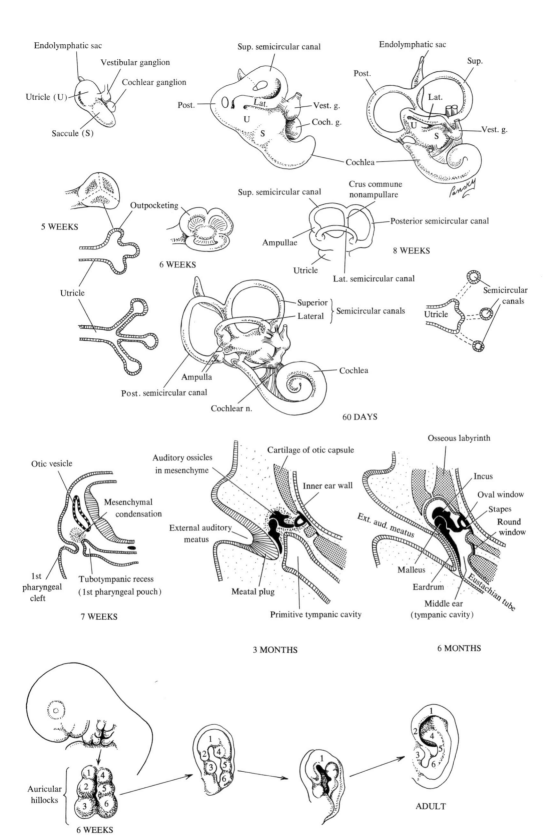

Endolymphatic sac
Vestibular ganglion
Cochlear ganglion
Utricle (U)
Saccule (S)

Sup. semicircular canal
Post.
Lat.
U
S
Vest. g.
Coch. g.
Cochlea

Endolymphatic sac
Post.
Sup.
Lat.
U
S
Vest. g.
Cochlea

Outpocketing
5 WEEKS
Utricle
6 WEEKS

Sup. semicircular canal
Crus commune nonampullare
Posterior semicircular canal
Ampullae
Utricle
Lat. semicircular canal
8 WEEKS

Superior
Lateral } Semicircular canals
Semicircular canals
Utricle
Ampulla
Cochlea
Post. semicircular canal
Cochlear n.
60 DAYS

Otic vesicle
Mesenchymal condensation
1st pharyngeal cleft
Tubotympanic recess (1st pharyngeal pouch)
7 WEEKS

Auditory ossicles in mesenchyme
Cartilage of otic capsule
Inner ear wall
External auditory meatus
Meatal plug
Primitive tympanic cavity
3 MONTHS

Osseous labyrinth
Incus
Oval window
Stapes
Round window
Ext. aud. meatus
Malleus
Eardrum
Middle ear (tympanic cavity)
Eustachian tube
6 MONTHS

Auricular hillocks
6 WEEKS
ADULT

-49-

24. THE MIDDLE EAR

I. The middle ear

A. TYMPANIC CAVITY AND EUSTACHIAN TUBE
1. The tympanic cavity is of entodermal origin and is derived from the 1st pharyngeal pouch, an outpocketing of the pharynx
   a. The pouch, lined with epithelium of entodermal origin, appears in embryos at about the 4th week
   b. The pouch grows rapidly in a lateral direction and temporarily comes in contact with the floor of the 1st ectodermal cleft
   c. The distal portion of the pouch, the tubotympanic recess, widens and gives rise to the primitive tympanic cavity; the proximal part remains narrow and forms the auditory or eustachian tube
2. The eustachian tube connects the tympanic cavity with the nasopharynx
   a. The tube's pharyngeal orifice is surrounded by a considerable amount of lymphoid tissue, the tubal or pharyngeal tonsil (adenoids)
      i. Nasal inflammation associated with the tubal tonsillar swelling often results in occlusion of the tube and inflammation of the tympanic cavity resulting in otitis media (seen in young children)
3. During late fetal life, the tympanic cavity expands dorsally and posteriorly to form the tympanic antrum
   a. Its walls are covered with epithelium of entodermal origin
4. After birth, the bone of the developing mastoid process is invaded by epithelium of the tympanic cavity and epithelial-lined air sacs are created (pneumatization)
   a. Later, most of the mastoid air cells come in contact with the antrum and tympanic cavity

B. THE OSSICLES
1. By the end of the 7th week, the mesenchyme above the primitive tympanic cavity demonstrates a number of condensations caused by proliferation of the dorsal tips of the 1st and 2nd pharyngeal arches
   a. The condensations become the cartilaginous precursors of the auditory ossicles (malleus, the incus, and the stapes)
      i. The malleus and incus are derived from the 1st pharyngeal arch; the stapes is derived from the 2nd arch
2. The ossicles appear during the 1st half of fetal life but remain embedded in mesenchyme until the 8th month, when the surrounding tissue dissolves
3. The entodermal epithelial lining of the primitive tympanic cavity gradually extends along the wall of the newly developed space, the tympanic cavity being at least twice as large as before
4. When the ossicles are entirely free from the surrounding mesenchyme, the entodermal epithelium connects them in a mesentery-like manner to the cavity wall; and the supporting ligaments of the ossicles develop in these mesenteries
5. Since the malleus is derived from the 1st pharyngeal arch, its muscle, the tensor tympani, is innervated by the mandibular branch of the trigeminal nerve (V)
6. Since the stapes is of the 2nd pharyngeal arch, its muscle, the stapedius, is innervated by the facial (VII) nerve

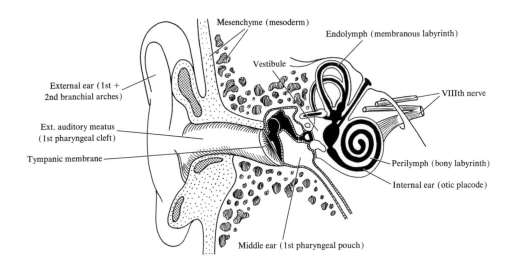

Mesenchyme (mesoderm)

Endolymph (membranous labyrinth)

Vestibule

External ear (1st + 2nd branchial arches)

VIIIth nerve

Ext. auditory meatus (1st pharyngeal cleft)

Tympanic membrane

Perilymph (bony labyrinth)

Internal ear (otic placode)

Middle ear (1st pharyngeal pouch)

ADULT EAR

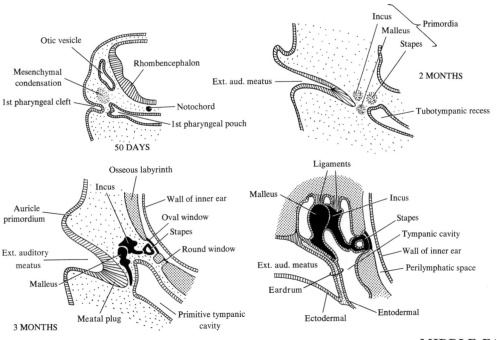

Otic vesicle

Incus
Malleus — Primordia
Stapes

Mesenchymal condensation

Rhombencephalon

Ext. aud. meatus

2 MONTHS

1st pharyngeal cleft

Notochord

Tubotympanic recess

1st pharyngeal pouch

50 DAYS

Osseous labyrinth

Ligaments

Incus

Malleus

Incus

Auricle primordium

Wall of inner ear

Stapes

Oval window

Stapes

Tympanic cavity

Stapes

Wall of inner ear

Ext. auditory meatus

Round window

Ext. aud. meatus

Perilymphatic space

Malleus

Eardrum

Meatal plug

Primitive tympanic cavity

Ectodermal

Entodermal

3 MONTHS

MIDDLE EAR

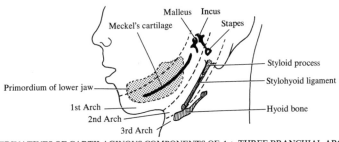

Malleus Incus

Meckel's cartilage

Stapes

Primordium of lower jaw

Styloid process

Stylohyoid ligament

1st Arch

2nd Arch

Hyoid bone

3rd Arch

DERIVATIVES OF CARTILAGINOUS COMPONENTS OF 1st THREE BRANCHIAL ARCHES

-51-

25. THE EAR: CONGENITAL MALFORMATIONS

I. **Malformations of the auditory system** can be classified in 3 different groups, corresponding to the specific embryologic characteristics of each of the ear's 3 parts:

A. LESIONS OF THE INTERNAL EAR
1. Cochlea
   a. Rubella in 2nd month of pregnancy is often a cause of lesions, since this is when internal ear differentiates
   b. Epithelium of cochlea and vestibule is altered, but organ of Corti has some intact regions so children affected perceive some deep frequencies
2. Canals: few malformations are known (seen with thalidomide)

B. LESIONS OF THE MIDDLE EAR: usually involve ossicles
1. After an infection, the mesenchymal plate separating the ossicles becomes sclerotic, impedes movement, and may cause total deafness

C. MALFORMATION OF THE EXTERNAL EAR
1. Results from absence or nonunion of tubercles
2. Major variations of the auricle have been associated with serious internal abnormalities such as kidney malformations
3. Abnormal position of the ear usually associated with abnormal mandibular development (agnathia or micrognathia)
   a. Instead of moving to sides of head, ears develop at site of primordia—at level of 1st branchial groove (seen in newborn at the angle of the missing jaw—octocephalus)
4. Auricular appendages or tags—common and due to accessory auricular hillocks.
5. Absence and hypoplasia of auricles are rare; associated with 1st arch syndrome; failure of hillocks to develop (anotia) or suppressed development (microtia)
6. Auricular sinus and fistulas: sinuses are usually preauricular; fistulas connect exterior with tympanic cavity
7. Atresia of external auditory meatus: failure of meatal plug to canalize

II. **CONGENITAL DEAFNESS** IS USUALLY ASSOCIATED WITH DEAF-MUTISM

A. MAY BE CAUSED BY
1. Abnormal development of the membranous and bony labyrinths
2. Malformations of the eardrum and ossicles
3. In extreme cases, tympanic cavity and external meatus are completely absent

B. OTHER CAUSES OF DEAFNESS
1. Heredity
2. Environmental and other factors affecting the mother early in pregnancy
   a. Rubella virus, affecting the embryo in the 7th and 8th week of development, will cause severe damage to the organ of Corti; diabetes; erythroblastosis fetalis; hypothyroidism; toxoplasmosis; x-radiation

C. MAJOR VARIATIONS of the auricle have been associated with serious internal abnormalities such as kidney malformations

AURICULAR APPENDAGES
OR TAGS

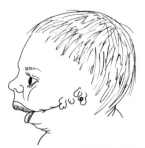

EXTERNAL EAR ANOMALY WITH
MAXILLOMANDIBULAR FISSURE

FISTULA AURIS
(PROBED)

MALFORMED AURICLE

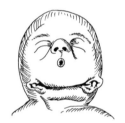

SYNOTIA (WITH MANDIBULAR
HYPOSTOMIA AND MACROSTOMIA)

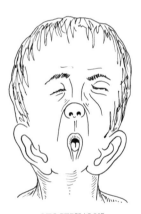

OTOCEPHALUS

HYPOPLASIA OF AURICLE
(WITH ABSENCE OF EXTERNAL
AUDITORY MEATUS)

26. CRITICAL OR SENSITIVE PERIODS OF HUMAN DEVELOPMENT*†

	Embryonic Period (in weeks)						Fetal Period (in weeks)		
1–2	3	4	5	6	7	8	9–16	20–36	38
Period of dividing zygote, implantation, and bilaminar embryo	Hrt. + CNS	Eye Leg Arm Hrt. CNS	Eye Leg Arm Hrt. CNS	Ear Teeth CNS Eye Hrt.	Palate CNS Ear Eye Hrt.	Ear Hrt. CNS Palate Genitalia	Genitalia	CNS Ear	Brain Brain

Period usually not susceptible to teratogens

Timeline bars (solid = very sensitive, dotted = less sensitive):
- HEART
- CNS
- ARMS
- EYES
- LEGS
- TEETH
- PALATE
- EXTERNAL GENITALIA
- EAR

Prenatal death period	Major morphologic abnormalities occur in these weeks	Physiologic defects and minor morphologic abnormalities occur in these weeks

*Organs listed refer to common sites of action of teratogens during weeks of development
†Dotted line indicates less sensitive periods of development (to teratogens)
Solid line indicates very sensitive periods of development (to teratogens)

Unit Two

GROSS AND
MICROSCOPIC
NEUROANATOMY

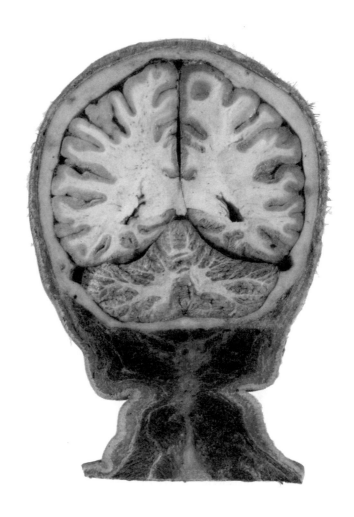

27. GENERAL HISTOLOGY OF NERVOUS SYSTEM AND NEURON CLASSIFICATION

I. Introduction: the nervous system may be defined as a specialized group of nerve cells (neurons) and supporting cells (glia) whose major functions are reception of information from external and internal environments; integration of incoming information; generation of new signals; and transformation and conduction of messages to special responding tissue (effectors)

A. NERVOUS TISSUE: characterized by its unique properties of excitation and conduction

B. NEURON is the anatomic and functional unit of the nervous system and chains of neurons comprise the nervous system
1. Each neuron is contiguous with the next cell in the chain, but the cells are not continuous with one another
2. Each neuron has a nerve cell body with all its protoplasmic processes, one axon, and usually one or more dendrites
    a. Dendrites are typically short, branching processes which form a major part of the receptor area of the neuron
    b. Each neuron has a single axon which varies greatly in length from one type of neuron to another and conducts impulses away from the soma or cell body

C. GANGLION is a collection of neurons outside the CNS

D. NUCLEUS is a cluster of neurons in the gray matter of the brain and spinal cord (nucleus in this sense refers to a functional aggregation of neurons and should not be confused with the nucleus of a cell)

II. Classification of neurons

A. ACCORDING TO FUNCTION
1. Motor neurons include neurons of the autonomic nervous system (ANS) and of anterior horns of the spinal cord. They carry impulses that produce movement
2. Sensory neurons carry impulses that produce sensation. Neurons of the dorsal root ganglia are of this type
3. Association (intercalated or internuncial) neurons serve to connect motor and sensory neurons

B. ACCORDING TO NUMBER OF PROCESSES FROM CELL BODY
1. Unipolar or pseudounipolar neurons have only 1 protoplasmic process
    a. This T-shaped process bifurcates and one branch functions as an axon, which enters the cord, the other as a dendrite, which extends to a peripheral tissue
    b. Neurons of this type are sensory and are found, e.g., in the dorsal root ganglia of the PNS and in the mesencephalic nucleus of the trigeminal (V) cranial nerve (the only example of unipolar neurons in the CNS)
2. Bipolar neurons have 2 processes, one from each pole of the cell body; one process serves as an axon, the other as a dendrite
    a. Neurons of this type are found in the retina, olfactory epithelium, and in the acoustic ganglion of cranial nerve VIII
3. Multipolar neurons have more than 2 processes from the cell body; only one process is an axon and the others are dendrites
    a. Neurons of the CNS are generally multipolar and include large motor neurons of the spinal cord, small association neurons of the cord, pyramidal cells of the cerebral cortex, Purkinje cells of the cerebellar cortex, and neurons of the ANS

C. ACCORDING TO LENGTH OF AXON
1. Golgi type 1 neurons are found in the CNS and have long axons which leave their site of origin and extend for a considerable distance before ending
    a. They form tracts, commissures, and projection and association fibers
2. Golgi type 2 neurons are found in the CNS and have short axons (0.5–5.0 mm)
    a. Their axons do not leave the gray matter where the cells are located
    b. Numerous in the cerebral and cerebellar cortices and spinal cord

UNIPOLAR NEURON
(SENSORY—AFFERENT)

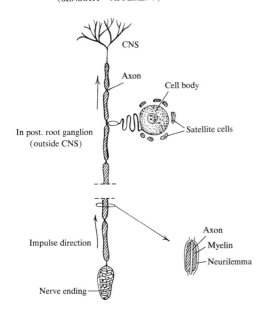

CNS

Axon

Cell body

Satellite cells

In post. root ganglion
(outside CNS)

Impulse direction

Axon
Myelin
Neurilemma

Nerve ending

BIPOLAR NEURON
(SENSORY—AFFERENT)

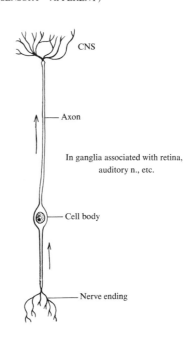

CNS

Axon

In ganglia associated with retina,
auditory n., etc.

Cell body

Nerve ending

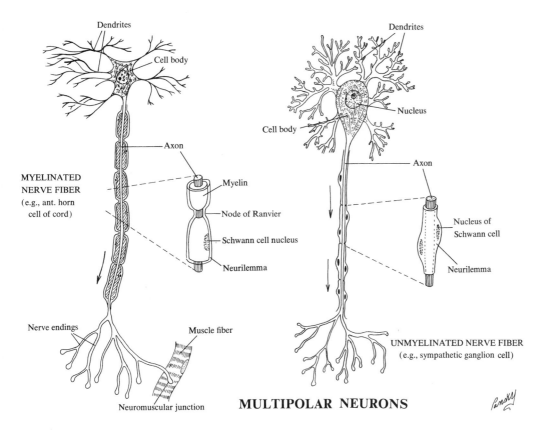

Dendrites

Cell body

Axon

MYELINATED
NERVE FIBER
(e.g., ant. horn
cell of cord)

Myelin

Node of Ranvier

Schwann cell nucleus

Neurilemma

Nerve endings

Muscle fiber

Neuromuscular junction

Dendrites

Nucleus

Cell body

Axon

Nucleus of
Schwann cell

Neurilemma

UNMYELINATED NERVE FIBER
(e.g., sympathetic ganglion cell)

MULTIPOLAR NEURONS

Pansky

28. GENERAL HISTOLOGY OF NERVOUS SYSTEM: NEUROGLIA—GENERAL

I. Neuroglia—general

A. SPECIALIZED NONNEURONAL SUPPORTING CELLS of the CNS are called neuroglia (nerve glue) or simply glia. These supporting cells are more numerous than neurons. They support neurons and provide axons with myelin sheaths that increase the speed of impulse conduction

B. UNLIKE NEURONS, glial cells have only 1 type process and do not form synapses

C. GLIAL CELLS can undergo cell division throughout adult life, particularly in response to damage to the CNS

D. CLASSIFICATION OF THE NEUROGLIA (three categories of glial cells can be recognized in the CNS)
1. Macroglia can be subdivided into astrocytes and oligodendrocytes
2. Microglia are not considered as true neuroglia by some authorities
3. Ependymal cells are not always classified as true neuroglia

E. EMBRYOLOGICALLY, all supportive neuroglia except microglia are derivatives of the neural tube (neuroectoderm); microglia arise from mesoderm

F. MOST KNOWLEDGE about the shapes and disposition of glial cells has been gained by examination of tissues stained by metallic impregnation because ordinary stains such as hematoxylin and eosin (H & E) show only nuclei of glial cells

II. Neuroglia—specific

A. ASTROCYTES can be separated into 2 types—fibrous and protoplasmic—by special staining techniques (one being the gold sublimate method of Cajal). Both types of astrocytes are stellate (star-shaped) and have long processes
1. Morphologic and functional characteristics of fibrous astrocytes
   a. Found primarily in white matter
   b. Processes are fewer in number, longer and straighter than those of protoplasmic astrocytes (resemble long-legged spiders)
   c. Some of their processes form expansions that are applied to the surfaces of blood vessels (perivascular end-feet or sucker feet)
      i. Other processes extend to surfaces of the CNS to form expansions known as the glial limiting membrane (secures pia mater to the CNS)
   d. Possess many cytoplasmic filaments which may occur in bundles (as seen in light microscopy) and provide rigidity and support
   e. They are the main supportive cells of the CNS and form firm links between capillaries
   f. They are important in the response of the CNS tissue to injury
      i. In the case of nonlethal injury to the astrocyte, the cell body swells, the processes elongate, the cell divides, and additional processes form
   g. They are often referred to as the scarring cells of the CNS because they fill in gaps after tissue is lost in disease processes
   h. When a lesion is extensive or when it is accompanied by liquefaction, the astrocytes are inadequate for the task and fibroblasts enter to assist in restoring supportive integrity
      i. Scarring results from the production of connective tissue fibers by fibroblasts (known as sclerosis) which differs from gliosis in which only astrocytes proliferate, as in response to chronic injury to the CNS
   i. Astrocytes are capable of limited phagocytosis (case of minimal CNS injury)

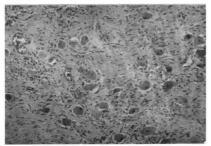

Sympathetic ganglion (low power)

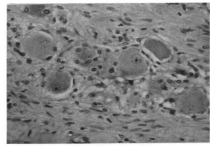

Sympathetic ganglion (high power)

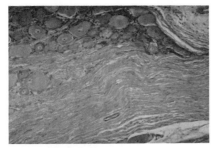

Dorsal root ganglion (low power)

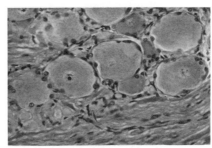

Dorsal root ganglion (high power)

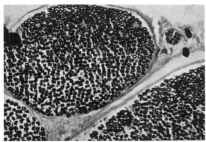

Spinal nerve in cross-section (0s04 preparation)

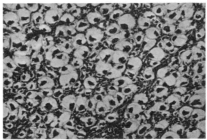

Myelinated nerve fibers in cross-section

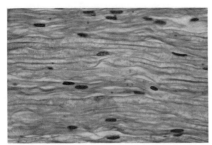

Spinal nerve in longitudinal-section showing myelinated fibers and Schwann cell nuclei

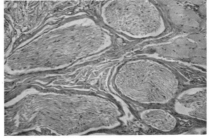

Spinal nerve in oblique-section

29. GENERAL HISTOLOGY OF NERVOUS SYSTEM: NEUROGLIA—SPECIFIC

I. Neuroglia—specific (cont.)

A. ASTROCYTES (CONT.)

1. Morphologic and functional characteristics of protoplasmic astrocytes
   a. Found primarily in gray matter
   b. Processes are far more numerous, much shorter, and much more extensively branched than those of fibrous astrocytes (resemble bushy shrubs)
   c. Processes cover nonsynaptic neuronal surfaces and surround synaptic zones
      i. Their pattern of arrangement suggests they serve to isolate the receptive surfaces of neurons and prevent the haphazard flow of impulses
   d. Processes also terminate on capillaries as perivascular end-feet
   e. Possess fewer cytoplasmic filaments than fibrous astrocytes

2. Identification (differentiation) of the two types of astrocytes
   a. The 2 types are indistinguishable histologically without special stains
   b. Identification based on location alone is unreliable (protoplasmic astrocytes predominate in gray matter and fibrous astrocytes in white matter)
   c. Nuclei of both types are large, pale, and lack nucleoli
   d. Nuclei are often indented and usually more irregular in outline than those of other glial cells
   e. Nuclei are smaller than those of neurons but larger than oligodendrocytes.

B. OLIGODENDROGLIA (OLIGODENDROCYTES)

1. A special neurohistologic technique (del Rio-Hortega's silver method) shows
   a. Cells are small with round or polygonal bodies; most numerous cell in CNS
   b. Possess few protoplasmic processes (no perivascular end-feet)
   c. Differ from each other only in location; produce myelin in the CNS

2. Classification of oligodendrocytes
   a. Interfascicular oligodendrocytes
      i. Found in white matter of brain and spinal cord
      ii. Elaborate myelin sheaths that surround axons of the CNS; comparable to Schwann cells that produce myelin of PNS
      iii. Each process has a diameter of about 0.1 μm
      iv. On reaching the axonal surface, the process enlarges into a wide, flat tongue which then wraps around the axon 1–20 times
      v. This wrapping material is the lipoprotein plasma membrane and constitutes the myelin sheath; the thickness of the sheath is determined by the number of times that the tongue wraps around the axon
      vi. Processes of one oligodendrocyte may form internodal myelin for 3–50 fibers
   b. Perineuronal oligodendrocytes
      i. Located in gray matter close to nerve cell bodies and dendrites
      ii. Often compared to satellite cells that surround the perikarya of peripheral ganglia and appear to have a similar relationship
   c. Perivascular oligodendrocytes
      i. Lined up along blood vessels in both gray and white matter
      ii. Their function is still not known

C. DIFFERENTIATION OF OLIGODENDROCYTES AND ASTROCYTES WITH ROUTINE STAINS

1. Nuclei of oligodendrocytes are round to ovoid and usually appear smaller and darker than nuclei of astrocytes
2. Oligodendrocyte nuclei contain denser clumps of chromatin
3. Oligodendrocytes are smaller in size but appear in far greater abundance

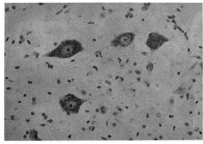

Typical neurons and their cytoplasmic processes surrounded by numerous glial cells

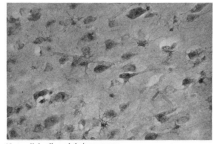

Neuroglial cells and their processes

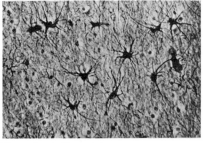

Fibrous astrocytes (low power)

A fibrous astrocyte (high power)

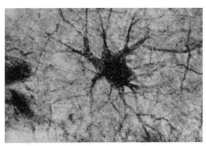

Protoplasmic astrocyte

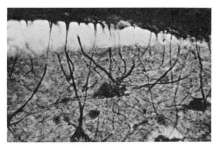

Astrocytes, their processes and perivascular footplates

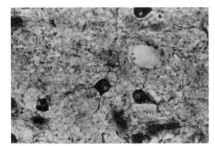

Oligodendrocytes

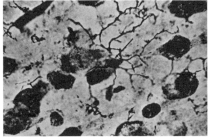

Microglia

30. SPINAL CORD: GENERAL GROSS APPEARANCE

I. Location and level

A. THE SPINAL CORD is that part of the CNS continuous with the medulla oblongata of the brainstem and located within the vertebral canal

B. IT EXTENDS, in the adult, from the foramen magnum cranially to the inferior border of the first lumbar vertebra caudally, where it tapers to a point (conus medullaris) and terminates

    1. From the conus, a slender, median, fibrous thread, the filum terminale, is prolonged as far as the back of the coccyx

C. IT USUALLY MEASURES 42–45 cm in length in the adult, and its diameter varies at the different levels

II. Spinal cord and vertebral column relationship

A. BECAUSE OF DIFFERENCES IN GROWTH RATE between the spinal cord and vertebral column, the cord in the adult ends at the lower border of the first lumbar vertebra

B. SINCE THE SPINAL CORD IS APPROXIMATELY 25 CM SHORTER than the column, lumbar and sacral nerves possess very long intervertebral roots which extend from the cord to the appropriate intervertebral foramina where dorsal and ventral roots join to form spinal nerves

    1. In the 3rd fetal month, the spinal cord fills the vertebral canal; at birth, the cord ends at a level between the first and second lumbar vertebrae (the lowest level is at the lower third of the third lumbar vertebra)

    2. Relationships in the adult are presented for various levels in the table

Spinal Nerve and Cord Segment	Vertebral Level	Spinous Process	Level of Nerve Exit
C1	C1	C1	Above C1
C8	C7	C6	Between C7–T1
T6	T5	T3	Between T6–T7
T12	T8	T8	Between T12–L1
L2	T10	T10	Between L1–L2
L5	T11	T12–L1	Below L5
S3	T12	L1–L2	Third sacral foramen

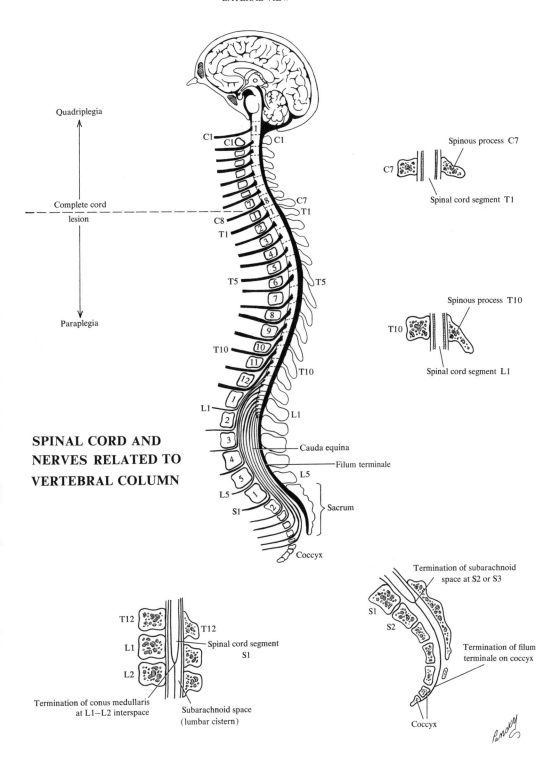

Quadriplegia

Complete cord
lesion

Paraplegia

C1 C1 C1

C7
T1

C8
T1

T5 T5

T10

T10

L1 L1

L5

L5
S1

Coccyx

**SPINAL CORD AND
NERVES RELATED TO
VERTEBRAL COLUMN**

Cauda equina

Filum terminale

L5

Sacrum

Spinous process C7

C7

Spinal cord segment T1

Spinous process T10

T10

Spinal cord segment L1

T12

L1

L2

T12

Spinal cord segment
S1

Termination of conus medullaris
at L1–L2 interspace

Subarachnoid space
(lumbar cistern)

Termination of subarachnoid
space at S2 or S3

S1

S2

Termination of filum
terminale on coccyx

Coccyx

31. SPINAL CORD: EXTERNAL MORPHOLOGY

I. External morphology

A. IT IS AN ELONGATED, UNSEGMENTED STRUCTURE intrinsically. Paired spinal nerves along its entire length produce an appearance of external segmentation

B. A SINGLE SPINAL CORD SEGMENT is defined as that part of the cord which receives or gives dorsal and ventral rootlets to a single pair of spinal nerves. There are 8 cervical segments, 12 thoracic segments, 5 lumbar segments, 5 sacral segments, and 1 coccygeal segment, or a total of 31 pairs of spinal nerves

C. INDIVIDUAL SPINAL SEGMENTS VARY IN LENGTH; those in the midthoracic region are usually much (at least $2\times$) longer than cervical and lumbar segments

D. GROSSLY, it has the appearance of a compressed cylinder with two enlargements
   1. Cervical enlargement corresponds to the nerves to the upper limb (brachial plexus) and is comprised of the lower four cervical and the first thoracic segments
   2. Lumbar enlargement corresponds to the area of nerves to the lower limbs (lumbosacral plexus) and is comprised of the five lumbar and the upper three sacral segments (i.e., acquires largest diameter at T12)

E. SINCE THERE IS NO SPINAL CORD WITHIN THE VERTEBRAL CANAL of most of the lumbar and all of the sacral vertebrae, the elongated dorsal and ventral roots are the only structures within the canal at these levels. The elongated mass of nerve roots suspended from the cord is known as the cauda equina (tail of a horse)

F. UP TO THE 3RD FETAL MONTH, the lengths of the cord and the vertebral canal are approximately equal. By the time of birth, the caudal extent of the cord is at the level between L1 and L2. This is one of the bases for the selection of the 3rd or 4th lumbar interspace for the introduction of the needle in lumbar puncture procedures

G. THE CORD IS ATTACHED TO SURROUNDING STRUCTURES, by extensions of its innermost meningeal covering, the pia mater; it intimately invests the cord (and brain) like a sausage skin does a sausage

H. THROUGHOUT THE LENGTH OF THE CORD, extensions of the pia mater are pulled out from the lateral aspects of the cord between dorsal and ventral roots, to be "tacked" to the walls of the outer investing meningeal covering, the dura mater. This horizontal shelflike ligament on either side of the cord, with its numerous toothlike attachment points is termed the dentate (toothed) or denticulate ligament. These ligaments serve to protect it from excessive shock and provide useful handles for manipulating the cord safely during surgery

I. THE INFERIOR PIAL EXTENSION that serves to anchor the cord is the filum terminale. This connective tissue structure is formed caudal to the conus medullaris, where the nerve tissue terminates, but the pial covering continues. This terminal filament is continued caudally to its point of attachment at the coccyx where it is known as the coccygeal ligament after the additional meningeal coverings (arachnoid and dura) are added

J. ON THE CORD THE DURA CONTINUES out along the dorsal and ventral roots and spinal nerves to the point where they pass through the intervertebral foramina. Here the dura becomes attached to the periosteum. It extends down to the level of S2, caudal to the termination of the conus medullaris, to "bag" the cauda equina. A sheathlike extension continues further caudally, wrapping the filum terminale to form the coccygeal ligament

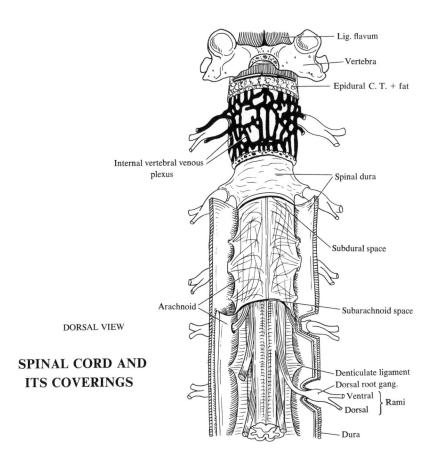

Lig. flavum

Vertebra

Epidural C. T. + fat

Internal vertebral venous
plexus

Spinal dura

Subdural space

Arachnoid

Subarachnoid space

DORSAL VIEW

**SPINAL CORD AND
ITS COVERINGS**

Denticulate ligament

Dorsal root gang.

Ventral
Dorsal } Rami

Dura

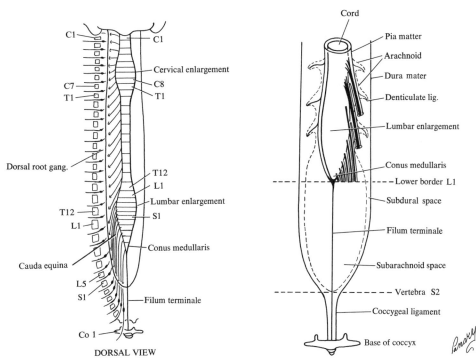

Cord

Pia matter

Arachnoid

Dura mater

Denticulate lig.

Lumbar enlargement

Conus medullaris

Lower border L1

Subdural space

Filum terminale

Subarachnoid space

Vertebra S2

Coccygeal ligament

Base of coccyx

C1
Cervical enlargement
C1
C7
C8
T1
T1
Dorsal root gang.
T12
L1
Lumbar enlargement
S1
T12
L1
Conus medullaris
Cauda equina
L5
S1
Filum terminale
Co 1

DORSAL VIEW

32. SPINAL CORD: MAJOR SUBDIVISIONS

I. Subdivisions of white and gray matter
A. Two BASIC DIVISIONS of the spinal cord are seen in cross section:
1. Gray matter is located in the central part and shaped like a butterfly or in the shape of an H. It is comprised of glia cells and cell bodies of efferent and internuncial neurons.
   a. The narrow midline connection (crossbar of the H) of the gray matter is divided into dorsal and ventral gray commissures
   b. Dorsal projections of the gray are known as posterior gray horns
   c. Ventral projections of the letter H are the ventral gray horns
   d. Intermediolateral gray horns are found only at segmental levels T1–L2 or L3. This enlargement of the central gray matter is intermediate and lateral in position between the ventral and posterior gray horns
2. White matter is arranged as the outer layer of the cord, and it is made up of nerve fibers and glia
   a. The surface reveals two longitudinally oriented grooves, located in the midline on the ventral and posterior surfaces
      i. The anterior median fissure is wide but shallow; it is found throughout the entire cord length
      ii. The posterior median sulcus is narrower and deeper and present at all levels
   b. Paired lateral sulci on the anterior and posterior surfaces
      i. Anterior lateral (ventrolateral) sulci are lines for the attachment of exiting ventral nerve rootlets
      ii. Posterior lateral (dorsolateral) sulci represent lines for attachment of entering dorsal nerve rootlets
B. In CROSS-SECTIONAL VIEW it is observed that components from the ventral and posterior gray horns extend to the surface of the cord at these sulci, and divide each half of white matter into three bundles of cell processes (fibers) termed funiculi. These funiculi are named according to position
1. The dorsal funiculi of the two sides are separated by the dorsal median sulcus and septum, and extend laterally to the dorsolateral sulcus
2. The ventral funiculi are separated by the anterior median fissure and extend laterally to the ventrolateral sulcus
3. The lateral funiculus of each half of white matter is bound medially by the gray horns and lies between the ventrolateral and dorsolateral sulci
C. In THE CERVICAL AND UPPER THORACIC PORTIONS of the cord (above T7), the dorsal funiculi are each subdivided by the dorsal intermediate sulcus and septum into smaller lateral and medial bundles termed fasciculi
1. Fasciculus cuneatus is the lateral portion of each dorsal funiculus and represents a tract carrying sensory impulses from upper half of body
2. Fasciculus gracilis is the medial portion of each dorsal funiculus and represents a tract carrying sensory impulses from lower half of body
D. In REFERENCE TO THE CENTRAL CANAL of the spinal cord, the white matter (like the gray) crossing the midline is separated into dorsal and ventral commissures. The ventral commissure is more prominent than the dorsal, and the former is present throughout the spinal cord at all levels. Both commissural and decussating fibers cross in the white commissure

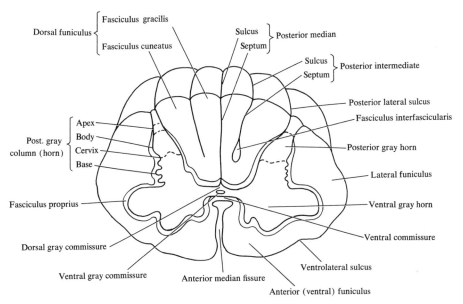

X-SECTION OF SPINAL CORD

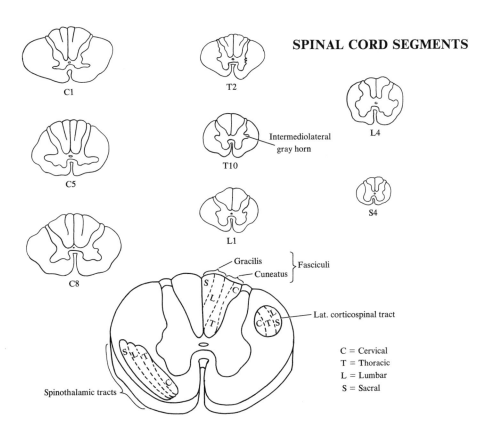

SPINAL CORD SEGMENTS

C = Cervical
T = Thoracic
L = Lumbar
S = Sacral

SOMATOTOPIC ARRANGEMENT OF FIBERS IN SPINAL CORD

33. SPINAL CORD: MICROSCOPIC ANATOMY

I. **Nuclei within the gray matter:** Neurons are not distributed uniformly within the gray substance of the spinal cord. Instead, their cell bodies are arranged somewhat into definitive columns or collections referred to as nuclei with the following distribution

A. NUCLEI OF THE POSTERIOR HORN
1. Intermediolateral nucleus is found at the levels of T1–L3 (sympathetic). The axons of neurons in this nucleus pass via the ventral roots and the white communicating rami to the sympathetic chain ganglia or to more peripheral sympathetic ganglia along the aorta
2. Sacral autonomic nuclei are found on the lateral surface of the gray matter at the base of the anterior horn in sacral segments S2, S3, and S4. Axons of these neurons make up the parasympathetic pelvic plexus
3. Posteromarginal nucleus is a thin layer of cells that cover the tip of the posterior horn. They are intersegmental association neurons and their axons ascend or descend in the lateral funiculus for several segments
4. Substantia gelatinosa forms the outer "caplike" portion of the posterior horn. The neurons of this area form an important part of the conscious and reflex pathways for pain, temperature, and some tactile impulses
5. Nucleus proprius of the posterior horn (nucleus centrodorsalis) sends axons across the midline in the anterior white commissure to form the anterior spinothalamic tract (pain and temperature)
6. Nucleus dorsalis (Clarke's column) is found in the medial portion of the base of the posterior horn. It is found from about C8 to L3 and axons from neurons in this nucleus pass into the posterior spinocerebellar tract of the same side.
7. Secondary visceral gray is an indistinct cell column that extends through the thoracic and upper lumbar levels. It receives incoming visceral afferent fibers

B. NUCLEI OF THE ANTERIOR HORN: The neurons (anterior horn cells) are the largest found within the spinal cord and have the following characteristics
1. Multipolar neurons with large, round vesicular nuclei and clumps of coarse Nissl substance within the cytoplasm
2. Long axons with distinctive axonal hillocks
3. Large anterior horn neurons are grouped into medial and lateral groups, which can be further subdivided into
   a. Dorsomedial nucleus
   b. Ventromedial nucleus
   c. Ventral nucleus
   d. Central nucleus
   e. Ventrolateral nucleus
   f. Dorsolateral nucleus
   g. Retrodorsolateral nucleus
4. These nuclei are most apparent at the level of cervical and lumbar enlargements
5. They supply motor innervation to somatic effectors (skeletal muscle) and are known as lower motor neurons

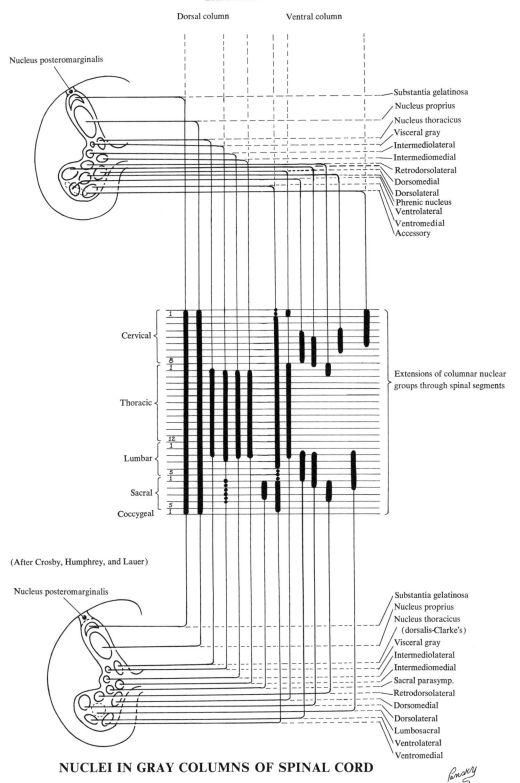

Lateral column

Dorsal column   Ventral column

Nucleus posteromarginalis

Substantia gelatinosa
Nucleus proprius
Nucleus thoracicus
Visceral gray
Intermediolateral
Intermediomedial
Retrodorsolateral
Dorsomedial
Dorsolateral
Phrenic nucleus
Ventrolateral
Ventromedial
Accessory

Cervical

Thoracic

Lumbar

Sacral

Coccygeal

Extensions of columnar nuclear
groups through spinal segments

(After Crosby, Humphrey, and Lauer)

Nucleus posteromarginalis

Substantia gelatinosa
Nucleus proprius
Nucleus thoracicus
  (dorsalis-Clarke's)
Visceral gray
Intermediolateral
Intermediomedial
Sacral parasymp.
Retrodorsolateral
Dorsomedial
Dorsolateral
Lumbosacral
Ventrolateral
Ventromedial

NUCLEI IN GRAY COLUMNS OF SPINAL CORD

Pansky

34. SPINAL CORD: REXED'S LAMINAE AND MAJOR TRACTS

I. Rexed's laminae within gray matter of spinal cord: The gray substance can be divided into nine layers or laminae of Rexed on the basis of the longitudinal arrangement of the neurons into columns and cytoarchitectonics. Rexed's laminae are more encompassing in area and are preferred by some authorities to the more traditional use of cell columns or nuclei because of problems with nomenclature and precise functions associated with the utilization of the latter. Basic facts essential for understanding laminae of Rexed include

A. SPINAL CORD is divided into nine cell laminae. Five are located in the posterior horn, two in the intermediate gray, and two in the anterior horn. An additional lamina surrounds the central canal

B. LAMINAE I THROUGH VI are found only in the posterior horn. They are concerned with sensory input to the spinal cord

C. LAMINA VI is found only in the cervical and lumbar enlargements

D. LAMINA VII is found in the intermediate gray area and extends into the anterior horn. It includes the nucleus dorsalis and the intermediolateral gray column

E. LAMINA VIII is located in the anterior horn. Many of the axons from the neurons located here cross to the opposite side (commissural)

F. LAMINA IX is found only in the anterior horn and is usually subdivided. It possesses alpha and gamma motor neurons whose axons leave the spinal cord via the anterior roots

G. THE FOLLOWING LAMINAE AND NUCLEI correspond in location
1. Lamina I with the posteromarginal nucleus
2. Lamina II with substantia gelatinosa
3. Lamina III and IV with nucleus proprius
4. Lamina IX with the subdivisions of the motor cell columns
5. Lamina VII with the nucleus dorsalis and the intermediolateral gray

II. Tracts within the white matter of the spinal cord: The position of fiber tracts (a tract is comprised of numerous axons within the CNS with the same origin, termination, and function) has been determined on the basis of experimental work with various mammals, stimulation of various areas during surgical procedures, and pathologic observations. A knowledge of the relative positions of the following tracts will form the basis for an understanding of the details of long ascending and descending tracts presented later

A. ASCENDING TRACTS include the following long* and short† tracts
1. Fasciculus gracilis*
2. Fasciculus cuneatus*
3. Anterior and posterior spinocerebellar*
4. Anterior and lateral spinothalamic*
5. Dorsolateral fasciculus†
6. Fasciculus proprius†

B. DESCENDING TRACTS include the following long* and short† tracts
1. Pyramidal (lateral and ventral corticospinal) tract*
2. Extrapyramidal tracts are comprised of all long descending tracts other than the lateral and ventral corticospinal tracts*
3. Dorsolateral fasciculus†
4. Fasciculus proprius†
5. Septomarginal fasciculus†
6. Fasciculus interfascicularis†

REXED'S LAMINAE IN GRAY MATTER OF CORD

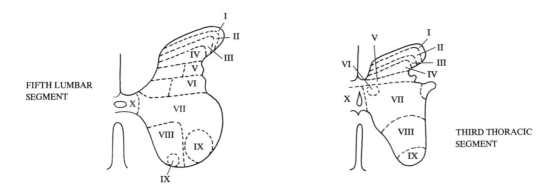

FIFTH LUMBAR
SEGMENT

THIRD THORACIC
SEGMENT

FIBER TRACTS

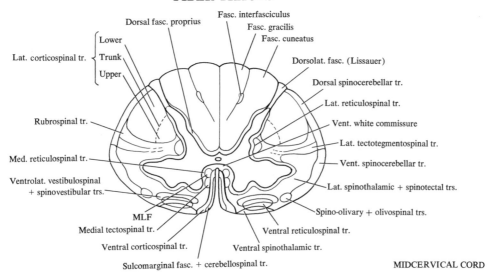

Fasc. interfasciculus

Dorsal fasc. proprius

Fasc. gracilis

Fasc. cuneatus

Lower
Trunk
Upper

Lat. corticospinal tr.

Dorsolat. fasc. (Lissauer)

Dorsal spinocerebellar tr.

Lat. reticulospinal tr.

Rubrospinal tr.

Vent. white commissure

Lat. tectotegmentospinal tr.

Med. reticulospinal tr.

Vent. spinocerebellar tr.

Ventrolat. vestibulospinal
+ spinovestibular trs.

Lat. spinothalamic + spinotectal trs.

Spino-olivary + olivospinal trs.

MLF

Medial tectospinal tr.

Ventral reticulospinal tr.

Ventral corticospinal tr.

Ventral spinothalamic tr.

Sulcomarginal fasc. + cerebellospinal tr.

MIDCERVICAL CORD

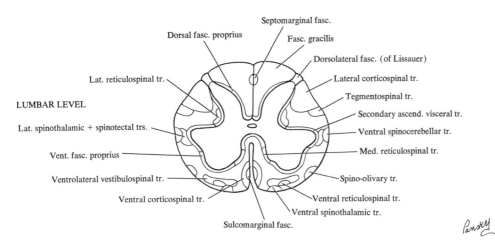

Septomarginal fasc.

Dorsal fasc. proprius

Fasc. gracilis

Dorsolateral fasc. (of Lissauer)

Lat. reticulospinal tr.

Lateral corticospinal tr.

LUMBAR LEVEL

Tegmentospinal tr.

Secondary ascend. visceral tr.

Lat. spinothalamic + spinotectal trs.

Ventral spinocerebellar tr.

Vent. fasc. proprius

Med. reticulospinal tr.

Ventrolateral vestibulospinal tr.

Spino-olivary tr.

Ventral corticospinal tr.

Ventral reticulospinal tr.

Ventral spinothalamic tr.

Sulcomarginal fasc.

-71-

35. SPINAL CORD: TYPICAL SPINAL NERVE

I. **Typical spinal nerve** is formed by a dorsal and a ventral root coming together inside the vertebral canal. Since the dorsal root, along with its dorsal root ganglion, is sensory and the ventral root carries motor fibers, the spinal nerve contains both modalities and is termed a mixed nerve

A. THE SPINAL NERVES, just after emerging from the intervertebral foramina, give off small recurrent meningeal branches (to meninges and their vessels) and filaments to nearby articular and ligamentous structures. They then divide into two major subdivisions

    1. Dorsal ramus splits into a medial and a lateral branch (one of which supplies both skin as well as muscle, and the other muscle only)

    2. Ventral ramus gives off a lateral or perforating branch (which is cutaneous) and then continues on around supplying motor and sensory fibers to muscles of the body wall. It terminates as an anterior perforating branch (also cutaneous)

B. EACH CUTANEOUS BRANCH (ANTERIOR, POSTERIOR, AND LATERAL) DIVIDES into two branches

    1. Anterior and posterior cutaneous branches each has a medial and lateral branch

    2. Lateral cutaneous branch has an anterior and a posterior branch

C. THE SKIN OF ONE ENTIRE SIDE OF THE BODY is innervated in this fashion, and the opposite side has a comparable set of nerves

D. MOTOR INNERVATION TO AND SENSORY FIBERS FROM skeletal muscles are carried by the anterior and posterior rami. Thus motor fibers from the anterior horn cells within the cord are carried by both anterior and posterior rami, and sensory fibers, whose cell bodies are located in the dorsal root ganglia, are carried by both anterior and posterior rami. In addition, both rami carry fibers from the lateral sympathetic chain ganglia to glands of the skin and blood vessels in both the skin and muscles

    1. The spinal nerves are connected with adjacent sympathetic trunk ganglia by rami communicantes. The latter contribute efferent and afferent sympathetic fibers to the spinal nerves

E. EACH OF THE 31 PAIRS OF SPINAL NERVES develops along this general pattern as described. This typical pattern is clearly seen, however, only on the trunk and particularly in the thoracic region

F. SEGMENTAL INNERVATION: the fact that each of the 31 pairs of nerves developmentally follows the typical pattern gives rise to this concept

    1. Parallel segments of skin innervated by a single nerve form bands. This resembles conceptually a loaf of bread, with each slice representing a segment

    2. Slices or segments (known as dermatomes) do not retain their absolute horizontal arrangement. The ventral portion of each segment tends to be located at a more caudal level than the dorsal

    3. Within each band or segment, the nerve supply is not absolute. Each nerve supplies not only its own segment but half of each adjacent segment. As a result, destruction of a single segmental nerve will not cause a band of anesthesia (total loss of sensation) but hypoesthesia (partial loss of sensation)

G. THE SKIN OF THE NECK where it joins the trunk is innervated by C5 which is immediately adjacent to a segment innervated by T1. This does not mean that C6, C7, and C8 are omitted. Instead, this is merely the cutaneous representation of the fact that these intervening nerves take part in the formation of a special plexus to supply the superior extremity

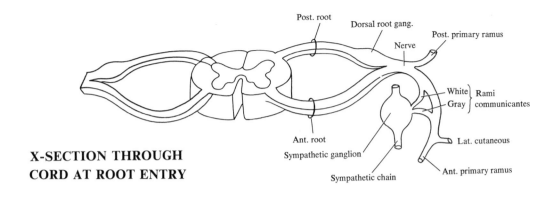

X-SECTION THROUGH CORD AT ROOT ENTRY

Post. root · Dorsal root gang. · Nerve · Post. primary ramus · White · Gray · Rami communicantes · Lat. cutaneous · Ant. primary ramus · Sympathetic ganglion · Sympathetic chain · Ant. root

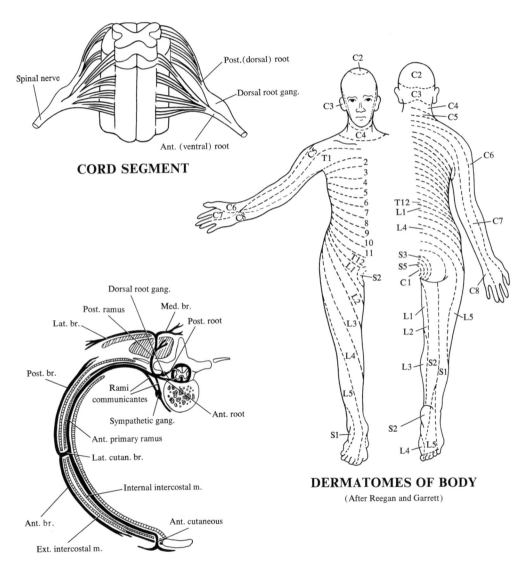

CORD SEGMENT

Spinal nerve · Post. (dorsal) root · Dorsal root gang. · Ant. (ventral) root

AN INTERCOSTAL SPINAL NERVE

Dorsal root gang. · Post. ramus · Med. br. · Post. root · Lat. br. · Post. br. · Rami communicantes · Ant. root · Sympathetic gang. · Ant. primary ramus · Lat. cutan. br. · Internal intercostal m. · Ant. br. · Ext. intercostal m. · Ant. cutaneous

DERMATOMES OF BODY
(After Reegan and Garrett)

Panoky

36. SPINAL CORD: CERVICAL AND BRACHIAL PLEXI

The three major plexi (cervical, brachial, and lumbosacral) are formed from only the ventral rami of the spinal nerves. The ventral rami supply the anterior and lateral parts of the neck and trunk, and make up the nerves of the perineum and limbs. Only in the thoracic region do they retain their separate identities as intercostal and subcostal nerves. The rest reunite as the following plexi (see below). Dorsal rami in all cases are not involved and follow the typical segmental distribution just described

I. **Cervical plexus** is formed from the ventral primary rami of spinal nerves C1–C4. Main branches: phrenic nerve, motor branches to prevertebral and infrahyoid muscles of neck, and cutaneous branches to the neck and head

II. **Brachial plexus** is formed from the ventral rami of cervical nerves C5, C6, C7, C8, and most of T1 and consists of
   A. ROOTS: used to identify the five spinal nerve components (ventral rami of C5–T1) until they continue to form distinct trunks
      1. Roots from C5 and C6 form the superior trunk
      2. Roots C8 and T1 form the inferior trunk
      3. Root from C7 continues alone to form the middle trunk
   B. THE THREE TRUNKS DIVIDE into an anterior and a posterior division which recombine to form three cords
      1. Posterior cord: formed by the posterior divisions of the three trunks
      2. Lateral cord: formed by anterior divisions of the upper and middle trunks
      3. Medial cord: formed by anterior division of the inferior trunk
   C. EACH CORD DIVIDES into two branches each of which either is a named nerve or contributes to a nerve as shown below
      1. Posterior cord gives rise to the axillary and the radial nerves which supply the extensor muscles of the upper extremity
      2. Lateral cord gives rise to musculocutaneous nerve and a branch to the median nerve
      3. Medial cord gives rise to the ulnar nerve and a branch to the median nerve
   D. TEN MORE SMALL NERVES take origin from the brachial plexus
      1. Dorsal scapular nerve arises directly from root C5
      2. Suprascapular nerve arises from a combination of C5 and C6
      3. Long thoracic nerve arises from C5, C6, and C7
      4. Lateral pectoral nerve arises from the lateral cord
      5. Medial pectoral nerve arises from the medial cord
      6. Medial brachial cutaneous nerve to the arm from the medial cord
      7. Medial antebrachial cutaneous nerve to the forearm from the medial cord
      8. Upper, middle, and lower subscapular nerves supply extensor muscles and arise from the posterior cord
   E. NERVES OF THE ARM, FOREARM, AND HAND
      1. Axillary nerve supplies the teres minor and deltoid muscles
      2. Radial nerve supplies all arm and forearm extensor muscles
      3. Musculocutaneous nerve supplies all the flexor muscles of the brachium or arm (coracobrachialis, biceps brachii, brachialis)
      4. Median nerve supplies nothing in the arm, but most of the flexor muscles of the forearm and part of the hand muscles
      5. Ulnar nerve supplies nothing in the arm, only two flexor muscles of the forearm (flexor carpi ulnaris and flexor digitorum profundus), but most of the muscles of the hand

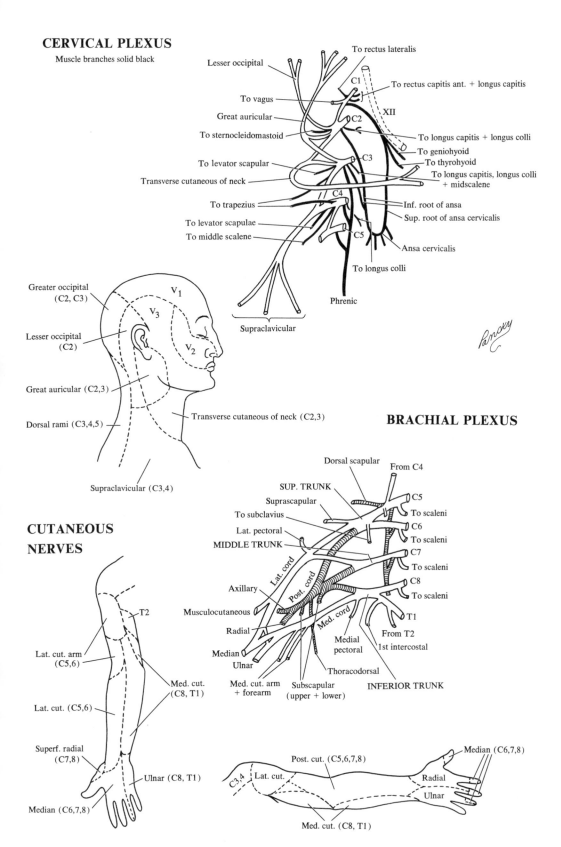

CERVICAL PLEXUS

Muscle branches solid black

To rectus lateralis

Lesser occipital

C1

To rectus capitis ant. + longus capitis

To vagus

XII

Great auricular

C2

To sternocleidomastoid

To longus capitis + longus colli

To geniohyoid

To levator scapular

C3

To thyrohyoid

Transverse cutaneous of neck

To longus capitis, longus colli + midscalene

C4

To trapezius

Inf. root of ansa

Sup. root of ansa cervicalis

To levator scapulae

To middle scalene

C5

Ansa cervicalis

To longus colli

Phrenic

Supraclavicular

Greater occipital (C2, C3)

V₁

V₃

Lesser occipital (C2)

V₂

Great auricular (C2,3)

Dorsal rami (C3,4,5)

Transverse cutaneous of neck (C2,3)

Supraclavicular (C3,4)

BRACHIAL PLEXUS

Dorsal scapular

From C4

SUP. TRUNK

Suprascapular

C5

CUTANEOUS NERVES

To subclavius

To scaleni

C6

Lat. pectoral

To scaleni

MIDDLE TRUNK

C7

To scaleni

Lat. cord

Post. cord

C8

To scaleni

Axillary

T1

Musculocutaneous

Med. cord

From T2

Radial

Medial pectoral

1st intercostal

T2

Median

Ulnar

Thoracodorsal

Lat. cut. arm (C5,6)

Med. cut. (C8, T1)

Med. cut. arm + forearm

Subscapular (upper + lower)

INFERIOR TRUNK

Lat. cut. (C5,6)

Superf. radial (C7,8)

Median (C6,7,8)

Post. cut. (C5,6,7,8)

Median (C6,7,8)

C3,4

Lat. cut.

Radial

Ulnar (C8, T1)

Ulnar

Median (C6,7,8)

Med. cut. (C8, T1)

37. SPINAL CORD: LUMBOSACRAL AND COCCYGEAL PLEXI

I. **Lumbosacral plexus** gives rise to the nerves that innervate the muscles of the hip, thigh, leg, and foot. In a very real sense, this plexus is actually comprised of two plexuses

A. LUMBAR PLEXUS is comprised of ventral rami of spinal nerves L1–L4 plus a contribution from T12. T12 and L1 supply cutaneous innervation primarily; therefore, it is simpler to think of the lumbar plexus as being the ventral rami of L2, L3, and L4. As in the case of the brachial plexus, muscles in both the original ventral (flexor) and original dorsal (extensor) groups are innervated by separate nerves

   1. Anterior divisions of the ventral rami of spinal nerves from L2–L4 combine to form the obturator nerve
   2. Posterior fibers of the anterior divisions of L2–L4 combine to form the femoral nerve

B. SACRAL PLEXUS is composed of ventral rami of spinal nerves L4–L5 and S1–S3, which combine into a single dissectable nerve, the sciatic nerve, characterized by the following

   1. The sciatic nerve is actually two nerves in a common wrapper which separate into the tibial and common peroneal nerves, which in turn supply muscles in the leg and foot
   2. As in the case of the lumbar plexus, two major nerves are formed from anterior and posterior divisions of the ventral rami of the spinal nerves
      a. Anterior divisions of L4–L5 and S1–S3 combine to form the tibial nerve, which supplies the original ventral or flexor muscle group of the leg (located posteriorly) and the plantar surface of the foot
      b. Posterior divisions combine to form the common peroneal nerve, which supplies muscles of the two original dorsal compartments of the leg, which are located anteriorly and laterally (extensors and muscles on the dorsum of the foot)

C. COCCYGEAL PLEXUS is composed of the ventral rami of the 4th and 5th sacral nerves and from the coccygeal nerve.

   1. It consists of two loops on the pelvic surface of the coccygeus and levator ani muscles. Twigs are given off to these muscles, and fine anococcygeal nerves supply the skin between the anus and coccyx

CUTANEOUS NERVES

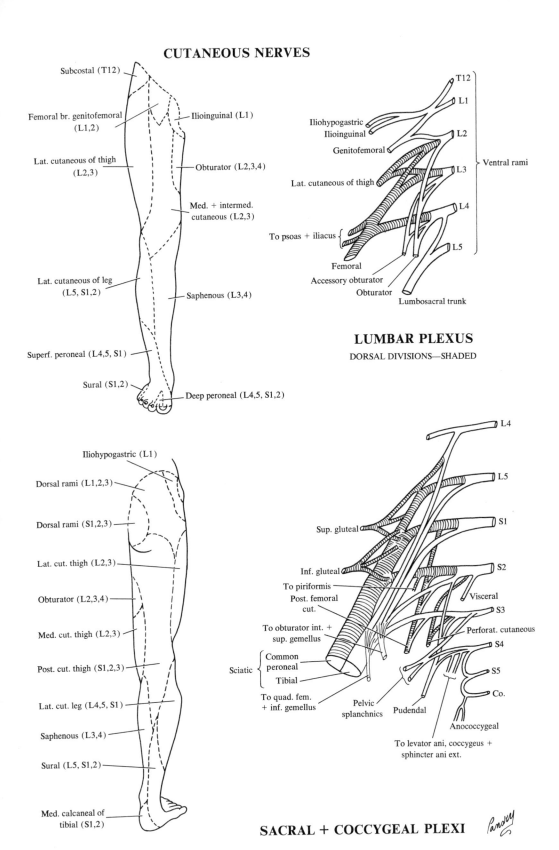

Subcostal (T12)

Femoral br. genitofemoral (L1,2)

Lat. cutaneous of thigh (L2,3)

Ilioinguinal (L1)

Obturator (L2,3,4)

Med. + intermed. cutaneous (L2,3)

Lat. cutaneous of leg (L5, S1,2)

Saphenous (L3,4)

Superf. peroneal (L4,5, S1)

Sural (S1,2)

Deep peroneal (L4,5, S1,2)

T12
L1

Iliohypogastric
Ilioinguinal
Genitofemoral

L2

Lat. cutaneous of thigh

L3

Ventral rami

L4

To psoas + iliacus

L5

Femoral
Accessory obturator
Obturator
Lumbosacral trunk

LUMBAR PLEXUS

DORSAL DIVISIONS—SHADED

Iliohypogastric (L1)

Dorsal rami (L1,2,3)

Dorsal rami (S1,2,3)

Lat. cut. thigh (L2,3)

Obturator (L2,3,4)

Med. cut. thigh (L2,3)

Post. cut. thigh (S1,2,3)

Lat. cut. leg (L4,5, S1)

Saphenous (L3,4)

Sural (L5, S1,2)

Med. calcaneal of tibial (S1,2)

L4

L5

Sup. gluteal

S1

Inf. gluteal

S2

To piriformis
Post. femoral cut.

Visceral

To obturator int. + sup. gemellus

S3

Perforat. cutaneous

Sciatic { Common peroneal / Tibial

S4

S5

Co.

To quad. fem. + inf. gemellus

Pelvic splanchnics

Pudendal

Anococcygeal

To levator ani, coccygeus + sphincter ani ext.

SACRAL + COCCYGEAL PLEXI

Pansky

-77-

38. SPINAL CORD: BLOOD SUPPLY

I. Arterial supply to the spinal cord: derived from two main sources

A. VERTEBRAL ARTERIES give rise to an anterior and a pair of posterior spinal arteries

   1. Anterior spinal artery: runs the entire length of the cord in the midline. From 6 to 10 anterior radicular arteries contribute to its entire length, branching up and down. As a result, there is a point between 2 anterior radicular arteries where, because of opposing blood flow, there is no flow in either direction. Occasionally, in the thoracic region, the anterior spinal artery narrows to a point where it will not function as an adequate anastomosis. The anterior $2/3$ of the substance of the cord is supplied via central and penetrating branches

     a. The cervical and first 2 thoracic segments are supplied by radicals arising from branches of the subclavian artery to join the anterior spinal at an angle of 60°–80°

     b. The middorsal region of cord (T3–T7) usually receives only 1 radicular artery. (This cord part has a poor afferent blood supply)

     c. The dorsolumbosacral cord (T8–conus) gets its supply from the artery of Adamkiewicz, which arises from a left-sided lumbar intercostal (80%)

     d. The cauda equina is supplied by lumbar, iliolumbar, and lateral and median sacral arteries

   2. Posterior spinal arteries: are paired, traverse the posterior surface of cord, and receive contributions from 10–23 posterior radicular arteries which supply the posterior $1/3$ of cord

B. SEGMENTAL ARTERIES (ascending and deep cervical, intercostal, lumbar, and sacrals) give rise to radicular arteries

   1. Each radicular artery follows the spinal nerve, passes through the intervertebral foramina, and separates into small anterior and larger posterior radicular branches. The radicular vessels supply the major blood supply to the cord segments and join the anterior and posterior spinal arteries

C. PERIPHERAL AND CENTRAL ARTERIES are both essentially end arteries (do not anastomose freely)

II. Venous drainage pattern of spinal cord: is similar to and closely parallels that of spinal arteries. Plexuses of veins, external and internal, extend along entire length of vertebral column and form a series of distinct rings around each vertebra. The plexuses freely anastomose and receive tributaries from vertebrae, ligaments, and spinal cord, and are relatively devoid of valves. Changes in intrathoracic or CSF pressure produce variations in blood volume, especially in the internal venous plexuses

A. EXTERNAL VERTEBRAL PLEXUS: consists of an anterior and a posterior plexus which communicate with the basivertebral veins. The posterior communicates with occipital veins, vertebral and deep cervical, and some via foramen magnum with dural sinuses

B. INTERNAL VERTEBRAL PLEXUS: lies in the epidural space in the vertebral canal and is arranged into anterior and posterior interconnected groups.

   1. Before draining into thoracic, abdominal and intercostal veins, the internal vertebral venous plexus receives a single anterior median vein, a single posterior median vein, paired anterior lateral veins, paired posterior lateral veins, and vasocoronal veins

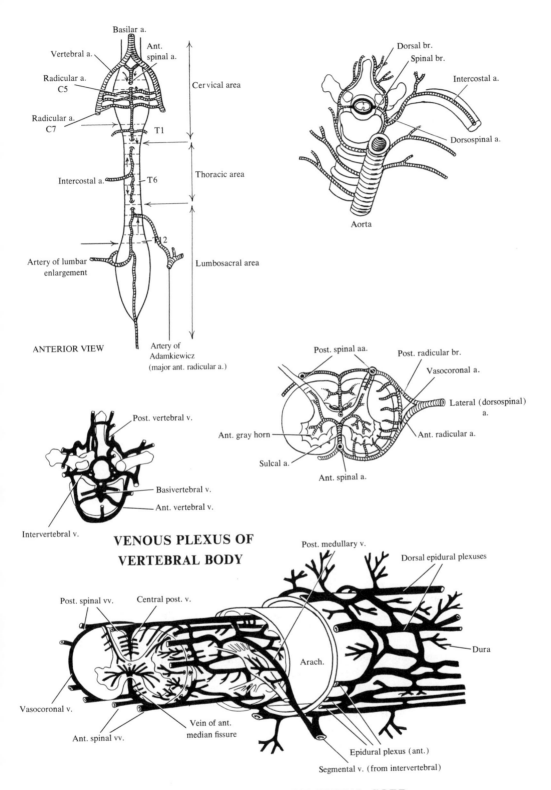

Basilar a.

Vertebral a.

Ant. spinal a.

Radicular a. C5

Cervical area

Radicular a. C7

T1

Intercostal a.

T6

Thoracic area

Artery of lumbar enlargement

L2

Lumbosacral area

ANTERIOR VIEW

Artery of Adamkiewicz (major ant. radicular a.)

Dorsal br.

Spinal br.

Intercostal a.

Dorsospinal a.

Aorta

Post. spinal aa.

Post. radicular br.

Vasocoronal a.

Lateral (dorsospinal) a.

Ant. gray horn

Ant. radicular a.

Sulcal a.

Ant. spinal a.

Post. vertebral v.

Basivertebral v.

Ant. vertebral v.

Intervertebral v.

VENOUS PLEXUS OF VERTEBRAL BODY

Post. medullary v.

Dorsal epidural plexuses

Post. spinal vv.

Central post. v.

Arach.

Dura

Vasocoronal v.

Vein of ant. median fissure

Ant. spinal vv.

Epidural plexus (ant.)

Segmental v. (from intervertebral)

ARTERIES AND VEINS OF SPINAL CORD

39. BRAINSTEM: DIENCEPHALON

I. **Introduction:** the brainstem is a collective term for the diencephalon, mesencephalon, metencephalon, and myelencephalon. It is that part of the brain remaining after the cerebral hemispheres and the cerebellum are removed. The definition of the term "brainstem" is somewhat variable. Some authors exclude the diencephalon, while others may include the deep structures of the telencephalon

II. **The diencephalon** lies between the cerebrum and midbrain and encloses the third ventricle. Its perimeter, viewed medially, includes the choroid plexus of the third ventricle; habenular commissure; pineal body (gland); posterior commissure; a line drawn through the posterior commissure and mamillary bodies (posterior boundary); mamillary bodies; tuber cinereum; hypophysis; optic chiasma; and a line drawn through the optic chiasma and the interventricular foramen (anterior boundary)

A. SUBDIVISIONS OF THE DIENCEPHALON (viewed in the midsagittal plane)

1. Epithalamus is the narrow band on the roof of the diencephalon and consists of
   a. Pineal body: a small mass lying in the depression between the superior colliculi
   b. Posterior commissure: comprised of myelinated fibers that cross the median plane on the dorsal aspect of the rostral end of the cerebral aqueduct. Some of its fibers interconnect the superior colliculi
   c. Habenula (habenular trigone) is a small triangular area anterior to the superior colliculus and contains the habenular nuclei (receives fibers from the stria medullaris)
   d. Habenular commissure: comprised of fibers interconnecting the nuclei
   e. Stria medullaris thalami runs parallel to the choroid plexus and is a fiber bundle from the septal area to the habenular nuclei
2. Thalamus is the largest subdivision of the diencephalon. It is a large ovoid gray mass located above the hypothalamic sulcus and on either side of the third ventricle
   a. It is the major relay (integrative) station interposed between many subcortical structures and the cerebral cortex
   b. The massa intermedia (interthalamic adhesion) is a site of fusion of the dorsal thalami across the midline (absent in 25–30% cases)
3. Hypothalamus is located below or ventral to the thalamus and forms the floor and the ventral half of the lateral walls of the third ventricle
   a. It contains the highest integrative centers of the ANS
   b. It is involved in the regulation of body temperature, endocrine gland functions, and the expression of emotions
   c. Hypothalamic structures visible on the ventral surface of the diencephalon include the optic chiasm and optic tracts, which delineate the floor of the diencephalon; mamillary bodies, which are paired white masses located inferior to the gray matter of the hypothalamic floor in front of the posterior perforated substance; tuber cinereum, which is a funnel-shaped eminence rostral to the mamillary bodies; median eminence, which is the most inferior extension of the tuber cinereum; and the infundibulum, a collective term for the median eminence and the infundibular stalk (a hollow process extending from the tuber cinereum to the posterior hypophysis)
4. Subthalamus lies ventral to the thalamus and flanks the hypothalamus laterally
   a. It has no visible or prominent external structures but is an important subcortical station for voluntary muscle activities

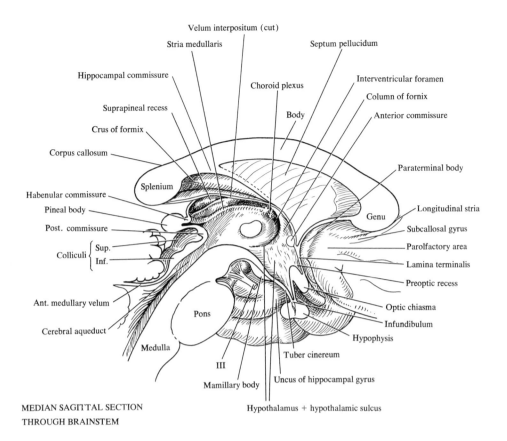

Velum interpositum (cut)

Stria medullaris

Hippocampal commissure

Suprapineal recess

Crus of formix

Corpus callosum

Splenium

Habenular commissure

Pineal body

Post. commissure

Colliculi { Sup. / Inf.

Ant. medullary velum

Cerebral aqueduct

Medulla

III

Mamillary body

Septum pellucidum

Choroid plexus

Body

Interventricular foramen

Column of fornix

Anterior commissure

Parateminal body

Genu

Longitudinal stria

Subcallosal gyrus

Parolfactory area

Lamina terminalis

Preoptic recess

Optic chiasma

Infundibulum

Hypophysis

Tuber cinereum

Pons

Uncus of hippocampal gyrus

Hypothalamus + hypothalamic sulcus

MEDIAN SAGITTAL SECTION
THROUGH BRAINSTEM

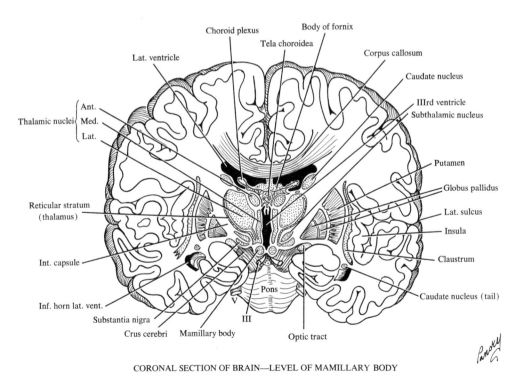

Choroid plexus

Body of fornix

Lat. ventricle

Tela choroidea

Corpus callosum

Caudate nucleus

Thalamic nuclei { Ant. / Med. / Lat.

IIIrd ventricle

Subthalamic nucleus

Putamen

Globus pallidus

Lat. sulcus

Insula

Claustrum

Reticular stratum (thalamus)

Int. capsule

Inf. horn lat. vent.

Substantia nigra

Crus cerebri

Mamillary body

Pons

III

V

Optic tract

Caudate nucleus (tail)

CORONAL SECTION OF BRAIN—LEVEL OF MAMILLARY BODY

–81–

40. BRAINSTEM: MESENCEPHALON

I. **The mesencephalon or midbrain** is the smallest of the four major subdivisions of the brainstem and is positioned between the pons and diencephalon. Its anterior boundary is formed by a line drawn through the posterior commissure dorsally and a point just caudal to the mamillary bodies ventrally

A. PROMINENT EXTERNAL FEATURES ON THE MIDBRAIN

    1. On the ventral surface, elevations on each side are formed by the cerebral peduncles (basis pedunculi, crus cerebri)

        a. The peduncles consist almost entirely of descending fibers. The corticospinal and corticonuclear fibers occupy approximately the middle $\frac{3}{5}$; the frontopontine fibers, the medial $\frac{1}{5}$; and the temporopontine and parietopontine fibers the lateral $\frac{1}{5}$

        b. They converge with one another on the pons as they emerge from the cerebrum in a position medial to the optic tracts

        c. The interpeduncular fossa lies between the cerebral peduncles

           i. Cranial nerve III (oculomotor) emerges from the side of this fossa

          ii. The floor of the fossa is known as the posterior perforated substance due to the many perforations produced by blood vessels that penetrate the midbrain

    2. On the dorsal surface, four rounded eminences, the corpora quadrigemina, are arranged in pairs and called the superior and inferior colliculi

        a. The colliculi form the roof or tectum of the midbrain which measures about 1.5 cm long

           i. The superior colliculi are larger and darker and are associated with the optic system. The brachium (bridgelike structure) of the superior colliculus connects with the lateral geniculate body and contains fibers from the cerebral cortex and retina to the superior colliculus. The latter is primarily a reflex center for movements of the eyes and head in response to visual and other stimuli

          ii. The inferior colliculi are usually more prominent and are important relay nuclei on the auditory pathway to the thalamus. The brachium of the inferior colliculus connects with the medial geniculate body. Some fibers reach the inferior colliculi from the lateral lemniscus, the acoustic part of the temporal cortex, and the spinotectal tracts

        b. The cruciate sulcus separates the superior and inferior colliculi

        c. Cranial nerve IV (trochlear) emerges from the brainstem just inferior to the inferior colliculus and curves around the brainstem

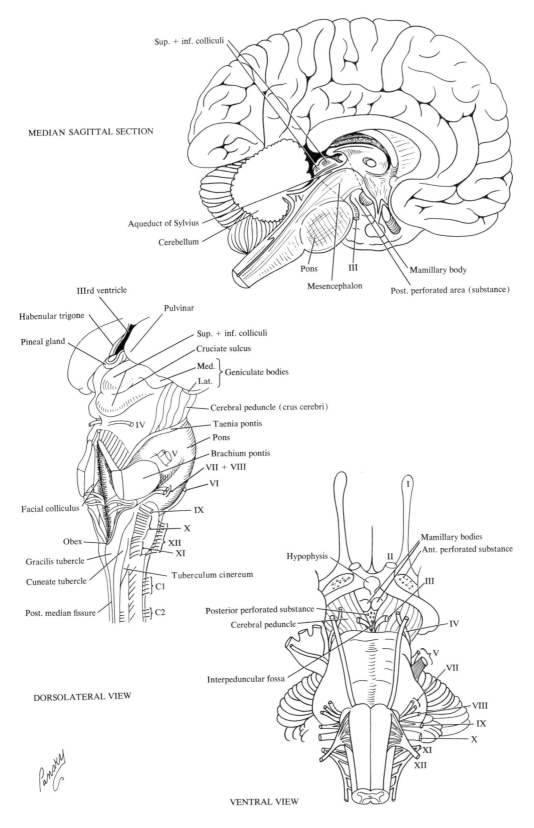

MEDIAN SAGITTAL SECTION

Sup. + inf. colliculi

Aqueduct of Sylvius

Cerebellum

Pons

III

Mesencephalon

Mamillary body

Post. perforated area (substance)

IV

DORSOLATERAL VIEW

IIIrd ventricle

Habenular trigone

Pineal gland

Pulvinar

Sup. + inf. colliculi

Cruciate sulcus

Med.
Lat. } Geniculate bodies

Cerebral peduncle (crus cerebri)

Taenia pontis

Pons

Brachium pontis

VII + VIII

VI

IX

X

XII

XI

Tuberculum cinereum

C1

C2

Facial colliculus

Obex

Gracilis tubercle

Cuneate tubercle

Post. median fissure

IV

V

VENTRAL VIEW

I

Mamillary bodies

Ant. perforated substance

Hypophysis

II

III

Posterior perforated substance

Cerebral peduncle

IV

Interpeduncular fossa

V

VII

VIII

IX

X

XI

XII

Pansky

41. BRAINSTEM: INTERNAL GROSS ANATOMY OF THE MESENCEPHALON

I. **The internal gross anatomy of the mesencephalon** is well demonstrated in transverse sections at the following levels

A. LEVEL OF THE SUPERIOR COLLICULUS: since the posterior part of the thalamus bulges into the midbrain during development and maturation of the brain, certain thalamic nuclei, medial and lateral geniculate bodies, and the pulvinar may be identified. Major structures also identified at this level and associated with the midbrain include

1. Nucleus of cranial nerve III in the periaqueductal gray
2. Fibers of nerve II as they course to their exit
3. Medial longitudinal fasciculi (MLF) form a V at this level with the oculomotor nucleus cradled in the arms of the V
4. Red nucleus and substantia nigra extend the length of the mesencephalon and into the diencephalon
   a. The tegmentum is separated anteriorly from the crura on each side by the substantia nigra
5. Cerebral aqueduct, medial lemniscus, brachium of the inferior colliculus, and the basis pedunculi are also present

B. LEVEL OF THE INFERIOR COLLICULUS

1. Lateral lemniscus and fibers from the lateral lemniscus form a covering for the inferior colliculus
2. Brachium of the inferior colliculus is formed by fibers that pass between the inferior colliculus and the medial geniculate body of the thalamus
3. Cerebral aqueduct connects the third and fourth ventricles
4. Periaqueductal gray, a rather extensive area of gray matter which surrounds the cerebral aqueduct
5. Nucleus of cranial nerve IV on each side, including a group of neurons located on the ventromedian aspect of the periaqueductal gray; the nucleus itself is located on and actually invaginates the dorsal surface of a fiber tract known as the MLF
6. Other structures that can be seen and identified at this level include
   a. The medial lemniscus
   b. The central tegmental tract
   c. The decussation of the superior cerebellar peduncles

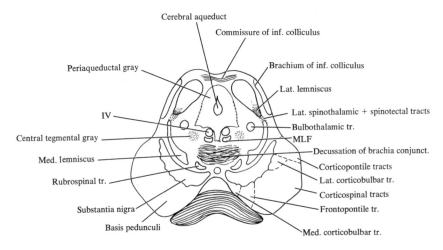

Cerebral aqueduct
Commissure of inf. colliculus
Brachium of inf. colliculus
Periaqueductal gray
Lat. lemniscus
Lat. spinothalamic + spinotectal tracts
IV
Bulbothalamic tr.
Central tegmental gray
MLF
Decussation of brachia conjunct.
Med. lemniscus
Corticopontile tracts
Rubrospinal tr.
Lat. corticobulbar tr.
Corticospinal tracts
Substantia nigra
Frontopontile tr.
Basis pedunculi
Med. corticobulbar tr.

C. MIDBRAIN: LEVEL OF INFERIOR COLLICULUS

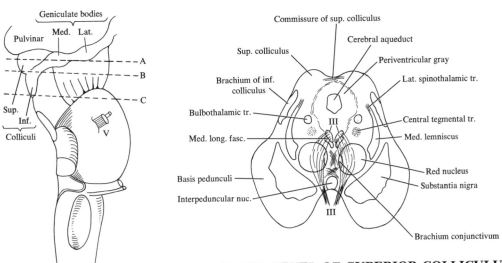

Geniculate bodies
Med. Lat.
Pulvinar
A
B
C
Sup.
Inf.
Colliculi
V

BRAINSTEM
LATERAL VIEW

Commissure of sup. colliculus
Sup. colliculus
Cerebral aqueduct
Periventricular gray
Brachium of inf. colliculus
Lat. spinothalamic tr.
Bulbothalamic tr.
III
Central tegmental tr.
Med. long. fasc.
Med. lemniscus
Basis pedunculi
Red nucleus
Interpeduncular nuc.
Substantia nigra
III
Brachium conjunctivum

B. MIDBRAIN: LEVEL OF SUPERIOR COLLICULUS

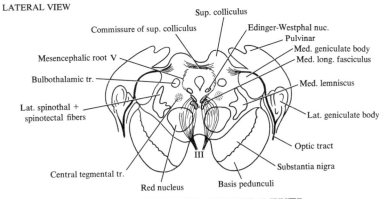

Sup. colliculus
Commissure of sup. colliculus
Edinger-Westphal nuc.
Pulvinar
Mesencephalic root V
Med. geniculate body
Med. long. fasciculus
Bulbothalamic tr.
Med. lemniscus
Lat. spinothal + spinotectal fibers
Lat. geniculate body
Optic tract
III
Central tegmental tr.
Substantia nigra
Red nucleus
Basis pedunculi

A. MIDBRAIN: UPPER LEVEL

42. BRAINSTEM: METENCEPHALON

I. **The metencephalon** is composed of both the pons and cerebellum in the adult. Since the latter is not defined as part of the brainstem proper, only the pons will be considered here

  A. LOCATION OF THE PONS: ventral to the cerebellum and anterior to the medulla. The term "pons" means bridge and refers to the prominent ventral bulge or bridge between the cerebellar hemispheres

  B. BOUNDARIES OF THE PONS: the rostral boundary is formed by a line through the superior pontine sulcus on the ventral surface and the root of cranial nerve IV on the dorsal surface; the caudal boundary is established by a line through the inferior pontine sulcus on the ventral surface and the stria medullares

  C. EXTERNAL GROSS FEATURES OF THE PONS: the dorsal pons is concealed by the cerebellum, while the ventral pons is comprised of a ventral bulge which is continuous with the cerebellum via the large, paired brachium pontis (middle cerebellar peduncles)

   1. Ventral surface of the pons has closely associated with it four cranial nerves
      a. Trigeminal (V) nerve is located on its lateral aspect
      b. Nerves VI, VII, and VIII are located in the inferior pontine sulcus, which marks the caudal boundary of the pons
      c. Nerves V and VII are comprised of two roots of unequal size. The larger root for nerve V is the sensory root (portio major) and the smaller root is the motor root (portio minor). The larger root of VII is motor, while the smaller root is the intermediate nerve
      d. Nerve VIII is comprised of two separate and distinct nerves, the vestibular nerve (ventromedial) and the cochlear nerve (dorsolateral)
   2. Dorsal surface of the pons is formed by the rostral half of the fourth ventricle. The lateral walls of this half of the fourth ventricle are comprised of the three pairs of cerebellar peduncles that form the lateral aspect of the pons and serve to connect the pons with the brainstem
      a. Median eminence lies on the floor of the fourth ventricle
      b. Facial colliculus is a prominent structure located on each side of the median eminence
      c. Locus ceruleus is a bluish area on each side of the median eminence positioned rostral to the vestibular area
      d. Median sulcus lies in the midsagittal plane
      e. Sulcus limitans is the more lateral sulcus separating the median eminence from the vestibular area

  D. MAJOR INTERNAL GROSS FEATURES OF THE PONS: at the midpontine level, the following structures are present (along with other structures which originate at other levels)
   1. Superior medullary velum is located in the roof of the rostral portion of the fourth ventricle
   2. Superior cerebellar peduncles form the lateral walls of the fourth ventricle
   3. Corticospinal (pyramidal) fasciculi and pontine nuclei are positioned in the basilar portion of the pons
   4. Transverse pontine fibers pass to the cerebellum via the brachium pontis
   5. Main sensory nucleus, motor nucleus, and fibers of the trigeminal (V) nerve
   6. Medial lemniscus and trapezoid body are situated in the dorsal pons or tegmentum
   7. Other internal gross features depicted at the levels of the facial colliculus and cranial nerve IV are considered elsewhere in the brainstem

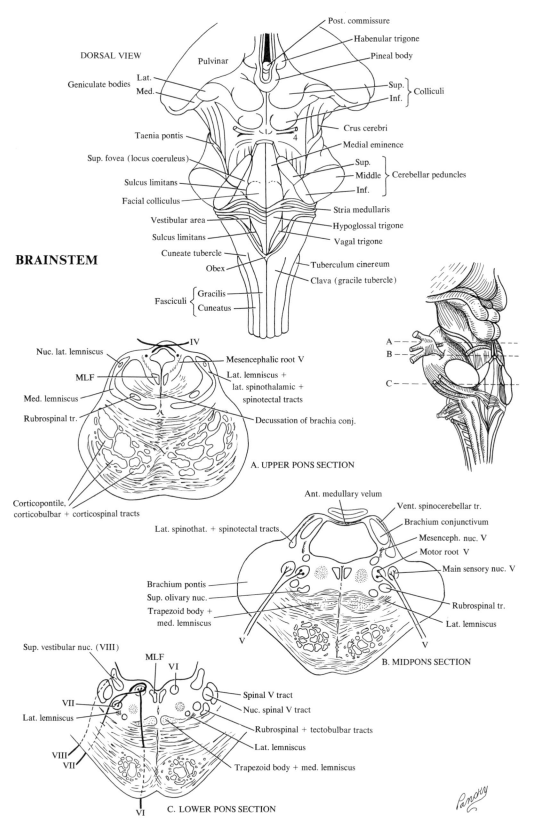

BRAINSTEM

DORSAL VIEW

Post. commissure
Habenular trigone
Pineal body
Pulvinar
Lat.
Geniculate bodies
Med.
Sup.
Inf. } Colliculi
Taenia pontis
Crus cerebri
Medial eminence
Sup. fovea (locus coeruleus)
Sup.
Middle } Cerebellar peduncles
Inf.
Sulcus limitans
Facial colliculus
Stria medullaris
Vestibular area
Hypoglossal trigone
Sulcus limitans
Vagal trigone
Cuneate tubercle
Obex
Tuberculum cinereum
Clava (gracile tubercle)
Fasciculi { Gracilis
Cuneatus

Nuc. lat. lemniscus
IV
Mesencephalic root V
MLF
Lat. lemniscus +
lat. spinothalamic +
spinotectal tracts
Med. lemniscus
Rubrospinal tr.
Decussation of brachia conj.
Corticopontile,
corticobulbar + corticospinal tracts
A. UPPER PONS SECTION

A
B
C

Ant. medullary velum
Vent. spinocerebellar tr.
Lat. spinothat. + spinotectal tracts
Brachium conjunctivum
Mesenceph. nuc. V
Motor root V
Main sensory nuc. V
Brachium pontis
Sup. olivary nuc.
Trapezoid body +
med. lemniscus
Rubrospinal tr.
Lat. lemniscus
V
V
B. MIDPONS SECTION

Sup. vestibular nuc. (VIII)
MLF
VI
Spinal V tract
VII
Nuc. spinal V tract
Lat. lemniscus
Rubrospinal + tectobulbar tracts
VIII
Lat. lemniscus
VII
Trapezoid body + med. lemniscus
VI
C. LOWER PONS SECTION

43. BRAINSTEM: MYELENCEPHALON

I. **The myelencephalon** refers to the medulla (medulla oblongata) in the adult. It is the pyramid-shaped portion of the brainstem which is located between the pons and the spinal cord. The rostral half of its dorsal portion forms the caudal half of the floor of the fourth ventricle.

A. BOUNDARIES OF THE MEDULLA are represented rostrally by a line through the stria medullares and the inferior pontine sulcus; caudally by a transverse plane through the rootlets of the first cervical spinal nerve

B. THE VENTRAL SURFACE OF THE MEDULLA reveals the pyramids, olives, pyramidal decussation, and rootlets of cranial nerves IX, X, XI, and XII. Cranial nerves IX, X, and XI exit in the postolivary sulcus, while XII exits in the preolivary sulcus.

C. THE DORSAL SURFACE OF THE MEDULLA (open portion) consists of the caudal portion of the floor of the fourth ventricle. Structures found on this portion of the floor inferior to the stria medullares include

1. Calamus scriptorius is the caudal portion (below the stria medullares) of the paired median eminences

2. Hypoglossal and vagal trigones, which represent prominences caused by underlying nuclei, comprise each half of the calamus scriptorius

D. LATERAL SURFACE OF THE MEDULLA is comprised of the following named prominences

1. Gracilis tubercle
2. Cunate tubercle
3. Tuberculum cinereum
4. Olive
5. Pyramid

E. MAJOR INTERNAL ANATOMIC FEATURES (at the level of the olive)

1. Fasciculus and nucleus gracilis form the clava or gracilis tubercle on the dorsal surface of the medulla

2. Fasciculus and nucleus cuneatus form the cuneate tubercle on the dorsal surface

3. Spinal nucleus and tract of the trigeminal (V) nerve form the tuberculum cinereum

4. Inferior olivary nuclear complex causes the bulges on the surface known as the olive

5. Corticospinal tracts comprise the pyramids at the level of the medulla

6. Other structures present include
   a. Nucleus of the hypoglossal (XII) nerve which forms the hypoglossal trigone
   b. Dorsal motor nucleus of the vagus (X) nerve which forms the vagal trigone
   c. Vestibular area
   d. Medial longitudinal fasciculus (MLF)
   e. Spinothalamic tract

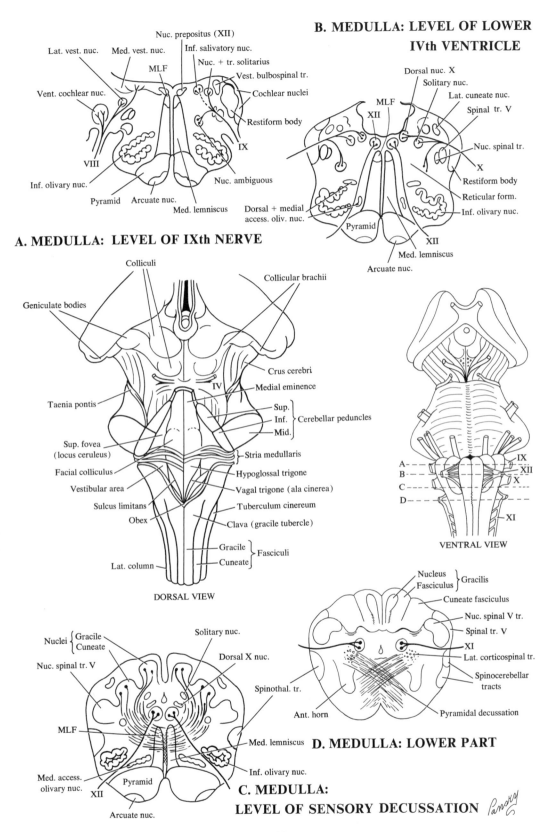

A. MEDULLA: LEVEL OF IXth NERVE

Lat. vest. nuc.
Med. vest. nuc.
Nuc. prepositus (XII)
Inf. salivatory nuc.
MLF
Nuc. + tr. solitarius
Vest. bulbospinal tr.
Vent. cochlear nuc.
Cochlear nuclei
Restiform body
IX
VIII
Inf. olivary nuc.
Pyramid
Arcuate nuc.
Med. lemniscus
Nuc. ambiguous

B. MEDULLA: LEVEL OF LOWER IVth VENTRICLE

Dorsal nuc. X
Solitary nuc.
Lat. cuneate nuc.
MLF
Spinal tr. V
XII
Nuc. spinal tr.
X
Restiform body
Reticular form.
Inf. olivary nuc.
Pyramid
XII
Med. lemniscus
Dorsal + medial access. oliv. nuc.
Arcuate nuc.

Colliculi
Collicular brachii
Geniculate bodies
Crus cerebri
Medial eminence
IV
Taenia pontis
Sup.
Inf. } Cerebellar peduncles
Mid.
Sup. fovea (locus ceruleus)
Stria medullaris
Facial colliculus
Hypoglossal trigone
Vestibular area
Vagal trigone (ala cinerea)
Sulcus limitans
Tuberculum cinereum
Obex
Clava (gracile tubercle)
Lat. column
Gracile } Fasciculi
Cuneate

DORSAL VIEW

A
B
C
D
IX
XII
X
XI

VENTRAL VIEW

Nuclei { Gracile / Cuneate
Solitary nuc.
Nuc. spinal tr. V
Dorsal X nuc.
Spinothal. tr.
Ant. horn
Med. lemniscus
MLF
Med. access. olivary nuc.
XII
Pyramid
Inf. olivary nuc.
Arcuate nuc.

C. MEDULLA: LEVEL OF SENSORY DECUSSATION

Nucleus / Fasciculus } Gracilis
Cuneate fasciculus
Nuc. spinal V tr.
Spinal tr. V
XI
Lat. corticospinal tr.
Spinocerebellar tracts
Pyramidal decussation

D. MEDULLA: LOWER PART

44. GROSS MORPHOLOGY OF BRAIN

I. **Introduction:** the brain (encephalon) is the greatly enlarged and modified superior portion of the CNS. It is ensheathed by three protective, connective tissue membranes (meninges) and is contained within the cranial vault of the skull. Division into cerebrum, cerebellum, and brainstem provides a useful basis for the study of gross structures of the brain

II. **Basic subdivisions of the brain:** the brain is generally divided into five major divisions
A. TELENCEPHALON (ENDBRAIN) gives rise to the bulk of the cerebrum in the adult and forms from the prosencephalon or forebrain
B. DIENCEPHALON (BETWEEN BRAINS) originates as the second vesicle of the original prosencephalon
C. MESENCEPHALON (MIDBRAIN) is the original middle vesicle and remains relatively unchanged. The term "mesencephalon" is retained
D. METENCEPHALON (AFTERBRAIN) forms when the original rhombencephalon divides into two segments. It forms the pons and cerebellum in the adult
E. MYELENCEPHALON (MEDULLA OBLONGATA) is the lowest division or region of the brainstem. It is the inferior segment of the original rhombencephalon

III. **Topography of the brain:** the two cerebral hemispheres, which make up the largest portion of the brain, are separated by the long, deep longitudinal fissure. The outer surface of each hemisphere is marked by a varied series of convolutions (elevated ridges) called gyri, and grooves termed sulci. Deeper furrows between some gyri are called fissures. The terms "sulci" and "fissures" are often used synonymously
A. LOBES AND MAJOR FISSURES OR SULCI of the cerebral hemispheres; each hemisphere is subdivided into six lobes
   1. Frontal lobe is located rostral to the central sulcus (fissure of Rolando) and antero-superior to the lateral fissure
   2. Parietal lobe lies between the central sulcus and the parieto-occipital groove or sulcus
   3. Occipital lobe is positioned posterior to the parieto-occipital line and sulcus
   4. Temporal lobe lies inferior to the lateral (sylvian) fissure and rostral to the parieto-occipital line
   5. Central (insular) lobe lies deep to the temporal lobe and can be seen only when this lobe is retracted
   6. Limbic lobe is located on the medial aspect of each hemisphere, along with medial portions of the frontal, parietal, and occipital lobes

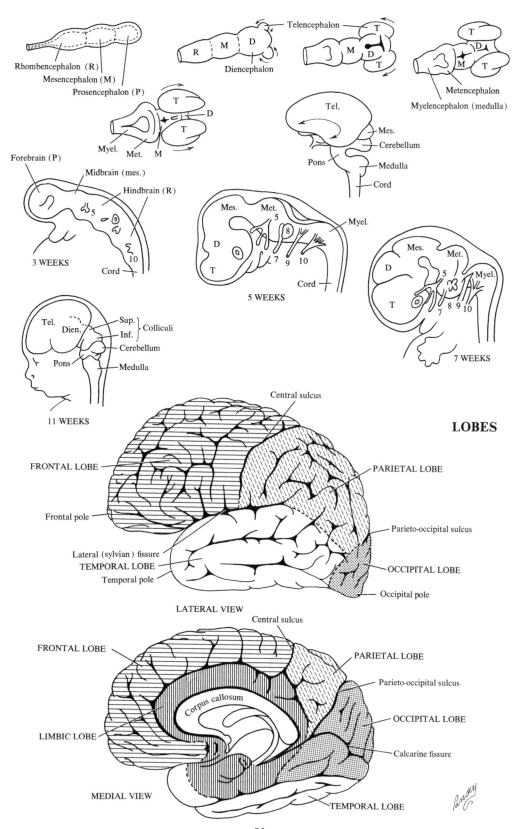

Rhombencephalon (R)
Mesencephalon (M)
Prosencephalon (P)

Telencephalon
R | M | D
Diencephalon

M
D
T

T
D
M
T
Metencephalon
Myelencephalon (medulla)

T
D
Myel. Met. M

Tel.
Mes.
Cerebellum
Pons
Medulla
Cord

Forebrain (P)
Midbrain (mes.)
Hindbrain (R)

3 WEEKS
Cord

Mes. Met.
D
T
5
8
7 9 10
Myel.
Cord
5 WEEKS

Mes. Met.
D
T
5
Myel.
7 8 9 10
7 WEEKS

Tel. Dien.
Sup. } Colliculi
Inf.
Cerebellum
Pons
Medulla

11 WEEKS

LOBES

Central sulcus

FRONTAL LOBE

Frontal pole

Lateral (sylvian) fissure
TEMPORAL LOBE
Temporal pole

PARIETAL LOBE

Parieto-occipital sulcus

OCCIPITAL LOBE

Occipital pole

LATERAL VIEW

Central sulcus

FRONTAL LOBE

Corpus callosum

LIMBIC LOBE

MEDIAL VIEW

PARIETAL LOBE

Parieto-occipital sulcus

OCCIPITAL LOBE

Calcarine fissure

TEMPORAL LOBE

Pansky

-91-

45. GYRI OF THE CEREBRAL HEMISPHERES— IMPORTANT LANDMARKS

I. **Frontal lobe:** recall that the central sulcus or fissure of Rolando separates the frontal and parietal lobes. Much of the frontal lobe is a vast unchartered ocean, but its function in higher cognitive processes, personality, etc., is likely

A. PRECENTRAL GYRUS is referred to as the motor cortex and contains neurons whose axons give rise to corticospinal and corticobulbar (brainstem) tracts

B. SUPERIOR AND MIDDLE FRONTAL GYRI are cortical areas in front of the precentral gyrus for the control of body and eye movements

C. INFERIOR FRONTAL GYRUS has a region referred to as Broca's area (44,45) which is important in the mechanics of speech production, especially in the dominant (usually left) hemisphere

D. ORBITAL GYRI of the frontal lobe (inferior surface of the hemisphere) give rise to pathways that are important in the expression of emotion

II. **Parietal lobe** has two major sulci, the postcentral sulcus and the interparietal sulcus. These divide the lobe into

A. POSTCENTRAL GYRUS is referred to as the sensory cortex and is the receiving area concerned with the appreciation of touch (somesthesis), the sense of position of the extremities (kinesthesis), and the vibratory sense and other fine tactile discriminatory processes

B. SUPERIOR AND INFERIOR PARIETAL LOBULES (LITTLE LOBES) are located posterior to the postcentral gyrus and are important in the synthetic aspects of multiple sensory experiences brought to consciousness

C. A SUPRAMARGINAL AND AN ANGULAR GYRUS are subdivisions of the inferior parietal lobule. These gyri are important in the reception and organization of language functions, especially in the dominant hemisphere

III. **Occipital lobe** is divided by secondary occipital sulci into several unnamed gyri. On the medial surface of the hemisphere, the calcarine sulcus separates gyri that contain neurons specialized to receive visual information

IV. **Temporal lobe** has two sulci that form the boundaries of three gyri. These include the superior and inferior sulci and the superior, middle, and inferior gyri

A. TRANSVERSE TEMPORAL GYRI (OF HESCHL) are found on the superior face of this lobe facing the lateral fissure and extending medially toward the insular (central) lobe

B. THIS REGION RECEIVES AUDITORY INFORMATION, while the inferior temporal gyrus is important in processing visual information

V. **Central (insular) lobe** is buried deep in the lateral fissure and surrounded by the circular sulcus

A. SHORT (IN THE ROSTRAL INSULA) AND LONG (IN THE POSTERIOR INSULA) OBLIQUE GYRI comprise most of the insular cortex

B. THE LIMEN INSULA is a tonguelike extension of the insula that projects medially to the anterior perforated substance

VI. **Limbic lobe and related gyri** form a ring of cortex and associated structures surrounding the central core of the medial aspect of the cerebrum. These structures are important in mechanisms of emotional expressions (see "Limbic System")

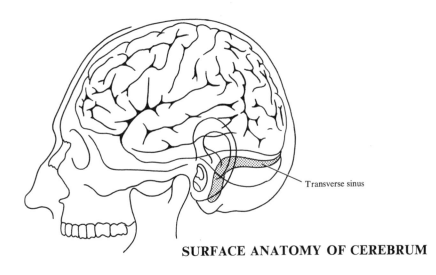

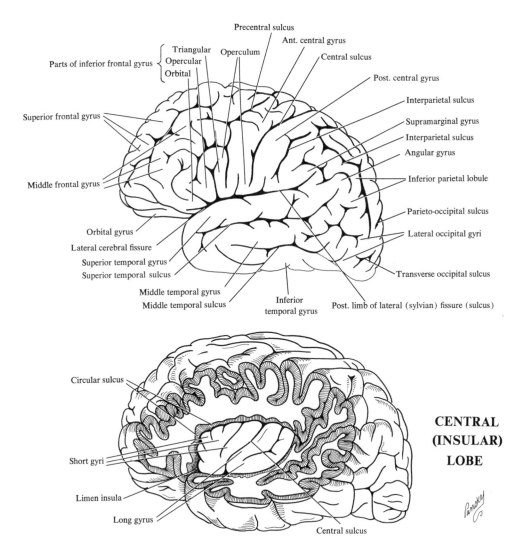

Transverse sinus

SURFACE ANATOMY OF CEREBRUM

Precentral sulcus

Ant. central gyrus

Triangular
Opercular
Orbital

Parts of inferior frontal gyrus

Operculum

Central sulcus

Post. central gyrus

Interparietal sulcus

Supramarginal gyrus

Superior frontal gyrus

Interparietal sulcus

Angular gyrus

Inferior parietal lobule

Middle frontal gyrus

Parieto-occipital sulcus

Lateral occipital gyri

Orbital gyrus

Lateral cerebral fissure

Superior temporal gyrus

Superior temporal sulcus

Transverse occipital sulcus

Middle temporal gyrus

Middle temporal sulcus

Inferior
temporal gyrus

Post. limb of lateral (sylvian) fissure (sulcus)

Circular sulcus

**CENTRAL
(INSULAR)
LOBE**

Short gyri

Limen insula

Long gyrus

Central sulcus

-93-

46. GROSS STRUCTURES OF BRAIN: MIDSAGITTAL ASPECT

I. The medial surface of the hemisphere consists of many named gyri and sulci or fissures

A. CINGULATE GYRUS is a crescentric convolution between the cingulate sulcus and the corpus callosum

B. THE UNCUS is wedge-shaped and lies between calcarine and parieto-occipital fissures

C. PARAHIPPOCAMPAL GYRUS lies between the hippocampal and collateral fissures

D. OCCIPITOTEMPORAL GYRUS (FUSIFORM) is medial to the inferior temporal sulcus

E. LINGUAL GYRUS is between the calcarine and collateral fissures

F. CUNEUS is the curved (hooked) anterior part of the parahippocampal gyrus

G. PRECUNEUS is separated from the cuneus gyrus by the parieto-occipital fissure

H. PARACENTRAL LOBULE is the quadrilateral gyrus around the end of the central sulcus

I. SUPERIOR FRONTAL GYRUS

II. The corpus callosum is a huge bundle of nerve fibers from neurons located entirely in the cortex. It is a major commissural (interhemispheric) pathway connecting the cerebral cortex on one side with its corresponding or homologous area on the other side.

A. THE ROSTRUM, GENU, BODY, AND SPLENIUM are divisions of the corpus callosum

B. THE FORCEPS MAJOR is formed by the splenium and the callosal fibers (U-shaped bundles) radiating to the occipital lobe

C. THE FORCEPS MINOR is formed by the genu and the U-shaped bundles of callosal fibers radiating to the frontal lobe

D. THE GENU blends with the rostrum, which is continuous with the lamina terminalis, a plate of neural tissue which forms the anterior boundary of the third cerebral ventricle

E. THE TAPETUM is a thin sheet of callosal fibers that radiate to line the outer border of the inferior and posterior horns of the lateral ventricles

F. THE INDUSIUM GRISEUM (SUPRACALLOSAL GYRUS) is a thin vestigial layer of gray matter found on the superior surface of the corpus callosum and parallel to its long axis. It contains two bilateral pairs of fiber bundles called the lateral and medial longitudinal striae. The striae are continuous with the septal region rostrally and with the fascicular gyrus caudally

III. The anterior commissure is a bundle of white fibers that crosses the midline to interconnect the cerebral hemispheres. The rostral portion joins both olfactory bulbs; the remainder connects the piriform areas of both hemispheres

IV. The interventricular foramina (of Monro) serve for the passage of CSF from the lateral ventricles to the third ventricle

V. The median telencephalon includes the following structures: anterior commissure; lamina terminalis; and neural tissue anterior to a line drawn through the foramen of Monro and the optic chiasma

VI. The fornix is a large bundle of axons forming an arc from the hippocampal formation to the mamillary body and part of the limbic system

VII. The septum pellucidum is a thin-walled paired structure (midplane) separating the lateral ventricles and extending between the fornix and corpus callosum. They are usually separated in the midline by a small cavity (cavum septum pellucidum)

VIII. The transverse cerebral fissure separates thalamus, mesencephalon, and cerebellum from the cerebral hemispheres

IX. The subcallosal (paraterminal) gyrus is cortical gray matter that covers the inferior aspect of the rostrum of the corpus callosum and is continuous above the genu with the indusium griseum

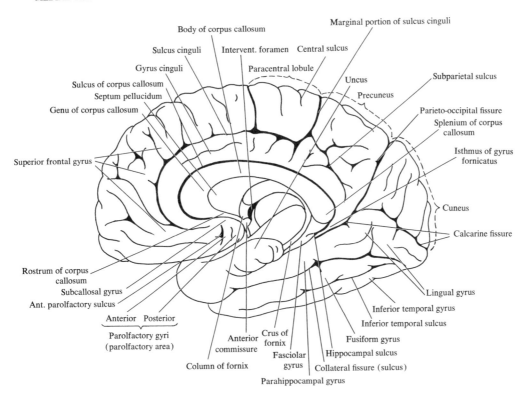

Marginal portion of sulcus cinguli

Body of corpus callosum

Sulcus cinguli

Intervent. foramen Central sulcus

Gyrus cinguli

Paracentral lobule

Sulcus of corpus callosum

Uncus

Septum pellucidum

Precuneus

Genu of corpus callosum

Subparietal sulcus

Parieto-occipital fissure

Splenium of corpus callosum

Superior frontal gyrus

Isthmus of gyrus fornicatus

Cuneus

Calcarine fissure

Rostrum of corpus callosum

Subcallosal gyrus

Ant. parolfactory sulcus

Lingual gyrus

Anterior Posterior

Inferior temporal gyrus

Inferior temporal sulcus

Parolfactory gyri (parolfactory area)

Anterior commissure

Crus of fornix

Fusiform gyrus

Fasciolar gyrus

Hippocampal sulcus

Column of fornix

Collateral fissure (sulcus)

Parahippocampal gyrus

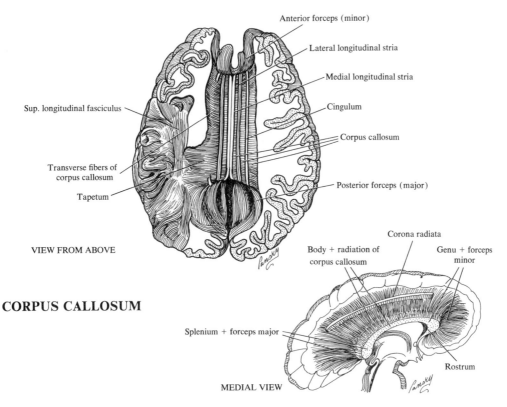

Anterior forceps (minor)

Lateral longitudinal stria

Medial longitudinal stria

Sup. longitudinal fasciculus

Cingulum

Corpus callosum

Transverse fibers of corpus callosum

Tapetum

Posterior forceps (major)

VIEW FROM ABOVE

Corona radiata

Body + radiation of corpus callosum

Genu + forceps minor

CORPUS CALLOSUM

Splenium + forceps major

Rostrum

MEDIAL VIEW

47. BASAL ASPECT OF THE BRAIN

I. **The basal aspect of the brain** includes both telencephalic and brainstem structures

A. TELENCEPHALIC STRUCTURES include portions of the frontal, temporal, and occipital lobes
1. The frontal lobe is demarcated by the olfactory sulcus (parallel to the medial border) and the orbital sulci (an irregular group) and gyri
2. The gyrus rectus lies medial to the olfactory sulcus
3. The rhinal sulcus and collateral sulcus serve to delineate the limbic lobe
4. The rhinencephalon is a phylogenetically old part of the cerebral hemisphere and includes portions associated with olfaction: olfactory bulb and tract, olfactory striae (medial, intermediate, and lateral), olfactory trigone, anterior perforated substance, and piriform area

B. DIENCEPHALIC STRUCTURES
1. Optic nerves
2. Optic tracts
3. Hypophysis
4. Tuber cinereum
5. Mamillary bodies
6. Anterior and posterior perforated substance

C. MESENCEPHALIC STRUCTURES
1. Interpeduncular fossa
2. Crura cerebri
3. Oculomotor nerve (III)
4. Trochlear nerve (IV)

D. METENCEPHALIC STRUCTURES
1. Pons
2. Trigeminal nerve (V)

E. MYELENCEPHALIC STRUCTURES
1. Olives (protuberance of the inferior olivary nucleus)
2. Pyramids and pyramidal decussation
3. Abducent nerve (VI)
4. Facial nerve (VII)
5. Vestibulocochlear nerve (VIII)
6. Glossopharyngeal nerve (IX)
7. Vagus nerve (X)
8. Spinal accessory nerve (XI)
9. Hypoglossal nerve (XII)
10. Posterolateral (postolivary) sulcus
11. Preolivary sulcus

F. ORBITAL SURFACE rests on the roofs of the orbit and nose and is marked by an orbital (demarcates the orbital gyri) and an olfactory sulcus (lodges the olfactory bulb and tract)

G. TENTORIAL SURFACE lies on the floor of the middle cranial fossa and partly on the tentorium cerebelli. It shows 2 grooves: collateral and occipitotemporal sulci. Anterior end of the collateral sulcus is the rhinal sulcus. Note the parahippocampal and lingual gyri, uncus, medial and lateral occipitotemporal gyri associated with this surface (the latter is continuous with the inferior temporal gyrus around the inferior margin of the hemisphere)

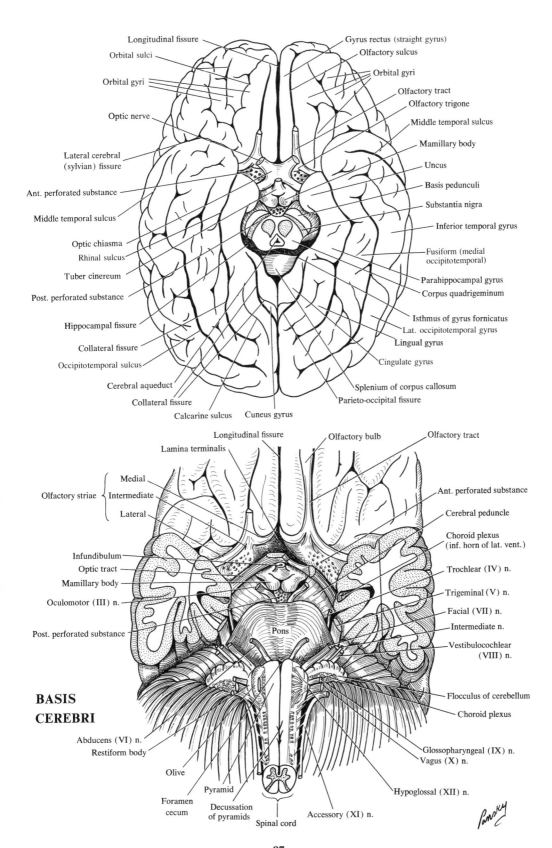

Longitudinal fissure
Orbital sulci
Orbital gyri
Optic nerve
Lateral cerebral (sylvian) fissure
Ant. perforated substance
Middle temporal sulcus
Optic chiasma
Rhinal sulcus
Tuber cinereum
Post. perforated substance
Hippocampal fissure
Collateral fissure
Occipitotemporal sulcus
Cerebral aqueduct
Collateral fissure
Calcarine sulcus
Cuneus gyrus

Gyrus rectus (straight gyrus)
Olfactory sulcus
Orbital gyri
Olfactory tract
Olfactory trigone
Middle temporal sulcus
Mamillary body
Uncus
Basis pedunculi
Substantia nigra
Inferior temporal gyrus
Fusiform (medial occipitotemporal)
Parahippocampal gyrus
Corpus quadrigeminum
Isthmus of gyrus fornicatus
Lat. occipitotemporal gyrus
Lingual gyrus
Cingulate gyrus
Splenium of corpus callosum
Parieto-occipital fissure

Longitudinal fissure
Lamina terminalis
Olfactory bulb
Olfactory tract

Olfactory striae { Medial / Intermediate / Lateral }

Infundibulum
Optic tract
Mamillary body
Oculomotor (III) n.
Post. perforated substance

Ant. perforated substance
Cerebral peduncle
Choroid plexus (inf. horn of lat. vent.)
Trochlear (IV) n.
Trigeminal (V) n.
Facial (VII) n.
Intermediate n.
Vestibulocochlear (VIII) n.
Flocculus of cerebellum
Choroid plexus

BASIS CEREBRI

Abducens (VI) n.
Restiform body
Olive
Pyramid
Foramen cecum
Decussation of pyramids
Spinal cord
Accessory (XI) n.
Pons
Glossopharyngeal (IX) n.
Vagus (X) n.
Hypoglossal (XII) n.

Pansky

-97-

48. MAJOR ASSOCIATION FIBERS OF THE CEREBRAL HEMISPHERES

I. Definitions: Association fibers are nerve fibers that interconnect cortical regions of the same cerebral hemisphere. Projection fibers connect cortical areas of the cerebrum with subcortical regions. Commissural fibers cross the midline and interconnect similar cortical regions in the two cerebral hemispheres

II. Association fibers on the lateral aspect of the cerebral hemispheres

A. UNCINATE FASCICULUS ("uncinate" means hook-shaped) interconnects the cortex of the uncus and temporal pole with the cortex of the inferior frontal region

B. INFERIOR OCCIPITOFRONTAL FASCICULUS is located along the inferior portion of the extreme capsule, dorsal to the uncinate fasciculus. It interconnects the cortex of the lateral or inferolateral portion of the frontal lobe and cortex of the occipital lobe, with connections along the way, including the inferior temporal and fusiform gyri of the temporal lobe

C. SUPERIOR LONGITUDINAL FASCICULUS is located along the dorsolateral border of the putamen, lateral to the internal capsule. It underlies and interconnects the cortices of the frontal, parietal, and occipital lobes and arches inferiorly and anteriorly with connections in the temporal lobe cortex

D. ARCUATE FASCICULUS curves over and around the posterior part of the insula to pass into the temporal lobe. It is a continuation of the superior longitudinal fasciculus (synonym for superior longitudinal fasciculus)

E. LATERAL OCCIPITAL FASCICULUS (also known as vertical or perpendicular occipital fasciculus and as the fasciculus of Wernicke) passes vertically through the occipital lobe and interconnects the fusiform gyrus of the temporal lobe and the posterior part of the parietal lobe

F. INFERIOR LONGITUDINAL FASCICULUS interconnects occipital lobe cortex and temporal lobe cortex in the inferior and lateral portion of the hemisphere

III. Association fibers on the medial aspect of the cerebral hemispheres

A. STRATUM CALCARIUM refers to a well-developed sheet of fibers curving around the bottom of the calcarine fissure from the cuneus above to the lingual gyrus

B. CINGULUM (means girdle) is an association bundle of the cerebrum located within the cingulate gyrus. It has connections all along its course with adjacent frontal, parietal, and temporal lobe cortex

C. SUPERIOR OCCIPITOFRONTAL FASCICULUS is located along the caudate nucleus medial to the interdigitating fibers of the internal capsule and corpus callosum. Its fibers interconnect the cortex of the occipital and temporal lobes with those of the frontal lobe and insula (synonym for subcallosal fasciculus)

IV. Association and commissural fibers in coronal section of cerebral hemispheres

A. CORPUS CALLOSUM (means hard body) is the thick band of commissural fibers interconnecting areas of the neopallium (cerebral cortex and underlying white matter)

B. CINGULUM

C. SUPERIOR AND INFERIOR OCCIPITOFRONTAL FASCICULI

D. SUPERIOR LONGITUDINAL FASCICULUS

E. ARCUATE FASCICULUS

F. UNCINATE FASCICULUS

G. ANTERIOR COMMISSURE

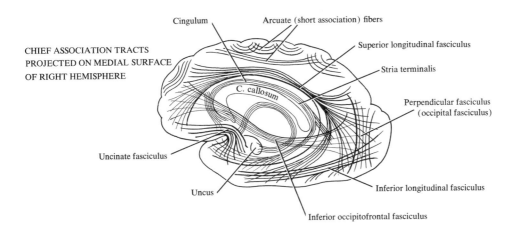

CHIEF ASSOCIATION TRACTS
PROJECTED ON MEDIAL SURFACE
OF RIGHT HEMISPHERE

Cingulum

Arcuate (short association) fibers

Superior longitudinal fasciculus

Stria terminalis

C. callosum

Perpendicular fasciculus
(occipital fasciculus)

Uncinate fasciculus

Uncus

Inferior longitudinal fasciculus

Inferior occipitofrontal fasciculus

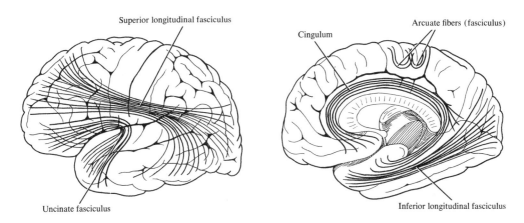

Superior longitudinal fasciculus

Uncinate fasciculus

Cingulum

Arcuate fibers (fasciculus)

Inferior longitudinal fasciculus

MAJOR ASSOCIATION FIBERS

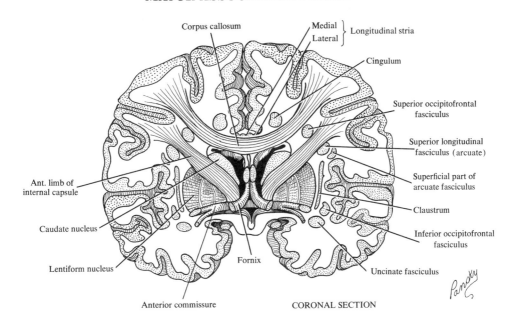

Corpus callosum

Medial } Longitudinal stria
Lateral }

Cingulum

Superior occipitofrontal
fasciculus

Superior longitudinal
fasciculus (arcuate)

Superficial part of
arcuate fasciculus

Claustrum

Inferior occipitofrontal
fasciculus

Uncinate fasciculus

Ant. limb of
internal capsule

Caudate nucleus

Lentiform nucleus

Fornix

Anterior commissure

CORONAL SECTION

-99-

49. MENINGES: THE DURA MATER

I. **Introduction:** the meninges are the three layers of nonneural, connective tissue coverings which enclose the brain and spinal cord and provide a shock absorber filled with CSF. The CSF is found in the ventricular system and the subarachnoid space

II. **The three meningeal layers** consist of the dura mater, arachnoid, and pia mater, each of which forms a separate continuous sheet to surround and protect the soft brain and the spinal cord

A. THE DURA MATER (PACHYMENINX) is the outer layer of the meninges and is a dense, tough, nonstretchable membrane composed of two layers around the brain and one layer around the spinal cord. Dura is relatively insensitive, except in the vicinity of its blood vessels and over the base areas of the skull

1. The outer layer of dura is the fibrous or periosteal (endosteal) layer and is actually the connective tissue membrane of the skull bone
2. The inner layer of dura is the meningeal layer, which folds into a doubled layer or partition in several regions of the skull. Four membranes formed by these folds or reflections are seen grossly in a sagittal plane
   a. The falx cerebri is the double fold of dura between the cerebral hemispheres in the midline; its free edge follows the corpus callosum
   b. The falx cerebelli is a similar midline partition between the cerebellar hemispheres
   c. The tentorium cerebelli is a doubled partition located within the transverse fissure separating the occipital lobe (cerebrum) and the cerebellum
   d. The diaphragma sellae is a double layer of dura that forms the roof of the hypophysis fossa (sella turcica). The infundibular stalk perforates this dural sheet
3. Branches of the trigeminal (V) nerve, vagus (X) nerve, and sympathetic nerves supply the dura mater
   a. Ethmoidal nerves of the ophthalmic division of V supply part of the falx cerebri and the anterior cranial fossa
   b. Recurrent tentorial branches of the ophthalmic division of V innervate the tentorium cerebelli and the posterior falx cerebri
   c. Branches of the maxillary and mandibular divisions of V innervate the middle cranial fossa
   d. Recurrent meningeal branches of the vagus (X) nerve innervate the posterior cranial fossa
   e. Sympathetic autonomic fibers follow the middle meningeal arteries to innervate the dura mater, but not the cerebral blood vessels
   f. Recurrent spinal nerve branches supply the dura mater of the spinal cord
4. The meningeal arteries and veins are located in the outer portion of the dura and groove the inner table of the calvaria
   a. Potential space between dural layers is the epidural space. Space between dura and arachnoid is the subdural space

III. **Clinical application:** dura mater has been used as a homograft in repair of thoracic wall and diaphragm defects, for the correction of transposition of the great arteries, and as a tracheal prosthesis. Since 1971, more than 5,000 bioprosthetic cardiac valves constructed of homologous dura mater have been implanted for the correction of acquired or congenital valvular disease

A. CEREBRAL VEINS crossing the subdural space have very little support and are vulnerable to injury

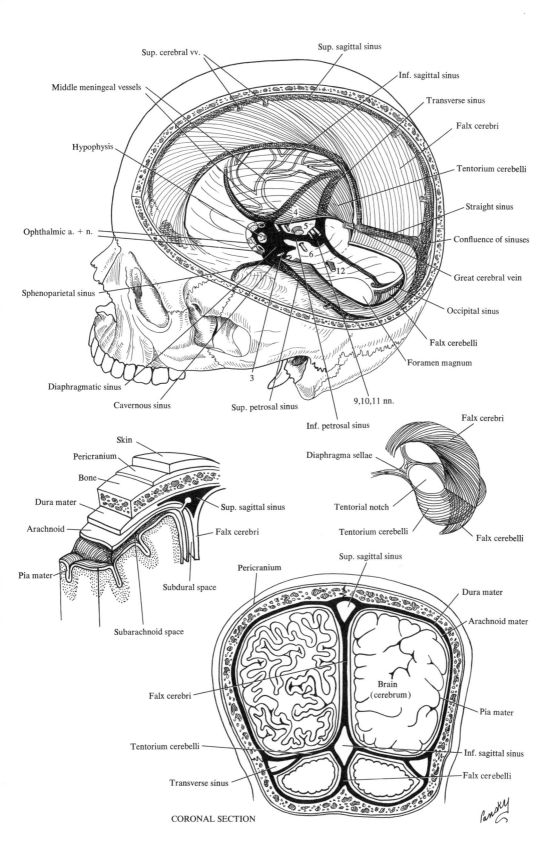

Sup. cerebral vv.

Sup. sagittal sinus

Inf. sagittal sinus

Middle meningeal vessels

Transverse sinus

Falx cerebri

Hypophysis

Tentorium cerebelli

Straight sinus

4

5

Confluence of sinuses

Ophthalmic a. + n.

6

12

Great cerebral vein

Occipital sinus

Sphenoparietal sinus

Falx cerebelli

Foramen magnum

3

Diaphragmatic sinus

9,10,11 nn.

Cavernous sinus

Sup. petrosal sinus

Inf. petrosal sinus

Falx cerebri

Skin

Pericranium

Bone

Diaphragma sellae

Dura mater

Arachnoid

Sup. sagittal sinus

Tentorial notch

Falx cerebri

Tentorium cerebelli

Falx cerebelli

Pia mater

Subdural space

Subarachnoid space

Pericranium

Sup. sagittal sinus

Dura mater

Arachnoid mater

Falx cerebri

Brain
(cerebrum)

Pia mater

Tentorium cerebelli

Inf. sagittal sinus

Falx cerebelli

Transverse sinus

CORONAL SECTION

50. MENINGES: ARACHNOID AND PIA MATER

I. **The arachnoid** is the thin, delicate middle layer of meninges which, like the dura mater, bridges the sulci and extends into only the major fissures. Arachnoid tends to follow dura

A. NUMEROUS FINE TRABECULAE extend from the arachnoid membrane to the pia mater to provide its spider-web appearance

B. THE SUBARACHNOID SPACE is the trabeculated space between the arachnoid and pia mater membranes and contains CSF

C. THE ARACHNOID GRANULATIONS are pia-arachnoid projections which extend into the superior sagittal sinus. The granulations (Pacchionian bodies) are each comprised of a number of arachnoid villi which appear to function as one-way valves allowing passage of substances from the CSF to venous blood in the sinuses

D. THE ARACHNOID, like the dura, sends prolongations over the roots of the spinal nerves

II. **The pia mater** is the inner layer of meninges which is intimately attached to the brain and spinal cord, following every contour (sulci and fissures), carrying with it small blood vessels that nourish the underlying nervous tissue

A. THE PIA-GLIAL MEMBRANE is formed by astrocyte end-feet which end in the pia

B. THE PIA MATER appears to serve as a barrier to the entrance of harmful materials and organisms

C. THE LEPTOMENINGES is a name given to the combination of the pia mater and the arachnoid membrane

D. MENINGITIS is an infectious disease of the nervous system that involves the meninges

E. THE PIA, by its various invaginations, forms the main substance of the tela choroidea and the choroid plexuses of the lateral, 3rd, and 4th ventricles

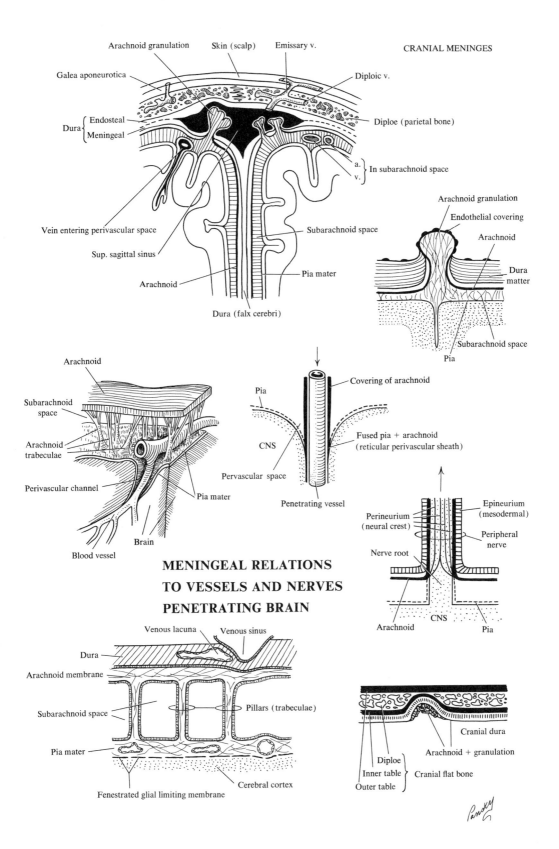

Arachnoid granulation
Skin (scalp)
Emissary v.
Galea aponeurotica
Diploic v.
Dura { Endosteal, Meningeal }
Diploe (parietal bone)
a. } In subarachnoid space
v. }
Vein entering perivascular space
Subarachnoid space
Sup. sagittal sinus
Pia mater
Arachnoid
Dura (falx cerebri)

Arachnoid granulation
Endothelial covering
Arachnoid
Dura matter
Subarachnoid space
Pia

Arachnoid
Subarachnoid space
Arachnoid trabeculae
Perivascular channel
Pia mater
Blood vessel
Brain

Pia
Covering of arachnoid
CNS
Pervascular space
Fused pia + arachnoid (reticular perivascular sheath)
Penetrating vessel

Epineurium (mesodermal)
Perineurium (neural crest)
Peripheral nerve
Nerve root
Arachnoid
CNS
Pia

MENINGEAL RELATIONS TO VESSELS AND NERVES PENETRATING BRAIN

Venous lacuna
Venous sinus
Dura
Arachnoid membrane
Subarachnoid space
Pillars (trabeculae)
Pia mater
Cerebral cortex
Fenestrated glial limiting membrane

Cranial dura
Arachnoid + granulation
Diploe
Inner table } Cranial flat bone
Outer table

51. MENINGES: VENOUS DURAL SINUSES

I. **The venous dural sinuses** are epithelially lined spaces or channels between the two layers of dura of the brain, and drainage of blood from the brain is chiefly into these sinuses. The dural sinuses do not have valves and are usually triangular in shape

A. THE SINUSES RECEIVE BLOOD from three major sources
   1. The cerebral veins are the major source of blood drainage
      a. The great anastomotic vein drains into the superior sagittal sinus
      b. The small anastomotic vein drains into the transverse sinus
      c. The deep cerebral vein drains the basal ganglia
      d. The internal cerebral vein drains the basal vein
      e. The great cerebral vein (vein of Galen) is formed at the junction of the two internal cerebral veins, is a short midline vein, and drains into the straight sinus
   2. The diploic veins lie between the layers of cranial bone. They anastomose freely with each other and also communicate with the meningeal veins internally and the superficial veins externally
   3. The emissary veins connect extracranial and intracranial veins

B. THE VENOUS SINUSES have little tendency to collapse as do most other veins because of the fibrous consistency of the dura mater. Because they have a tendency to remain filled even against gravity, they are subjected to profuse bleeding when ruptured by a bone fragment

C. THE DURAL SINUSES are both paired and unpaired
   1. The unpaired dural venous sinuses include
      a. Superior longitudinal (sagittal) sinus begins as a continuation of the nasal vein at the foramen cecum and receives blood from cerebral, diploic, and emissary veins. It also receives CSF from arachnoid granulations
      b. Inferior longitudinal (sagittal) sinus receives blood from veins on the medial surface of the brain
      c. Straight sinus occupies the area between the falx cerebri and tentorium cerebelli. It is the posterior continuation of the great cerebral vein of Galen and joins the superior sagittal sinus to form the confluence of sinuses
      d. Occipital sinus: smallest of the sinuses. Begins near the foramen magnum and ascends in the attached margin of the falx cerebelli to end in the confluence of sinuses. It also receives the inferior cerebellar veins
   2. The paired dural venous sinuses include
      a. Transverse sinus receives blood from the superior sagittal sinus and straight sinus and drains into the internal jugular veins
      b. Cavernous sinuses are located lateral to the body of the sphenoid bone, and receive blood from the middle cerebral veins and superior ophthalmic veins. They communicate with each other via intercavernous sinuses as well as with the petrosal sinuses and pterygoid plexus of veins. These sinuses are the largest and most clinically important of the deep interconnecting sinuses and drain into the internal jugular veins and transverse sinuses
      c. Sigmoid sinuses are a continuation of the transverse sinuses. In their S-shaped course toward the jugular foramen, they receive blood from the inferior cerebrum, cerebellum, and emissary veins
      d. Petrosal sinuses: the superior and inferior petrosal sinuses drain from the cavernous sinuses into the transverse sinuses and internal jugular veins, respectively
      e. Sphenoparietal sinuses extend along the crest of the lesser wing of the sphenoid bone and drain into the cavernous sinuses

CEREBRAL VENOUS SINUSES

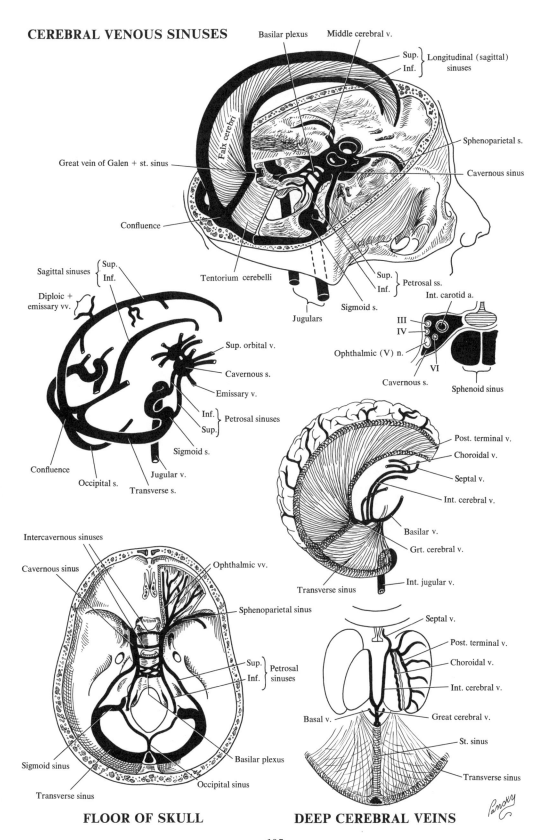

Basilar plexus

Middle cerebral v.

Sup. ⎫
Inf. ⎭ Longitudinal (sagittal) sinuses

Sphenoparietal s.

Falx cerebri

Great vein of Galen + st. sinus

Cavernous sinus

Confluence

Sagittal sinuses ⎰ Sup.
⎱ Inf.

Diploic + emissary vv.

Tentorium cerebelli

Sup. ⎫
Inf. ⎭ Petrosal ss.

Int. carotid a.

Sigmoid s.

Jugulars

III
IV

Ophthalmic (V) n.

VI

Cavernous s.

Sphenoid sinus

Sup. orbital v.

Cavernous s.

Emissary v.

Inf. ⎱
Sup. ⎰ Petrosal sinuses

Sigmoid s.

Confluence

Occipital s.

Jugular v.

Transverse s.

Post. terminal v.

Choroidal v.

Septal v.

Int. cerebral v.

Basilar v.

Grt. cerebral v.

Transverse sinus

Int. jugular v.

Intercavernous sinuses

Cavernous sinus

Ophthalmic vv.

Sphenoparietal sinus

Sup. ⎫
Inf. ⎭ Petrosal sinuses

Septal v.

Post. terminal v.

Choroidal v.

Int. cerebral v.

Basal v.

Great cerebral v.

St. sinus

Sigmoid sinus

Basilar plexus

Occipital sinus

Transverse sinus

Transverse sinus

FLOOR OF SKULL

DEEP CEREBRAL VEINS

Panoky

52. VENTRICULAR SYSTEM

I. **The ventricular system** is a series of four communicating cavities within the brain, lined by ependyma and filled with CSF which is elaborated from blood by the choroid plexuses

A. THE CAVITIES OF THE SYSTEM consist of a pair of lateral ventricles (right and left), the third ventricle, and the fourth ventricle with the latter two connected by the cerebral aqueduct (of Sylvius)

1. The two lateral ventricles are contained within the cerebral hemispheres, each of which is connected with the third ventricle through the interventricular foramen of Monro. Each lateral ventricle has four distinct parts

   a. The anterior horn lies anterior to the interventricular foramina and is located in the frontal lobe
      i. Its anterior border and roof are surrounded by the corpus callosum
      ii. Its medial wall is formed by the septum pellucidum
      iii. Its floor and lateral wall are formed by the caudate nucleus (head)

   b. The body (corpus) lies posterior to the foramina of Monro and is located in the parietal lobe
      i. It extends posteriorly to a point adjacent to the splenium
      ii. Its roof is also surrounded by the corpus callosum
      iii. Its medial wall is formed by the septum pellucidum
      iv. Structures found on the floor of the body include (medial to lateral): fornix, choroid plexus, dorsal thalamus, stria terminalis, vena terminalis, and caudate nucleus

   c. The inferior (temporal) horn is contained in the temporal lobe
      i. Its roof is comprised of white matter of the cerebral hemispheres
      ii. Its medial wall and floor are formed by the fimbria of the fornix and the hippocampus; the stria terminalis and caudate nucleus (tail) are found along its medial wall
      iii. The rostral end is partially surrounded by the amygdala

   d. The posterior horn is located in the occipital lobe
      i. Its roof is formed by the corpus callosum
      ii. The calcar avis, an eminence produced by the calcarine fissure, is found on its medial wall

   e. The trigone (atrium) of the lateral ventricle is the cavity or chamber at the junction of the body, posterior horn, and inferior horn

2. The third ventricle is a thin vertical chamber or cavity located in the midline between the paired lateral ventricles
   a. Its roof is formed by a thin layer of ependyma
   b. Its lateral walls are formed primarily by the paired thalami
   c. Its floor and lower lateral walls are formed by the hypothalamus and subthalamus.
   d. Its anterior boundary is formed by the lamina terminalis and anterior commissure

3. The fourth ventricle is a broad, rhomboid-shaped cavity overlying the pons and medulla and is bound dorsally by the cerebellum and medullary veli
   a. It is continuous with the central canal of the upper cervical spinal cord below and the cerebral aqueduct (of Sylvius) of the mesencephalon above. The latter connects the third and fourth ventricles
   b. The lateral boundaries of the floor are the cerebellar peduncles, the cuneate tubercles, and the clava
   c. The calamus scriptorius is the most inferior portion of the rhomboid fossa or floor of the ventricle
   d. This ventricle also communicates with the subarachnoid space via the two foramina of Luschka and the foramen of Magendi

VENTRICULAR SYSTEM

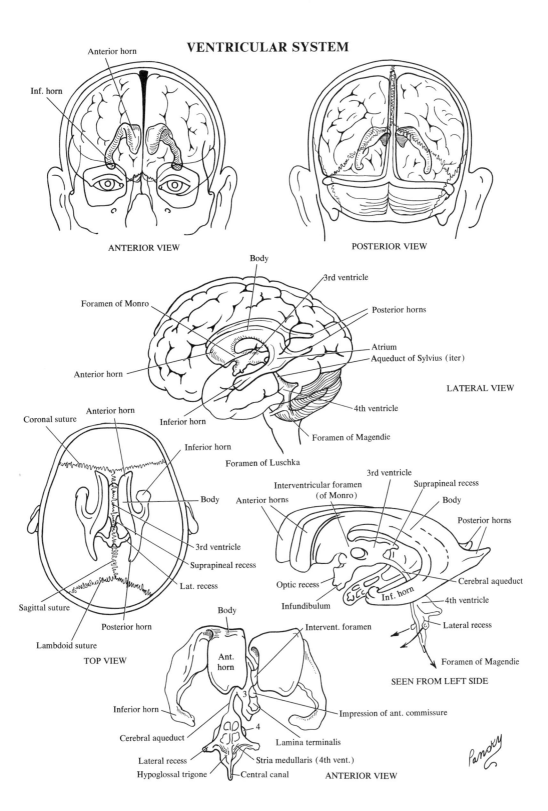

Anterior horn

Inf. horn

ANTERIOR VIEW

POSTERIOR VIEW

Body

3rd ventricle

Foramen of Monro

Posterior horns

Atrium

Aqueduct of Sylvius (iter)

Anterior horn

LATERAL VIEW

Anterior horn

Inferior horn

4th ventricle

Coronal suture

Foramen of Magendie

Foramen of Luschka

Anterior horn

Inferior horn

3rd ventricle

Interventricular foramen
(of Monro)

Suprapineal recess

Anterior horns

Body

Posterior horns

Body

3rd ventricle

Suprapineal recess

Cerebral aqueduct

Optic recess

Lat. recess

Sagittal suture

Infundibulum

Inf. horn

4th ventricle

Lambdoid suture

Posterior horn

Lateral recess

TOP VIEW

Body

Intervent. foramen

Foramen of Magendie

SEEN FROM LEFT SIDE

Ant.
horn

Inferior horn

Impression of ant. commissure

Cerebral aqueduct

Lamina terminalis

Lateral recess

Stria medullaris (4th vent.)

Hypoglossal trigone

Central canal

ANTERIOR VIEW

Panoky

53. CHOROID PLEXUS AND CEREBROSPINAL FLUID

I. **The choroid plexus** is a rich network of blood vessels of the pia mater which projects into each ventricular cavity to form a semipermeable filter between arterial blood and the CSF

A. AN EPITHELIAL LAYER OF EPENDYMA covers each choroid plexus

B. THE TELA CHOROIDEA of the lateral ventricle is a thin membrane resembling a web which is continuous with the choroid plexus of the third ventricle through the interventricular foramina. It is formed by vascular folds invaginating the ependyma. Thus, the tela choroidea is pia plus ependyma

C. THE CHOROID PLEXUS EXTENDS from the interventricular foramen to the rostral end of the inferior horn of the ventricular system

   1. The anterior and posterior choroidal arteries, which are branches of the internal carotids and posterior cerebral arteries, respectively, supply the plexuses of the lateral ventricles

   2. The choroid plexus of the 3rd ventricle is found in its roof and extends on each side from the interventricular foramina to the caudal extent of the roof

   3. The choroid plexus of the fourth ventricle is found in the roof of the medulla and is T-shaped. Right and left halves extend in the direction of the lateral recesses

II. **Cerebrospinal fluid** is a clear, colorless, almost protein-free filtrate of blood which forms in the ventricles and circulates through the subarachnoid space

A. FUNCTION: serves to support and cushion the CNS against injury

   1. May serve nutritive functions and remove waste products of neuronal metabolism

   2. Its pressure measured in the lumbar cistern is 100–150 mm H_2O when the subject is lying down and 200–300 mm H_2O in the sitting position

   3. It contains small amounts of protein, glucose, and potassium and relatively large amounts of sodium chloride

   4. Very few cellular components are found in CSF (1–5 cells/mm^3 is normal)

   5. Bears close resemblance to blood plasma ultrafiltrate except for great differences in protein concentration (CSF, 25 mg/100 gm; plasma, 6500 mg/100 gm)

   6. It has higher Mg^{3+}, Na^+, and Cl^- concentrations and lower K^+, Ca^{2+}, and glucose concentrations than plasma

   7. It is a secretory product produced by active transport mechanisms; Na^+, K^+, and Cl^- are actively transported from choroid plexi into the ventricles, while water follows passively to maintain equilibrium

   8. There is considerable evidence that the choroid plexi are the chief sites for production of CSF, but small amounts are believed to be produced from the ependyma, glial elements, and pia-arachnoid capillary beds

   9. It is constantly being formed at the rate of 600–700 ml/day

   10. The total volume in the ventricles and subarachnoid space is about 140 ml, with an average ventricular capacity of about 20–25 ml

B. CIRCULATION OF CEREBROSPINAL FLUID is through the ventricles and subarachnoid spaces to literally float the brain and spinal cord

   1. From the lateral ventricles it flows through the interventricular foramina (of Monro) into the 3rd ventricle; then it flows through the aqueduct of the midbrain (of Sylvius) into the 4th ventricle; from the latter it flows through the single (Magendie) and two lateral (Luschka) foramina and recesses to circulate in the subarachnoid space around the brain and spinal cord

   2. It returns to the venous system through small membranous villi, the arachnoid granulations, located in the superior sagittal sinus

   3. The flow of CSF is unidirectional because of the pressure gradient between its source, arterial blood and venous blood

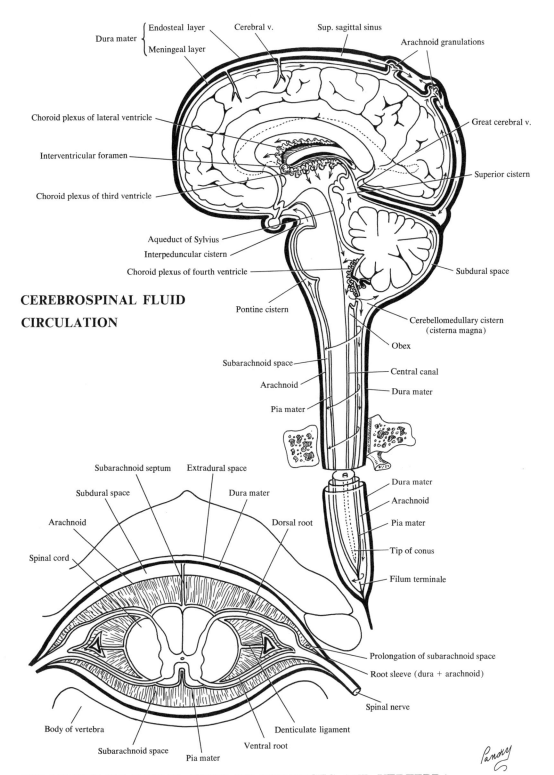

Dura mater { Endosteal layer
{ Meningeal layer

Cerebral v.

Sup. sagittal sinus

Arachnoid granulations

Choroid plexus of lateral ventricle

Great cerebral v.

Interventricular foramen

Superior cistern

Choroid plexus of third ventricle

Aqueduct of Sylvius

Interpeduncular cistern

Choroid plexus of fourth ventricle

Subdural space

**CEREBROSPINAL FLUID
CIRCULATION**

Pontine cistern

Cerebellomedullary cistern
(cisterna magna)

Obex

Subarachnoid space

Central canal

Arachnoid

Dura mater

Pia mater

Subarachnoid septum

Extradural space

Dura mater

Arachnoid

Subdural space

Dura mater

Pia mater

Arachnoid

Dorsal root

Tip of conus

Spinal cord

Filum terminale

Body of vertebra

Prolongation of subarachnoid space

Root sleeve (dura + arachnoid)

Spinal nerve

Subarachnoid space

Denticulate ligament

Pia mater

Ventral root

RELATIONS OF SPINAL CORD TO MENINGES AND VERTEBRA

54. SUBARACHNOID SPACE CISTERNS

I. The subarachnoid space surrounding the brain and spinal cord shows local variations. Over the convexities of the cerebral hemispheres, this space is small except in the depths of the sulci. At the base of the brain, around the brainstem, and at the termination of the spinal cord, the pia and arachnoid are separated, creating several large spaces, called subarachnoid cisternae

A. THREE NAMED CISTERNAE (CISTERNS) LOCATED ON VENTRAL ASPECT OF BRAINSTEM
   1. The chiasmatic cistern is located in the region of the optic chiasma. The cistern continues upward as the cistern of the lamina terminalis, which in turn continues into the cistern of the corpus callosum above the commissure
   2. The interpeduncular cistern is located in the interpeduncular fossa of the mesencephalon
   3. The pontine cistern is located at the pontomedullary junction

B. TWO NAMED CISTERNAE (CISTERNS) POSITIONED ON POSTERIOR ASPECT OF BRAINSTEM
   1. The cerebellomedullary cistern (cisterna magna), one of the largest cisterns, is located between the choroid plexus of the medulla and the cerebellum. The foramina of the 4th ventricle open into this cistern
   2. The superior cistern (cisterna ambiens) surrounds the superior and lateral surfaces of the mesencephalon. This cistern contains the great cerebral vein of Galen and the posterior cerebral and superior cerebellar arteries, which are of great clinical importance. The superior cistern lies between the splenium and superior surfaces of the midbrain and the cerebellum. The superior cistern and interpeduncular cistern are continuous with the cisterna ambiens on the sides of the midbrain

C. THE CISTERN OF THE LATERAL FOSSA is on the lateral side of the hemisphere where the arachnoid bridges over the lateral fissure

D. THE LUMBAR (SPINAL) CISTERN is located between the lumbar and upper sacral levels of the spinal column (L1 to S2)
   1. This cistern contains the filum terminale and dorsal and ventral nerve roots of the cauda equina
   2. CSF is withdrawn here in a lumbar spinal tap

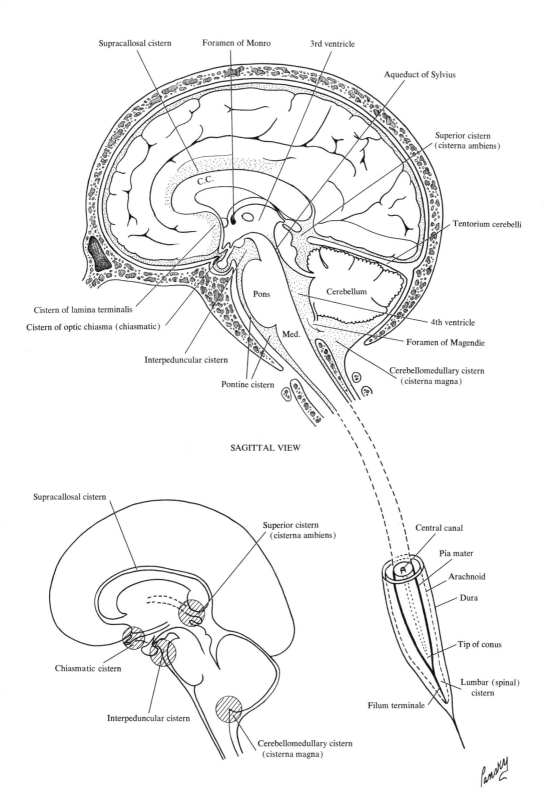

Supracallosal cistern   Foramen of Monro   3rd ventricle

Aqueduct of Sylvius

Superior cistern
(cisterna ambiens)

C.C.

Tentorium cerebelli

Cistern of lamina terminalis

Cistern of optic chiasma (chiasmatic)

Pons

Cerebellum

4th ventricle

Interpeduncular cistern

Med.

Foramen of Magendie

Pontine cistern

Cerebellomedullary cistern
(cisterna magna)

SAGITTAL VIEW

Supracallosal cistern

Superior cistern
(cisterna ambiens)

Central canal

Pia mater

Arachnoid

Dura

Chiasmatic cistern

Tip of conus

Interpeduncular cistern

Lumbar (spinal)
cistern

Filum terminale

Cerebellomedullary cistern
(cisterna magna)

Pansky

55. BRAIN BARRIERS

I. **Ultrastructural studies on the brain, choroid plexus, and meninges** have helped to modify the concept of the hematoencephalic (blood-brain) barrier demonstrated in 1887 by Ehrlich. This original concept was based on the fact that certain aniline dyes, which stain most tissues of the body readily, when injected intravascularly, failed to stain the brain and spinal cord. Thus, a barrier was said to exist between the CNS and the vascular system. The concept has been altered to include the various morphologic features as well as functional systems which serve to subdivide the CNS and associated structures into named compartments separated by specific brain barriers

A. THE BLOOD-BRAIN BARRIER exists between the brain and vascular compartments. The perivascular components of this barrier consist of two continuous layers, the endothelial cells of the blood vessels and the basement membrane. The barrier functions as a differential filter that controls the exchange of various substances from the blood to interstitial fluid

B. THE BLOOD-CSF BARRIER exists between the brain and the CSF compartments. It is an effective one-way entry into the CSF compartment. The epithelium (ependyma) and adnexa of the choroid plexi are responsible for the actively secreted CSF

C. THE BRAIN-CSF BARRIER exists between the brain and the CSF compartments. The ependyma and the subjacent glial cells form a potential barrier between CSF and the interstitial fluid of the brain

D. COMPARTMENTS: CNS is usually divided into two compartments
   1. The intracellular fluid compartment of the brain is comprised of neurons and glial cells. Substances pass into and out of neurons and glial cells from the extracellular space and through plasma membranes instead of by direct exchange from the plasma
   2. The intercellular (interstitial or extracellular) compartment is comprised of the space between nerve cells (neurons and glia) and capillaries. Electron microscopic studies of the brain suggest a total extracellular space of about 4%. Neurochemical studies, using brain chloride as a measure of the extent of the extracellular volume, give values between 25 and 40%

E. ABSENCE OF BLOOD-BRAIN BARRIER: certain areas of the brain appear to be devoid of a blood-brain barrier. These areas are highly vascular and some are secretory in function. These areas include area postrema, hypophysis, pineal body, hypothalamic regions, subfornical organ, supraoptic crest, and choroid plexi
   1. The structure of the blood vessels in these areas differs in that their endothelial cells have fenestrations or pores
   2. There may be an extensive perivascular space containing basement membrane material and other connective tissue elements (connective tissue stroma of the perivascular space in choroid plexi may be 50–100 μm thick)
   3. The boundary between such an extensive perivascular space and the brain parenchyma may be lined with a definitive basement membrane as well as the basal part of the endothelial cells
   4. On the brain parenchymal side of the basement membrane, astrocytic end-feet fit close together up against the membrane

F. BREAKDOWN OF THE BLOOD-BRAIN BARRIER
   1. When large molecules pass from the blood to the brain parenchyma during induced cerebral edema (anoxia, cold-induced, etc.), gross structural damage to small blood vessels has been demonstrated
   2. Histamine, bacterial endotoxins, and other agents that increase extravasation in most tissues do not appear to affect the vascular endothelium of the brain
   3. Systemically induced inflammation does not appear to affect the blood-brain barrier

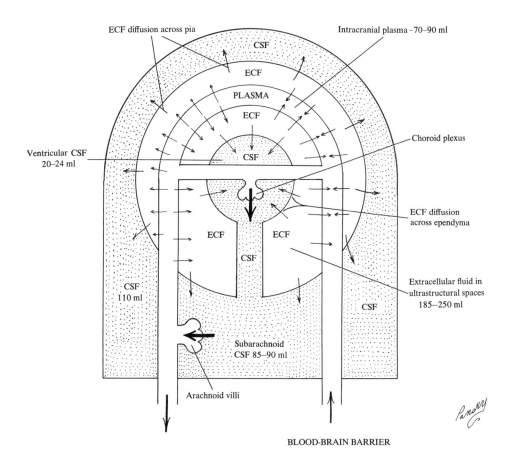

ECF diffusion across pia

Intracranial plasma –70–90 ml

CSF

ECF

PLASMA

ECF

CSF

Choroid plexus

Ventricular CSF 20–24 ml

ECF diffusion across ependyma

ECF

ECF

CSF

Extracellular fluid in ultrastructural spaces 185–250 ml

CSF 110 ml

CSF

Subarachnoid CSF 85–90 ml

Arachnoid villi

Panoky

BLOOD-BRAIN BARRIER

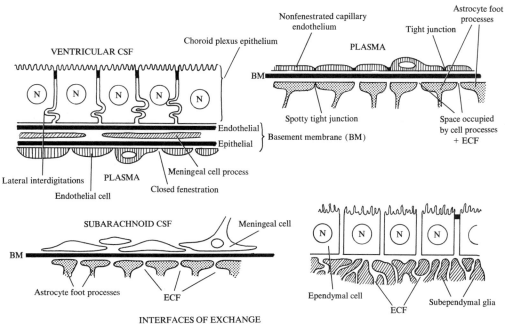

VENTRICULAR CSF

Choroid plexus epithelium

Nonfenestrated capillary endothelium

Astrocyte foot processes

Tight junction

PLASMA

BM

N   N   N   N   N

Endothelial
Epithelial
} Basement membrane (BM)

Spotty tight junction

Space occupied by cell processes + ECF

Lateral interdigitations

Endothelial cell

PLASMA

Meningeal cell process

Closed fenestration

SUBARACHNOID CSF

Meningeal cell

BM

Astrocyte foot processes

ECF

N   N   N   N

Ependymal cell

ECF

Subependymal glia

INTERFACES OF EXCHANGE

56. BLOOD SUPPLY OF THE BRAIN:
ARTERIAL SUPPLY

I. The arterial supply to intracranial structures is basically derived from branches of the two internal carotid arteries and two vertebral arteries

A. THE INTERNAL CAROTID ARTERIES arise at the bifurcation of the common carotids in the neck below the angle of the mandible. They ascend to the base of the skull, extend anteromedially through the carotid canal, curve as a sigmoid-shaped vessel within the cavernous sinus, and at the lesser wing of the sphenoid bone turn between the anterior and middle clinoid processes to reach the brain. The internal carotid arteries do not give off any branches in the neck; all branches are within the cranial cavity and consist of

  1. Ophthalmic artery is given off just as the internal carotid leaves the cavernous sinus. It has an important branch, the central retinal artery, which runs in the center of the optic nerve to end in the retina
  2. Anterior choroidal artery is given off just before the carotid divides into its terminal branches
     a. It runs along the optic tract, around the cerebral peduncle to the level of the lateral geniculate body, and then becomes part of the choroid plexus of the lateral ventricle
     b. It sends branches to cerebral peduncles, internal capsule, caudate nucleus, hippocampus, and optic tract
  3. Anterior and middle cerebral arteries are the terminal divisions of the internal carotid as it reaches the ventral surface of the brain lateral to the optic chiasma
     a. Anterior cerebral artery runs medially over the optic tract to the longitudinal cerebral fissure to supply blood to the frontal lobes
        i. Anterior communicating artery is located in the fissure
        ii. Branches pass to the medial aspect of the frontal and parietal lobes; anterior perforated substance, septum pellucidum, and part of the corpus callosum
        iii. Medial striate artery, when present, supplies the caudate nucleus, putamen, and internal capsule (anterior limb)
     b. Middle cerebral artery extends laterally (in lateral fissure) over the insula, giving off branches to the lateral aspect of temporal and parietal lobes
        i. Lateral striate artery supplies basal ganglia and internal capsule
        ii. Posterior communicating artery runs dorsally to join posterior cerebral branch of basilar artery. In its course, it gives off branches to the internal capsule and thalamus

B. VERTEBRAL ARTERIES are branches of the subclavian artery and ascend in the neck through the transverse foramina of cervical vertebrae. They enter the foramen magnum and join to form the basilar artery at the caudal end of pons. Branches include

  1. Anterior spinal artery descends on ventral aspect of cord to supply it
  2. Posterior spinal arteries descend on dorsal aspect of cord to supply it
  3. Posterior inferior cerebellar arteries are given off laterally to supply part of the cerebellum and the choroid plexus of the 4th ventricle

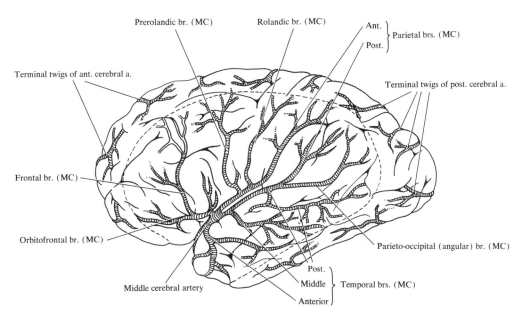

Prerolandic br. (MC)

Rolandic br. (MC)

Ant. } Parietal brs. (MC)
Post.

Terminal twigs of ant. cerebral a.

Terminal twigs of post. cerebral a.

Frontal br. (MC)

Orbitofrontal br. (MC)

Parieto-occipital (angular) br. (MC)

Middle cerebral artery

Post.
Middle } Temporal brs. (MC)
Anterior

ARTERIAL SUPPLY— LATERAL VIEW

CEREBRAL CORTEX

Anterior cerebral

Anterior cerebral

Posterior cerebral

Middle cerebral

Middle cerebral

Posterior cerebral

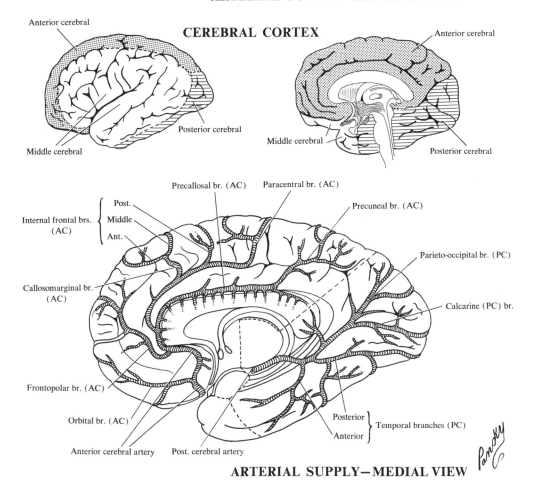

Precallosal br. (AC)

Paracentral br. (AC)

Precuneal br. (AC)

Post.
Internal frontal brs. { Middle
(AC) Ant.

Parieto-occipital br. (PC)

Callosomarginal br.
(AC)

Calcarine (PC) br.

Frontopolar br. (AC)

Orbital br. (AC)

Posterior } Temporal branches (PC)
Anterior

Anterior cerebral artery

Post. cerebral artery

ARTERIAL SUPPLY—MEDIAL VIEW

57. BLOOD SUPPLY OF THE BRAIN: BASILAR ARTERY AND CIRCLE OF WILLIS

I. Arterial supply (cont.)

A. BASILAR ARTERY is formed by the union of the two vertebral arteries, and it courses along the ventral aspect of the pons. Its branches include
1. Pontine arteries supply the pons
2. Anterior inferior cerebellar arteries originate at the level of the abducens, facial, and vestibulocochlear nerves, between the pons and medulla to supply the cerebellum
3. Labyrinthine arteries follow the facial and vestibulocochlear nerves through the internal acoustic meatus to the inner ear
4. Superior cerebellar arteries branch off the basilar artery near its termination at the level of the upper pons. They are in close proximity to the oculomotor and trochlear nerves and extend superiorly around the pons. They supply the upper cerebellum, corpora quadrigemina, choroid plexus of the third ventricle, and the pineal body
5. Posterior cerebral arteries are the terminal branches of the basilar artery. They receive the posterior communicating arteries from the internal carotid.
   a. Supply the medial and inferior aspects of the occipital and temporal lobes
   b. Give off posterior choroidal branches to the choroid plexi of the 3rd and lateral ventricles

II. The circle of Willis is formed by the anterior and posterior communicating arteries and proximal portions of the anterior, middle (internal carotid), and posterior cerebral arteries

A. FUNCTION OF CIRCLE OF WILLIS makes possible an adequate blood supply to the brain in case of occlusion of either the carotid or vertebral arteries
1. The circle surrounds the optic chiasma and pituitary stalk
2. The circle of Willis is subject to frequent anatomic variation and a "normal" circle is seen in only about 50% of the population
   a. Common variations occur when the posterior cerebral arteries arise directly from the internal carotid via an enlarged posterior communicating artery, when one or both posterior communicating arteries are absent, and when there are multiple small anterior communicating arteries
3. Numerous arteries arise from the circle, penetrate the brain substance, and are very important because not only are they small and easily plugged but are also end arteries (no collateral circulation) and supply vital areas
4. The initial segment of the anterior cerebral artery also gives rise to the recurrent artery of Heubner, which supplies the rostral limb of the internal capsule, the anteroinferior part of the head of the caudate nucleus, and the putamen

III. Collateral circulation: symptoms of a cerebrovascular accident (CVA) are determined by the location and size of the pathologic process. The effect on brain tissue of a gradually developing occlusion depends on available anastomotic channels and collateral circulation. A thrombosis in the internal carotid artery is likely to cause a massive cerebral infarct resulting in death, but may go unnoticed if there is efficient collateral circulation in the circle of Willis. Collateral circulation via superficial pial anastomoses between the major cerebral arteries can also be effective. The perforating arteries to deep areas of the cerebrum, cerebellum, and brainstem, however, are in effect "end arteries," and occlusions here usually result in severe effects

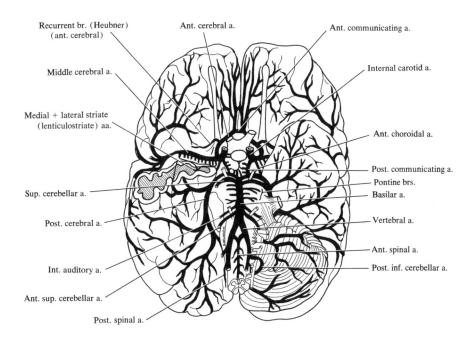

Recurrent br. (Heubner) (ant. cerebral)

Ant. cerebral a.

Ant. communicating a.

Middle cerebral a.

Internal carotid a.

Medial + lateral striate (lenticulostriate) aa.

Ant. choroidal a.

Sup. cerebellar a.

Post. communicating a.

Pontine brs.

Basilar a.

Post. cerebral a.

Vertebral a.

Int. auditory a.

Ant. spinal a.

Post. inf. cerebellar a.

Ant. sup. cerebellar a.

Post. spinal a.

ARTERIES AT BASE OF BRAIN

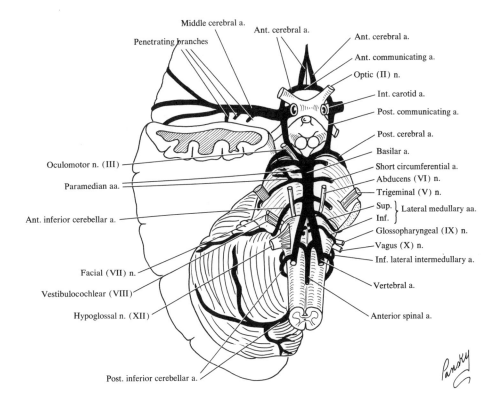

Middle cerebral a.

Ant. cerebral a.

Ant. cerebral a.

Penetrating branches

Ant. communicating a.

Optic (II) n.

Int. carotid a.

Post. communicating a.

Post. cerebral a.

Oculomotor n. (III)

Basilar a.

Short circumferential a.

Paramedian aa.

Abducens (VI) n.

Trigeminal (V) n.

Sup. } Lateral medullary aa.
Inf. }

Ant. inferior cerebellar a.

Glossopharyngeal (IX) n.

Vagus (X) n.

Inf. lateral intermedullary a.

Facial (VII) n.

Vertebral a.

Vestibulocochlear (VIII)

Hypoglossal n. (XII)

Anterior spinal a.

Post. inferior cerebellar a.

Pansky

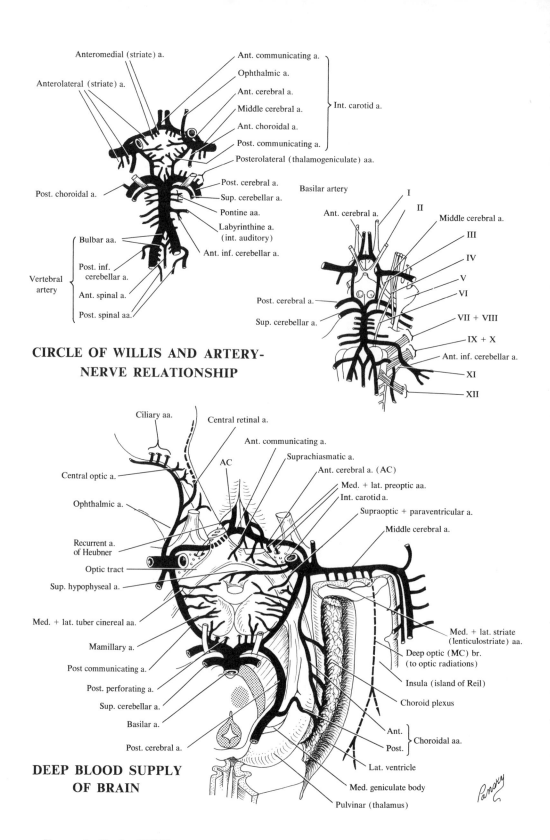

CIRCLE OF WILLIS AND ARTERY-NERVE RELATIONSHIP

DEEP BLOOD SUPPLY OF BRAIN

FIGURE 1. **Circle of Willis.**

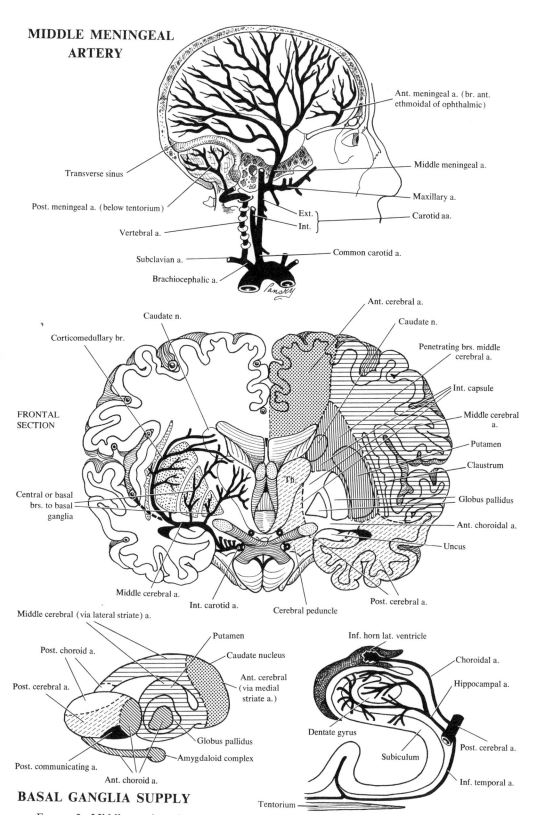

MIDDLE MENINGEAL ARTERY

Ant. meningeal a. (br. ant. ethmoidal of ophthalmic)

Transverse sinus

Middle meningeal a.

Post. meningeal a. (below tentorium)

Maxillary a.

Vertebral a.

Carotid aa.

Ext.
Int.

Subclavian a.

Common carotid a.

Brachiocephalic a.

Pansky

Ant. cerebral a.

Caudate n.

Caudate n.

Corticomedullary br.

Penetrating brs. middle cerebral a.

Int. capsule

**FRONTAL
SECTION**

Middle cerebral a.

Putamen

Claustrum

Central or basal brs. to basal ganglia

Th.

Globus pallidus

Ant. choroidal a.

Uncus

Middle cerebral a.

Int. carotid a.

Cerebral peduncle

Post. cerebral a.

Middle cerebral (via lateral striate) a.

Putamen

Inf. horn lat. ventricle

Choroidal a.

Post. choroid a.

Caudate nucleus

Hippocampal a.

Post. cerebral a.

Ant. cerebral (via medial striate a.)

Post. communicating a.

Globus pallidus

Amygdaloid complex

Dentate gyrus

Post. cerebral a.

Subiculum

Ant. choroid a.

Inf. temporal a.

BASAL GANGLIA SUPPLY

Tentorium

FIGURE 2. **Middle meningeal artery and basal ganglia arterial supply.**

58. BLOOD SUPPLY OF THE BRAIN:
VENOUS DRAINAGE

I. Veins of the brain: all terminate in the dural sinuses and are classified as external and internal cerebral veins

A. EXTERNAL CEREBRAL VEINS

    1. Superior cerebral veins (8–12) drain the superior, lateral, and medial surface of the hemisphere and terminate in the superior sagittal sinus

    2. Middle cerebral veins begin on the lateral surface of the hemispheres and end directly or indirectly in the cavernous sinuses (there are also intercommunications with the superior sagittal and transverse sinuses)

    3. Medial cerebral veins drain the medial surface of the hemispheres and terminate in the inferior sagittal sinus

    4. Inferior cerebral veins are small and drain the inferior surface of the hemispheres

        a. Those on the frontal lobe connect via the superior cerebral tributaries with the superior sagittal sinus

        b. Those on the temporal lobe join either the middle cerebral or basilar venous system and end in the cavernous or superior petrosal sinuses

    5. Basal veins (of Rosenthal) are formed by the confluence of 3 small veins: anterior cerebral draining the frontal lobe; deep middle cerebral, draining the insula and adjoining gyri; and inferior striate, draining the corpus striatum

        a. Receive tributaries from the interpeduncular fossa, hippocampal gyrus, inferior horn of the lateral ventricle, and mesencephalon

        b. Terminate in the great cerebral vein (of Galen)

B. INTERNAL CEREBRAL VEINS: each formed by a confluence of two veins

    1. The terminal (thalamostriate) vein lies in the groove between the thalamus and corpus striatum and receives tributaries from both

    2. The anterior septal vein is formed by the union of multiple intramedullary veins, which drain the deep white matter of the anterior portion of the frontal lobe. These unite in the anterior part of the anterior horn of the lateral ventricle and form 1 or 2 veins. The anterior septal vein runs medially and joins the ipsilateral thalamostriate vein to form the internal cerebral vein

    3. The internal cerebral veins receive the superior choroidal veins, roof veins of the lateral ventricle, veins of the posterior horn of the lateral ventricle, and thalamic veins

C. CEREBELLAR VEINS (veins of the posterior cranial fossa) are of 3 groups

    1. Superior (Galenic) group drains the superior portion of the cerebellum and upper part of the brainstem, and include precentral, superior vermian, superior cerebellar hemispheric, posterior mesencephalic, anterior pontomesencephalic, lateral mesencephalic, and quadrigeminal veins. Major drainage is to the great cerebral vein and straight, transverse, and petrosal sinuses

    2. Anterior or petrosal group receives venous drainage from the anterior aspect of the brainstem, superior and inferior surfaces of the cerebellar hemispheres, region of the cerebellomedullary fissure, and lateral recess of the 4th ventricle. It drains into the superior petrosal sinus and has several named tributaries

    3. Posterior or tentorial group of veins drains the inferior portion of the cerebellar vermis and medial part of the superior and inferior cerebellar hemispheres. Major drainage is to the transverse, straight, or lateral sinuses

    4. When the internal cerebral veins reach the splenium of the corpus callosum, they are joined by the basal veins to form the great cerebral vein (of Galen)

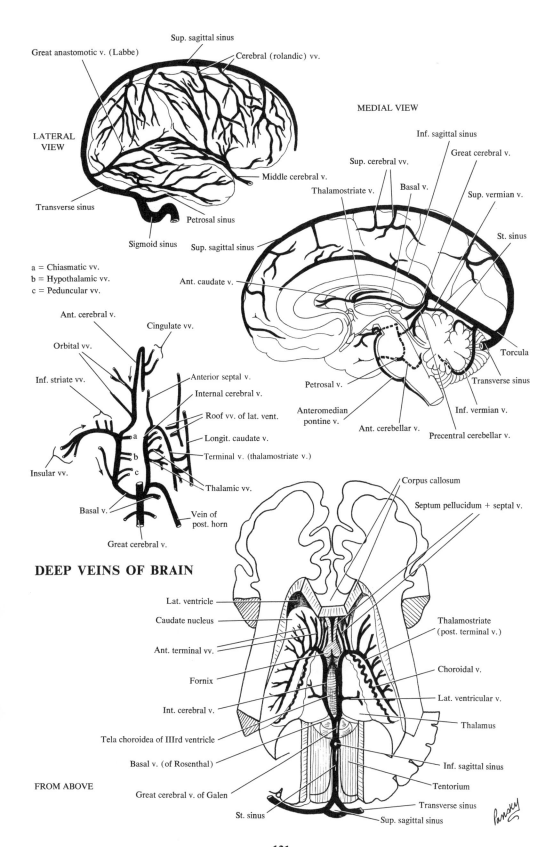

Sup. sagittal sinus

Great anastomotic v. (Labbe)

Cerebral (rolandic) vv.

MEDIAL VIEW

LATERAL VIEW

Inf. sagittal sinus

Great cerebral v.

Sup. cerebral vv.

Thalamostriate v.

Basal v.

Sup. vermian v.

Middle cerebral v.

St. sinus

Transverse sinus

Petrosal sinus

Sup. sagittal sinus

Ant. caudate v.

Sigmoid sinus

a = Chiasmatic vv.
b = Hypothalamic vv.
c = Peduncular vv.

Torcula

Transverse sinus

Ant. cerebral v.

Cingulate vv.

Petrosal v.

Anteromedian pontine v.

Inf. vermian v.

Orbital vv.

Anterior septal v.

Internal cerebral v.

Ant. cerebellar v.

Precentral cerebellar v.

Inf. striate vv.

Roof vv. of lat. vent.

Longit. caudate v.

Terminal v. (thalamostriate v.)

Thalamic vv.

Insular vv.

Corpus callosum

Septum pellucidum + septal v.

Basal v.

Vein of post. horn

Great cerebral v.

DEEP VEINS OF BRAIN

Lat. ventricle

Caudate nucleus

Thalamostriate (post. terminal v.)

Ant. terminal vv.

Choroidal v.

Fornix

Lat. ventricular v.

Int. cerebral v.

Thalamus

Tela choroidea of IIIrd ventricle

Basal v. (of Rosenthal)

Inf. sagittal sinus

Tentorium

FROM ABOVE

Great cerebral v. of Galen

Transverse sinus

St. sinus

Sup. sagittal sinus

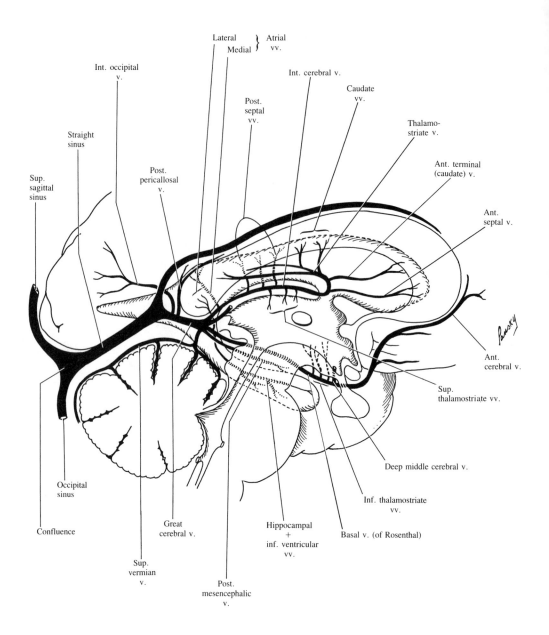

Lateral
Medial } Atrial vv.

Int. occipital v.

Int. cerebral v.

Caudate vv.

Post. septal vv.

Thalamo-striate v.

Straight sinus

Ant. terminal (caudate) v.

Sup. sagittal sinus

Post. pericallosal v.

Ant. septal v.

Ant. cerebral v.

Sup. thalamostriate vv.

Ant. cerebral v.

Deep middle cerebral v.

Occipital sinus

Inf. thalamostriate vv.

Confluence

Great cerebral v.

Hippocampal + inf. ventricular vv.

Basal v. (of Rosenthal)

Sup. vermian v.

Post. mesencephalic v.

FIGURE 3. **Subependymal veins.**

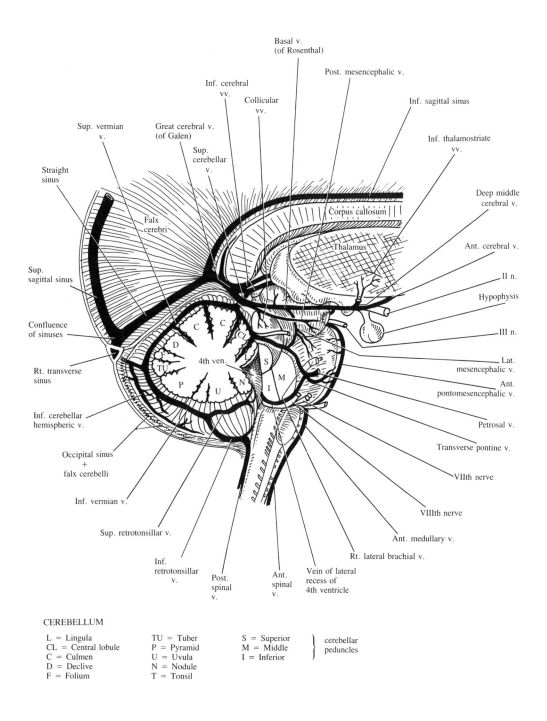

Basal v.
(of Rosenthal)

Post. mesencephalic v.

Inf. cerebral vv.

Collicular vv.

Inf. sagittal sinus

Inf. thalamostriate vv.

Sup. vermian v.

Great cerebral v. (of Galen)

Sup. cerebellar v.

Deep middle cerebral v.

Straight sinus

Falx cerebri

Corpus callosum

Thalamus

Ant. cerebral v.

Sup. sagittal sinus

II n.

Hypophysis

Confluence of sinuses

III n.

Rt. transverse sinus

C   C   CL

D

F

TU

4th ven.

P   U   N   S   I   M

Lat. mesencephalic v.

Ant. pontomesencephalic v.

Inf. cerebellar hemispheric v.

Petrosal v.

Transverse pontine v.

Occipital sinus
+
falx cerebelli

VIIth nerve

Inf. vermian v.

VIIIth nerve

Sup. retrotonsillar v.

Ant. medullary v.

Inf. retrotonsillar v.

Post. spinal v.

Ant. spinal v.

Vein of lateral recess of 4th ventricle

Rt. lateral brachial v.

CEREBELLUM

L = Lingula
CL = Central lobule
C = Culmen
D = Declive
F = Folium

TU = Tuber
P = Pyramid
U = Uvula
N = Nodule
T = Tonsil

S = Superior
M = Middle
I = Inferior

} cerebellar peduncles

FIGURE 4. **Posterior cranial fossa veins.**

59. COMPUTERIZED TOMOGRAPHY (CT) OF THE BRAIN

I. Introduction
The development of the CT scanner has had a great impact on all fields of neuroscience. At the same time, computerized cerebral scanning has clearly demonstrated and emphasized the need to understand the anatomic minutiae of the brain and its coverings in order to interpret the details and relationships now possible with the CT scanner. In addition to revealing the extent and exact location of intracranial space-occupying lesions, cerebral hemorrhage, cerebral edema, infarcts, malformations, and demyelinating diseases, it can serve also as an invaluable tool for studying normal neuroanatomy. The student of neuroanatomy is able to obtain a better three-dimensional image when coronal and horizontal CT scans of the same brain are compared

II. Some characteristics and major advantages of the CT scanner
A. NONINVASIVE CHARACTER makes it practically free of morbidity
B. REPEATED SCANS without exposure to x-radiation are possible
C. SAFER THAN ANGIOGRAPHY AND PNEUMOENCEPHALOGRAPHY
D. PAINLESS, COMFORTABLE, AND RELATIVELY FAST PROCEDURE
E. AMBULATORY UTILIZATION makes it more practical

III. Basis of computer tomography
A. CT UTILIZES DIFFERENTIAL ABSORPTION of head and neck (or body) obtained by axial circumferential measurements which are mathematically arranged by a computer into a matrix (grid) of picture elements on an oscilloscope
B. EVEN SOFT TISSUES have slightly different absorption coefficients, and by taking a number of measurements of the amount of x-ray absorption and systematically altering the angle of the passage of the x-ray beam through the head, the absorption coefficient can be calculated
C. THE IMAGE that is formed can be manipulated by using the computer's data bank to suit the particular organ or pathology in the area of interest which includes magnification, density, and area of measurements
D. HIGHER DENSITIES appear light (bone), while lower densities are darker
E. CERTAIN TISSUES (abscess margins, tumors, edematous fluids, etc.) and intravascular blood can show increased CT number density when intravenous contrast materials (diatrizoate, iothalamate ions, etc.) are utilized
F. THE DEGREE OF INCREASE IN BLOOD CT DENSITY is related to the dose and rate of intravenous infusion (e.g., fast high dose causes more intravascular CT density enhancement)

IV. Procedure used for neuroradiologic application
A. X-RAY BEAM that is finely focused is rotated around the head numerous times
B. AMOUNT OF X-RADIATION EMITTED is known, and that passing through the skull is determined by means of scintillation crystals
C. SECTIONS (slices) of the head are scanned in this fashion
D. THE BRAIN CAN BE PROGRESSIVELY SCANNED, in this manner, from its base to its vertex when the positions of the slices are varied
E. DATA FROM THE SCANS are processed by the computer, which in turn displays the absorption coefficients of the tissues in each section on a cathode ray tube. The intensity of the beam is proportional to the absorption coefficient
F. SOME OF THE STRUCTURES (both normal and pathologic) that can be recognized are shown on the CT scans that are included herein

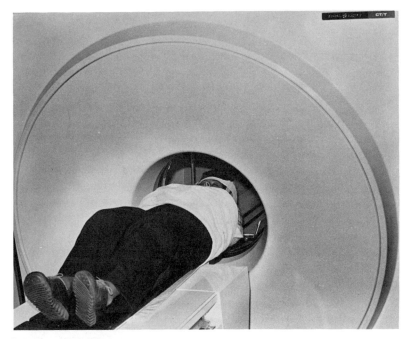

Patient in position in CT scanner

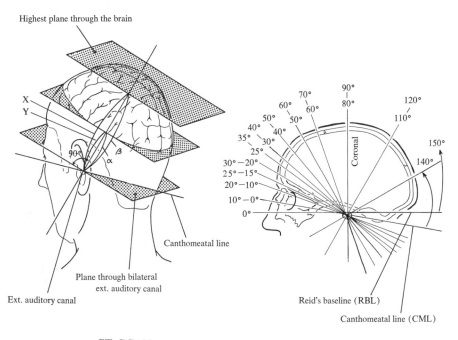

Highest plane through the brain

X
Y

90°

α
β

Canthomeatal line

Plane through bilateral
ext. auditory canal

Ext. auditory canal

90°
80°
70°
60°   60°
50°   50°
40°   40°
35°   30°
25°
30° — 20°
25° — 15°
20° — 10°
10° — 0°
0°

120°
110°

Coronal

150°
140°

Reid's baseline (RBL)

Canthomeatal line (CML)

CT SCAN

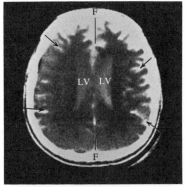

Horizontal CT scan of normal adult brain showing lateral ventricles (LV), falces (F), and sulci (arrows). Radiopaque medium used for contrast.

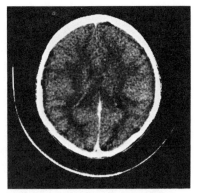

Horizontal CT scan of normal 7-yr-old brain. Note difference between white and gray matter.

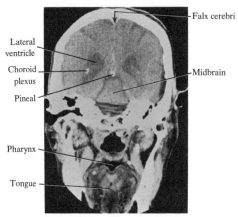

Coronal CT scan of normal adult brain.

Falx cerebri

Lateral ventricle

Choroid plexus

Pineal

Midbrain

Pharynx

Tongue

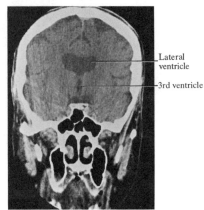

Coronal CT scan of normal adult brain.

Lateral ventricle

3rd ventricle

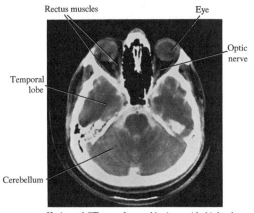

Horizontal CT scan of normal brain at midorbit level.

Rectus muscles

Eye

Optic nerve

Temporal lobe

Cerebellum

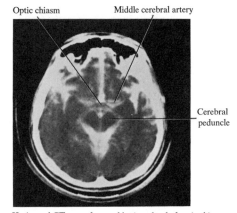

Horizontal CT scan of normal brain at level of optic chiasm.

Optic chiasm

Middle cerebral artery

Cerebral peduncle

FIGURE 5. **Normal CT scans.**

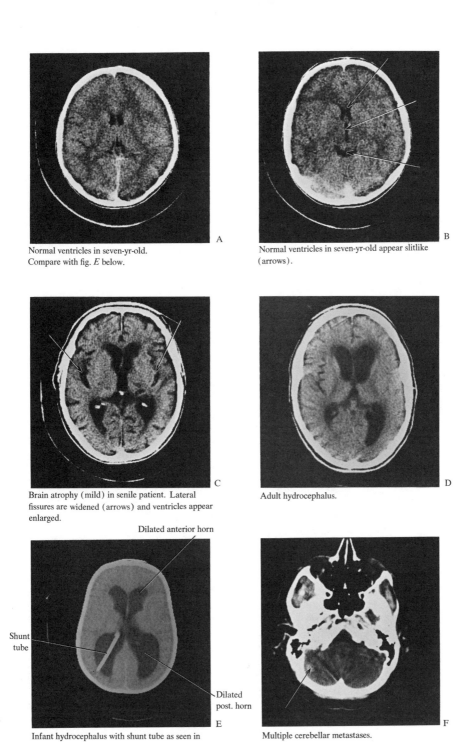

Normal ventricles in seven-yr-old. Compare with fig. *E* below.

A

Normal ventricles in seven-yr-old appear slitlike (arrows).

B

Brain atrophy (mild) in senile patient. Lateral fissures are widened (arrows) and ventricles appear enlarged.

C

Adult hydrocephalus.

D

Dilated anterior horn

Shunt tube

Dilated post. horn

Infant hydrocephalus with shunt tube as seen in CT scan.

E

Multiple cerebellar metastases.

F

FIGURE 6. **CT Scans.** **A** and **B** normal; **C**, brain atrophy; **D** and **E**, hydrocephalus; **F**, cerebellar metastases.

60. MAGNETIC RESONANCE IMAGING (MRI) OF THE BRAIN

I. Characteristics: MRI reflects both the chemical environment and concentration of hydrogen atoms in the body. These atoms, which behave like tiny spinning magnets, align with the axis of the magnetic field in the scanner. The ancillary equipment beams a certain radiofrequency at the atoms, and they pick up energy and change alignment. When the signal stops, the atoms ''relax'' into their original orientation with the magnetic field, giving off radio signals, and a computer transforms the signals into an image of body tissue. Thus, a picture of the body's tissues is formed quickly, harmlessly, and more accurately than with any other imaging method. The current major clinical application of MRI is for diagnosing diseases of the CNS. MRI can discriminate among overlapping structures and has no ionizing radiation

II. Basic fundamentals

A. ATOMIC NUCLEI have an angular momentum arising from their inherent property of rotation
   1. The rotation or spin, being electrically charged, corresponds to a current flowing on a spin axis which generates a small magnetic field
   2. Only nuclei with an odd number of protons or neutrons have a net spin and lend themselves to MRI spectroscopy
   3. The magnetic moments or dipoles of the nuclei with spin are pointed in random directions but orient, in a magnetic field, with the field's lines of induction or force
      a. The magnetic behavior of the entire population of nuclei is predictable and can be controlled in a magnetic field
B. A ROTATING MAGNETIC FIELD is applied by surrounding the patient with a coil connected to a source of radiofrequency power
   1. The frequency of the applied electromagnetic radiation must match the natural precessional frequency of the nuclei [term used: nuclear magnetic resonance (NMR)]
      a. The frequencies used are in the radiofrequency band for the electromagnetic spectrum and do not disrupt molecules of living systems
C. THE PATIENT is inserted into a body-size chamber, in a large magnet, which also contains radiofrequency coils
   1. He is immersed in a magnetic field of up to 3,000 gauss, which causes a magnetic alignment of the H^+ nuclei in the body, as well as other nuclei with an odd number of protons or neutrons (thus, a net spin)
   2. When subjected to energy of the correct phase and radiofrequency, the spins of these nuclei change direction and absorb energy in the process
   3. The energy is subsequently reemitted at a characteristic radiofrequency, and the signal is picked up by a receiver. Signal characteristics depend on the strength of the field and density of the nuclei in the plane of the body, and correspond to the number of H^+ or other nuclei present
D. DATA FROM THE RECEIVER are fed into a computer that plots the distribution of the nuclei responsible for the emission and shows the abundance of nuclei, as well as the molecular environment in which they reside
   1. Resolution is achieved in the 2- to 4-mm range, improving on standard CT images
   2. MRI is inherently a 3-dimensional phenomenon, and signals are obtained from the total volume of material enclosed in the transmitter and receiver coils
   3. MRI imaging systems can be designed to receive data from a single point, from a line, from a plane, and from a complete 3-dimensional volume all at once

MRI OF PATIENT WITH HYDROCEPHALUS

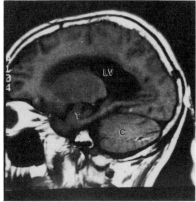

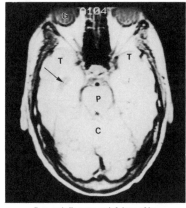

Parasagittal
section

Horizontal
section

C = cerebellum
LV = enlarged lateral ventricle
T = temporal lobe

G = eyeball; arrow = inf. horn of lat.
ventricle; C = cerebellum
T = temporal lobes; P = pons

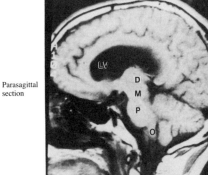

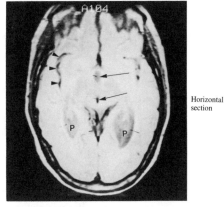

Parasagittal
section

Horizontal
section

M = mesencephalon; LV = enlarged
ventricle
P = pons
O = medulla oblongata
D = diencephalon

P = posterior horns of lateral ventricle
arrows = 3rd ventricle
arrowheads = inf. horn of temporal lobe

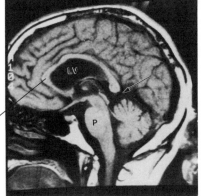

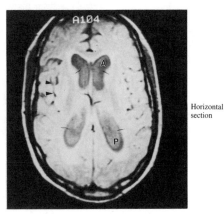

Midsagittal
section

Corpus
callosum

Horizontal
section

P = pons; arrow = superior cistern

A = anterior horn; P = posterior horn
arrowheads = insular gyri

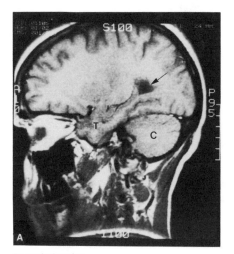

Parasagittal section

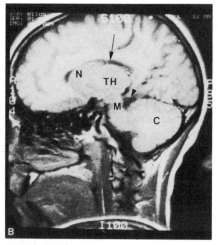

Parasagittal section

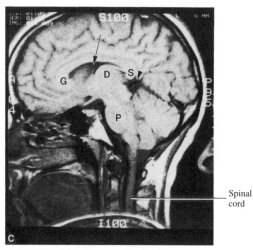

Spinal
cord

Sagittal section (near midsagittal
plane)

FIGURE 7. **Normal magnetic resonance images (MRI).** **A,** through cerebrum and cerebellum *(C),* showing temporal lobe *(T)* and lateral ventricle near antrum (arrow); **B,** through cerebrum and cerebellum *(C),* including thalamus *(TH),* mesencephalon *(M),* head of caudate nucleus *(N),* and lateral ventricle (arrow); **C,** through cerebrum and brainstem, showing diencephalon *(D),* corpus callosum *(G =* genu and *S =* splenium), pons *(P),* lateral ventricle (arrow), and superior cistern (arrowhead).

-130-

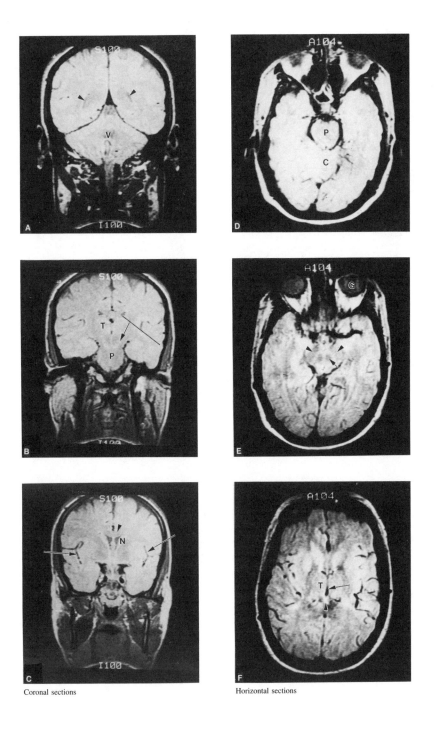

Coronal sections

Horizontal sections

FIGURE 8. **Normal magnetic resonance images (MRI).** *Coronal views.* **A,** through cerebrum, cerebellum and cord, lateral ventricles (arrowheads), and fourth ventricle *(V)*; **B,** thalamus *(T)*, cerebral peduncle (arrow), and pons *(P)*; **C,** insula (arrows), corpus callosum (arrowhead), and caudate nucleus *(N)*. *Horizontal views.* **D,** through cerebrum, cerebellar vermis *(C)*, and pons/ midbrain junction *(P)*; **E,** at level of midbrain and eyeball *(G)*, note red nucleus (arrow) and substantia nigra (arrowheads); **F,** at level of diencephalon, note thalamus *(T)*, third ventricle (arrow), and pineal gland (arrowhead).

Unit Three

CRANIAL NERVES

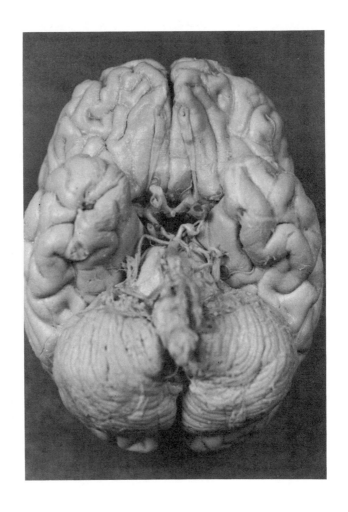

61. FUNCTIONAL CLASSIFICATION OF SYSTEMS

I. Neurons

A. AFFERENT (A): conduction current is traveling toward brain or spinal cord. The word "sensory" implies consciousness and not all afferent impulses reach consciousness

1. Somatic (S) pertains to the framework of the body as distinguished from the viscera or organs. Embryologically is concerned with those parts derived from somatopleure. SA fibers carry information about changes in the external environment

   a. General (G) refers to impulses that begin at or near the body surface. In the GSA system are 3 sensory modalities of sensation: pain; temperature (both warm and cold); and touch (light touch and deep touch (synonymous with pressure)

   b. Special (S) receptors are highly specialized and are concentrated in relatively small areas or organs. Usually contained in a framework of nonnervous tissue and often associated with accessory structures (eye and ear)

      i. Vision—stimuli are short, high-frequency waves called light

      ii. Hearing—stimuli are long, low-frequency waves called sound

2. Visceral (V) implies afferent impulses arising in or around the viscera

   a. General (G) receptors are in or on mucous membranes and in organ walls

      i. The GVA system carries impulses due to physical or chemical composition of substances in the organ and degree of distention of their walls

B. EFFERENT (E) impulses originate in the CNS and flow away from the brain or spinal cord. The term "motor" is used since its conduction currents result in activity

1. Somatic (S) tissue activated is skeletal muscle (embryonic somites)

   a. General (G): Most of the efferent fibers (GSE) originate in the ventral gray columns of the cord, pass via ventral roots, join spinal nerve trunks and run to the myoneural junction (motor end plate) of skeletal muscle

      i. Fibers also originate in cranial nuclei of nerves III, IV, VI, and XII

2. Visceral (V): system of efferent fibers that activate organs

   a. General (G) pass to all smooth muscle, to cardiac muscle, and to all glands

      i. The GVE system is concerned with body homeostasis: regulation of heart rate and blood pressure; temperature, glandular secretion, peristalsis, sphincter tension, and pupil size

   b. Special (S): relatively small distribution and tissue activated is typical striated muscle but not from somites but from the embryonic branchial arches

      i. Innervate muscles of facial expression, of mastication, of pharynx and larynx

      ii. Cells of origin of the SVE fibers are in groups or nuclei in the SVE column of the brainstem and use nerves V, VII, IX, X, and XI

   c. Special (S): the receptors do answer, in part, the structural and functional requirements of organs of special sense. The stimuli are chemical, and the impulses have profound effects on the activity of certain organs

      i. The SVA category also includes taste and smell

C. Proprioception (P): although its receptors lie deep in the body tissue, this system is concerned with the position and movement of the body

1. General (G) receptors are widely dispersed: in muscles, in tendons, and in and around articular capsules

   a. The GP system processes receptors of the pressure or tension type activated by changes in tension of muscles or tendons and by movements at joints

   b. Activation of system is outside consciousness: make possible normal walking, in part, balancing, regulation of head position, muscle contraction adjustment

2. Special (S) receptors are in inner ear (ampullae of canals, utricle, and saccule)

   a. SP system consists of a group of cells related to a fluid medium so that head movement causes current flow, stimulating the receptors

CRANIAL NERVE NUCLEI
& COMPONENTS

DORSAL VIEW

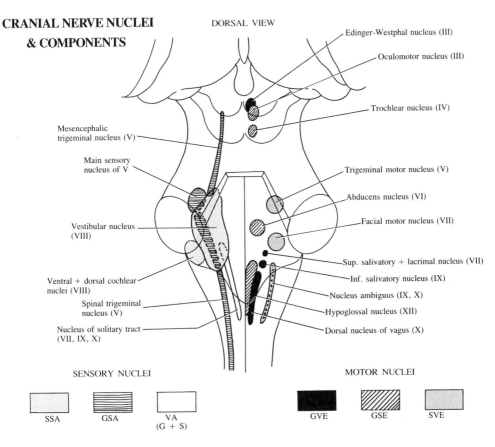

Edinger-Westphal nucleus (III)

Oculomotor nucleus (III)

Trochlear nucleus (IV)

Mesencephalic trigeminal nucleus (V)

Main sensory nucleus of V

Trigeminal motor nucleus (V)

Abducens nucleus (VI)

Facial motor nucleus (VII)

Vestibular nucleus (VIII)

Sup. salivatory + lacrimal nucleus (VII)

Inf. salivatory nucleus (IX)

Ventral + dorsal cochlear nuclei (VIII)

Nucleus ambiguus (IX, X)

Spinal trigeminal nucleus (V)

Hypoglossal nucleus (XII)

Nucleus of solitary tract (VII, IX, X)

Dorsal nucleus of vagus (X)

SENSORY NUCLEI

MOTOR NUCLEI

SSA   GSA   VA (G + S)

GVE   GSE   SVE

Edinger-Westphal nucleus

Red nucleus

Oculomotor nucleus

Mesencephalic nucleus of V

Trochlear nucleus

Motor nucleus of V

Main sensory nucleus of V

Ventral cochlear nucleus

Abducens nucleus

Sup. + inf. salivatory nuclei

Vestibular nucleus

Inferior olivary nucleus

Dorsal cochlear nucleus

Nucleus ambiguus

Dorsal vagal nucleus

Spinal tract + nucleus of V

Hypoglossal nucleus

Spinal nucleus of accessory n.

Solitary tract nucleus

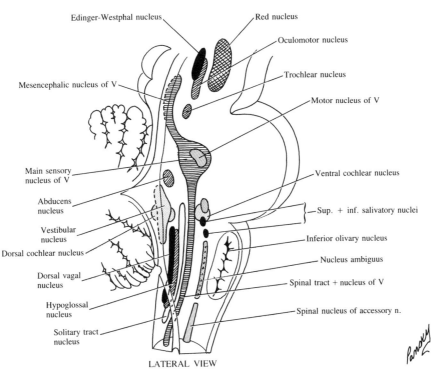

LATERAL VIEW

62. OLFACTORY (I) AND OPTIC (II) NERVES

I. Olfactory nerve: concerned with special sense of smell

A. FUNCTIONAL COMPONENTS: special visceral afferent (SVA)
  1. Receptors: neuroepithelial cells of the olfactory mucosa
  2. Location: nasal mucous membrane
  3. Distribution: upper $\frac{1}{3}$ of nasal cavity (septum and superior conchae)
  4. Cell bodies: in the nasal mucous membrane
  5. Termination: olfactory bulb

B. SYMPTOMS OF DISEASE: anosmia or the loss of the sense of smell

C. CLINICAL TESTING: odor is applied in one nostril at a time

II. Optic nerve

A. FUNCTIONAL COMPONENTS: special somatic afferent (SSA)
  1. Review the anatomy of the eye (see Unit One)
  2. The gross and functional aspects of the visual pathways
    a. The primary neurons are the receptor cells of the retina, the rods and cones
    b. The secondary neurons are the bipolar cells of the retina
    c. The tertiary neurons are the ganglionic cells of the retina and the axons of these neurons which form the optic nerve
      i. The axons run over the inner surface of the retina, in contact with the vitreous humor, and converge toward the optic papilla or optic disk
    d. The optic nerve passes through the orbit via the optic foramen, into the cranial cavity, and the 2 nerves converge to form the optic chiasma. The fibers from the nasal retina (temporal field of vision) cross in the chiasma and join the uncrossed fibers from the temporal retina (nasal field of vision) to form the optic tracts
    e. The optic tracts course posteriorly and end at the lateral geniculate bodies
      i. Most of the optic fibers terminate by synapsing with neurons in the lateral geniculate bodies
      ii. Some fibers bypass the bodies and course through the brachium of the superior colliculus or in the pretectal region to synapse in the colliculus and pretectal region, thus forming reflex pathways
    f. Axons of cells in the geniculate bodies project through the sublenticular and retrolenticular parts of the internal capsule to enter the geniculocalcarine tract (optic radiations) and terminate in the visual cortex around the calcarine fissure
    g. The fibers synapsing in the superior colliculus and the pretectal region are primarily concerned with visual reflexes
      i. Fibers from the superior colliculus enter the tectobulbar and tectospinal tracts, pass through the dorsal tegmental decussation, and terminate in the various motor nuclei of the cranial nerves and in the anterior gray column of the cervical segments of the spinal cord
      ii. Axons of the neurons in the pretectal region pass to the Edinger-Westphal nuclei (parasympathetic portion of the oculomotor nucleus) to establish a reflex that will constrict the pupil in response to strong light (loss of the light reflex without loss of vision indicates pathology in the pretectal region)

B. CLINICAL TESTING: visual acuity (uncorrected and corrected for errors of refraction) using Snellen's chart, visual fields (tangent screen, confrontation, perimeter), fundi (ophthalmoscopy), and light reflexes (direct and consensual)

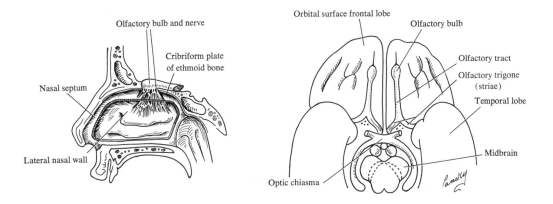

OLFACTORY NERVE (I)

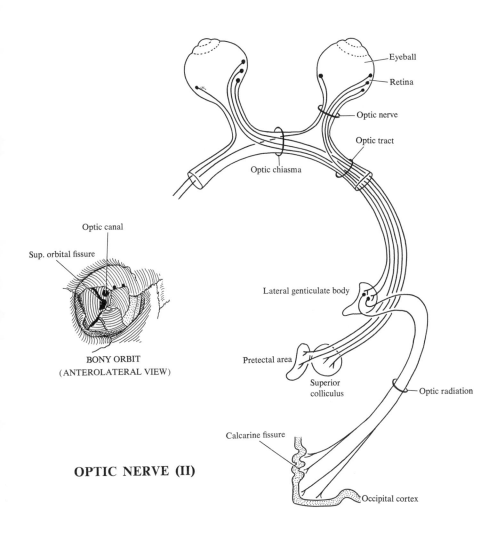

BONY ORBIT
(ANTEROLATERAL VIEW)

OPTIC NERVE (II)

63. OCULOMOTOR (III) NERVE

I. Functional components contain 2 functionally distinct types of motor fibers

A. GENERAL SOMATIC EFFERENT (GSE): to all the extraocular muscles except the lateral rectus and superior oblique

B. GENERAL VISCERAL EFFERENT (GVE): parasympathetic preganglionic fibers to the ciliary and episcleral ganglia

 1. Postganglionic fibers terminate in sphincter pupillae and ciliary muscles and function to constrict pupil by causing contraction of the iris sphincter muscle and relaxation of the suspensory ligament of the lens by contraction of the ciliary muscle.

II. Oculomotor nuclear complex is located in the mesencephalon at the level of the superior colliculus, near the midline, ventral to the aqueduct, in a V-shaped trough formed by the medial longitudinal fasciculus (MLF)

A. THE NUCLEUS can be divided into

 1. A paired lateral nuclear group

 2. An unpaired paramedian caudal central nucleus

 a. Both 1 and 2 give rise to SE fibers, and both contain typical large to medium-sized somatic motor neurons

 3. The Edinger-Westphal nucleus, which gives rise to GVE fibers, is located dorsomedial to the rostral ⅔ of the lateral nuclei, and consists of small neurons similar to those seen in the dorsal motor nucleus of the vagus nerve

B. SOME FIBER CONNECTIONS of the oculomotor complex

 1. The somatic efferent nuclei in the complex receive corticobulbar fibers, probably via relay by intercalated neurons in the reticular formation; fibers from the MLF, most of which originate in the vestibular nuclei and reflexly correlate eye position with that of the head (some fibers in the MLF, however, are intranuclear and associate the abducens and oculomotor nuclei for conjugate eye movement); fibers from the superior colliculus, probably via the intercalated neurons, placing eye movements under control of the optic reflex centers in the colliculus and thus under control of eye movement centers in the cerebrum that project to the colliculus; and fibers from the reticular formation

 2. The Edinger-Westphal nucleus receives

 a. Fibers from the pretectum that participate in the pupillary light reflex

 b. The visceral efferent components of the oculomotor nucleus also possess a convergence center responsible for the accommodation-convergence reaction

 i. This center receives fibers from the visual cortex or via the frontal eye centers

III. The internal course of the oculomotor root fibers

A. THE AXONS COURSE VENTRALLY from the oculomotor nucleus in bundles that pass medial to, lateral to, and through the red nucleus to emerge in the interpeduncular fossa medial to the basis pedunculi

IV. Intracranial course of the oculomotor nerve

A. THE NERVE COURSES FORWARD between the posterior cerebral and superior cerebellar arteries before piercing the dura and entering the cavernous sinus

B. IN THE SINUS, THE NERVE LIES SUPERIOR to cranial nerves IV and VI, and bends down and laterally to enter orbit through the medial aspect of the superior orbital fissure

V. Symptoms of disease

A. GSE: eye looks downward and outward with no convergence

B. GVE: pupils do not constrict to light or for accommodation

VI. Clinical testing of parasympathetic fibers (GVE): pupillary response to light and accommodation; observe shape, size, and equality of pupils; test pupillary reflex to accommodation by having patient look at near and then far object (for GSE, see cranial nerves IV and VI)

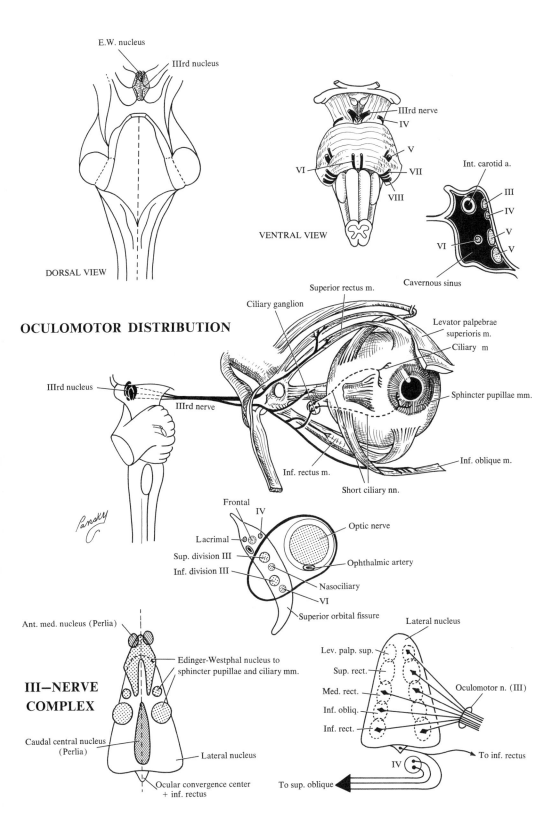

E.W. nucleus

IIIrd nucleus

DORSAL VIEW

IIIrd nerve

IV

V

VI

VII

VIII

VENTRAL VIEW

Int. carotid a.

III

IV

V

V

VI

Cavernous sinus

OCULOMOTOR DISTRIBUTION

Superior rectus m.

Ciliary ganglion

Levator palpebrae superioris m.

Ciliary m

IIIrd nucleus

IIIrd nerve

Sphincter pupillae mm.

Inf. oblique m.

Inf. rectus m.

Short ciliary nn.

Frontal

IV

Lacrimal

Optic nerve

Sup. division III

Inf. division III

Ophthalmic artery

Nasociliary

VI

Superior orbital fissure

Ant. med. nucleus (Perlia)

Lateral nucleus

Lev. palp. sup.

Edinger-Westphal nucleus to sphincter pupillae and ciliary mm.

Sup. rect.

Med. rect.

III–NERVE COMPLEX

Inf. obliq.

Inf. rect.

Oculomotor n. (III)

Caudal central nucleus (Perlia)

Lateral nucleus

To inf. rectus

IV

Ocular convergence center + inf. rectus

To sup. oblique

64. TROCHLEAR (IV) AND ABDUCENS (VI) NERVES

I. Trochlear (IV) nerve

A. FUNCTIONAL COMPONENTS
1. Somatic efferent fibers (GSE): fibers supply the superior oblique eye muscle

B. TROCHLEAR NUCLEUS is found in the mesencephalon at the level of the inferior colliculus in the ventromedian portion of the central gray
1. The nucleus contains typical somatic motor neurons and appears to indent the MLF

C. INTERNAL COURSE OF THE TROCHLEAR ROOT FIBERS
1. The trochlear nerve is the only cranial nerve that exits from the dorsal aspect of the brainstem
2. Axons of motor neurons in the trochlear nucleus sweep dorsolaterally and caudally, in the outer margin of the central gray, toward the dorsal surface of the brainstem where they decussate completely
3. The nerves exit the brainstem at the dorsal junction of the mesencephalon and pons, just lateral to the anterior medullary velum

D. INTRACRANIAL COURSE OF THE TROCHLEAR NERVE
1. The thin IVth nerve swings around the lateral surface of the brainstem, passes between the posterior cerebral and superior cerebellar arteries, and pierces the dura in the anterior attaching fold of the tentorium cerebelli
2. The nerve then enters the cavernous sinus inferior to the IIIrd nerve and enters the orbit through the superior orbital fissure

E. SYMPTOMS OF DISEASE: eye tends to look slightly upward

II. Abducens (VI) nerve

A. FUNCTIONAL COMPONENTS
1. Somatic efferent fibers (GSE): fibers supply the lateral rectus muscle of eyeball

B. ABDUCENS NUCLEUS is located in the lateral part of the median eminence in the floor of the 4th ventricle
1. Fibers of the facial nerve loop (internal genu of the facial nerve) around the nucleus
2. The facial colliculus marks the location of the genu of the facial nerve and the VIth nucleus in the floor of the 4th ventricle

C. CONNECTIONS OF THE ABDUCENS NUCLEUS
1. Bilateral corticobulbar fibers via intercalated neurons
2. Vestibular fibers via the MLF that may be concerned with control of eye movements
   a. Fibers in the MLF also serve to interconnect the abducens and oculomotor nuclei

D. INTERNAL COURSE OF THE ABDUCENS ROOT FIBERS
1. Root fibers emerge from medial side of the abducens nucleus and pass ventrally and somewhat caudally to exit from the brainstem at the caudal border of the pons
   a. As it exits, the abducens root fibers pass just lateral to the corticospinal tract

E. INTRACRANIAL COURSE OF THE ABDUCENS NERVE
1. The nerve passes forward to pierce the dura and enters the lateral aspect of the cavernous sinus, lying inferior to the trochlear nerve
2. The abducens nerve enters the orbit via the superior orbital fissure

F. SYMPTOMS OF DISEASE: the eye turns inward

III. Clinical testing of nerves III, IV, and VI to extrinsic muscles: range of ocular movements and convergence, lid movements; look for nystagmus (particularly lateral and upward gaze), ptosis of upper lid, and diplopia

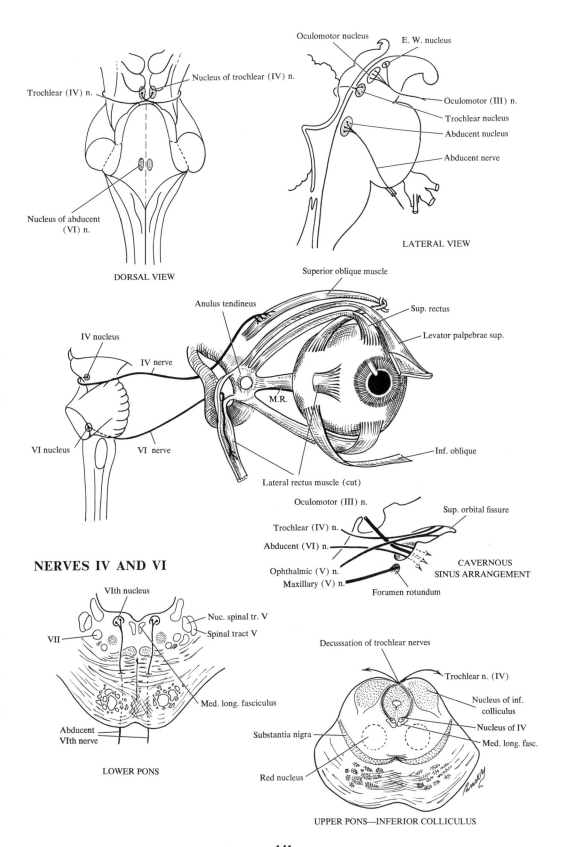

Oculomotor nucleus

E. W. nucleus

Nucleus of trochlear (IV) n.

Trochlear (IV) n.

Oculomotor (III) n.

Trochlear nucleus

Abducent nucleus

Abducent nerve

Nucleus of abducent (VI) n.

LATERAL VIEW

DORSAL VIEW

Superior oblique muscle

Anulus tendineus

Sup. rectus

IV nucleus

Levator palpebrae sup.

IV nerve

M.R.

VI nucleus

VI nerve

Inf. oblique

Lateral rectus muscle (cut)

Oculomotor (III) n.

Sup. orbital fissure

Trochlear (IV) n.

Abducent (VI) n.

NERVES IV AND VI

Ophthalmic (V) n.

Maxillary (V) n.

CAVERNOUS
SINUS ARRANGEMENT

Foramen rotundum

VIth nucleus

Nuc. spinal tr. V

Spinal tract V

VII

Decussation of trochlear nerves

Trochlear n. (IV)

Med. long. fasciculus

Nucleus of inf. colliculus

Nucleus of IV

Abducent
VIth nerve

Substantia nigra

Med. long. fasc.

LOWER PONS

Red nucleus

UPPER PONS—INFERIOR COLLICULUS

-141-

65. TRIGEMINAL (V) NERVE—PART I

I. Functional components: contains both sensory and motor components

A. GENERAL SOMATIC AFFERENT (GSA) FIBERS convey the impulses of touch, pain, and temperature from the head and face, from the teeth and mucous membrane of the oral cavity, and from the mucous membrane of the nasal cavity and sinuses as well as from the dura mater (portio major)

B. GENERAL PROPRIOCEPTION (GP) FIBERS: fibers of deep pressure and kinesthetic sense
   1. Convey impulses from the teeth, temporomandibular joint, hard palate, and from stretch receptors in the muscles of mastication

C. SENSORY PORTION OF THE TRIGEMINAL NERVE
   1. The somatic sensory column forms only the sensory nucleus of the Vth nerve and is a long column of gray matter in the dorsolateral areas extending from the upper midbrain to the upper cervical segments of the cord
      a. The sensory nucleus is often referred to as a series of contiguous nuclei (nucleus of the mesencephalic tract of the Vth nerve, principal sensory nucleus of the Vth nerve, and the nucleus of the spinal Vth tract)
      b. The cells of the nucleus receive, mainly via the Vth nerve, somatic sensory impulses from the head, expect for the scalp in back of the head, which is related to the 2nd and 3rd cervical dermatomes
         i. Some cutaneous branches of the VIIth, IXth, and Xth cranial nerves also send pain and temperature fibers to the spinal sensory nucleus of the Vth nerve
      c. Neurons that send afferent fibers to the trigeminal sensory nucleus have their cell bodies in the semilunar (gasserian) ganglion outside the CNS
         i. Proprioceptive fibers from the face, however, have their cell bodies within the mesencephalic nucleus of the Vth nerve

D. SPECIAL VISCERAL EFFERENT (SVE) FIBERS: motor portion of the trigeminal nerve
   1. Motor impulses pass to the muscles of mastication, the anterior belly of the digastric muscle, the tensor tympani muscle, the tensor veli palatini muscle, and the mylohyoid muscle
   2. Because the muscles of mastication are of branchial arch origin, the motor component is classified as SVE and the fibers are called branchiomotor

II. Trigeminal nuclear complex consists of 4 named nuclei (3 sensory, 1 motor)

A. THE 3 SENSORY NUCLEI (see above for details): the spinal nucleus, the main sensory nucleus, and the mesencephalic nucleus

B. THE MOTOR NUCLEUS: composed of typical multipolar motor neurons and located just medial to the incoming trigeminal root fibers and the main sensory nucleus
   1. Some connections of the motor nucleus
      a. The motor neurons of the Vth nerve are acted on by fibers from higher centers, as are most motor neurons; the most abundant of these are the corticobulbar fibers, many of which terminate on reticular neurons which in turn project to the motor neurons
      b. Bilateral secondary trigeminal fibers, sensory fibers from other cranial nerves, and fibers from the mesencephalic nucleus of the Vth nerve apparently activate the motor nucleus to provide reflex control of jaw muscles (especially to superficial stimuli from the tongue and oral cavity)

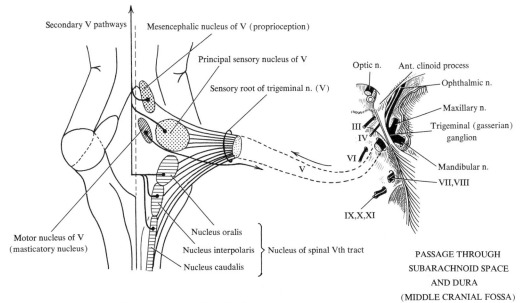

Secondary V pathways

Mesencephalic nucleus of V (proprioception)

Principal sensory nucleus of V

Sensory root of trigeminal n. (V)

Optic n. Ant. clinoid process

Ophthalmic n.

Maxillary n.

III

IV

VI

Trigeminal (gasserian) ganglion

V

Mandibular n.

VII,VIII

IX,X,XI

Motor nucleus of V (masticatory nucleus)

Nucleus oralis

Nucleus interpolaris } Nucleus of spinal Vth tract

Nucleus caudalis

PASSAGE THROUGH
SUBARACHNOID SPACE
AND DURA
(MIDDLE CRANIAL FOSSA)

NUCLEI OF TRIGEMINAL NERVE (V)

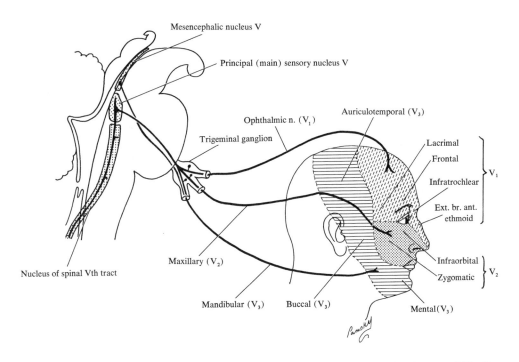

Mesencephalic nucleus V

Principal (main) sensory nucleus V

Auriculotemporal (V_3)

Ophthalmic n. (V_1)

Trigeminal ganglion

Lacrimal

Frontal

Infratrochlear

Ext. br. ant. ethmoid

} V_1

Nucleus of spinal Vth tract

Maxillary (V_2)

Infraorbital

Zygomatic

} V_2

Mandibular (V_3) Buccal (V_3)

Mental(V_3)

CUTANEOUS SENSORY AREAS

66. TRIGEMINAL (V) NERVE—PART II

I. Internal course of trigeminal root fibers
A. COURSE OF THE GSA COMPONENT: see Part I (portio major)
B. AXONS OF NEURONS IN THE MOTOR NUCLEUS pass on the medial side of the incoming sensory fibers, pass through the lateral side of the pontine tegmentum and the middle cerebellar peduncle at a midpontine level, and emerge as the portio minor, internal to the incoming sensory fibers

II. Intracranial course of the trigeminal nerve
A. THE PONS lies in the most anterior part of the posterior cranial fossa and only the Vth nerve attaches to it. This leaves laterally and has 2 roots: a very large sensory root and a small motor root. The two roots run anterolaterally from the posterior into the middle cranial fossa across a notch on the petrous bone which houses the trigeminal ganglion in the middle fossa
   1. The sensory nerves enter the ganglion via the ophthalmic nerve (V_1) through the superior orbital fissure, the maxillary nerve (V_2) via the foramen rotundum, and the mandibular nerve (V_3) via the foramen ovale
      a. V_1 is sensory but is joined by filaments from the internal carotid sympathetic plexus. Its major divisions are the lacrimal, frontal, and nasociliary branches.
      b. V_2 is larger than V_1 and also acts as a carrier for autonomic fibers
      c. V_3 is the largest branch of V and is a mixed nerve
   2. The motor root passes beneath the semilunar ganglion to join the mandibular division and exits the cranium through the foramen ovale, where it is distributed to the face via its inferior alveolar and lingual branches
   3. The GSA fibers have their cell bodies in the ganglion and terminate in the spinal V nucleus (pain and temperature) and in the chief sensory nucleus (touch)

III. Symptoms of disease
A. GSA: anesthesia in the area of distribution
B. SVE: paralysis of the jaws as related to the muscles of mastication
C. NEURALGIAS, strictly speaking, are pains in the distribution of a nerve, but the term is often vaguely used with some recurring severe pains which do not fall into a definite pattern that can be recognized. Some definitive types are
   1. Trigeminal neuralgia or tic douloureux is an intense spasmotic pain affecting the areas supplied by the branches of the Vth cranial nerve
      a. May affect persons of all ages, but is less common below the age of 50
      b. Symptoms include intense stabs of pain, lasting only a few seconds but often frequently repeated, that shoot upward from the lower lip, or upward from the eyebrows, or downward into the point of the jaw and side of the tongue. Pain may also start from the same trigger point and can be brought on by touching this point, brushing the face, shaving, exposure to cold wind, eating, drinking, or talking
   2. Atypical facial neuralgia: is seen in many patients, particularly women, who complain of constant burning pain that lasts for months or years, affecting one side, but often spreads to the other side of the face
      a. Its onset is often dated to some dental treatment
      b. Patients are depressed and there is usually some deficiency in their lives, e.g., a loss of a husband or a failure to have children
   3. Postherpetic neuralgia
      a. After an attack of herpes zoster, neuralgic pain may persist for months or years
      b. This condition is rare in the young but common in the aged, especially affecting those who have had ophthalmic herpes (herpes of V_1)
      c. The pain seems to become part of one's personality and to control patients' lives
      d. All forms of treatment appear to fail, and the greatest danger is drug addiction

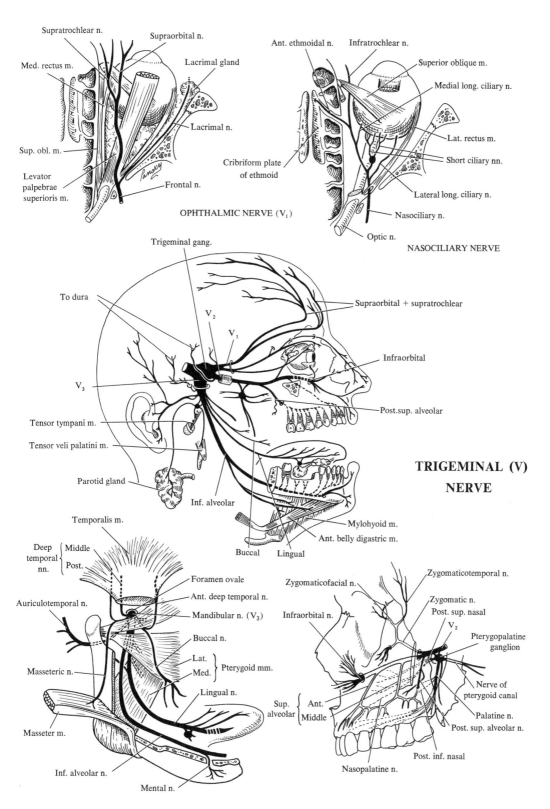

Supratrochlear n.
Supraorbital n.
Med. rectus m.
Lacrimal gland
Sup. obl. m.
Lacrimal n.
Levator palpebrae superioris m.
Cribriform plate of ethmoid
Frontal n.

OPHTHALMIC NERVE (V₁)

Ant. ethmoidal n.
Infratrochlear n.
Superior oblique m.
Medial long. ciliary n.
Lat. rectus m.
Short ciliary nn.
Lateral long. ciliary n.
Nasociliary n.
Optic n.

NASOCILIARY NERVE

Trigeminal gang.
To dura
V₂
V₁
V₃
Supraorbital + supratrochlear
Infraorbital
Post.sup. alveolar
Tensor tympani m.
Tensor veli palatini m.
Parotid gland
Inf. alveolar
Mylohyoid m.
Ant. belly digastric m.
Buccal
Lingual

TRIGEMINAL (V) NERVE

Temporalis m.
Deep temporal nn. { Middle Post.
Auriculotemporal n.
Foramen ovale
Ant. deep temporal n.
Mandibular n. (V₃)
Buccal n.
Lat. } Med. } Pterygoid mm.
Masseteric n.
Lingual n.
Masseter m.
Inf. alveolar n.
Mental n.

MANDIBULAR (V₃) NERVE

Zygomaticofacial n.
Zygomaticotemporal n.
Zygomatic n.
Post. sup. nasal
Infraorbital n.
V₂
Pterygopalatine ganglion
Nerve of pterygoid canal
Sup. { Ant. alveolar { Middle
Palatine n.
Post. sup. alveolar n.
Post. inf. nasal
Nasopalatine n.

MAXILLARY (V₂) NERVE

67. FACIAL (VII) NERVE—PART I

I. Functional components

A. GENERAL SOMATIC AFFERENT (GSA) FIBERS arise from neurons in the geniculate ganglion and innervate the external auditory meatus and skin behind the ear
   1. Centrally, these sensory fibers probably synapse in the trigeminal sensory nuclei and are conveyed to the thalamus with secondary trigeminal fibers before they are relayed to the somesthetic cortex

B. SPECIAL VISCERAL AFFERENT (SVA) FIBERS have their cell bodies in the geniculate ganglion and convey taste from the anterior $\frac{2}{3}$ of tongue via the chorda tympani
   1. These fibers form a major portion of the nervus intermedius
   2. Centrally, taste fibers enter the tractus solitarius to synapse in the upper portion of the nucleus of the tractus solitarius
      a. The central course of secondary gustatory fibers is not well known, but they are somehow transmitted to the thalamus and then to the cortical taste area located close to the area for somatic sensibility of the tongue
      b. Some gustatory secondary fibers also reach the hypothalamus and presumably influence autonomic function
      c. Taste fibers of the IXth and Xth cranial nerves are also distributed to the nucleus of the solitary tract

C. GENERAL VISCERAL AFFERENT (GVA) FIBERS have their cell bodies in the geniculate ganglion and convey impulses from the deep tissues of the face. Centrally, the fibers terminate in the nucleus of the tractus solitarius

D. GENERAL PROPRIOCEPTION (GP)
   1. Receptors lie in the muscles of facial expression
   2. Cell bodies are in the mesencephalic nucleus of the trigeminal
   3. Termination is in the reticular formation of the brainstem by way of the mesencephalic nucleus

E. GENERAL VISCERAL EFFERENT (GVE) FIBERS
   1. Supply parasympathetic preganglionic impulses to the submandibular and pterygopalatine ganglia. Postganglionic fibers from the ganglia innervate the submandibular, sublingual, and lacrimal glands, along with glands in the mucous membranes of the nasal and oral cavities

F. SPECIAL VISCERAL EFFERENT (SVE) FIBERS: branchiomotor fibers
   1. Innervate the superficial skeletal muscles of the face and scalp (muscles of facial expression), the platysma, the stylohyoid, the stapedius, and the posterior belly of the digastric muscles

II. The facial nuclei

A. THE SENSORY COMPONENTS OF THE VIITH NERVE utilize the trigeminal sensory nuclei (GSA, GP) centrally and the nucleus of the solitary tract (SVA)

B. THE SVE FIBERS originate in the facial motor nucleus located in the caudal pons just medial to the spinal nucleus of the Vth nerve. The facial nucleus is composed of several distinct cellular groups, but it is sufficient to know that
   1. The dorsal portion of the nucleus supplies motor innervation to muscles of the upper face and forehead
   2. The ventral portion supplies the lower facial muscles

C. THE GVE FIBERS originate in the visceral motor nucleus called the superior salivatory nucleus, which is difficult to discern but is apparently continuous caudally with the inferior salivatory nucleus of the IXth cranial nerve and lacrimal nucleus

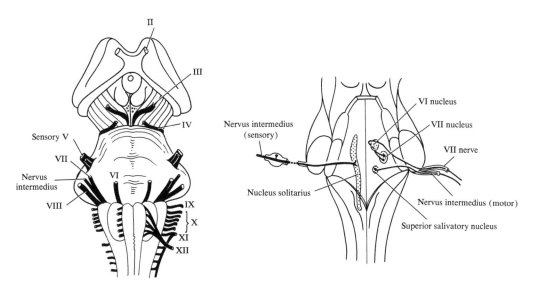

BRAINSTEM (VENTRAL SURFACE) CRANIAL NERVES

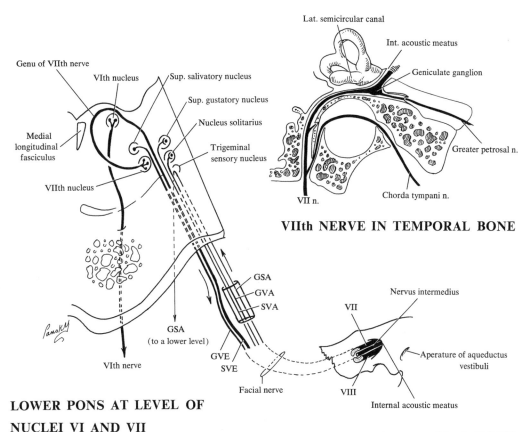

VIIth NERVE IN TEMPORAL BONE

LOWER PONS AT LEVEL OF NUCLEI VI AND VII

POSTERIOR PETROUS TEMPORAL BONE

68. FACIAL (VII) NERVE—PART II

I. Connections of the facial motor nucleus: many sources send fibers to the facial motor nucleus

A. SEVERAL IMPORTANT REFLEXES are initiated by optic, auditory, and other stimuli from around the face and mouth (some are clinically useful)

1. Optic reflexes are conveyed to the 7th motor nucleus via the superior colliculus to mediate such reflexes as blinking or closing the eye in response to bright light
2. Fibers from the superior olive mediate impulses of acoustic origin and produce reflexes such as blinking in response to a loud sound, or contraction or relaxation of the stapedius muscles in response to varying sound intensities
3. Strong projections from sensory nuclei of nerve V pass to the facial nucleus
   a. The corneal reflex is an important reflex mediated by these projections
      i. When the cornea is touched, both eyes normally blink
      ii. Corneal reflex tests the integrity of the Vth sensory complex as well as the motor fibers and nuclei of the VIIth nerve
4. Descending fibers from higher centers also influence the 7th motor nucleus
   a. Direct and indirect (impulses relayed by reticular neurons) corticobulbar fibers from the motor cortex presumably mediate voluntary control of face
      i. Clinical evidence indicates that corticobulbar fibers pass bilaterally to the dorsal part of the facial nucleus, but those to the ventral part of the nucleus arise in the contralateral motor cortex only; thus, unilateral damage to the corticobulbar fibers above the facial nucleus produces paralysis of the contralateral lower facial muscles only
   b. Other higher centers projecting to the 7th motor nucleus are red nucleus, mesencephalic reticular formation, globus pallidus, and perhaps the thalamus
      i. Many project indirectly to the facial nucleus via the reticular formation
      ii. Projections from the globus pallidus are probably involved in emotional (mimetic) facial movement. The ability of the lower facial muscles to respond to emotional stimuli even in the presence of a central-type facial paralysis (lesion above the 7th motor nucleus) is well documented

II. Internal course of facial root fibers

A. THE BRANCHIOMOTOR FIBERS have a complex course to their point of exit: emerge from the dorsal surface of the facial nucleus and pass dorsomedially to approach the floor of the 4th ventricle; the fibers then turn to ascend in a compact bundle and pass medial to the VIth nucleus; at the rostral pole of the latter, the fibers loop dorsolaterally over the nucleus and pass ventrolaterally between the spinal trigeminal complex and the motor nucleus of VII to exit from the brainstem near the pontomedullary junction. The loop around the VIth nucleus is the internal genu of the VIIth nerve

III. Intracranial course of the facial nerve

A. TOGETHER WITH the nervus intermedius and the VIIIth cranial nerve (bound in a common meningeal sheath), the facial nerve passes laterally from its point of exit from the brainstem to enter the internal acoustic meatus

B. TO REACH THE FACE, fibers pass through the wall of the ear cavity (giving off the great petrosal nerve, nerve to stapedius muscle, and the chorda tympani) and exit via the stylomastoid foramen just behind the parotid gland and spread on face

VI. Symptoms of disease

A. SVE: facial paralysis (Bell's palsy) (test muscles of facial expression)

B. GVE: poor secretion from gland of termination (test lacrimation)

C. SVA: ageusia on anterior ⅔ of tongue (test for taste)

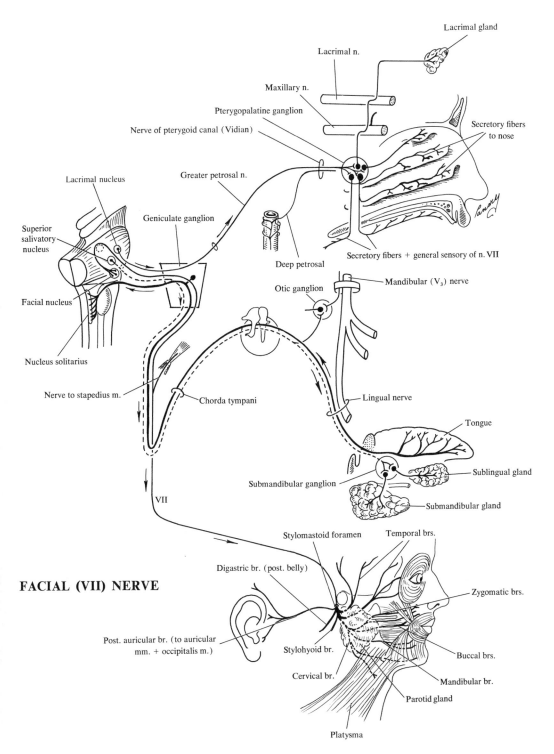

Lacrimal gland

Lacrimal n.

Maxillary n.

Pterygopalatine ganglion

Nerve of pterygoid canal (Vidian)

Secretory fibers
to nose

Lacrimal nucleus

Greater petrosal n.

Geniculate ganglion

Superior
salivatory
nucleus

Deep petrosal

Secretory fibers + general sensory of n. VII

Facial nucleus

Mandibular (V₃) nerve

Otic ganglion

Nucleus solitarius

Nerve to stapedius m.

Chorda tympani

Lingual nerve

Tongue

Submandibular ganglion

Sublingual gland

Submandibular gland

VII

Stylomastoid foramen

Temporal brs.

Digastric br. (post. belly)

FACIAL (VII) NERVE

Zygomatic brs.

Post. auricular br. (to auricular
mm. + occipitalis m.)

Stylohyoid br.

Buccal brs.

Cervical br.

Mandibular br.

Parotid gland

Platysma

MUSCLES OF FACIAL EXPRESSION

69. COCHLEAR (VIII) NERVE: AUDITORY SYSTEM

I. Functional components

A. SPECIAL SOMATIC AFFERENT (SSA) FIBERS
1. Receptors: the hair cells located in the organ of Corti
2. Distribution: cochlear duct of the inner ear
3. Cell bodies: the spiral ganglion
4. Terminations: dorsal and ventral cochlear nuclei

II. Functional auditory system

A. THE AUDITORY PATHWAY TO THE CORTEX consists of at least a 4-neuron chain; in addition, the auditory pathways are complicated by other small relay nuclei
1. The superior olivary complex and the nucleus of the lateral lemniscus are two such small relay nuclei (omitted here to simplify the initial study)
2. The primary neurons are bipolar cells located in the spiral ganglion within the bony spiral lamina
   a. The distal processes terminate in relation to the hair cells of the organ of Corti
   b. The central processes pass via the auditory portion of the VIIIth cranial nerve to terminate in the dorsal and ventral cochlear nuclei
3. The secondary neurons are located in the dorsal and ventral cochlear nuclei
   a. These axons pass rostrally to either decussate in and form the trapezoid body or synapse with neurons in the superior olivary nucleus
      i. The fibers that decussate enter the lateral lemniscus and ascend to the inferior colliculus
      ii. Some fibers in the lateral lemniscus pass through the commissure of the inferior colliculus of the opposite side
4. Axons of the 3rd-order neurons in the inferior colliculus pass via the brachium of the inferior colliculus to the medial geniculate body
5. The fourth neuron of the chain begins in the medial geniculate body. Fibers pass from here through the sublenticular portion of the internal capsule to end in the auditory area of the cerebral cortex

III. Auditory reflexes are affected by the inferior colliculus, which gives fibers to the tectospinal and tectobulbar tracts, which both cross to the opposite side via the dorsal tegmental decussation and run downward, giving off collateral and terminal branches to various cranial motor nuclei and to the anterior gray columns in the cervical segments of the spinal cord (fibers from these 2 tracts probably terminate in the reticular formation, as well)

IV. Symptoms of disease

A. THE SENSATION OF HEARING has bilateral cortical representation; thus, unilateral lesions in the CNS do not cause a complete loss of hearing in either ear
B. SYMPTOMS OF DISEASE in this system, however, could be deafness or a decrease in the ability to hear

V. Clinical testing: qualitative test for hearing (watch ticking); Rinne and Weber tuning fork tests (for air and bone conduction); audiometry; lateralization of sound; examine eardrum with otoscope for signs of middle ear infection

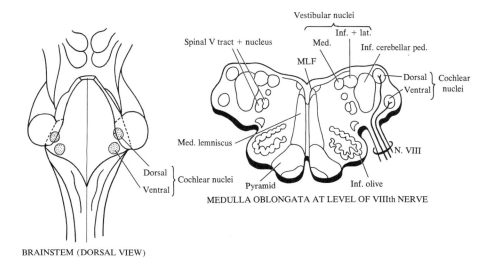

BRAINSTEM (DORSAL VIEW)

MEDULLA OBLONGATA AT LEVEL OF VIIIth NERVE

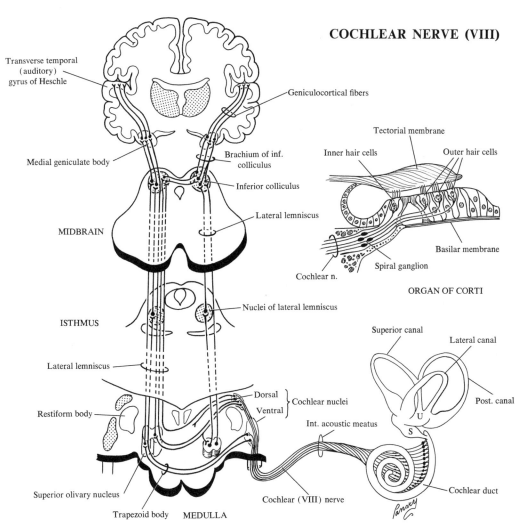

COCHLEAR NERVE (VIII)

ORGAN OF CORTI

MEDULLA

70. VESTIBULAR (VIII) NERVE: VESTIBULAR SYSTEM—PART I

I. Functional components

A. SPECIAL SOMATIC AFFERENT (SSA) FIBERS
1. Receptors: hair cells of the cristae and maculae
2. Distribution: ampullae of the semicircular canals of the inner ear, utriculus of the inner ear, and the sacculus of the inner ear
3. Cell bodies: in the vestibular ganglion
4. Terminations: vestibular nuclei and the cerebellum

II. Functional vestibular system consists primarily of a 3-neuron chain

A. THE PRIMARY NEURON is a bipolar cell in the vestibular ganglion in the internal auditory meatus
1. The distal processes pass to the crista ampullaris of the semicircular canals and to the maculae of the utricle and saccule
2. The central processes pass in the vestibular portion of the VIIIth nerve to enter the brainstem at the junction of the pons and medulla
3. Majority of vestibular fibers terminate in the vestibular nuclei, but some pass directly into the cerebellum via the inferior cerebellar peduncle

B. THE SECONDARY NEURONS are located in 1 of 4 vestibular nuclei: superior nucleus of Bechterew, lateral nucleus of Deiters, medial nucleus of Schwalbe, and the inferior nucleus (spinal or descending nucleus)
1. Secondary connections are numerous
   a. Fibers from the superior nucleus enter the MLF of the same side and turn rostrally to end among the cells of cranial nerves III, IV, and VI
   b. Fibers from the lateral nucleus (the majority) descend, on the same side, through the brainstem, then down the cord in the anterior funiculus as the vestibulospinal tract, and terminate among cells of the ventral gray column
   c. Fibers from the inferior nucleus pass medially, cross the midline, and descend in the MLF, to terminate in the nucleus of the accessory nerve and in the ventral columns of the cervical cord
   d. Fibers from the medial nucleus cross the midline, enter the opposite MLF, and bifurcate into fibers that ascend and descend in that fascicle to terminate just as other fibers do in it
      i. Some fibers from the medial nucleus enter the reticular formation of both sides to be distributed to motor nuclei of the autonomic system in the brainstem, then go to the intermediolateral column of the cord via the reticulospinal tract

C. THE 3RD-ORDER NEURONS (EFFERENT NEURONS)
1. Axons from the ventral columns of the cord, from lower cervical to lower sacral levels, are distributed to skeletal muscles of trunk and extremities
   a. Completes a typical somatic reflex and is important in relating body movements with those of the head; significant in maintaining balance
2. Fibers from the ventral columns of cord in the cervical region and from nucleus of the spinal portion of accessory nerve are distributed to muscles of neck (sternocleidomastoid and trapezius); important in maintaining the normal position of the head
3. Axons from somatic efferent columns of the brainstem, particularly nuclei of the IIIrd, IVth, and VIth nerves, supply extrinsic muscles of eyeball and help coordinate eye movements with those of head
4. Fibers from autonomic centers, especially the dorsal motor nucleus of X and intermediolateral cell columns of the cord, account for nausea, vomiting, and pallor manifested after overstimulation of the vestibular system

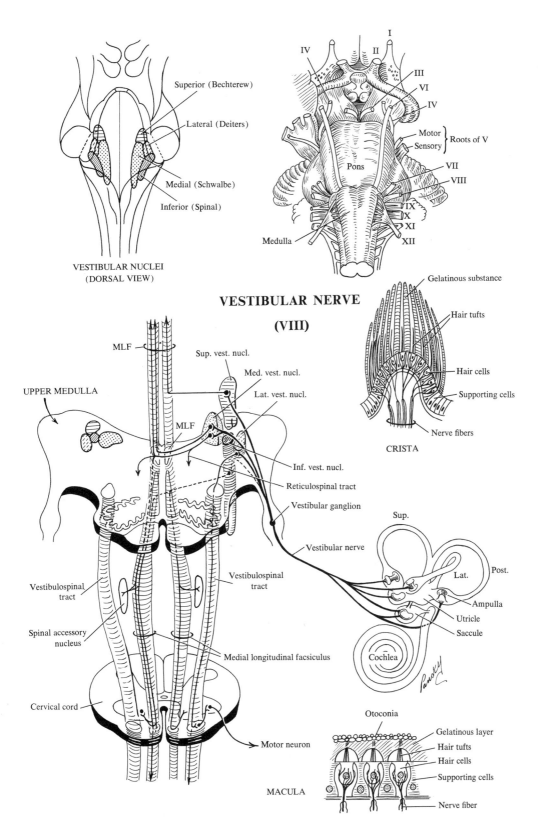

VESTIBULAR NUCLEI
(DORSAL VIEW)

Superior (Bechterew)

Lateral (Deiters)

Medial (Schwalbe)

Inferior (Spinal)

I

IV

II

III

VI

IV

Motor
Sensory } Roots of V

VII

VIII

IX

X

XI

XII

Pons

Medulla

VESTIBULAR NERVE

(VIII)

MLF

UPPER MEDULLA

MLF

Sup. vest. nucl.

Med. vest. nucl.

Lat. vest. nucl.

Inf. vest. nucl.

Reticulospinal tract

Vestibular ganglion

Vestibular nerve

Vestibulospinal
tract

Vestibulospinal
tract

Spinal accessory
nucleus

Medial longitudinal facsiculus

Cervical cord

Motor neuron

Gelatinous substance

Hair tufts

Hair cells

Supporting cells

Nerve fibers

CRISTA

Sup.

Lat.

Post.

Ampulla

Utricle

Saccule

Cochlea

Otoconia

Gelatinous layer

Hair tufts

Hair cells

Supporting cells

Nerve fiber

MACULA

-153-

71. VESTIBULAR (VIII) NERVE: VESTIBULAR SYSTEM—PART II

I. Cerebellar influence on the vestibular system: cerebellum does exert control over the activities initiated by the stimulation of the vestibular receptors

A. DIRECT PATHWAYS

  1. Afferent neurons travel through ascending fibers to terminate in the nodulus and flocculus of the cerebellum

  2. From cells in the cortex of the nodulus and flocculus, fibers descend to terminate in the fastigial nucleus of the cerebellum

    a. From the fastigial nucleus, axons descend via the medial segment of the inferior cerebellar peduncle to end in the vestibular nuclei

    b. Other axons from the fastigial nucleus cross the midline and descend to the vestibular nuclei of the opposite side via the uncinate bundle (of Russell)

    c. Additional fibers from the vestibular nuclei are carried by the pathways already described (vestibulospinal tract, MLF, and reticular formation) to the same terminations

B. INDIRECT PATHWAYS

  1. The afferent neurons terminate in the superior and lateral vestibular nuclei

  2. From cells mainly in the superior nucleus (but also lateral), fibers ascend in the medial segment of the inferior cerebellar peduncle to terminate both in the flocculonodular lobe and in the lingula and uvula of the cerebellum

  3. From cells in the cerebellar cortex of the above regions, fibers descend both to the fastigial and to the globose nuclei of the cerebellum

    a. Axons from the above descend by the same pathways and have the same terminations as described above for the direct route

II. Clinical considerations

A. DISEASE IN THE VESTIBULAR MECHANISM expresses itself in disturbances of equilibrium, balance, and coordination

  1. Simple observation of the patient standing or walking becomes an integral part of the examination

  2. Since the system is so closely tied to reflex movements of the eye, such reflexes are used to localize a lesion

B. SPECIAL PATHOLOGY

  1. Vertigo: most characteristic symptom of vestibular disease

  2. Inflammation of the labyrinth

  3. Ménière's disease: paroxysmal attacks of vertigo associated with unilateral tinnitus and deafness

  4. Basal or pontine tumors or hemorrhage can involve the extra- or intramedullary course of the afferent fibers of the vestibular nerve

III. Clinical testing

A. POSTURE AND COORDINATION TESTS (rotation test of Bárány)

B. CALORIC TESTS

C. OCULONYSTAGMOGRAPHY: documenting nystagmus via amplifiers and recorders

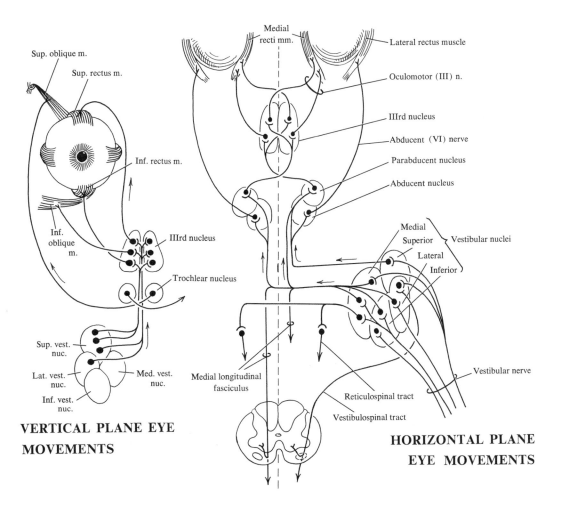

Sup. oblique m.

Sup. rectus m.

Inf. rectus m.

Inf. oblique m.

IIIrd nucleus

Trochlear nucleus

Sup. vest. nuc.

Lat. vest. nuc.

Inf. vest. nuc.

Med. vest. nuc.

VERTICAL PLANE EYE MOVEMENTS

Medial recti mm.

Lateral rectus muscle

Oculomotor (III) n.

IIIrd nucleus

Abducent (VI) nerve

Parabducent nucleus

Abducent nucleus

Medial

Superior

Lateral

Inferior

Vestibular nuclei

Medial longitudinal fasciculus

Reticulospinal tract

Vestibulospinal tract

Vestibular nerve

HORIZONTAL PLANE EYE MOVEMENTS

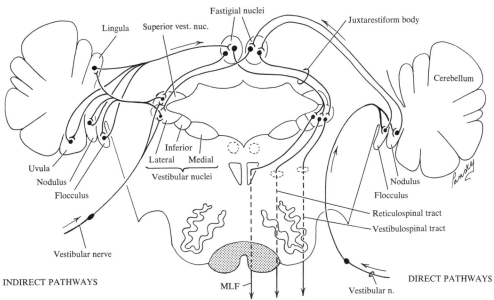

Fastigial nuclei

Lingula

Superior vest. nuc.

Juxtarestiform body

Cerebellum

Inferior

Lateral Medial

Vestibular nuclei

Uvula

Nodulus

Flocculus

Nodulus

Flocculus

Reticulospinal tract

Vestibulospinal tract

Vestibular nerve

INDIRECT PATHWAYS

MLF

Vestibular n.

DIRECT PATHWAYS

VESTIBULOCEREBELLAR + CEREBELLOVESTIBULAR CONTROL

–155–

72. GLOSSOPHARYNGEAL (IX) AND VAGUS (X) NERVES—PART I

I. **Functional components:** the nerves are discussed together because of their close relationship. Both nerves are mixed and contain the same functional components

A. GENERAL SOMATIC AFFERENT (GSA) FIBERS: for vagus (X) only
1. Receptors: external auditory meatus
2. Cell bodies: jugular ganglion
3. Termination: spinal V nucleus
4. Symptoms of disease: anesthesia in the area of distribution

B. GENERAL VISCERAL AFFERENT (GVA) FIBERS
1. In the IXth and Xth nerves, arise from cell bodies in the petrosal and inferior nodose ganglion, respectively
   a. Those of the IXth nerve convey touch, pain, and temperature from the posterior portion of the tongue, tonsil, and eustachian tube
   b. Those from the Xth convey sensory sensations from the pharynx, larynx, trachea, and the viscera of the thorax and abdomen

C. SPECIAL VISCERAL AFFERENT (SVA) FIBERS: fibers of taste
1. The IXth nerve from the posterior $\frac{1}{3}$ of tongue; Xth fibers from the epiglottis and tongue adjacent to it
2. Their cell bodies are in the petrosal and nodose ganglia, respectively
3. Centrally, all visceral afferent fibers of IX, X (and also taste fibers of VII) will enter the solitary nucleus
   a. Secondary pathways to the thalamus and cerebral cortex are not well known, but apparently do exist
   b. There is a definitive gustatory receptive cortex
   c. Other visceral sensations are not definitively localized
   d. The solitary nucleus does send numerous projections to the centers in the reticular formation
   e. The respiratory center is located dorsal to the olivary complex and responds to vagal and other neural impulses, as well as to changes in CO_2 concentration in the blood
   f. There are also cardiovascular and vomiting centers in the reticular formation that receive secondary visceral afferent fibers
   g. Numerous fibers also pass from the solitary nucleus to cranial nerve nuclei, especially to the dorsal motor nucleus of nerve X
4. Of special interest is the SVA component of the IXth nerve from the carotid sinus to the CNS
   a. Afferent impulses are initiated by increased carotid arterial pressure and conveyed centrally by the sinus nerve
      i. Centrally, collaterals are given off to the dorsal motor nucleus of nerve X
      ii. The impulses are transmitted to preganglionic neurons of the vagus nerve which send their axons to postganglionic neurons in the atria of the heart
      iii. Stimulation of the sinus causes a decrease in heart rate and blood pressure

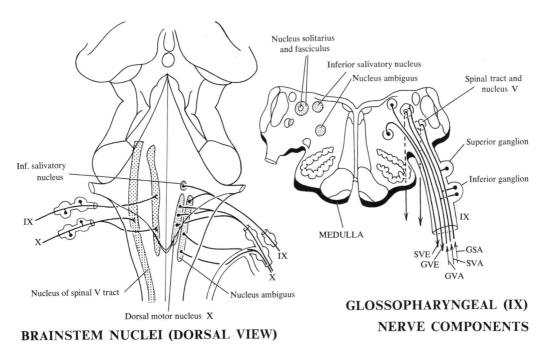

Nucleus solitarius
and fasciculus

Inferior salivatory nucleus

Nucleus ambiguus

Spinal tract and
nucleus V

Superior ganglion

Inferior ganglion

Inf. salivatory
nucleus

IX

X

IX

X

MEDULLA

IX

SVE
GVE

GSA
SVA

GVA

Nucleus of spinal V tract

Nucleus ambiguus

Dorsal motor nucleus X

BRAINSTEM NUCLEI (DORSAL VIEW)

**GLOSSOPHARYNGEAL (IX)
NERVE COMPONENTS**

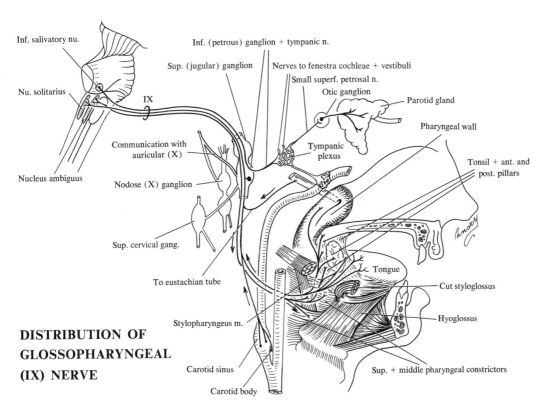

Inf. salivatory nu.

Inf. (petrous) ganglion + tympanic n.

Sup. (jugular) ganglion

Nerves to fenestra cochleae + vestibuli

Small superf. petrosal n.

Otic ganglion

Parotid gland

Nu. solitarius

IX

Pharyngeal wall

Communication with
auricular (X)

Tympanic
plexus

Tonsil + ant. and
post. pillars

Nucleus ambiguus

Nodose (X) ganglion

Sup. cervical gang.

Tongue

Cut styloglossus

Hyoglossus

To eustachian tube

**DISTRIBUTION OF
GLOSSOPHARYNGEAL
(IX) NERVE**

Stylopharyngeus m.

Carotid sinus

Carotid body

Sup. + middle pharyngeal constrictors

73. GLOSSOPHARYNGEAL (IX) AND VAGUS (X) NERVES—PART II

I. Functional components (cont.)

A. GENERAL VISCERAL EFFERENT (GVE) FIBERS

1. Those of the IXth nerve arise from neurons in the inferior salivatory nucleus and pass as parasympathetic fibers to the otic ganglion via the tympanic nerve. Postganglionic fibers innervate the parotid gland (lesser petrosal nerve)
2. Those of the Xth nerve arise in the dorsal motor nucleus of nerve X and pass as preganglionic parasympathetic fibers to terminate in parasympathetic ganglia innervating the thoracic and abdominal viscera

B. SPECIAL VISCERAL EFFERENT (SVE) FIBERS

1. Those of the IXth nerve arise in the rostral pole of the nucleus ambiguus and innervate the stylopharyngeus muscle
2. Those of the Xth nerve arise from the middle portion of the nucleus ambiguus and innervate the voluntary striated muscles of the larynx and pharynx
3. The nucleus ambiguus is a column of motor neurons found in the medulla about halfway between the spinal V nucleus and the inferior olive
   a. Axons in its rostral portion contribute fibers to the IXth nerve
   b. Axons from its middle portion pass obliquely dorsally and medially to join the other exiting fibers of the Xth nerve
   c. Neurons in its caudal portion contribute to the XIth nerve

II. Symptoms of disease

A. THE IXTH NERVE

1. GVA: interference with the afferent limb of the swallowing reflex
2. SVA: ageusia on the posterior $\frac{1}{3}$ of the tongue
3. GVE: decreased secretion of the parotid gland
4. SVE: some difficulty in swallowing

B. THE XTH NERVE

1. GSA: anesthesia in the area of distribution (external auditory meatus)
2. GVA
   a. Loss of reflex control of circulation (heart rate, etc.)
   b. Disturbance in the depth of respiration
3. SVA: anesthesia in the area of distribution (epiglottis and adjacent tongue)
4. GVE
   a. Loss of reflex control of circulatory system (heart rate, blood pressure)
   b. Poor digestion due to decreased secretion of digestive enzymes and a slowing of peristalsis
5. SVE: difficulty in swallowing and speaking

III. Clinical testing

A. MOTOR FIBERS: swallowing, gag reflex, movement of palate and uvula (IX), voice (X)
B. SENSORY FIBERS: gag reflex, pharyngeal sensation, taste on posterior $\frac{1}{3}$ of tongue (IX); look at vocal cords if indicated by abnormal voice (X)

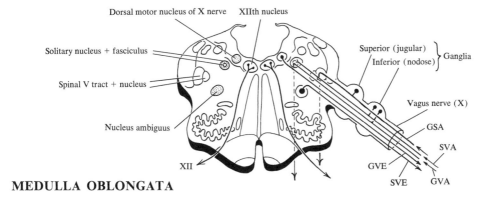

Dorsal motor nucleus of X nerve
XIIth nucleus
Solitary nucleus + fasciculus
Spinal V tract + nucleus
Nucleus ambiguus
Superior (jugular)
Inferior (nodose) } Ganglia
Vagus nerve (X)
GSA
SVA
GVE
SVE
GVA
XII

MEDULLA OBLONGATA

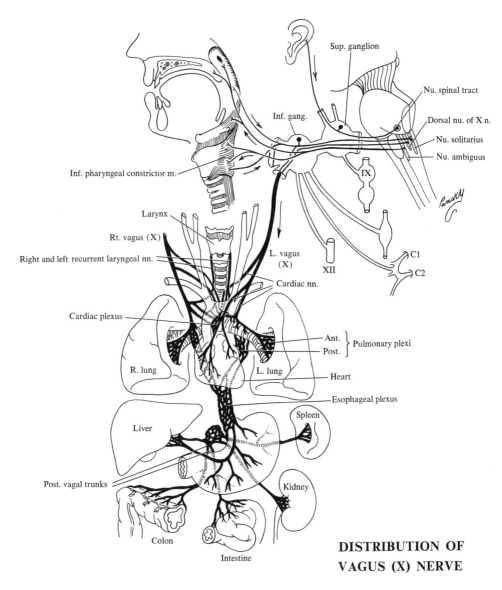

Sup. ganglion
Nu. spinal tract
Inf. gang.
Dorsal nu. of X n.
Nu. solitarius
Nu. ambiguus
Inf. pharyngeal constrictor m.
IX
Larynx
Rt. vagus (X)
Right and left recurrent laryngeal nn.
L. vagus (X)
XII
C1
C2
Cardiac nn.
Cardiac plexus
Ant.
Post. } Pulmonary plexi
R. lung
L. lung
Heart
Esophageal plexus
Spleen
Liver
Post. vagal trunks
Kidney
Colon
Intestine

**DISTRIBUTION OF
VAGUS (X) NERVE**

74. ACCESSORY (XI) NERVE

I. Functional components
A. THE XITH NERVE is generally considered to be an efferent nerve, although some sensory fibers (GP) probably exist in muscles supplied by its SVE component
B. SPECIAL VISCERAL EFFERENT (SVE) FIBERS: to the intrinsic muscles of the larynx and the sternocleidomastoid and trapezius muscles

II. Nuclei of the XIth nerve
A. THE NERVE HAS TWO DIVISIONS, spinal and cranial
   1. The spinal portion arises from the spinal nucleus of XI in the anterior horn, extending from the 5th or 6th cervical vertebra to the middle of the pyramidal decussation
   2. The cranial portion arises from the caudal pole of the nucleus ambiguus to supply the intrinsic muscles of the larynx

III. Connections of the nuclei
A. BILATERAL CORTICOBULBAR
B. COLLATERAL from other cranial nuclei and incoming sensory systems
C. RETICULAR FORMATIONS, etc.

IV. Internal and intracranial course of the XIth nerve root fibers
A. FIBERS OF THE SPINAL PORTION
   1. Exit from the lateral aspect of the spinal cord between the dorsal and ventral nerve roots
   2. The rootlets of this portion of the XIth nerve unite to form the external branch of XI, which ascends in the spinal canal just posterior to the dentate ligaments and enters the skull through the foramen magnum to exit eventually from the skull via the jugular foramen with nerves IX and X. The external branch of the spinal division runs downward and backward through the sternocleidomastoid muscle (supplying it), crosses the posterior triangle of the neck, and ends in the trapezius muscle (supplies it). It also communicates with branches of spinal nerves C2, C3, and C4
B. FIBERS OF THE CRANIAL PORTION pass from the caudal pole of the nucleus ambiguus to join the fibers of the vagus nerve and to exit with that nerve through the jugular foramen
   1. They later form the motor portion of the recurrent laryngeal nerves which supply the intrinsic muscles of the larynx

V. Symptoms of disease: difficulty in turning the head and possibly in swallowing

VI. Clinical testing: test sternocleidomastoid and trapezius muscles for function

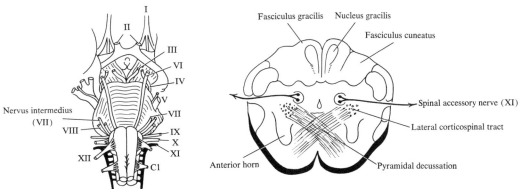

Nervus intermedius (VII)

Fasciculus gracilis   Nucleus gracilis   Fasciculus cuneatus

Spinal accessory nerve (XI)

Lateral corticospinal tract

Anterior horn

Pyramidal decussation

TRANSVERSE SECTION—
LOWER MEDULLA

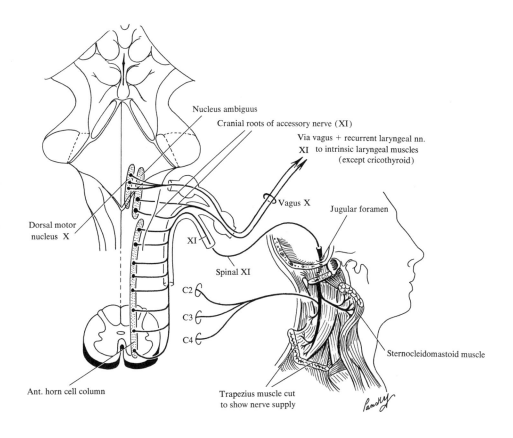

Nucleus ambiguus

Cranial roots of accessory nerve (XI)

Via vagus + recurrent laryngeal nn.
XI to intrinsic laryngeal muscles
(except cricothyroid)

Vagus X

Jugular foramen

Dorsal motor
nucleus X

XI

Spinal XI

C2

C3

C4

Sternocleidomastoid muscle

Ant. horn cell column

Trapezius muscle cut
to show nerve supply

DISTRIBUTION OF ACCESSORY NERVE (XI)

75. HYPOGLOSSAL (XII) NERVE

I. Functional components
A. THE XIITH NERVE is classified primarily as a GSE nerve to the skeletal musculature (extrinsic and intrinsic) of the tongue, but it probably carries some impulses (GP) originating from muscle spindles
B. GENERAL SOMATIC EFFERENT (GSE) FIBERS
    1. Origin: hypoglossal nucleus
    2. Termination: all extrinsic and intrinsic muscles of the tongue

II. Nucleus of the XIIth nerve
A. THE NUCLEUS IS LOCATED near the floor of the 4th ventricle just beneath the hypoglossal trigone
    1. It is a longitudinal cell column of typical somatic motor neurons extending from the caudalmost portion of the medulla to the level of the stria medullaris
    2. The nucleus consists of several distinct cell groups, each group supplying certain muscles of the tongue
        a. The medially placed groups probably innervate the intrinsic muscles of the tongue

III. Connections of the XIIth nucleus
A. CORTICOBULBAR FIBERS are involved where voluntary movement of the tongue is involved
    1. Some of these apparently end directly on neurons in the nucleus
    2. Many of the corticobulbar impulses are apparently relayed by crossed and uncrossed reticular neurons
        a. Some uncrossed corticobulbar fibers end predominantly in the medially placed cell groups that control the intrinsic tongue muscles, which work symmetrically
B. OTHER FIBERS, primarily from the sensory nuclei of V and the nucleus solitarius, enter the hypoglossal nucleus to activate the reflex actions of sucking, swallowing, chewing, etc.

IV. Internal course of the hypoglossal root fibers
A. THE AXONS of the hypoglossal neurons gather in bundles and leave the ventral surface of the nucleus
B. THE ROOTLETS course in a ventral and slightly lateral direction, traverse the medullary reticular formation just lateral to the medial lemniscus, pierce the most medial part of the inferior olive, and exit the brainstem in the ventrolateral sulcus (preolivary) between the pyramid and the inferior olive
C. THE CLOSE RELATIONSHIP of the emerging fibers of XII and the pyramidal tract is the anatomic basis for the syndrome called alternating hypoglossal hemiplegia

V. Intracranial course of the XIIth nerve: the large to medium-sized, myelinated fibers of the XIIth nerve exit as 10–15 rootlets that soon join to leave this skull via the hypoglossal canal in the occipital bone

VI. Symptoms of disease: paralysis of the tongue

VII. Clinical testing: tongue function (look for evidence of wasting or fibrillation); have patient stick out tongue and look for deviation to either side

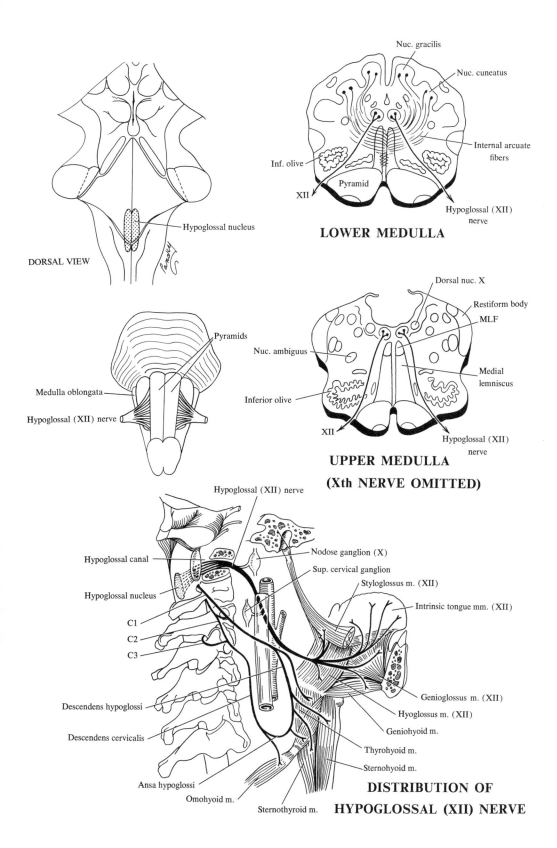

Nuc. gracilis

Nuc. cuneatus

Inf. olive

Pyramid

XII

Internal arcuate fibers

Hypoglossal (XII) nerve

LOWER MEDULLA

Hypoglossal nucleus

DORSAL VIEW

Dorsal nuc. X

Restiform body

MLF

Nuc. ambiguus

Inferior olive

XII

Medial lemniscus

Hypoglossal (XII) nerve

UPPER MEDULLA

(Xth NERVE OMITTED)

Pyramids

Medulla oblongata

Hypoglossal (XII) nerve

Hypoglossal (XII) nerve

Hypoglossal canal

Hypoglossal nucleus

C1

C2

C3

Descendens hypoglossi

Descendens cervicalis

Ansa hypoglossi

Omohyoid m.

Sternothyroid m.

Nodose ganglion (X)

Sup. cervical ganglion

Styloglossus m. (XII)

Intrinsic tongue mm. (XII)

Genioglossus m. (XII)

Hyoglossus m. (XII)

Geniohyoid m.

Thyrohyoid m.

Sternohyoid m.

DISTRIBUTION OF

HYPOGLOSSAL (XII) NERVE

76. SUMMARY OF CRANIAL NERVES

Nerve	Chief Components	Origin or Termination in CNS	Peripheral Ending (Sensory or Motor)	Major Functions
I	SVA	Olfactory bulb	Olfactory epithelium	Smell
II	SSA	Mainly in lateral geniculate nucleus	Ganglion cells of retina	Vision
III	GSE	Oculomotor nucleus	Superior, inferior, and medial recti; inferior oblique and levator palpebral superioris muscles	Eye movements and lid elevation
	GVE	Edinger-Westphal nucleus	Sphincter pupillae and ciliary muscles	Miosis and accommodation
IV	GSE	Trochlear nucleus	Superior oblique muscle	Eye movements
V	GSA	Spinal and main sensory nuclei Mesencephalic nucleus	Muscle spindles and other mechanoreceptors	Sensation from head
	SVE	Trigeminal motor nucleus	Muscles of mastication, tensor tympani and tensor veli palatini and others	Major movements of mandible
VI	GSE	Abducens nucleus	Lateral rectus muscle	Eye movements
VII	GSA	Spinal trigeminal nucleus	Outer ear	General skin sensation
	SVA	Solitary nucleus	Taste buds—anterior $\frac{2}{3}$ of tongue	Taste
	GVA	Solitary nucleus	Part of nasopharynx	Sensation from nasal cavity and soft palate
	GVE	Superior salivatory nucleus	Submandibular, sublingual, and lacrimal glands	Secretion of tears and saliva
	SVE	Facial motor nucleus	Muscles of facial expression, stapedius, stylohyoid, and posterior belly of digastric	Facial expression
VIII	SSA	Cochlear (spiral) nuclei	Organ of Corti	Hearing
		Vestibular nuclei	Cristae of canals Ampullae of utricle and saccule	Equilibrium

SUMMARY OF CRANIAL NERVES *(continued)*

Nerve	Chief Components	Origin or Termination in CNS	Peripheral Ending (Sensory or Motor)	Major Functions
IX	GSA	Spinal trigeminal nucleus	Outer ear sensation	General skin sensation
	SVA	Solitary nucleus	Taste buds	Taste—posterior $\frac{1}{3}$ of tongue
	GVA	Solitary and spinal Vth nuclei	Carotid body and sinus; mucous membranes of oral and nasal pharynx	Sensation in tongue and pharynx; visceral reflexes
	GVE	Inferior salivatory nucleus	Parotid gland	Secretion of saliva
	SVE	Nucleus ambiguus	Stylopharyngeus muscle	Elevation of pharynx
X	GSA	Spinal trigeminal nucleus	Outer ear	General skin sensation
	SVA	Solitary nucleus	Taste buds of epiglottis	Taste
	GVA	Solitary and spinal Vth nuclei	Thoracic and abdominal viscera; mucous membranes of larynx and laryngeal pharynx	Sensation in pharynx, larynx, and thoracic and abdominal viscera; visceral reflexes
	GVE	Dorsal motor nucleus of X	Thoracic and abdominal viscera	Movements and secretion of thoracic and abdominal viscera
	SVE	Nucleus ambiguus	Larynx and pharynx	Movements of larynx and pharynx
XI	SVE (cranial)	Nucleus ambiguus	Larynx and pharynx	Movements of larynx and pharynx
	SVE (spinal)	Accessory nucleus of cervical cord	Sternocleidomastoid and trapezius muscles	Movements of head and shoulders
XII	GSE	Hypoglossal nucleus	Muscles of tongue (extrinsic and intrinsic)	Movements of tongue

Proprioceptive (P) fiber (e.g., from muscle spindles) pathways from extraocular muscles and muscles of the tongue are still uncertain. Those from extraocular muscles probably travel via nerves III, IV, and VI in the orbit, then join the Vth nerve for the remainder of their course in the brainstem. The XIIth nerve does contain lingual proprioceptive fibers for part or all of its course. Many texts classify proprioception (P) under GSA.

77. PARASYMPATHETIC GANGLIA ASSOCIATED WITH CRANIAL NERVES

Ganglion	Location	Parasympathetic Root	Sympathetic Root	Chief Distribution
Ciliary	Lateral to optic nerve	Oculomotor nerve	Internal carotid plexus	Ciliary muscle and sphincter pupillae Dilator pupillae and tarsal muscles
Pterygo-palatine	In pterygopala-tine fossa	Greater petrosal (VII) and n. of pterygoid canal	Internal carotid plexus	Lacrimal gland
Otic	Below foramen ovale	Lesser petrosal (IX)	Plexus on middle meningeal artery	Parotid gland
Subman-dibular	On hyoglossus muscle	Chorda tympani (VII) by way of lingual nerve	Plexus on facial artery	Submandibular and sublingual glands

THE DIENCEPHALON

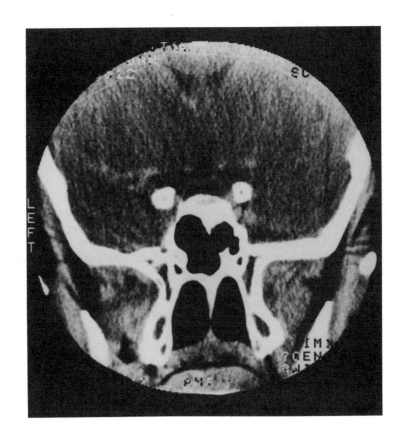

78. DIENCEPHALON: EPITHALAMUS

I. **Introduction:** the diencephalon lies rostral to the brainstem, begins at the level of the posterior commissure and extends to the interventricular foramina, includes the structures surrounding the 3rd ventricle, and is subdivided into four major parts: epithalamus, thalamus, hypothalamus, and subthalamus

A. EPITHALAMUS is located dorsomedial to the thalamus in the dorsal wall of the posterior portion of the 3rd ventricle, inferior to the splenium of the corpus callosum. This division consists of the pineal body, posterior commissure, habenular commissure, habenula (habenular nuclei), tela choroidea, and the stria medullaris thalami

  1. The membranous tela choroidea stretches between the stria medullaris thalami of the opposite sides to form most of the roof of the third ventricle. Remaining structures form only a small posterior part of the roof

  2. Habenular trigone is the small, triangular, depressed area just anterior to the superior colliculi and consists of

    a. Habenular nuclei receive nerve fibers from the stria medullaris thalami

    b. Habenular commissure joins the habenular nuclei

    c. Habenulopeduncular tract (retroflex tract of Meynert) is comprised of nerve fibers from the habenular nucleus to the interpeduncular ganglion of the mesencephalon

  3. Posterior commissure is a bundle of white fibers that crosses the midsagittal plane immediately rostral to the superior colliculi at the point where the 3rd ventricle becomes continuous with the cerebral aqueduct. It contains fibers

    a. Connecting the two superior colliculi

    b. From the pretectal nuclei

    c. From nuclei of the posterior commissure

    d. From the olivary nuclei

  4. Pineal body is the main structure of the epithalamus

    a. Located in the depression between the superior colliculi

    b. Formed by an evagination of the roof of the diencephalon

    c. Attached by a stalk that is continuous with both the habenular and posterior commissures

    d. Its recess (pineal recess) is the cavity of the evagination formed by the separation of the dorsal and ventral laminae of the stalk

    e. In man, it secretes melatonin, which aids in the regulation of the secretion of gonadotropins

    f. This gland also secretes melanocyte-stimulating hormones which contain endorphin molecules, important in pain control

    g. Clinical significance: the pineal body is located in the midline, and upon calcification (after 20 years) it serves as a useful marker of midline structures on a roentgenogram. A shift of this structure to one side may represent a shift of brain structures due to a mass lesion on the opposite side or from an atropic lesion on the same side

    h. In fish, amphibians, and reptiles, the gland is found beneath the dorsum of the head. It contains photoreceptors and nerve cells. The former register changes in light intensity due to translucence of tissues overlying the gland

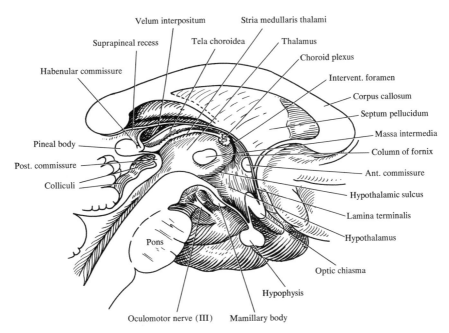

Velum interpositum

Suprapineal recess

Stria medullaris thalami

Tela choroidea

Habenular commissure

Thalamus

Choroid plexus

Intervent. foramen

Corpus callosum

Septum pellucidum

Massa intermedia

Pineal body

Column of fornix

Post. commissure

Ant. commissure

Colliculi

Hypothalamic sulcus

Lamina terminalis

Hypothalamus

Pons

Optic chiasma

Hypophysis

Oculomotor nerve (III)   Mamillary body

DIENCEPHALON

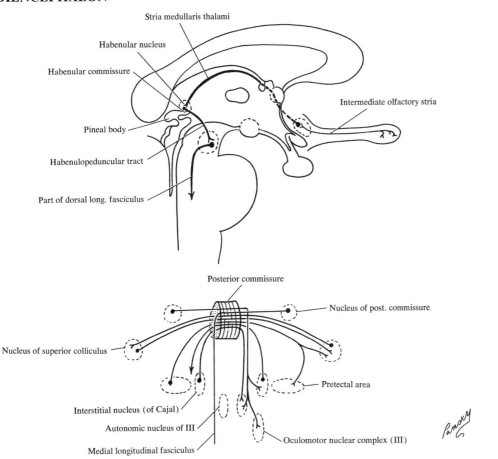

Stria medullaris thalami

Habenular nucleus

Habenular commissure

Intermediate olfactory stria

Pineal body

Habenulopeduncular tract

Part of dorsal long. fasciculus

Posterior commissure

Nucleus of post. commissure

Nucleus of superior colliculus

Pretectal area

Interstitial nucleus (of Cajal)

Autonomic nucleus of III

Oculomotor nuclear complex (III)

Medial longitudinal fasciculus

Pansky

79. DIENCEPHALON: PINEAL GLAND/BODY

I. **The pineal gland/body** is composed of cell groups, pinealocytes, arranged as cords separated by connective tissue and supported by glial cells. The gland is rich in blood and is innervated by postganglionic sympathetic nerve fibers whose cell bodies are in the superior cervical ganglia. The postganglionics contain several neurotransmitters, of which norepinephrine is critical for melatonin production. Concretions of calcified material (''brain sand'') accumulate in its parenchyma after the age of 16 years, consisting of calcium and magnesium salts. The mammalian gland contains neither photoreceptors nor nerve cells, having changed its histophysiology from that of submammalian vertebrates

A. SYNTHESIS OF MELATONIN (indole metabolism in the gland): tryptophan is taken up from the blood and converted into serotonin by pinealocytes, which then metabolize it to melatonin. Pineal serotonin concentration is higher than that of any other tissue due to the high level of tryptophan hydroxylase in the gland. Melatonin has the property of clumping melanin pigment granules in cutaneous melanocytes and influencing the lightness or darkness of the animal

1. Norepinephrine is released from nerve endings at night; it interacts with postjunctional adrenergic receptors on pinealocyte membranes, which enhances adenylate cyclase activity, resulting in formation of cyclic AMP and, in turn, increasing melatonin production

B. ACTIONS OF MELATONIN: are diverse

1. Melatonin aggregates in the skin (dermal melanocytes) of amphibians
2. Inhibits the reproductive physiology of mammals (gonad inhibition)
3. It is metabolized, after secretion, predominantly in the liver

C. PINEAL MELATONIN RHYTHM (reduction of melatonin in hamsters)

1. Radioimmunoassay of melatonin over a 24-hour period (light:dark cycle of 14:10 hours) was compared in young and old female and male hamsters
   a. Young animals (both sexes) showed an eightfold rise in melatonin during the dark phase
   b. Old animals exhibited little or no melatonin increase during the dark phase
   c. The marked decrease in pineal biosynthetic activity in the aging animal may be due to (any or all of the following)
      i. Catecholamine receptors on pinealocytes lose sensitivity to norepinephrine neurotransmitters
      ii. Postganglionic sympathetic fibers do not release norepinephrine
      iii. Precursors of norepinephrine are not available

D. CLINICAL OBSERVATIONS suggest an antigonadotropic function for the pineal gland in man, since tumors developing near puberty may alter the age of onset of pubertal changes—may be precocious if the parenchymatous cells are destroyed or delayed (if the tumor is derived from the functional cells of the gland)

1. Pinealectomy in experimental animals stimulates the genital system, with resulting genital hypertrophy, precocious opening of the vagina in immature females, and estrous cycle changes
2. Pineal extract administration in animals inhibits the gonads

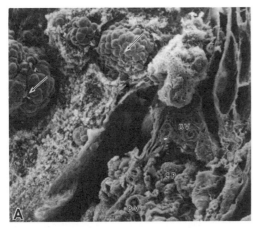

SEM of pineal body and stalk (low power). Concretions (arrows), choroid plexus (CP), and blood vessels (BV).

Large acervuline conglomerate of many concretions surrounded by pinealocytes and glial cells (low power).

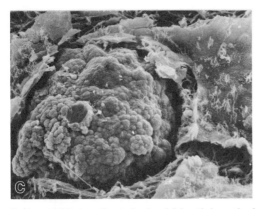

Conglomerate of "brain sand" protruding through 3rd ventricular ependymal lining (low power).

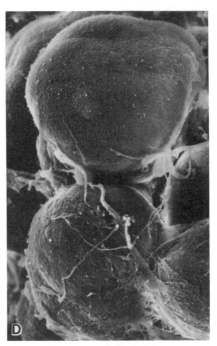

Parenchymal cells in pineal body surrounded by vascular and C.T. elements (high power).

Third ventricular luminal surface with cilia and blebs on ependymal lining (high power).

80. THALAMUS: ANATOMY AND NUCLEI

I. **Introduction:** the thalamus is the most dorsal of the four subdivisions of the diencephalon and forms the dorsal half of each lateral wall of the 3rd ventricle. Major ascending pathways of the auditory, visual, and somatosensory systems have their final subcortical relays here. The large, ovoid (egg-shaped) gray thalami of both sides often fuse (70%) in the midline of the 3rd ventricle to form the interthalamic adhesion or massa intermedia

A. THE THALAMIC RADIATION: comprised of tracts emerging from the lateral thalamic surface, entering the internal capsule and terminating in the cerebral cortex. The thalamus is the major source of afferent fibers to the cortex

B. THE THALAMUS IS ALSO THE RELAY CENTER for both cerebellar projections to the motor cortex and major efferent fibers from the basal ganglia

C. THALAMIC ANATOMY is best seen by dividing its more than 30 nuclei into 3 gray areas: anterior, medial, and lateral, formed by a vertical sheet of white matter, the internal medullary lamina, which divides each of the thalami

   1. Intralaminar nuclei are made up of several small groups of nuclei which are enclosed within the medullary lamina
   2. The reticular nucleus is separated from the main nuclear mass by the external medullary lamina, a layer of myelinated fibers on the lateral surface of the 2 thalami adjacent to the internal capsule

D. FOUR EXTERNAL PROMINENCES are present on the surface of each of the thalami: the anterior tubercle, pulvinar, medial, and lateral geniculate bodies

II. **Four major groups of the more than 30 thalamic nuclei**

A. ANTERIOR NUCLEAR GROUP corresponds to and is visualized as the anterior tubercle. Afferent fibers from the mamillary bodies run in the mamillothalamic tract; efferent fibers project to the cingulate gyrus (cortex)

B. MEDIAL NUCLEAR GROUP includes the gray substance medial to the internal medullary lamina; the dorsal medial nucleus, which projects to the frontal cortex lying anterior to the precentral gyrus (motor cortex); and the centromedian nucleus (intralaminar nucleus) which connects with the corpus striatum

C. MIDLINE NUCLEAR GROUP includes groups of cells lying just beneath the ependymal lining of the 3rd ventricle and in interthalamic adhesion. These nuclei connect with nuclei in the hypothalamus and with periaqueductal gray matter

D. LATERAL NUCLEAR GROUP: located anterior to and includes the pulvinar. It lies between the internal and external medullary laminae. It is further subdivided into a dorsal and ventral tier of nuclei

   1. Pulvinar is the largest nucleus of the dorsal tier and connects with cortices of the parietal and temporal lobes
   2. Other nuclei of the dorsal tier
      a. Lateral dorsal nucleus: caudal to the anterior nucleus. Has fiber connections with the cingulate gyrus, precuneus, and mamillary bodies
      b. Lateral posterior nucleus receives fibers from medial and lateral geniculate bodies, from the ventral posterior nuclear complex, and has reciprocal connections with cortices of precuneus and superior parietal lobule
   3. Lateral and medial geniculate bodies: nuclei are part of the ventral tier
      a. Lateral geniculate body: an oval elevation on lateral part of posterior end of thalamus. It receives fibers from the optic tract and projects fibers to visual cortex as the geniculocalcarine (optic) radiations
      b. Medial geniculate body: located under the pulvinar, lateral to midbrain. It receives auditory fibers from the lateral lemniscus and inferior colliculus, and projects fibers to Heschl's gyri (temporal cortex) as the auditory radiations

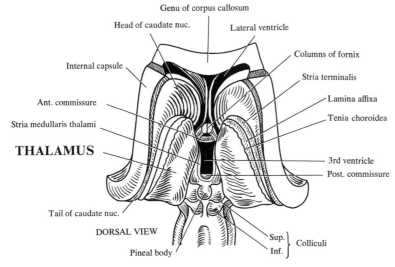

Genu of corpus callosum

Head of caudate nuc.

Lateral ventricle

Internal capsule

Columns of fornix

Stria terminalis

Ant. commissure

Lamina affixa

Stria medullaris thalami

Tenia choroidea

THALAMUS

3rd ventricle

Post. commissure

Tail of caudate nuc.

DORSAL VIEW

Sup. } Colliculi
Inf. }

Pineal body

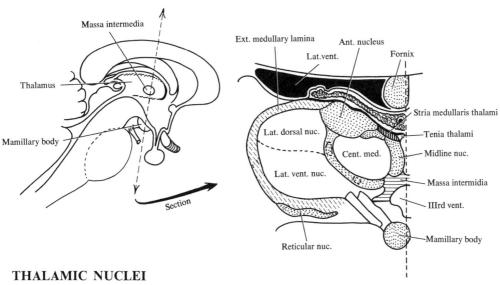

Massa intermedia

Ext. medullary lamina

Ant. nucleus

Lat.vent.

Fornix

Thalamus

Stria medullaris thalami

Tenia thalami

Lat. dorsal nuc.

Mamillary body

Cent. med.

Midline nuc.

Lat. vent. nuc.

Massa intermidia

IIIrd vent.

Reticular nuc.

Mamillary body

Section

THALAMIC NUCLEI

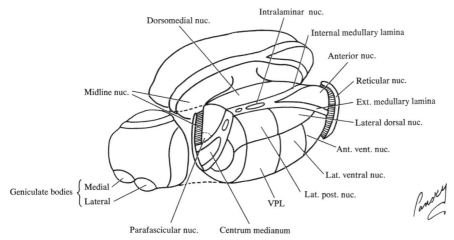

Dorsomedial nuc.

Intralaminar nuc.

Internal medullary lamina

Anterior nuc.

Reticular nuc.

Midline nuc.

Ext. medullary lamina

Lateral dorsal nuc.

Ant. vent. nuc.

Lat. ventral nuc.

Geniculate bodies { Medial
Lateral

Lat. post. nuc.

VPL

Parafascicular nuc.

Centrum medianum

-173-

81. THALAMUS: VENTRAL TIER NUCLEI

I. Four major groups of thalamic nuclei (cont.)

A. LATERAL NUCLEAR GROUP (CONT.)

    1. Other nuclei of the ventral tier (from anterior to posterior)

        a. Ventral anterior nucleus receives input from substantia nigra, globus pallidus, other thalamic nuclei (midline and intralaminar), reticular formation, and cerebral cortex. Its output goes to the caudate nucleus and premotor cortex (area 6)

        b. Ventral lateral nucleus receives input from the globus pallidus, substantia nigra, other thalamic nuclei, reticular formation, cerebral cortex, cerebellum (dentatothalamic tract), and the red nucleus (rubrothalamic tract). It has reciprocal connections with the precentral gyrus (area 4)

        c. Both of the above nuclei are specific motor nuclei for the integration and transmission of information from the motor control pathways in the motor system to the motor cortex of the cerebrum. Both send efferent fibers to widespread areas in the frontal cortex

    2. The ventral posterior nuclear complex consists of a ventral posteromedial nucleus (VPM), which is also called the semilunar or arcuate nucleus, and a ventral posterolateral nucleus (VPL)

        a. These nuclei relay somatic sensory system information to the cortex

        b. The VPL nucleus is where second-order neurons of the spinothalamic and medial lemniscus pathways from the trunk and limbs terminate

        c. The VPM nucleus is the site of termination of neurons of the trigeminal pathways from the head

        d. Axons from the VPL and VPM nuclei project to the postcentral gyrus (primary somatosensory cortex) via the middle thalamic radiation in the posterior limb of the internal capsule

II. Summary of nuclear areas and thalamic nuclei

Anterior nucleus ———————————————————	ANTERIOR AREA
Midline nucleus ———————	
Dorsomedial (medial nucleus)	} MEDIAL AREA
Centromedian (intralaminar) nucleus —————	

Lateral dorsal nucleus —————		
Lateral posterior nucleus	} DORSAL TIER	
Pulvinar ———————		
Ventral anterior nucleus —————		LATERAL AREA
Ventral lateral nucleus		
Ventral posterolateral nucleus		
Ventral posteromedial nucleus	} VENTRAL TIER	
Lateral geniculate body		
Medial geniculate body		
Reticular nuclei ———————		

III. Clinical considerations

A. SPECIFIC NEUROSURGICAL LESIONS in the ventral anterior thalamic nuclei interrupt connections between the basal ganglia and the cerebral cortex and serve to decrease rigidity and tremor in patients with Parkinson's disease

B. LESIONS THAT DESTROY THE VENTRAL POSTERIOR NUCLEUS OF THE THALAMUS produce a contralateral hemianesthesia with loss of sensory modalities of the face, trunk, and limbs. This is often caused by an infarction of a branch of the posterior cerebral artery

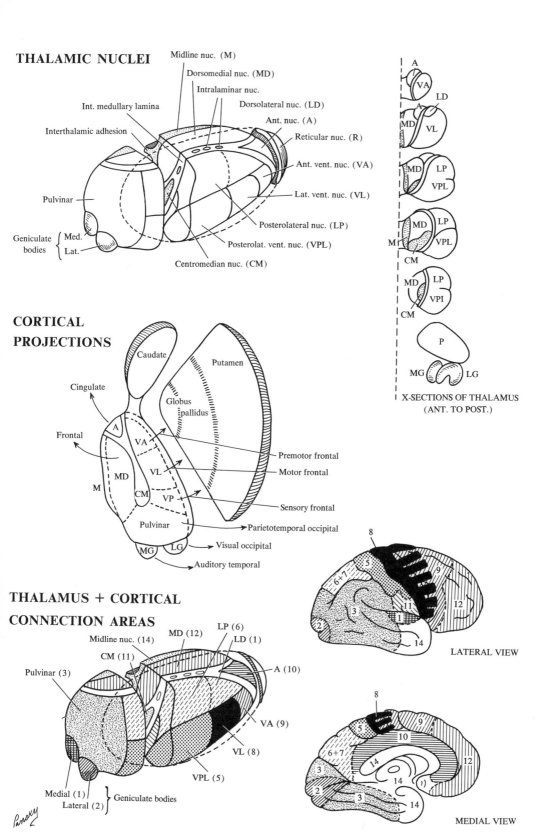

THALAMIC NUCLEI

Midline nuc. (M)
Dorsomedial nuc. (MD)
Intralaminar nuc.
Int. medullary lamina
Dorsolateral nuc. (LD)
Interthalamic adhesion
Ant. nuc. (A)
Reticular nuc. (R)
Ant. vent. nuc. (VA)
Lat. vent. nuc. (VL)
Pulvinar
Posterolateral nuc. (LP)
Geniculate bodies { Med. Lat.
Posterolat. vent. nuc. (VPL)
Centromedian nuc. (CM)

X-SECTIONS OF THALAMUS
(ANT. TO POST.)

A
VA
LD
A
MD
VL
MD
LP
VPL
MD
LP
M
VPL
CM
MD
LP
VPI
CM
P
MG
LG

CORTICAL PROJECTIONS

Caudate
Putamen
Cingulate
Globus pallidus
Frontal
A
VA
Premotor frontal
Motor frontal
MD
VL
M
CM
VP
Sensory frontal
Pulvinar
Parietotemporal occipital
Visual occipital
MG
LG
Auditory temporal

THALAMUS + CORTICAL CONNECTION AREAS

Midline nuc. (14)
MD (12)
LP (6)
LD (1)
CM (11)
A (10)
Pulvinar (3)
VA (9)
VL (8)
VPL (5)
Medial (1)
Lateral (2) } Geniculate bodies

LATERAL VIEW

MEDIAL VIEW

Panexy

-175-

82. FUNCTIONAL THALAMUS AND SUBTHALAMUS

I. Functions of the thalamus
A. INTEGRATION, CORRELATION, AND RELAY of information for the sensory, motor, consciousness, limbic, and visual systems
B. FOCUSING ATTENTION: by making certain cortical sensory areas more receptive and other areas less receptive
C. CONSCIOUS PERCEPTION AND INTERPRETATION OF PAIN: the only peripheral sensory stimulus to be interpreted at the level of the thalamus

II. Functional classification of thalamic nuclei
A. SPECIFIC NUCLEI possess reciprocal connections with specific areas of the cerebral cortex that have specific motor or sensory functions: these include most of the ventral tier nuclei of the lateral area, namely
   1. Anterior nucleus (some include these with the association nuclei)
   2. Ventral anterior (in part)
   3. Ventral lateral and ventral posterior nuclei
   4. Lateral and medial geniculates
B. SUBCORTICAL NUCLEI have no direct fiber connections with cerebral cortices and include thalamic reticular nucleus, intrathalamic nucleus, dorsomedial nucleus (in part), and the ventral anterior nucleus (in part)
C. ASSOCIATION NUCLEI have reciprocal fiber connections with association areas of the cerebral cortex and include the lateral dorsal nucleus, the lateral posterior nucleus, the pulvinar, and the dorsomedial nucleus (in part)

III. Subthalamus (ventral thalamus) lies between the tegmentum of the midbrain and the dorsal thalamus
A. THE SUBTHALAMUS IS LOCATED ventral to the thalamus and posterolateral to the hypothalamus. Lateral to it lies the internal capsule
B. THE SUBTHALAMUS CONTAINS nuclei and pathways associated with the control circuits of the basal ganglia
C. THE RED NUCLEUS AND SUBSTANTIA NIGRA extend into the subthalamus from the midbrain
D. THE SUBTHALAMUS NUCLEUS (BODY OF LUYS) is the major structure within the ventral thalamus. It is a lens-shaped mass of gray matter dorsolateral to the upper end of the substantia nigra
E. THE ANSA LENTICULARIS is the major pathway through this region and conveys fibers from the globus pallidus to the ventral anterior thalamic nucleus
F. THE FIELDS OF FOREL lie anterior to the red nucleus at this level and are formed by fibers from the globus pallidus
   1. Field H is the ventromedial portion
   2. Field H_1 is the dorsomedial portion
   3. Field H_2 (fasciculus lenticularis) is the ventrolateral portion which runs medially
   4. The ansa lenticularis joins field H_2 as it bends acutely in field H
   5. The thalamic fasciculus (Forel's field H_1) extends through field H_1 to reach the anterior ventral nucleus of the thalamus
   6. The zona incerta is a thin gray zone (reticular formation) above the fasciculus lenticularis

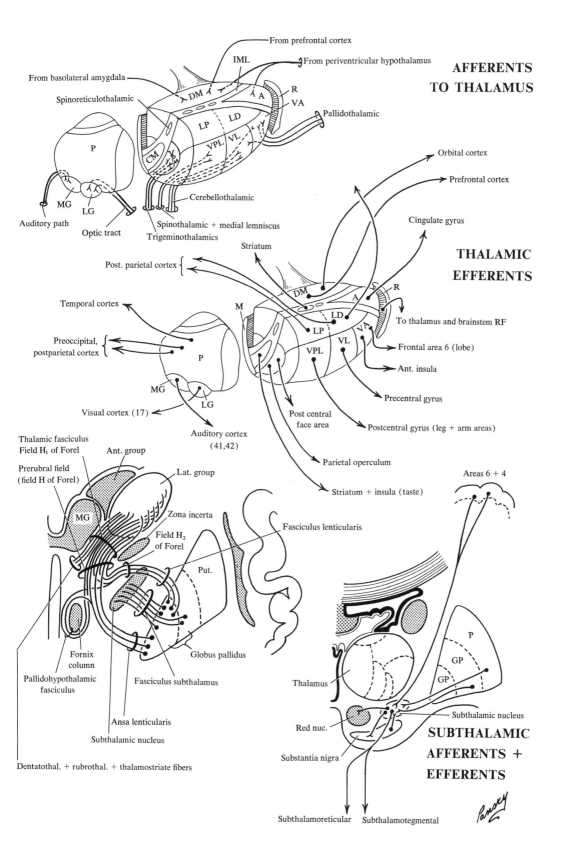

AFFERENTS TO THALAMUS

From prefrontal cortex
From periventricular hypothalamus
IML
From basolateral amygdala
Spinoreticulothalamic
DM
R
A
VA
Pallidothalamic
LP
LD
VPL
VL
CM
P
MG
LG
Auditory path
Optic tract
Cerebellothalamic
Spinothalamic + medial lemniscus
Trigeminothalamics

THALAMIC EFFERENTS

Orbital cortex
Prefrontal cortex
Cingulate gyrus
Striatum
Post. parietal cortex
DM
A
R
LD
To thalamus and brainstem RF
Temporal cortex
M
LP
VA
Frontal area 6 (lobe)
VPL
VL
Ant. insula
Preoccipital, postparietal cortex
P
Precentral gyrus
MG
LG
Postcentral gyrus (leg + arm areas)
Visual cortex (17)
Auditory cortex (41,42)
Post central face area
Parietal operculum
Striatum + insula (taste)

Thalamic fasciculus Field H₁ of Forel
Ant. group
Areas 6 + 4
Prerubral field (field H of Forel)
Lat. group
MG
Zona incerta
Fasciculus lenticularis
Field H₂ of Forel
Put.
P
GP
GP
Fornix column
Globus pallidus
Thalamus
Subthalamic nucleus
Pallidohypothalamic fasciculus
Fasciculus subthalamus
Red nuc.
Ansa lenticularis
Subthalamic nucleus
Substantia nigra
Dentatothal. + rubrothal. + thalamostriate fibers

SUBTHALAMIC AFFERENTS + EFFERENTS

Subthalamoreticular   Subthalamotegmental

Pansky

-177-

83. HYPOTHALAMUS

I. **Introduction:** hypothalamus is located on the inferior aspect of the diencephalon and bordered by lamina terminalis anteriorly; midbrain, posteriorly; hypothalamic sulcus, dorsally; third ventricle, medially; and subthalamic nucleus, laterally

II. **Hypothalamic structures seen on the ventral aspect of the diencephalon**
A. MAMILLARY BODIES: paired, spherical (pea-sized), white masses inferior to the 3rd ventricle, caudal to the tuber cinereum, and rostral to the posterior perforated substance
B. TUBER CINEREUM: that part of the hypothalamic floor between the mamillary bodies and the optic chiasm
C. MEDIAN EMINENCE: funnel-like extension of the tuber cinereum which is a hypothalamic derivative of the pituitary (hypophysis) gland
D. INFUNDIBULUM: a hollow, ventral extension of the median eminence and includes the infundibular stalk to the posterior lobe (pars nervosa) of the hypophysis
E. NEUROHYPOPHYSIS consists of the median eminence, the infundibular stalk, and the neural lobe (posterior lobe) of the hypophysis
F. OPTIC CHIASM AND OPTIC TRACTS delineate the floor of the hypothalamus
G. TWO HYPOTHALAMIC AREAS are seen on each side of the 3rd ventricle on either side of a sagittal plane through the fornix and mamillothalamic tract
   1. Lateral nuclear area: positioned in the lateral part of the tuber cinereum and includes the following nuclei
      a. Lateral preoptic nucleus: between anterior commissure and supraoptic nucleus
      b. Supraoptic nucleus straddles the lateral part of the optic chiasm
      c. Lateral nucleus: in the floor of hypothalamus lateral to the arcuate nucleus
      d. Tuberal nucleus: in a position dorsal to the tuber cinereum
   2. Medial nuclear area is subdivided into three regions based on external anatomic features, as indicated
      a. Supraoptic (anterior) region: located dorsal to optic chiasm. Includes
         i. Medial preoptic nucleus: medial extension of the preoptic nucleus; medial and lateral preoptic nuclei are part of the telencephalon developmentally, but usually are included with the hypothalamic nuclei because it is hypothalamic functionally
         ii. Anterior nucleus: in the plane with and between supraoptic and paraventricular nuclei
         iii. Paraventricular nucleus: a group of cells dorsal in position to the anterior nucleus and lying close to the lining of the 3rd ventricle
         iv. Supraoptic nucleus closely parallels the optic chiasm. Occupies both lateral nuclear area and anterior region of medial nuclear area
      b. Middle (tuberal) region: located dorsal to the tuber cinereum and includes
         i. Dorsomedial nucleus: a mass lying above the ventromedial nucleus
         ii. Ventromedial nucleus: an ovoid mass of cells lying anterior to the mamillary bodies and posterior to the supraoptic nucleus
         iii. Arcuate (infundibular) nucleus: located in the floor of the hypothalamus ventral to the ventromedial nucleus
      c. Posterior (mamillary) region: positioned dorsal and immediately anterior to the mamillary bodies and includes
         i. Mamillary nuclei: found in mamillary bodies. Consist of a medial nucleus, larger of the two and responsible for protuberance of mamillary body, and a lateral nucleus, located between lateral border of the medial mamillary nucleus and base of brain
         ii. Posterior nucleus: dorsal to the mamillary nuclei and located in the lateral wall of the hypothalamus

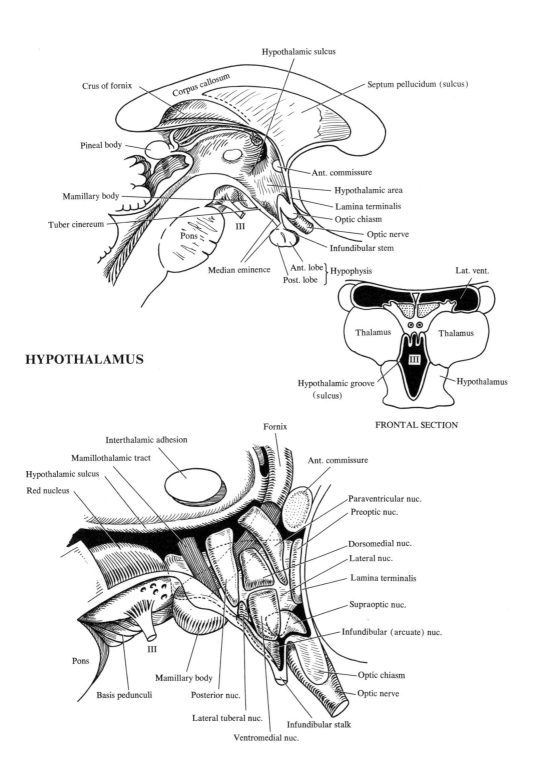

Hypothalamic sulcus

Crus of fornix

Corpus callosum

Septum pellucidum (sulcus)

Pineal body

Ant. commissure

Hypothalamic area

Mamillary body

Lamina terminalis

Optic chiasm

Tuber cinereum

III

Optic nerve

Pons

Infundibular stem

Median eminence

Ant. lobe }
Post. lobe } Hypophysis

Lat. vent.

Thalamus   Thalamus

III

Hypothalamic groove
(sulcus)

Hypothalamus

HYPOTHALAMUS

FRONTAL SECTION

Interthalamic adhesion

Fornix

Mamillothalamic tract

Ant. commissure

Hypothalamic sulcus

Paraventricular nuc.

Red nucleus

Preoptic nuc.

Dorsomedial nuc.

Lateral nuc.

Lamina terminalis

Supraoptic nuc.

Infundibular (arcuate) nuc.

Optic chiasm

III

Optic nerve

Pons

Mamillary body

Basis pedunculi

Posterior nuc.

Lateral tuberal nuc.

Infundibular stalk

Ventromedial nuc.

HYPOTHALAMIC NUCLEI— SCHEMATIC

84. HYPOTHALAMIC FIBER CONNECTIONS AND NUCLEAR SUMMARY

I. Summary of hypothalamic nuclei

A. THE NUCLEI ARE LOOSELY ORGANIZED GROUPS OF CELLS in the walls of the 3rd ventricle. They are not very well defined or distinct but appear as more or less well-circumscribed cell groups in a more diffuse conglomerate of cells

B. THE HYPOTHALAMUS MAY BE DIVIDED into medial and lateral areas
1. The medial hypothalamus has many short axonal connections among cells which allow excitation to spread in any given direction from any focus to establish many self-exciting networks
2. Large neurons of the lateral hypothalamus have more specific axonal connections. The many neuronal connections of the medial forebrain bundle traverse this lateral hypothalamic area

II. Fiber connections of the hypothalamus include those to the spinal cord, brainstem, and cerebrum, as well as intrahypothalamic nuclear fibers

A. AFFERENT FIBER CONNECTIONS
1. Stria terminalis: a small fiber bundle running in the terminal sulcus (between thalamus and caudate nucleus) and carrying fibers from the amygdaloid nucleus to the anterior, ventromedial, and preoptic nuclei of the hypothalamus
2. Pallidohypothalamic fibers pass from the lentiform nucleus to the ventromedial hypothalamic nucleus
3. Thalamohypothalamic fibers originate in the medial and midline thalamic nuclei
4. Fornix (corticomamillary fibers) conveys fibers from the hippocampus to the mamillary bodies and possibly the preoptic nuclei
5. Medial forebrain bundle (olfactohypothalamic fibers) is a phylogenetically ancient fiber system comprised of many neuronal connections interposed in the bundle
   a. Extends throughout the entire lateral hypothalamus
   b. Continues rostrally through the preoptic area to the olfactory region and caudally to the midbrain
   c. Carries fibers from nuclei in the paraolfactory area and corpus striatum to the hypothalamus
   d. Composed of short relay and long fibers
   e. Conduction is from the lateral to the more medial areas, with short axon neurons traversing the medial forebrain bundle or collaterals arising from fibers in the bundle

B. EFFERENT FIBER CONNECTIONS begin as a conspicuous bundle of fibers, the fasciculus mamillaris princeps, in the mamillary nuclei and divide to form
1. Mamillothalamic fasciculus (hypothalamic-thalamic tract), which ends in the anterior thalamic nucleus
2. Mamillotegmental fasciculus ends in the dorsal tegmental and interpeduncular nuclei of the midbrain
3. Dorsal longitudinal fasciculus (periventricular system) is located in the periventricular gray of the brainstem and connects the hypothalamus with the reticular formation of the mesencephalon and parasympathetic nuclei
4. Hypothalamicohypophyseal tract is made up of fibers that arise in the supraoptic and paraventricular nuclei and carry neurosecretory products to the neurohypophysis (pars nervosa of the pituitary)
5. Medial forebrain bundle receives fibers from the hypothalamus and continues to the paramedian cell groups of the midbrain tegmentum and in the mesencephalon reticular formation and raphe nuclei

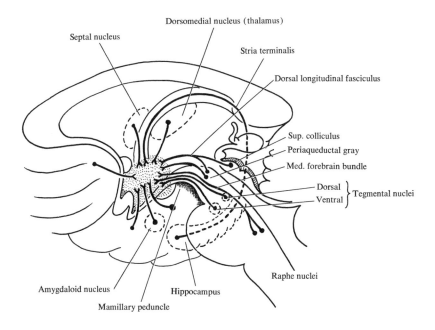

HYPOTHALAMIC AFFERENTS (INPUT)

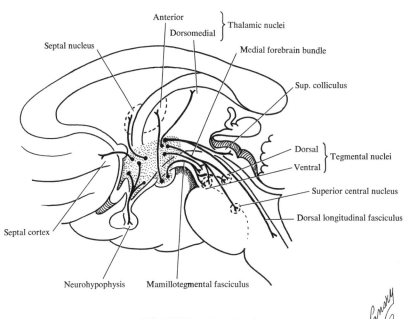

HYPOTHALAMIC EFFERENTS (OUTPUT)

85. HYPOTHALAMIC FUNCTIONAL CONSIDERATIONS

I. **The hypothalamus** functions as the primary control center for the visceral system and serves to integrate activity for consciousness, visceral, limbic, and endocrine systems. It is considered to be diversified in activities and includes regulatory influence on the autonomic nervous system. Stimulation of the anterior area nuclei is excitatory to the restorative (parasympathetic) activities, while stimulation of the posterior nuclei is preparative (sympathetic). Except for the supraoptic and paraventricular nuclei, specific functions have not been determined for individual nuclei. The following functions are controlled by the hypothalamus

  A. THE SUPRAOPTIC AND PARAVENTRICULAR NUCLEI contain neurons whose axons convey neurosecretory droplets in the hypothalamohypophyseal tract to the pars nervosa (posterior lobe of pituitary)

    1. Supraoptic nucleus is associated with the release of antidiuretic hormone (ADH)

    2. Paraventricular nucleus is associated with the release of oxytocin

  B. THE HYPOTHALAMUS ALSO INFLUENCES THE SECRETORY ACTIVITY of the pars distalis (anterior lobe of the pituitary)

  C. THE HYPOTHALAMUS HAS CERTAIN NEURONAL ORGANIZATIONS, referred to as centers, which subserve other centers in the brainstem via complex synaptic pathways. These centers in the brainstem are concerned with the autonomic regulation of respiration, vomiting, micturition, and cardiovascular activity

  D. TEMPERATURE IS CONTROLLED in part by the hypothalamus. Thermoreceptor-sensitive neurons (center) in the hypothalamus initiate the loss or conservation of heat by vasodilatation, vasoconstriction, sweating, shivering, etc.

    1. The anterior hypothalamic center regulates loss of heat

    2. The posterior hypothalamic center regulates the conservation of heat

  E. WATER METABOLISM AND DRINKING are influenced by the stimulation of the lateral hypothalamus

  F. FOOD INTAKE AND FEEDING BEHAVIOR are controlled in the lateral hypothalamic areas, while the satiety of feeding is controlled in the medial (ventromedial) areas. Lesions in the ventromedial area usually lead to hyperphagia and obesity; lateral hypothalamic lesions cause aphagia and ultimate emaciation

  G. SLEEP is regulated by the consciousness system which has connections with the posterior hypothalamus. Lesions here may cause excessive sleep (hypersomnia)

  H. SEXUAL FUNCTIONS are influenced by the hypothalamus and its complex interconnections with other parts of the brain

  I. EMOTIONS are considered in the realm of hypothalamic activities since it is part of the limbic system

II. **Horner's syndrome:** The hypothalamus gives rise to many fibers concerned with autonomic regulation. Many involved in sympathetic control traverse the brainstem and reach the ipsilateral (some contralateral) intermediolateral cell column of the cord. Interruption of the descending sympathetic pathway causes ipsilateral Horner's syndrome: miosis (small pupil), ptosis (drooping eyelid), and enophthalmos (recession of the eyeball; more apparent than real). May be accompanied by flushing and lack of sweating in the ipsilateral skin of the face and part of the body

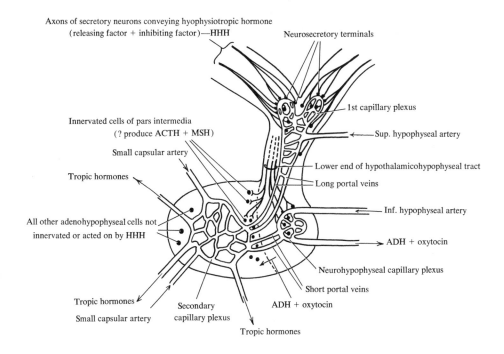

Axons of secretory neurons conveying hyophysiotropic hormone
(releasing factor + inhibiting factor)—HHH

Neurosecretory terminals

1st capillary plexus

Innervated cells of pars intermedia
(? produce ACTH + MSH)

Sup. hypophyseal artery

Small capsular artery

Lower end of hypothalamicohypophyseal tract

Tropic hormones

Long portal veins

Inf. hypophyseal artery

All other adenohypophyseal cells not
innervated or acted on by HHH

ADH + oxytocin

Neurohypophyseal capillary plexus

Short portal veins

Tropic hormones

ADH + oxytocin

Secondary
capillary plexus

Small capsular artery

Tropic hormones

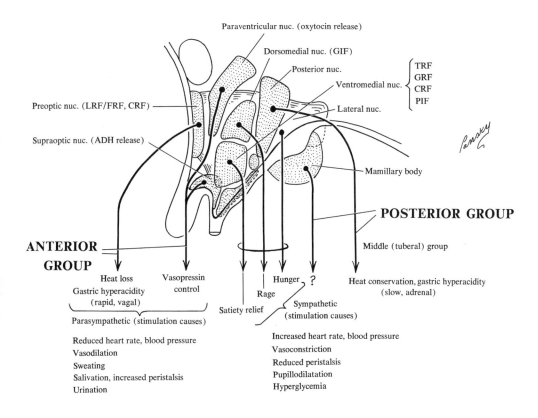

Paraventricular nuc. (oxytocin release)

Dorsomedial nuc. (GIF)

Posterior nuc.

Ventromedial nuc.

{ TRF
GRF
CRF
PIF

Preoptic nuc. (LRF/FRF, CRF)

Lateral nuc.

Supraoptic nuc. (ADH release)

Mamillary body

POSTERIOR GROUP

ANTERIOR
GROUP

Middle (tuberal) group

Heat loss

Vasopressin
control

Hunger   ?

Heat conservation, gastric hyperacidity
(slow, adrenal)

Gastric hyperacidity
(rapid, vagal)

Rage

Satiety relief

Sympathetic
(stimulation causes)

Parasympathetic (stimulation causes)

Reduced heart rate, blood pressure
Vasodilation
Sweating
Salivation, increased peristalsis
Urination

Increased heart rate, blood pressure
Vasoconstriction
Reduced peristalsis
Pupillodilatation
Hyperglycemia

86. SUMMARY OF THALAMIC NUCLEAR CONNECTIONS

I. Specific relay nuclei: receive well-defined fiber bundles and project to specific functional cortical areas

Nuclei	Input	Output
Lateral geniculate	Optic tract	Visual cortex
Medial geniculate	Inferior brachium	Auditory cortex
VPL	Medial lemniscus, spinothalamic tract	Somatosensory cortex
VL/VA	Cerebellum, basal ganglia	Motor/premotor cortex

II. Association nuclei: receive inputs from a variety of sources and project to broad areas of the association cortex

Nuclei	Input	Output
Pulvinar	Superior colliculus and other thalamic nuclei	Parietal, occipital, and temporal association cortex
LP	Other thalamic nuclei	Cingulate gyrus
LD	Amygdala, hypothalamus, and other thalamic nuclei	Prefrontal cortex

III. Nonspecific nuclei: project to widespread cortical areas (often via fibers that are collaterals of those going to other places)

Nuclei	Input	Output
Part of VA	Other thalamic nuclei	Collaterals to prefrontal cortex
Intralaminar	Reticular formation, basal ganglia, and other thalamic nuclei	Collaterals to widespread cortical areas

IV. Subcortical nuclei: these nuclei have no cerebral cortical projections

Nuclei	Input	Output
Reticular	Thalamus	Thalamus

THE BASAL
GANGLIA

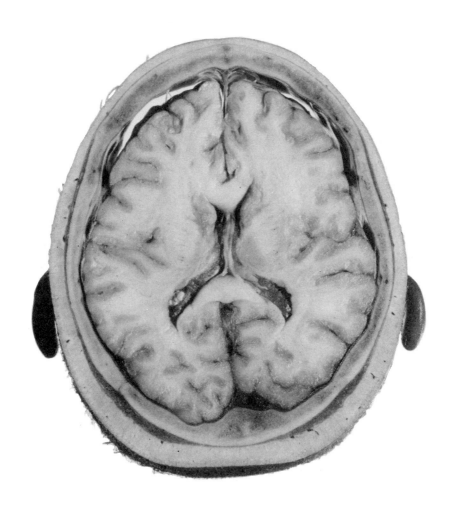

87. BASAL GANGLIA: GENERAL CONSIDERATIONS AND DEFINITIONS

I. **Definitions:** the basal ganglia are the three large subcortical gray (nuclear) masses derived from the telencephalon

A. THESE SUBCORTICAL COLLECTIONS are comprised of neurons and intermixed finely myelinated axons

B. THE GANGLIA ARE PHYSICALLY SEPARATED from the diencephalon by the internal capsule, but intimately related to the cerebral cortex, thalamus, subthalamus, and substantia nigra

C. THE TERM "BASAL GANGLIA" was originally used to include all gray matter in the diencephalon and telencephalon, but is currently used to designate only the following structures

   1. Caudate nucleus: made up of a head which forms the lateral wall of the anterior horn of the lateral ventricle, a body overlying the lateral part of the dorsal thalamus, and a tail located above temporal horn of lateral ventricle

   2. Lenticular (lentiform) nucleus: divided into an outer larger portion, the putamen, and an inner portion, the globus pallidus, by the lateral medullary lamina (a vertical plate of white matter). The whole nucleus has the size and shape of a Brazil nut. Unlike the caudate nucleus, it has no ventricular surface and lies deeply buried in the white matter of the cerebral hemisphere. It is closely applied to the lateral surface of the internal capsule which separates it from the caudate nucleus rostrally and thalamus caudally. In current usage, the individual terms ("putamen" and "globus pallidus") are preferred over "lenticular nucleus"

      a. Putamen (shell): the largest and most lateral part of the basal ganglia. It underlies the insula and is continuous anteriorly with the head of the caudate nucleus. It is composed primarily of small to medium-sized nerve cells

      b. Globus pallidus (pale globe): the nucleus that comprises the medial portion of the lenticular nucleus located between the internal capsule and putamen. Fibers from this nucleus enter the ansa lenticularis, lenticular fasciculus, and other tracts associated with the extrapyramidal system

D. "STRIATUM" is a collective term for caudate nucleus and putamen because of the striated appearance, while "corpus striatum" refers to the combination of caudate nucleus, putamen, and globus pallidus

   1. Corpus striatum is a single gray mass during early development, but becomes separated by the fibers of the internal capsule into two distinct cellular masses, the caudate and the lenticular nuclei

   2. The corpus striatum is the part of the basal ganglia considered to be concerned with somatic motor functions. Neurologists often use the term "extrapyramidal motor system" to group together the corpus striatum and certain related brainstem nuclei considered to subserve these somatic motor functions

E. THE AMYGDALA (AMYGDALOID NUCLEAR COMPLEX) AND CLAUSTRUM: included by some authorities with the basal ganglia, but for clinical purposes they are not considered to be part of the basal ganglia

F. THE NEOSTRIATUM comprises the caudate and putamen

G. THE PALEOSTRIATUM consists of the globus pallidus (lateral and medial portions)

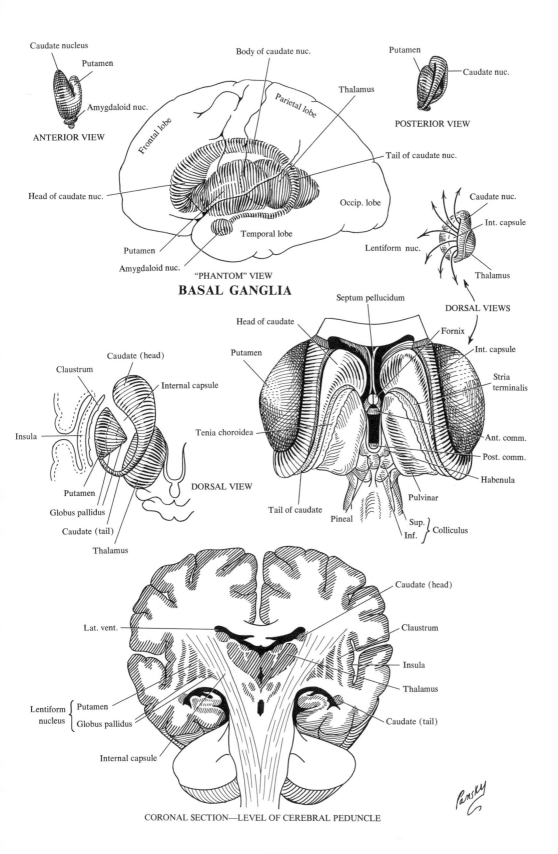

BASAL GANGLIA

ANTERIOR VIEW

Caudate nucleus
Putamen
Amygdaloid nuc.

POSTERIOR VIEW

Putamen
Caudate nuc.

"PHANTOM" VIEW

Body of caudate nuc.
Parietal lobe
Thalamus
Frontal lobe
Head of caudate nuc.
Putamen
Amygdaloid nuc.
Temporal lobe
Occip. lobe
Tail of caudate nuc.

DORSAL VIEWS

Caudate nuc.
Int. capsule
Lentiform nuc.
Thalamus

DORSAL VIEW

Claustrum
Caudate (head)
Internal capsule
Insula
Putamen
Globus pallidus
Caudate (tail)
Thalamus

Head of caudate
Putamen
Tenia choroidea
Tail of caudate
Pineal
Septum pellucidum
Fornix
Int. capsule
Stria terminalis
Ant. comm.
Post. comm.
Habenula
Pulvinar
Sup.
Inf. } Colliculus

CORONAL SECTION—LEVEL OF CEREBRAL PEDUNCLE

Lat. vent.
Lentiform { Putamen
nucleus { Globus pallidus
Internal capsule
Caudate (head)
Claustrum
Insula
Thalamus
Caudate (tail)

Pansky

88. BASAL GANGLIA: ANATOMIC RELATIONSHIPS

I. Anatomic relationships

A. THE CORPUS STRIATUM is shown to be interrupted by massive fiber bundles of the internal capsule when viewed in horizontal and coronal sections of the brain

    1. In a rostral coronal section, it is revealed that the heads of the caudate and putamen are fused along the ventromedial surface of the caudate, thus indicating their common origin. There is a topographic relationship, with rostral cortex projecting to the head of the caudate

    2. In a slightly more posterior coronal section, the caudate and lenticular nuclei are completely separated by the fiber bundles of the internal capsule

B. THE INTERNAL CAPSULE in horizontal section is V-shaped, with the apex directed medially, with the following spatial relationships

    1. The anterior limb or arm of the internal capsule separates caudate and lenticular nuclei. The anterior limb is comprised of

        a. Thalamocortical and corticothalamic fibers which are reciprocal connections between the lateral thalamic nucleus and the frontal lobe of the cerebral cortex

        b. Frontopontine fibers from the frontal lobe to the pontine nuclei

        c. Fibers from the caudate nucleus to the putamen. (*Note:* the existence of these fibers is currently in question.)

    2. The posterior limb of the internal capsule separates the thalamus and the lenticular nucleus and is subdivided into three portions which contain the following designated fibers or fiber tracts

        a. Lenticulothalamic portion

            i. Corticobulbar tract and corticospinal tract (nerve fibers to superior limb lie anterior to those to inferior limb)

            ii. Corticorubral tract: comprised of nerve fibers from cortex of the frontal lobe to red nucleus which run closely with those in the corticospinal tract

        b. Retrolenticular part

            i. Fibers from lateral nucleus of thalamus to postcentral gyrus

        c. Sublenticular (below lenticular nucleus) part

            i. Parietotemporopontine fibers from parietal and temporal lobe cortex to pontine nuclei

            ii. Auditory radiations from medial geniculate body to transverse temporal gyrus or Heschl's gyrus (areas 41, 42)

            iii. Optic fibers from lateral geniculate body to area 17

C. THE CAUDATE NUCLEUS is an elongated subcortical nuclear mass

    1. Head portion is contiguous with anterior perforated substance

    2. Lies adjacent to inferior border of anterior horn of lateral ventricle

    3. Tapers as it continues posteriorly and inferiorly as the tail, which enters the roof of the inferior (temporal) horn of lateral ventricle to end at level of the amygdala

D. THE LENTICULAR NUCLEUS is wedge-shaped and located within the limbs of the internal capsule between insula (laterally), caudate nucleus, and thalamus (medially). It is subdivided into the putamen, the larger convex portion lying medial to the insular cortex, and the globus pallidus, the smaller medial, triangular-shaped portion, which appears lighter in color because of its many myelinated fibers

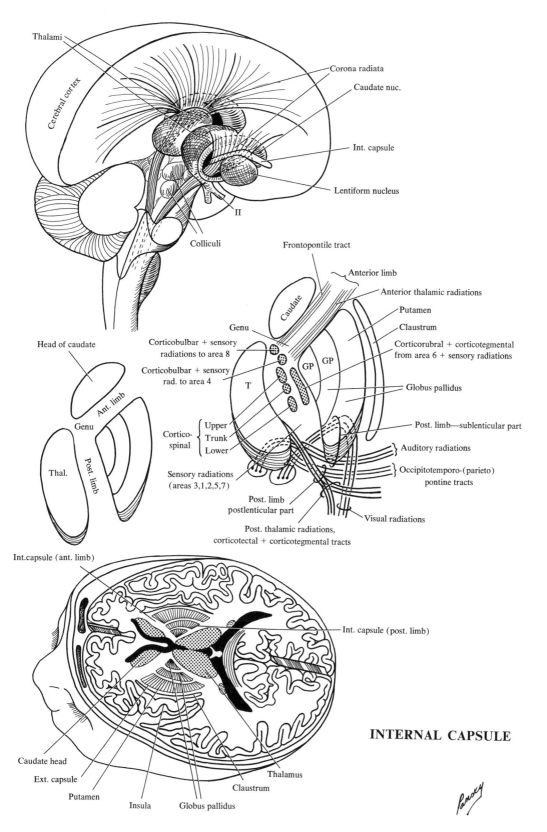

Thalami

Cerebral cortex

Corona radiata

Caudate nuc.

Int. capsule

Lentiform nucleus

II

Colliculi

Frontopontile tract

Anterior limb

Anterior thalamic radiations

Putamen

Claustrum

Corticorubral + corticotegmental from area 6 + sensory radiations

Caudate

Genu

Corticobulbar + sensory radiations to area 8

Corticobulbar + sensory rad. to area 4

GP GP

T

Globus pallidus

Head of caudate

Ant. limb

Genu

Post. limb—sublenticular part

Cortico-spinal { Upper / Trunk / Lower

Auditory radiations

Occipitotemporo-(parieto) pontine tracts

Thal.

Post. limb

Sensory radiations (areas 3,1,2,5,7)

Post. limb postlenticular part

Visual radiations

Post. thalamic radiations, corticotectal + corticotegmental tracts

Int.capsule (ant. limb)

Int. capsule (post. limb)

Caudate head

Ext. capsule

Putamen

Insula

Globus pallidus

Claustrum

Thalamus

INTERNAL CAPSULE

Panoky

– 189 –

89. BASAL GANGLIA: FIBER CONNECTIONS AND PATHWAYS

I. Principal afferent connections

A. CEREBRAL CORTEX contributes majority of afferent projections to caudate and putamen (neostriatum)
   1. Fibers from the rostral cortex project ipsilaterally to the head of the caudate
   2. Sensorimotor cortex (areas 3, 2, 1) sends fibers mainly to the tail of the caudate
B. THALAMUS is the most important source of subcortical afferents to the caudate nucleus and putamen
   1. Centromedian nucleus sends afferent fibers to the putamen
   2. Intralaminar nuclei project to the caudate nucleus
C. SUBSTANTIA NIGRA sends afferents to the caudate nucleus and globus pallidus
D. NEOSTRIATUM (CAUDATE AND PUTAMEN) sends afferents to the globus pallidus
E. SUBTHALAMIC NUCLEUS sends fibers to the medial portion of the globus pallidus

II. Principal efferent connections

A. CAUDATE AND PUTAMEN (NEOSTRIATUM) project efferents to both medial and lateral portions of the globus pallidus and the substantia nigra
B. GLOBUS PALLIDUS (PALEOSTRIATUM) provides the major outflow system for the corpus striatum
   1. Lateral portion gives rise to majority of pallido-subthalamic fibers
   2. Medial portion gives rise to three fiber bundles
      a. Lenticular fasciculus originates from the dorsal aspect of the medial portion of the globus pallidus and runs rostral to the subthalamic nucleus and ventral to the zona incerta as Forel's field H_2
      b. Ansa lenticularis fibers merge with the fibers of the lenticular fasciculus in the prerubral field (field H of Forel). These two bundles then project dorsolaterally as a component of the thalamic fasciculus
      c. Thalamic fasciculus (Forel's field H_1) is a complex bundle which contains pallidothalamic, rubrothalamic, and dentatothalamic fibers
         i. It projects fibers to VA, VL, centromedian, and intralaminar thalamic nuclei
         ii. The region where the thalamic fasciculus enters the thalamus is perhaps the most strategic single location in the brain
            (a) Fibers from the corpus striatum and cerebellum converge at this point to enter the specific thalamic nuclei
            (b) Fibers exit the thalamus at this point to project to motor and premotor cerebral cortex
C. SUBSTANTIA NIGRA sends efferent fiber projections to the medial portion of VA and VL nuclei of the thalamus and the neostriatum

PRINCIPAL AFFERENT CONNECTIONS— BASAL GANGLIA

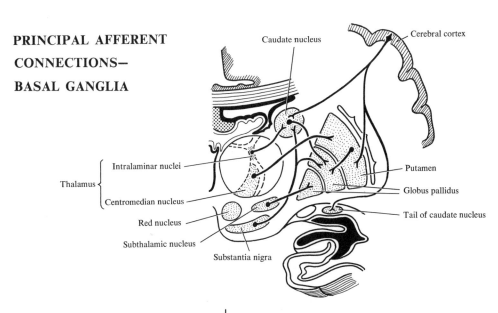

Caudate nucleus

Cerebral cortex

Thalamus
- Intralaminar nuclei
- Centromedian nucleus

Red nucleus

Subthalamic nucleus

Substantia nigra

Putamen

Globus pallidus

Tail of caudate nucleus

PRINCIPAL EFFERENT CONNECTIONS

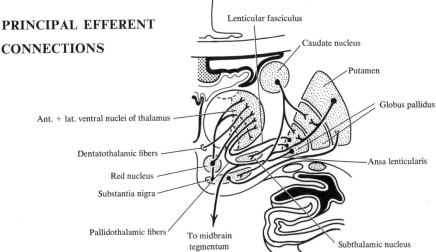

Lenticular fasciculus

Caudate nucleus

Putamen

Globus pallidus

Ant. + lat. ventral nuclei of thalamus

Dentatothalamic fibers

Red nucleus

Substantia nigra

Ansa lenticularis

Pallidothalamic fibers

To midbrain tegmentum

Subthalamic nucleus

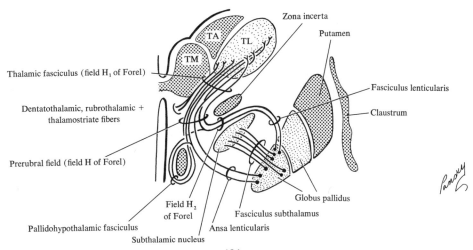

Zona incerta

Putamen

Thalamic fasciculus (field H₁ of Forel)

TA

TL

TM

Dentatothalamic, rubrothalamic + thalamostriate fibers

Fasciculus lenticularis

Claustrum

Prerubral field (field H of Forel)

Field H₂ of Forel

Fasciculus subthalamus

Globus pallidus

Pallidohypothalamic fasciculus

Ansa lenticularis

Subthalamic nucleus

90. BASAL GANGLIA: FUNCTIONAL AND CLINICAL CONSIDERATIONS—PART I

I. **Introduction:** the basal ganglia control system includes putamen, globus pallidus, caudate nucleus, ventral anterior nucleus of the thalamus, subthalamic nucleus, substantia nigra, and their connections in the cerebral hemispheres and diencephalon

A. THIS CONTROL SYSTEM OR CIRCUIT exerts a profound influence on motor activities, but the precise mechanism of action for specific areas has not been demonstrated

B. THE BASAL GANGLIA CONTROL CIRCUIT is a modulating system that functions to either facilitate or inhibit motor activities
   1. Excessive discharge (electrical stimulation) of the caudate nucleus slows or suppresses motor function
   2. Electrical stimulation of the globus pallidus causes contraversive turning and postural changes
   3. Lesions in the subthalamic nucleus usually suppress motor activity
   4. Lesions in the striatum (caudate nucleus and putamen) cause deficits in the more complex motor functions

II. **Pathologic conditions:** when the delicate balance in this modulating system (facilitation vs. suppression) is disrupted, the following pathologic symptoms or signs of abnormalities in motor function are observed

A. LOSS OR REDUCTION OF MOVEMENT (HYPOKINESIA): the term "hypokinesia" refers to the reduction in initiation, implementation, and facility of execution of movement and is characterized by
   1. Tone: there is usually an increase in tone (hypertonia) which occurs as a direct effect on alpha motor neurons and is known as rigidity. Depending upon the degree of hypertonia, some or all of the following clinical signs may be observed
      a. Movements of limbs or the entire body are initiated slowly and stopped with difficulty
      b. Conscious and emotional movements may be suppressed
      c. Face is masklike (fixed facial expression)
      d. Abnormal postures may be assumed
      e. Patient may appear paralyzed due to exaggerated tone
      f. Arm swing during walking is absent
      g. Tendon reflexes usually appear normal
   2. Rigidity: when hypertonicity is widespread over most of the body, the condition is called rigidity and manifests itself in one of the following
      a. Cogwheel rigidity: increased resistance to passive movement, usually with "cogwheel jerks" but without the clasp-knife phenomenon of spasticity
      b. Plastic rigidity: increased resistance to passive movement is constant, continuous, and smooth (not jerky)

B. INVOLUNTARY MOVEMENTS (HYPERKINESIA): greatly increased activity in the dopaminergic pathway may result in the following types of involuntary movements
   1. Tremor may occur when the body is at rest (postural, static, or alternating) or during voluntary muscular activity (intention or action)
      a. Alternating tremor is the result of the alternating contraction of opposing muscle groups such as adductors-abductors and flexors-extensors. The "pill-rolling" phenomenon is an example of tremor in the hand. All alternating tremors are usually suppressed or absent during voluntary activity and may be completely absent during sleep
      b. Intention tremor occurs during voluntary muscular activity and becomes more exaggerated near the end of the voluntary act

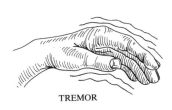

TREMOR

RIGIDITY

TREMOR WITH RIGIDITY—
EXAGGERATED WITH
VOLUNTARY ACT

MASKLIKE "FIXED"
FACIAL EXPRESSION

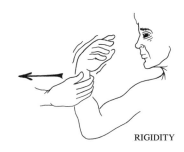

RIGIDITY

UNCONTROLLED
CONTINUED
MOVEMENT

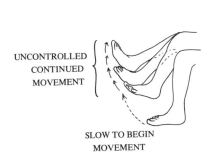

SLOW TO BEGIN
MOVEMENT

RIGIDITY

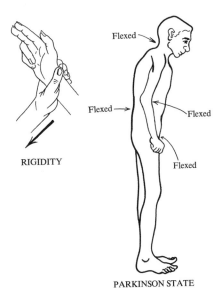

Flexed

Flexed

Flexed

Flexed

PARKINSON STATE

91. BASAL GANGLIA: FUNCTIONAL AND CLINICAL CONSIDERATIONS—PART II

I. Pathologic conditions (cont.)
A. INVOLUNTARY MOVEMENTS (CONT.)
   1. Athetosis is the disorder in which there are slow, twisting, involuntary movements of the extremities, particularly of the fingers and hands. It may occur at rest or during voluntary muscle movements and is referred to as serpentine movement
   2. Dystonia is the disorder that involves abnormal posturing of the trunk or an extremity. Movements are slow, involuntary, and accompanied by bizarre, grotesque twisting of the shoulder, pelvic girdle, and trunk
   3. Chorea is characterized by sudden, jerky, involuntary movements, along with grimacing, twitching of facial muscles, and faulty vocalization. These movements occur at rest or during voluntary acts
   4. Ballism is characterized by involuntary movements of an entire limb which begin in the proximal portion and pass in wavelike movements distally. There may be gross, rapid, flinging movements when the subthalamic nucleus is involved. Hemiballism is the condition when both limbs on the same side are involved. Monoballism denotes only one limb is involved. This disorder is the only one involving the basal ganglia whose onset may occur suddenly

II. Specific hypokinetic and hyperkinetic disorders
A. PARALYSIS AGITANS (PARKINSON'S DISEASE), the most common type of basal ganglia disease, is characterized by the following symptoms, clinical signs, and traits
   1. Massive degeneration of the cerebral cortex, globus pallidus, and substantia nigra (usually accompanied by lesions in the substantia nigra)
   2. Alternating tremor (resting tremor), characteristically involving the hands in a "pill-rolling" movement; diminishes during voluntary movement and increases during emotional stress. Occurs at the rate of 3–5 movements per second
   3. Decreased striatal dopamine level which results in rapid firing of cells of the striatum; normally dopamine of the substantia nigra inhibits this rapid firing of cells of the striatum
   4. Loss of the normal modulating influence of the basal ganglia on motor activities of the pyramidal and extrapyramidal systems
   5. Cogwheel and plastic rigidity due to increased tone in all muscles; may be observed along with akinesia
   6. Flexed posture throughout the body may result
   7. Difficulty in moving (bradykinesia, or slow movements; hypokinesia, or few movements); decreased blinking; expressionless face; and absence of arm movements with walking
   8. Cause is unknown (idiopathic) but often follows encephalitis, carbon monoxide poisoning, manganese poisoning, or other drug toxicity
B. HUNTINGTON'S CHOREA is a progressive, dominantly inherited disorder of adults with onset usually between the ages of 35 and 40, with the following signs and symptoms
   1. Moderate to severe choreiform motor disability; chorea is random, with continuous occurrence of brief, abrupt, jerky movements of body parts
   2. Diffuse degeneration with widespread loss of neurons in the cerebral cortex and basal ganglia; atrophy of caudate nucleus and putamen allows dilatation of the lateral ventricles
   3. General and progressive mental deterioration (dementia)
   4. Decreased muscle tone
   5. Reflexes and sensations are usually normal
C. SYDENHAM'S CHOREA is a disorder in children or adolescents often associated with rheumatic fever with the following symptoms and signs
   1. Sudden, irregular, involuntary, and purposeless movements
   2. Petechial (small hemorrhagic spots) lesions may be present in corpus striatum

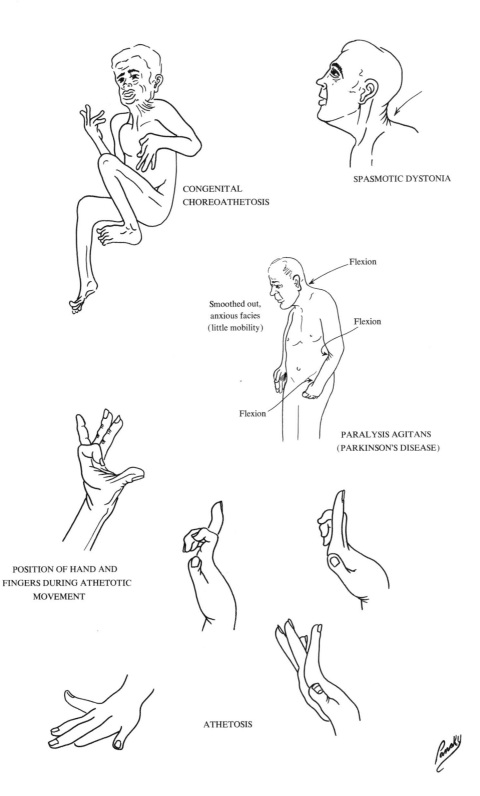

CONGENITAL
CHOREOATHETOSIS

SPASMOTIC DYSTONIA

Flexion

Smoothed out,
anxious facies
(little mobility)

Flexion

Flexion

PARALYSIS AGITANS
(PARKINSON'S DISEASE)

POSITION OF HAND AND
FINGERS DURING ATHETOTIC
MOVEMENT

ATHETOSIS

Unit Six

FUNCTIONAL
CEREBRAL CORTEX

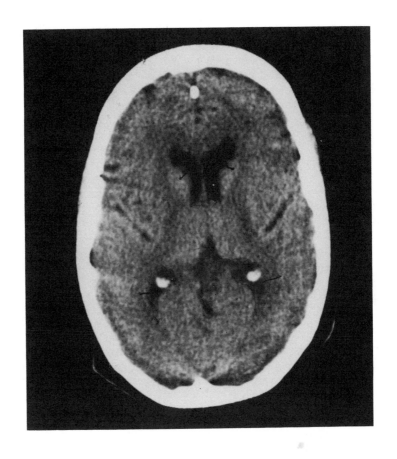

92. CEREBRAL CORTEX: GENERAL CONSIDERATIONS

I. **Definition:** "cerebral cortex" or "pallium" is the term applied to the thin layer of gray matter which covers the white matter of the convoluted cerebral hemispheres. This gray matter or cortex is comprised of cell bodies of neurons, neuroglia, nerve fibers, and blood vessels

A. GRAY APPEARANCE: characteristic of the cerebral cortex; imparted by its rich vascular supply and relative absence of myelinated nerve fibers

B. NEARLY 14 BILLION NEURONS have been estimated to be in the 2200 cm^2 of cortex within both cerebral hemispheres

C. THICKNESS OF THE CORTEX: averages 2.5 mm. It varies considerably in different areas: 4.5 mm in the motor area of the precentral gyrus; 1.5 mm in the calcarine fissure

II. **Summary of the development of the cerebral cortex** (also see p. 30)

A. DEVELOPS FROM THE TELENCEPHALON, the most rostral of the three original brain vesicles

B. THE EARLY TELENCEPHALIC VESICLE reveals three zones or concentric layers
   1. Matrix (germinal or ependyma) layer surrounds the lateral ventricles
   2. Mantle (intermediate) layer gives rise to the white matter of the cerebrum
   3. Marginal (outer) layer gives rise to the cortical layer of the cerebrum

C. COLUMNAR EPITHELIAL CELLS, originally present in the most primitive neural tube, extend through the above three layers

D. CELLS MIGRATE FROM THE MANTLE LAYER into the marginal layer to form the cerebral cortex by the end of the second month
   1. The cortex increases in thickness as cells migrate into and become differentiated in the marginal zone

E. CELLS IN THE CORTICAL ZONE become organized into layers (horizontally arranged) between 6 and 8 months
   1. Cells formed at the same time tend to migrate to the same layer and remain together
   2. Newly formed cells migrate through these previously formed layers to form more superficial layers—"inside-out" neuronal migration

III. **Summary of cerebral function:** based upon known information on cortical organization in terms of cells and fibers, the following general statements can be made about cortical functions

A. CORTICAL LAYERS I–III (MOLECULAR, EXTERNAL GRANULAR, AND EXTERNAL PYRAMIDAL) have numerous stellate cells, which indicate that these three layers are important for association and higher functions such as memory, interpretation of sensory input, and certain discriminative functions

B. INTERNAL GRANULAR LAYER (IV) is mainly a receptive layer (thalamocortical fibers end here)

C. LAYERS V AND VI (GANGLIONIC AND MULTIFORM) are primarily efferent layers that contain nerve cell bodies whose axons enter the corticospinal tract

IV. **Interrelation of cortical nerve cells:** histologic information concerning these neurons, their processes, and intracortical relationships has been obtained by using special silver impregnation methods such as the Golgi method. The latter is especially useful when studying neuronal relationships and their axonal and dendritic arborizations. In Nissl-stained sections, cell bodies can be studied; with the Weigert method, the course and distribution of myelin fibers can be investigated

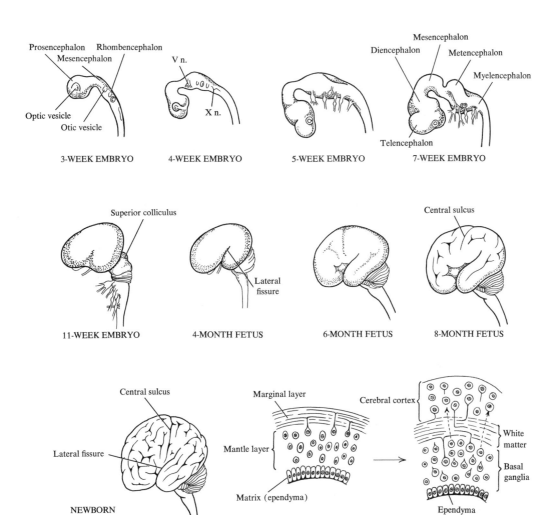

Prosencephalon   Rhombencephalon
    Mesencephalon

Optic vesicle

Otic vesicle

3-WEEK EMBRYO

V n.

X n.

4-WEEK EMBRYO

5-WEEK EMBRYO

Mesencephalon
Diencephalon   Metencephalon
                  Myelencephalon

Telencephalon

7-WEEK EMBRYO

Superior colliculus

11-WEEK EMBRYO

Lateral
fissure

4-MONTH FETUS

6-MONTH FETUS

Central sulcus

8-MONTH FETUS

Central sulcus

Lateral fissure

NEWBORN

Marginal layer

Mantle layer

Matrix (ependyma)

Cerebral cortex

White
matter

Basal
ganglia

Ependyma

HEMISPHERE WALL

CEREBRAL CORTEX

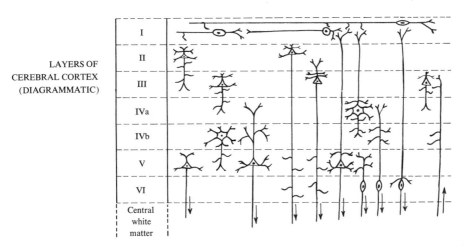

LAYERS OF
CEREBRAL CORTEX
(DIAGRAMMATIC)

I

II

III

IVa

IVb

V

VI

Central
white
matter

93. CEREBRAL CORTEX: CORTICAL CELLS

I. **Cortical cells:** five types of neurons are found in the cerebral cortex. These cells have been demonstrated by several neurohistologic stains (Golgi, Nissl, and Weigert)

A. PYRAMIDAL CELLS: most characteristic and numerous cells of the cortex

    1. Cell bodies are pyramidal or triangular in shape and possess prominent Nissl granules in their cytoplasm

    2. They vary in size from 10 to more than $100 \mu m$ in height

       a. Small pyramidal cells measure $10-20 \mu m$; medium-sized cells $20-30 \mu m$; large cells $30-70 \mu m$; and giant cells $70-100 \mu m$ (and larger)

    3. One or more types (sizes) of pyramidal cells may be found in all but the outermost cortical (molecular) layer

    4. Giant pyramidal cells (Betz cells) are characteristic of the precentral gyrus (motor area 4, layer V)

    5. All pyramidal cells are usually oriented with their apices directed toward the cortical surface

    6. Apical dendrites branch profusely within the cortical layers that lie superficial to the location of the cell body

    7. Other dendrites sprout from the base and body of these cells and usually extend horizontally within the same layer of the cortex

    8. The axon typically arises from the base of the cell body and enters the underlying white matter as one of the efferent cortical fibers

B. GRANULE CELLS (STELLATE) are small star- or polygon-shaped neurons with scant cytoplasm (second most numerous type)

    1. Cell bodies vary in size from 4 to 8 μm

    2. Dendrites are numerous, short, and highly branched in all directions

    3. The axon is short and branches near the parent cell body

    4. These cells are present in all cortical layers, but most numerous in layer IV

    5. They are classified as Golgi type II neurons and are correlative in function

C. FUSIFORM (POLYMORPHIC MULTIFORM) CELLS are usually spindle-shaped, but they may show wide variation in morphology (such as pyramidal or ovoid)

    1. Limited to the deepest layer of the gray cortex

    2. Their long axis is perpendicular to the cortical surface

    3. Dendrites sprout from the two poles of the cell, with more superficial dendrites ascending toward the surface and deeper ones branching in the same cortical layer. Their axons leave the cell body to enter the white matter

D. HORIZONTAL CELLS (OF CAJAL) are small neurons with fusiform cell bodies (usually spear- or spindle-shaped) with large nuclei and scant cytoplasm

    1. Found only in the outermost layer (molecular) of the cortex; their axis is parallel to the surface

    2. Dendrites spread out from each end of the cell body in a plane that is also parallel to the surface of the cortex

    3. Axon leaves the cell body and spreads out parallel to the surface in the outermost layer of the cortex

    4. These cells are correlative in function

E. CELLS OF MARTINOTTI (ASCENDING AXON CELLS) are small and usually polygonal in shape but may be round or triangular

    1. Cell bodies are found in all cortical layers, with possible exception of the outer (molecular) layer. These cells are associative in function

    2. Dendrites are short and spread out in all directions from the cell body

    3. Axon ascends from cell body toward the cortical surface and branches in the outermost layer of cortex

HUMAN CEREBRAL CORTEX

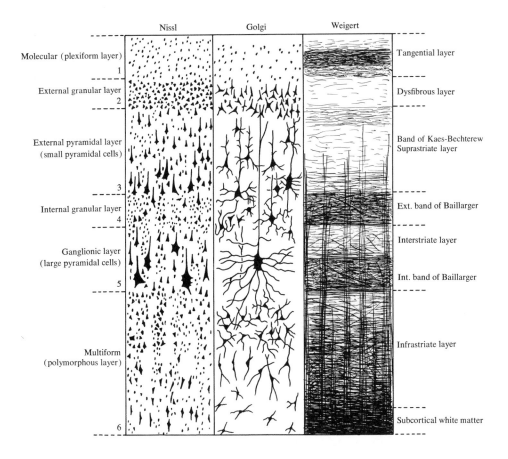

Nissl Golgi Weigert

Molecular (plexiform layer) — 1 — Tangential layer

External granular layer — 2 — Dysfibrous layer

External pyramidal layer
(small pyramidal cells) — Band of Kaes-Bechterew
Suprastriate layer

— 3 —

Internal granular layer — Ext. band of Baillarger
— 4 — Interstriate layer

Ganglionic layer
(large pyramidal cells) — Int. band of Baillarger
— 5 —

Infrastriate layer

Multiform
(polymorphous layer)

Subcortical white matter
— 6 —

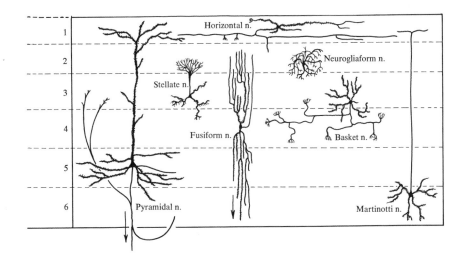

Horizontal n.

1

2 Neurogliaform n.

Stellate n.

3

Fusiform n. Basket n.

4

5

6 Pyramidal n. Martinotti n.

Parsky

94. CEREBRAL CORTEX: CORTICAL LAYERS

I. Cortical layers: the major portion (90%) of the cerebral cortex in man is characterized by 6 layers of cells in the adult. This 6-layer cellular arrangement is found in the neopallium (new cortex), which is also termed "neocortex," "isocortex," and "homogenetic cortex." Olfactory cortex or paleopallium (old cortex) and the hippocampal formation (archipallium) usually have 3 basic cellular layers and are termed the "allocortex" or "heterogenetic cortex." The 6 basic layers seen in the neocortex are, from the pial surface

A. MOLECULAR LAYER I (200 μm): often called the plexiform layer because of the fairly dense fiber plexus of axons of horizontal cells and dendrites of deeper lying pyramidal and granule cells
   1. Horizontal cells of Cajal are distinguishing cells (also see granule cells)
B. EXTERNAL GRANULAR LAYER II (200 μm) is made up of some small pyramidal cells and numerous small stellate cells
   1. Round cell bodies of these stellate cells, seen in Nissl preparations, have led to the designation of these cells as granule cells
   2. Apical dendrites enter the molecular layer
   3. Axons of granule cells descend to either terminate within the cortex or extend into the white matter
C. EXTERNAL PYRAMIDAL LAYER III (450 μm) is made up mainly of pyramidal cells, but it also contains some small granule and Martinotti cells
   1. Apical dendrites of pyramidal cells ascend into the molecular layer
   2. Axons from pyramidal cells descend into the multiform layer
D. INTERNAL GRANULAR LAYER IV (200 μm): homogenetic cortex has a preponderance of small stellate or granular cells. The latter are more numerous in the granular heterogenetic cortex of the postcentral sensory cortex. In contrast, granule cells are obscured by large pyramidal cells in the agranular heterogenetic cortex of the precentral motor area, so that the motor cortex is called agranular
   1. Axons of granule cells are short and usually remain within layer IV (some may descend into deeper layers)
   2. Layer IV has a large number of horizontal, myelinated nerve fibers (external band of Baillarger) from the thalamus which branch and end here
E. INTERNAL PYRAMIDAL (GANGLIONIC) LAYER V (250 μm): made up mainly of cell bodies of medium- and large-sized pyramidal neurons
   1. Stellate and Martinotti cells are present but not in abundance
   2. Apical dendrites of large pyramidal cells extend into the molecular layer; those of smaller pyramidal cells penetrate into layer IV and branch
   3. Axons of pyramidal cells descend into the white matter mainly as projection fibers (some association fibers)
   4. Internal band of Baillarger is formed by horizontal myelinated fibers in the deeper part of this layer
   5. Precentral motor cortex contains giant pyramidal cells of Betz in this layer to make its identification easy
F. FUSIFORM (MULTIFORM) LAYER VI (250 μm): predominant cell is the fusiform cell
   1. Spindle-shaped cells oriented with their long axis vertical to surface
   2. Some Martinotti, stellate, and small pyramidal cells may be present here
   3. Dendrites from large fusiform cells ascend into the molecular layer; those from small cells pass into layer IV
   4. Fusiform cell axons descend as projection and association fibers
   5. All fibers entering or leaving the cortex must pervade this layer
   6. Motor cortex may contain a few giant pyramidal cells of Betz in this layer, making it difficult to distinguish from the overlying ganglionic layer

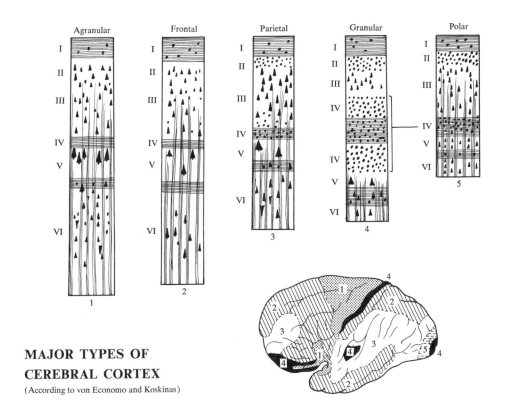

**MAJOR TYPES OF
CEREBRAL CORTEX**

(According to von Economo and Koskinas)

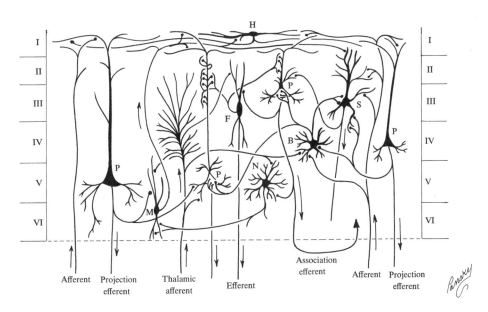

SOME INTRACORTICAL CIRCUITS OF NEOCORTEX

95. CEREBRAL CORTICAL AREAS: PREMOTOR AREAS

I. Introduction: the distribution pattern of cerebral cortical cells varies in different regions of the brain. Based on morphologic characteristics (some obvious, others very subtle), several cytoarchitectural maps have been drawn, dividing the cerebrum into areas. Some areas have known specialized functions, while for others no clear function has been demonstrated. These stated functions have been arrived at primarily by electrically stimulating the area and observing the response elicited. Histologic surveys based on differences in the arrangement and types of cells, and in the pattern of myelinated fibers, have resulted in several fundamental cortical maps (Campell, 1905, recognized some 20 cortical areas; Brodmann, 1909, extended the number of fields to 52; Economo, 1929, increased the number to 109; and Vogts, 1919, subdivided the human cerebral cortex into more than 200 fields). Brodmann's cortical maps with 47–52 areas are still used for descriptive purposes

II. Motor functions of specific cortical areas

A. PRIMARY MOTOR AREA (BRODMANN'S AREA 4)
   1. Located in precentral gyrus, along the anterior wall of the central sulcus, and in the most anterior part of the paracentral lobule
   2. Cortex is unusually thick (4.5 mm)
   3. This area is classified as heterogenetic, agranular in structure (layer VI is obscured by an increase in pyramidal cells)
   4. Ganglionic layer (V) contains giant pyramidal cells of Betz
   5. Corticospinal (pyramidal) tract arises in part from area 4
      a. This tract carries impulses for highly skilled voluntary movements to lower motor neurons
      b. Largest corticospinal fibers may be axons of Betz cells found in layer V
      c. Hemispherectomy or complete decortication results in degeneration of corticospinal fibers
   6. Area 4 controls fine, highly skilled, voluntary movements of the body, especially the facial muscles and flexor musculature of the extremities
   7. Though the point of location of cortical centers for specific movements of body parts varies in individuals, the sequence of these centers seems to be constant. In Penfield's homunculus, the center for the pharynx (swallowing) is in the most inferior portion of the precentral gyrus followed upward (superiorly) by the tongue, jaw, lips, face, fingers, hand, etc.

B. PREMOTOR AREA (AREA 6)
   1. Located in front of area 4 and continues on the medial surface (superior frontal gyrus) to the cingulate gyrus
   2. Histologically, area 6 resembles area 4
      a. Giant pyramidal cells of Betz are not present, however, in area 6
      b. Internal granular layer (layer IV) is obscured by pyramidal cells (agranular heterogenetic cortex)
   3. This area is concerned with the development of motor skills
   4. Ablation of this specific area alone does not result in paresis, hypertonia, or grasp reflexes

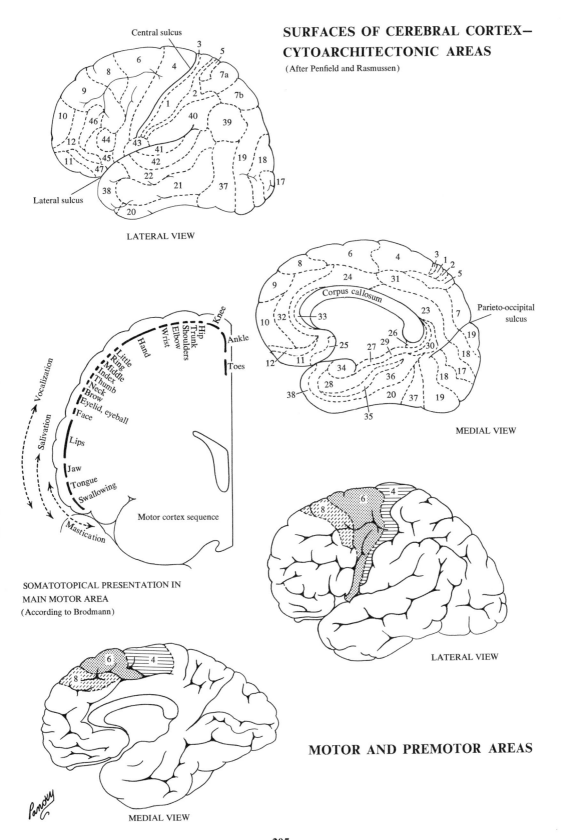

Central sulcus

7a

7b

40

39

19 18

17

41

42

37

22

21

38

20

Lateral sulcus

LATERAL VIEW

Knee

Hip
Trunk
Shoulders
Elbow
Wrist
Hand

Ankle

Little
Ring
Middle
Index
Thumb
Neck
Brow
Eyelid, eyeball
Face

Toes

Vocalization

Salivation

Lips

Jaw

Tongue

Swallowing

Mastication

Motor cortex sequence

**SOMATOTOPICAL PRESENTATION IN
MAIN MOTOR AREA**
(According to Brodmann)

Corpus callosum

Parieto-occipital
sulcus

33

23

7

24

31

32

25

26

29

30

19

27

18

11

34

36

18

17

28

38

20

37

19

35

MEDIAL VIEW

8

6

4

LATERAL VIEW

6

4

8

MOTOR AND PREMOTOR AREAS

Panoky

MEDIAL VIEW

96. CEREBRAL CORTICAL AREAS: SUPPLEMENTARY MOTOR, FRONTAL, AND OCCIPITAL EYE FIELDS AND PREFRONTAL AREAS

I. Motor functions (cont.)

A. SUPPLEMENTARY MOTOR AREAS: no assigned number
1. Located in the superior frontal gyrus on its medial surface just rostral to primary motor area 4
2. Somatotopic pattern (sequence for representation of body parts) is reversed from that of motor area 4
3. This area is concerned with the following movements as determined by electrically stimulating the area
   a. Assumption of posture
   b. Complex patterned muscle movements
   c. Rapid, infrequent, incoordinate muscle activities
4. This area deals with bilateral, synergistic muscle activities

B. FRONTAL EYE FIELD (AREA 8)
1. Located primarily in the caudal aspect of the middle frontal gyrus
2. Histology within area 8 varies (more than one cytoarchitectural plan)
3. This area is concerned with voluntary eye movements; these movements are independent of visual stimuli
4. Stimulation of the area results in conjugate deviation of the eyes to the opposite side. These same movements can be made by asking the patient to look right/left
5. The exact pathway for eliciting these responses is unknown, but the superior colliculus is probably involved

C. OCCIPITAL EYE FIELD: no assigned number
1. Located in the occipital lobe, but not well localized
2. Stimulation of the occipital cortex results in conjugate deviation of the eyes to the opposite side
3. This area is concerned with involuntary eye movements; these movements are induced by visual stimuli
4. These eye fields in the two cerebral hemispheres are interconnected by commissural fibers through the splenium, but the pathway is unknown

D. PREFRONTAL CORTEX (AREAS 9–12)
1. Located rostral to motor and premotor cortical areas and comprises the remaining part of the frontal lobe
2. Receives fibers from the dorsomedial nucleus of the thalamus
3. Receives long association fibers from all lobes of the cerebrum
4. This region is concerned with abstract thinking, mature judgment, foresight, tactfulness, and self-control
5. Patients with large lesions (tumors or cortical atrophy from syphilis) of the frontal lobe often reveal the following traits
   a. Poor personal habits
   b. Vulgar speech
   c. General euphoria (clownish behavior)
   d. Poor self-control

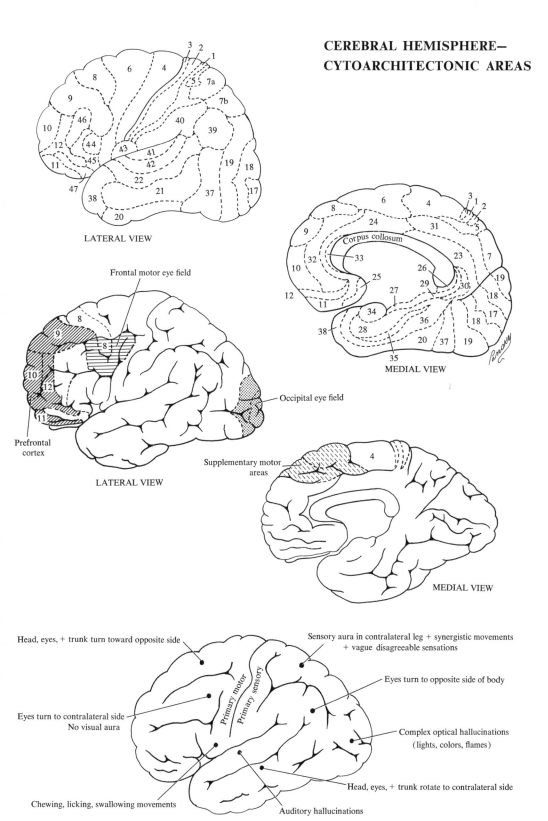

CEREBRAL HEMISPHERE—
CYTOARCHITECTONIC AREAS

LATERAL VIEW

Frontal motor eye field

Occipital eye field

Prefrontal cortex

LATERAL VIEW

Corpus collosum

MEDIAL VIEW

Supplementary motor areas

MEDIAL VIEW

Head, eyes, + trunk turn toward opposite side

Sensory aura in contralateral leg + synergistic movements + vague disagreeable sensations

Eyes turn to opposite side of body

Primary motor

Primary sensory

Eyes turn to contralateral side
No visual aura

Complex optical hallucinations
(lights, colors, flames)

Head, eyes, + trunk rotate to contralateral side

Chewing, licking, swallowing movements

Auditory hallucinations

FUNCTIONAL AREAS OF CORTEX BY ELECTRIC STIMULATIONS

97. CEREBRAL CORTEX: SENSORY FUNCTIONS OF CERTAIN CORTICAL AREAS

I. **Primary sensory or first somesthetic cortex (areas 3, 1, 2):** called the first sensory area because it has the highest density of points that can produce localized sensations on electrical stimulation

A. THE PRIMARY SENSORY CORTEX is necessary to recognize the source, quality, and quantity (severity) of pain and temperature; appreciate simple touch, light pressure, and vibrations from bony prominences; and discern fine discriminating touch and position and movement of the body parts (proprioception and kinesthetic sense)
  1. Located in the postcentral gyrus on lateral surface of hemisphere, including its medial extension into the paracentral lobule (posterior portion)
  2. Histologically the postcentral gyrus consists of 3 narrow strips of cortex (areas 3, 1, 2) that differ in structure: area 3 is granular heterogenetic cortex, while areas 1 and 2 are homogenetic and thicker
  3. The postcentral gyrus receives fibers from VPM and VPL nuclei of the thalamus
     a. These nuclei serve as relays for fibers from the medial lemniscus, spinothalamic tracts, and trigeminothalamic tracts
     b. Sensory modalities are detected and recognized at the level of the thalamus
  4. The various regions of the contralateral half of the body are represented as inverted in the postcentral gyrus, corresponding to that of the motor cortex (area 4)
B. SECONDARY OR SECOND SOMESTHETIC CORTEX
  1. Located around the most ventral aspect of the pre- and postcentral gyri; continues posteriorly along the superior lip of the lateral fissure for a limited distance; and may extend onto insula
  2. Receives impulses from several sensory modalities, but pain appears to predominate. Input is from the intralaminar nuclei and from the posterior complex of nuclei of the thalamus. The ascending afferent fibers to these nuclei arrive via spinothalamic and trigeminothalamic tracts and the reticular formation. Thus, it is involved in less discriminative aspects of sensation
  3. The sequence for the location of body parts is the reverse of that given for the primary sensory cortex (no cortical areas identified for face, mouth, and swallowing mechanism)
  4. Representation of most body parts (extremities) is mainly contralateral, although parts of the body are bilaterally represented
  5. No clinical disorder has been identified for selective destruction of the second somesthetic area

II. **The somesthetic association cortex** is predominantly in the superior parietal lobule on the lateral surface and in the precuneus on the medial surface, and coincides with areas 5 and 7 (of Brodmann)

A. THESE CORTICAL AREAS receive fibers from the first somesthetic area and have connections with the dorsal tier nuclei in the lateral mass of the thalamus
  1. Data relating to general senses are integrated (e.g., one can assess objects held in the hand without actually seeing them)
  2. A lesion in the association cortex (assuming the somesthetic area is intact) results in awareness of general senses, but the significance of the information received is difficult to interpret—a condition called agnosia. Several forms exist, depending on the sense that is affected
     a. Tactile agnosia and astereognosis (loss of awareness of the spatial relations of parts of the body) are related and are the result of a lesion that destroys somesthetic association cortex. The symptoms combine when a person cannot identify a common object held in the hand while the eyes are closed—cannot correlate surface texture, shape, size, and object weight with previous experience

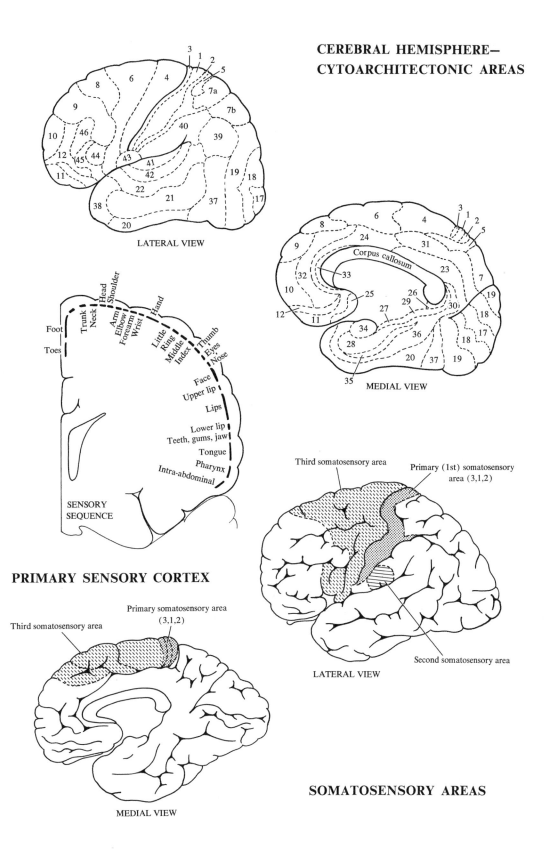

CEREBRAL HEMISPHERE–
CYTOARCHITECTONIC AREAS

LATERAL VIEW

MEDIAL VIEW

SENSORY SEQUENCE

Foot
Toes
Trunk
Neck
Head
Shoulder
Arm
Elbow
Forearm
Wrist
Hand
Little
Ring
Middle
Index
Thumb
Eyes
Nose
Face
Upper lip
Lips
Lower lip
Teeth, gums, jaw
Tongue
Pharynx
Intra-abdominal

PRIMARY SENSORY CORTEX

Corpus callosum

Third somatosensory area

Primary (1st) somatosensory area (3,1,2)

Second somatosensory area

LATERAL VIEW

Third somatosensory area

Primary somatosensory area (3,1,2)

MEDIAL VIEW

SOMATOSENSORY AREAS

98. CEREBRAL CORTEX: VISUAL AND AUDITORY CORTEX

I. Primary visual cortex (area 17)
A. AREA 17 surrounds the calcarine fissure and is known as the striate area
  1. This cortical area is very thin (1.5 mm)
  2. Classified as granular, heterogenetic cortex
    a. Layer IV is thick and packed with stellate (granule) cells
  3. Outer band of Baillarger (Gennari) appears in layer IV of area 17
  4. This area gives rise to projection fibers to
    a. The pons (corticopontine)
    b. The tectum of the midbrain (corticotegmental)
    c. The oculomotor complex and abducens nuclei
      i. Conjugate deviation of the eyes (involuntary) is initiated through these centers
      ii. Fixation on objects (both stationary and moving) is possible through these fiber connections
      iii. Accommodation reflexes necessary for the adjustment to near/far vision may be initiated through cortical connections with these nuclei
  5. Primary visual cortex receives the geniculocalcarine tract from the lateral geniculate body

II. Secondary visual cortex (areas 18 and 19)
A. THESE VISUAL PERCEPTION AREAS (18 AND 19) deal with the interpretation of the printed word and the spatial localization of an object
  1. Located in cortical regions that surround the primary visual area on its medial and lateral surfaces and account for visual perception
  2. Area 18 receives fibers from area 17 and sends fibers to area 19; areas 18 and 19 both send fibers to area 17 (reverberating circuits)

III. Primary auditory cortex (areas 41 and 42)
A. AREAS 41 AND 42 are concerned with the perception of simple auditory impulses
  1. Located in the floor of the lateral fissure on the transverse temporal gyri (Heschl's convolutions or gyri)
  2. Auditory input into the medial geniculate nucleus is mainly from the contralateral organ of Corti (some input from the ear of the same side), and it is the medial geniculate nucleus that is the main source of fibers (auditory radiations) into these auditory cortical areas
  3. A lesion in the primary auditory cortex on one side results in diminished auditory acuity in both ears, with the greater loss in the opposite ear
  4. A unilateral lesion in the auditory areas makes it more difficult to detect the direction from which sound comes

IV. Auditory association cortex (area 22)
A. AREA 22 is essential for the interpretation and appreciation of intricate sounds in the dominant hemisphere
  1. It is located on the lateral surface of the superior temporal gyrus (Wernicke's area)
  2. It has reciprocal fiber connections with areas 41 and 42
  3. Fibers from areas 41 and 42 into this area are responsible for the turning of the head and eyes to follow sounds
  4. Area 22 receives and synthesizes language data from sensory areas into a comprehensive sensory language pattern

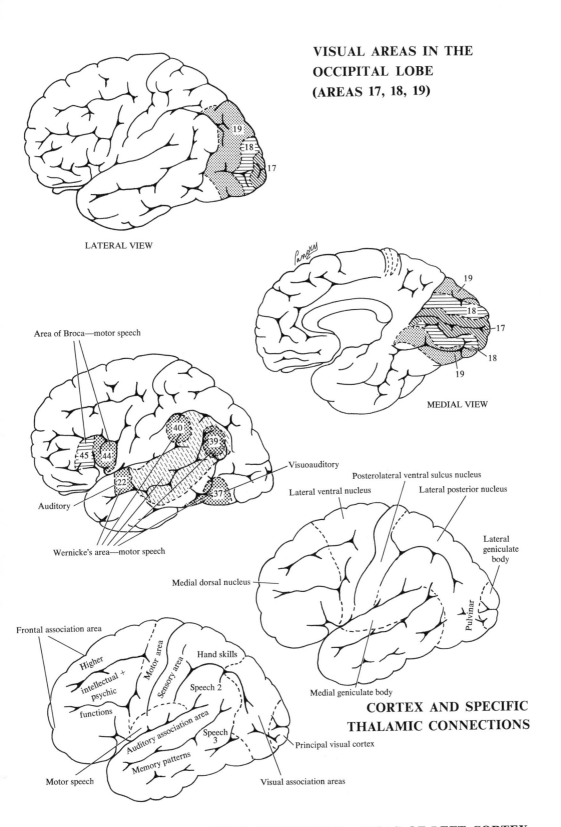

VISUAL AREAS IN THE OCCIPITAL LOBE (AREAS 17, 18, 19)

19
18
17

LATERAL VIEW

Pansky

19
18
17
18
19

MEDIAL VIEW

Area of Broca—motor speech

40
39
45 44
22
37

Visuoauditory

Posterolateral ventral sulcus nucleus
Lateral ventral nucleus
Lateral posterior nucleus

Auditory

Lateral geniculate body

Wernicke's area—motor speech

Medial dorsal nucleus

Pulvinar

Frontal association area

Higher intellectual + psychic functions

Motor area
Sensory area
Hand skills
Speech 2

Auditory association area
Memory patterns
Speech 3

Medial geniculate body

CORTEX AND SPECIFIC THALAMIC CONNECTIONS

Motor speech

Principal visual cortex

Visual association areas

GENERALIZED FUNCTIONAL AREAS OF LEFT CORTEX

99. THE APHASIAS AND OTHER CORTICAL SYNDROMES

I. **Aphasia** is the inability to use language and involves cortical language areas or their connections. Types are described according to the location of lesions

   A. EXPRESSIVE (BROCA'S, NONFLUENT, MOTOR) APHASIA: associated with a lesion in Broca's frontal lobe area, which contains motor programs for language generation

      1. Characterized by hesitant and distorted speech but relatively good comprehension. The patient tends to leave out all but meaningful words and speaks or writes in a telegraphic way

      2. The lesion deprives the motor cortex of input to generate language even though the muscles involved are normal

   B. RECEPTIVE (WERNICKE'S, FLUENT, SENSORY) APHASIA: auditory and visual comprehension of language, naming of objects, and repetition of a sentence spoken by an examiner are all defective. Caused by lesion in sensory language area (Wernicke's). Broca's area is intact so words are produced without regard to meaning

      1. The patient produces written and spoken words, but the words or sequences in which they are used are defective. One may see substitution of one word for another (paraphasia) and insertion of new and meaningless words (neologisms)

   C. JARGON APHASIA: stringing words and phrases together with little meaning; involves Wernicke's area, superior longitudinal and/or arcuate fasciculi

   D. CONDUCTION APHASIA: result of interruption of arcuate fasciculus connecting Broca's and Wernicke's areas. There is good comprehension and spontaneous speech but poor repetition. Broca's area is unchecked, leading to fluent aphasia, but Wernicke's area is undamaged and comprehension is intact

   E. ANOMIC APHASIA: result of destruction of angular gyrus in dominant hemisphere. A type of fluent aphasia in which major deficit is inability to recall names of persons and things

   F. ALEXIA (WORD BLINDNESS): loss of ability to read. Occurs with or without other aspects of aphasia. Pure type may be due to lesion of occipital lobe of dominant hemisphere and splenium of corpus callosum

   G. DYSLEXIA: incomplete alexia characterized by a level of reading ability far below that expected on the basis of overall intelligence or ability in skills

   H. GLOBAL APHASIA: complete loss of verbal communication except for stereotyped sayings. Results from destruction of cortex on either side of lateral sulcus

II. **Parietal lobe syndromes:** The parietal association cortex (areas 5 and 7) responds to complicated stimuli. Large lesions of the right lobe can be complex. Patients have trouble with spatial orientation to everything on the left and may ignore objects to the left, even the left half of the body. Lesion is rarely confined and often involves hemiplegia, hemisensory loss and other defects such as

   A. AGNOSIAS (LACK OF KNOWLEDGE): inability to recognize objects when using a given sense (e.g., visual agnosia is inability to recognize common objects by sight even with visual fields intact)

   B. APRAXIAS (LACK OF ACTION): inability to perform an action even though muscles needed are sound (e.g., unable to touch nose with finger on request, but able to do so spontaneously)

III. **Prefrontal syndromes:** are complex; related to areas 4, 6, and 8; they are difficult to qualify or relate anatomically, since these areas are related to activity of other cerebral lobes via long association bundles. There are emotional changes, intellectual deficits, and often loss of affect

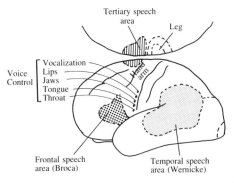

Tertiary speech area

Leg

Vocalization
Lips
Jaws
Tongue
Throat

Voice Control

Frontal speech area (Broca)

Temporal speech area (Wernicke)

(after Penfield and Roberts, 1959)

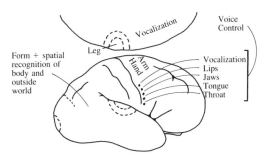

Voice Control

Vocalization

Leg

Arm

Hand

Form + spatial recognition of body and outside world

Vocalization
Lips
Jaws
Tongue
Throat

CONTROL OF VOCAL MUSCLES—LOCALIZED BILATERALLY
IN PRECENTRAL GYRI

CEREBRAL CORTEX: LOCALIZATION OF FUNCTION
& ASSOCIATION PATHWAYS

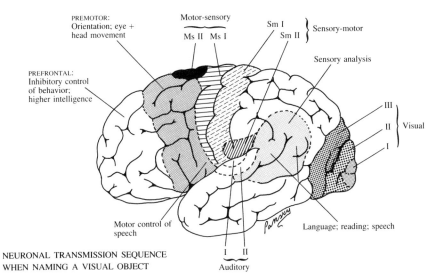

PREMOTOR: Orientation; eye + head movement

Motor-sensory

Ms II Ms I

Sm I
Sm II

Sensory-motor

PREFRONTAL: Inhibitory control of behavior; higher intelligence

Sensory analysis

III

II

Visual

I

Motor control of speech

I II

Auditory

Language; reading; speech

NEURONAL TRANSMISSION SEQUENCE
WHEN NAMING A VISUAL OBJECT

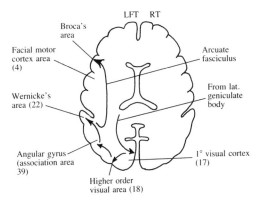

LFT RT

Broca's area

Facial motor cortex area (4)

Wernicke's area (22)

Angular gyrus (association area 39)

Arcuate fasciculus

From lat. geniculate body

1° visual cortex (17)

Higher order visual area (18)

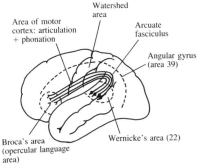

Watershed area

Area of motor cortex: articulation + phonation

Arcuate fasciculus

Angular gyrus (area 39)

Broca's area (opercular language area)

Wernicke's area (22)

PRIMARY BRAIN LANGUAGE AREAS

100. LANGUAGE

I. Introduction
A. LANGUAGE is the only fundamental process in which man differs from the animals
B. SINCE NO EXPERIMENTAL ANIMAL known has highly developed language skills, the study of language is difficult
C. THERE ARE NO SIMPLE ANATOMIC DIFFERENCES between the brains of man and other animals to account for language, yet subtle differences between the two hemispheres of man's brain do exist and are related to the fact that, in adults, language functions occur predominantly in the left dominant hemisphere

II. General
A. LANGUAGE IS SEPARABLE INTO TWO COMPONENTS: conceptualization and expression
   1. In forms of aphasia (language deficits not due to mental defects) related to conceptualization, the patient cannot understand spoken or written language, even though hearing and vision are not impaired
   2. In expressive aphasia the patient is unable to carry out the coordinated oral and respiratory movements needed for language, even though he can move his tongue and lips, understand the spoken language, and knows what he wants to say. This is often associated with inability to write
B. Different cortical areas have been related to specific aspects of language
   1. Areas in the frontal lobe near the motor cortex are involved in the articulation of speech (Broca's area)
   2. Areas in the parietal and temporal lobes are involved in sensory functions and language interpretations (Wernicke's area)
   3. The cortical specializations are not present at birth but develop gradually in childhood during language acquisition
   4. Localization of language functions in the left hemisphere (in 96% of the population) is not true during early childhood, when both hemispheres function
      a. Damage to left hemisphere of a child results in no impediment of future language development; develops in intact right hemisphere
      b. Even if the left hemisphere is traumatized after language onset, it develops in the right hemisphere after transient period of loss
      c. Prognosis becomes worse as the age at which the damage occurs increases, so that after early teens, language interference is permanent
      d. The change in the possibility of establishing language in the teens is probably related to the fact that the brain reaches its final structural, biochemical, and functional maturity at that time
         i. With maturation of the brain, language functions are apparently irrevocably assigned and language utilization by the right hemisphere is no longer possible
C. Because language resides in the left hemisphere, only that hemisphere can communicate orally or in writing about its conscious experience
   1. If vision is limited so that only the part of the retina whose fibers pass to the right side is excited, the left hemisphere, which controls speech, is unaware of the visual experience and the patient cannot describe it orally or in writing
   2. If the portion of return projecting to the left hemisphere is activated, the patient can easily describe the experience
   3. The right silent hemisphere does understand language and has linguistic functions, but not those involved in graphic or vocal expression
D. The capacity for language is characteristic of man, and language develops provided there is some stimulation or opportunity
   1. The capacity is partially inherited, although environmental circumstances are important and can limit language activity
   2. Behavioral characteristics and anatomic correlates of language can be identified, but mechanisms of neuronal operations resulting in language are unknown

AREAS INVOLVED IN LANGUAGE

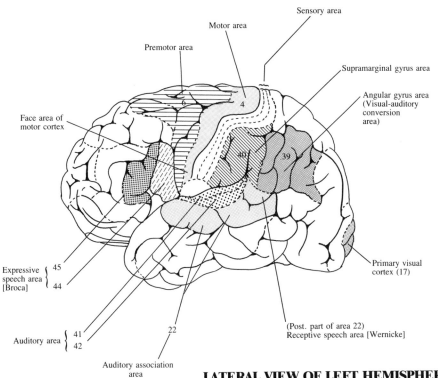

Sensory area

Motor area

Premotor area

Supramarginal gyrus area

Angular gyrus area
(Visual-auditory
conversion
area)

6

4

Face area of
motor cortex

40

39

Primary visual
cortex (17)

Expressive
speech area
[Broca] { 45
          44

Auditory area { 41
                42

22

(Post. part of area 22)
Receptive speech area [Wernicke]

Auditory association
area

LATERAL VIEW OF LEFT HEMISPHERE OF
CEREBRAL CORTEX

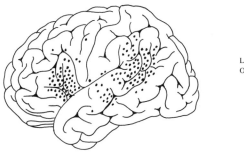

LEFT HEMISPHERE
OF CEREBRAL CORTEX

ELECTRICAL STIMULATION POINTS THAT ARREST SPEECH
Anterior cluster overlies Broca's area
Posterior cluster overlies Wernicke's area

101. LEARNING AND MEMORY

I. **Knowledge** is acquired by learning. Memory is the retention of learned knowledge. Nonassociative learning is obtained by repeated exposure to a single stimulus type; associative learning comes from comparison of one stimulus to another (classical conditioning) or comparison of a stimulus to the organism's behavior (operant conditioning). Knowledge can also be acquired by repeated practice (reflexive learning) or by evaluation and comparison with previous experience (declarative learning)

  A. Only a small part of the information entering the consciousness is stored. Concepts and ideas can be stored as words (verbal memory) or as nonverbal codes

II. **Four levels of permanence** of information in memory storage sites are recognized

  A. SENSORY MEMORY: sensory information (e.g., visual images received by retinal receptors) stored automatically for a few hundred milliseconds is evaluated and immediately processed (verbally or nonverbally coded) or extinguished (forgotten) by replacement with other information

  B. PRIMARY MEMORY (SHORT-TERM MEMORY): verbally coded information is temporarily stored in the order received. Primary memory is forgotten when the stored information is replaced with new information. Repetition ensures the transfer of the contents of primary memory to secondary memory

  C. SECONDARY MEMORY: information in long-term memory is stored in order of significance as engrams. Retrieval from secondary memory is slower than from primary memory. Previous experience or information acquired subsequently (proactive or retroactive information) may interfere with the learning process, so that partial or complete forgetting may occur. Proactive inhibition is important in causing memory loss.

  D. TERTIARY MEMORY: important information is stored indefinitely, is rarely forgotten and can be rapidly retrieved (e.g., an individual's name)

III. **Mechanisms of learning:** the engram may exist as information in a reverberating circuit (dynamic engram) or as modifications of synaptic connections (structural engram). There may be changes in the proportions of nucleotides in RNA during learning, but attempts to extract and transfer these changes (as memory traces) between one organism and another have given inconsistent results

IV. **Autonomic nervous system learning:** the coupling of an ineffective (conditioned) stimulus with an effective (unconditioned) stimulus may cause changes in the behavior of organs which are innervated by autonomic neurons. Utilizing a form of conditioning (operant conditioning) in which the change in behavior is reinforced by a reward, it is possible to change heart rate, glandular secretion, smooth muscle tone, etc.

  A. BIOFEEDBACK OPERANT CONDITIONING is useful clinically in the treatment of hypertension and headache. Changes in blood pressure can be effected with audible and/or visual signals, especially when coupled with relaxation

V. **Memory disturbances**

  A. ANTEGRADE AMNESIA (AMNESTIC SYNDROME; KORSAKOFF'S SYNDROME): inability to learn new long-term information. Failure to transfer from primary to secondary memory

  B. RETROGRADE AMNESIA: individuals are unable to retrieve some information in primary and secondary memory due to disruption of access to secondary memory. Memory loss is proportional to degree of damage. Forgotten material may gradually be recalled

  C. HYSTERICAL AMNESIA: individuals cannot recollect who they are and have no memory of prior experience. New information can easily be remembered

  D. DEMENTIAS: loss of recent memory, impaired learning, intellectual decline, and other aspects of mental decline characterize the dementias. In Alzheimer's disease, there are tangled neurofibrillary deposits in the cortex and other areas

HUMAN MEMORY PROCESSING

(after Waugh & Norman, 1965;
Ervin & Anders, 1970)

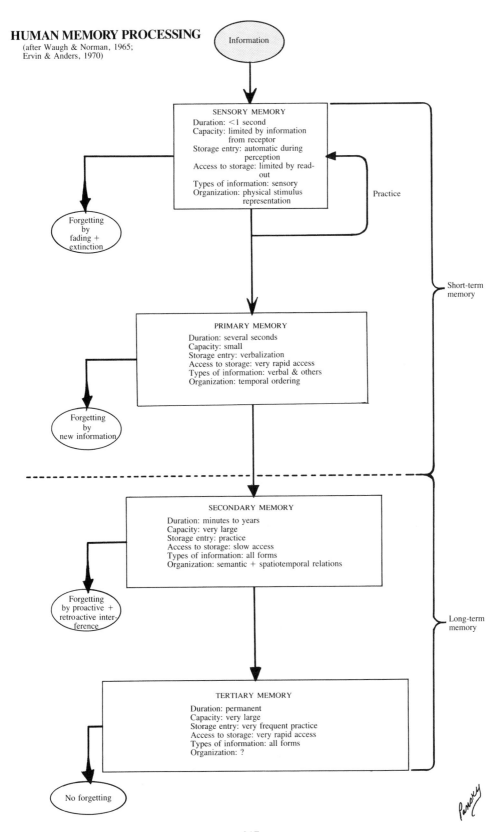

Information

SENSORY MEMORY

Duration: <1 second
Capacity: limited by information
　　　　　from receptor
Storage entry: automatic during
　　　　　perception
Access to storage: limited by read-
　　　　　out
Types of information: sensory
Organization: physical stimulus
　　　　　representation

Practice

Forgetting
by
fading +
extinction

Short-term
memory

PRIMARY MEMORY

Duration: several seconds
Capacity: small
Storage entry: verbalization
Access to storage: very rapid access
Types of information: verbal & others
Organization: temporal ordering

Forgetting
by
new information

SECONDARY MEMORY

Duration: minutes to years
Capacity: very large
Storage entry: practice
Access to storage: slow access
Types of information: all forms
Organization: semantic + spatiotemporal relations

Forgetting
by proactive +
retroactive inter-
ference

Long-term
memory

TERTIARY MEMORY

Duration: permanent
Capacity: very large
Storage entry: very frequent practice
Access to storage: very rapid access
Types of information: all forms
Organization: ?

No forgetting

Panoky

102. CONSCIOUSNESS AND SLEEP—PART I

I. **Consciousness includes two distinct concepts:** states of consciousness and conscious experience (of which one is aware, such as thoughts, feelings, ideas, dreams, etc.)
A. A PERSON'S STATE OF CONSCIOUSNESS (awake, asleep, drowsy) is defined by his behavior, from coma to maximum attentiveness and by the pattern of brain activity that can be electrically recorded as the electroencephalogram (EEG)
B. Wavelike patterns of the EEG change in frequency and amplitude as behavior changes
  1. The EEG is produced by the intermittent synchronization of the electric activity of small groups of neurons of the cerebral cortex
  2. The basic units of electric activity are considered to be individual synaptic potentials (or groups of them) rather than action potentials
C. The EEG is a useful tool clinically because the normal patterns are altered over the brain areas that are diseased, damaged, or injured
D. The EEG is also useful in defining states of consciousness, but one does not know what function, if any, this electric activity serves in the processing of information

II. **States of consciousness**
A. THE WAKING STATE AND AROUSAL are far from homogeneous
  1. The prominent EEG wave pattern of an awake relaxed adult with eyes closed is a slow oscillation of 8–13 Hz, known as the alpha rhythm
    a. Each person and brain region has a characteristic pattern of alpha rhythm
    b. The alpha rhythms are always larger at the back of the head over the visual cortex and are also larger when a person has his eyes closed and is not thinking
  2. When a person is attentive to external stimuli, the alpha rhythm is replaced by lower, faster oscillations. The change is called arousal and is associated with the act of attending to stimuli rather than perception itself
B. SLEEP is an active process and not a mere absence of wakefulness. There are 2 states of sleep characterized by different EEG and behavior patterns
  1. As a person becomes drowsy, the alpha rhythm is gradually replaced by irregular, low-voltage potential differences, and as sleep deepens, the EEG waves become slower, larger, and more irregular. This slow wave is periodically interrupted by episodes of paradoxical sleep during which the subject still seems asleep but has an EEG pattern similar to that of EEG arousal or of an awake, alert person
    a. In slow-wave sleep there is considerable tonus in postural muscles and only a small change in cardiovascular or respiratory activity. The person can be awakened, and rarely reports dreaming (thoughts rather than dreams)
    b. During paradoxical sleep the behavior criteria are precise. At the onset, there is an abrupt and complete inhibition of tone in postural muscles, with periodic episodes of twitching of facial muscles and limbs, and rapid eye movements behind closed lids occur. This sleep is called rapid eye movement or REM sleep
      i. Respiration & heart rate are irregular; blood pressure may go up or down
      ii. When awakened from REM sleep, 80–90% of people report dreams
  2. The two states of sleep follow a regular 30- to 90-minute cycle, each episode of paradoxical sleep lasting 10 to 15 minutes. Thus, slow sleep takes up about 80% of total sleeping time and paradoxical sleep about 20%
    a. Normally it is not possible to pass directly from the waking state to REM sleep; at least 30 minutes of slow sleep are required
    b. If deprived of REM sleep for several nights, all subjects spend a greater portion of time in REM sleep the next time they sleep; thus, the total number of hours spent in REM sleep tends to remain constant

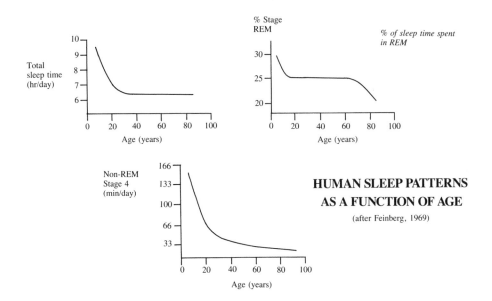

**HUMAN SLEEP PATTERNS
AS A FUNCTION OF AGE**

(after Feinberg, 1969)

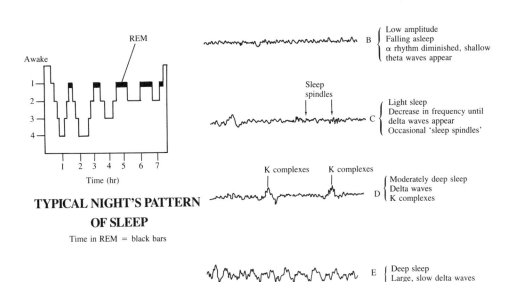

**TYPICAL NIGHT'S PATTERN
OF SLEEP**

Time in REM = black bars

A { Relaxed waking
    α rhythm prevails

B { Low amplitude
    Falling asleep
    α rhythm diminished, shallow
    theta waves appear

C { Light sleep
    Decrease in frequency until
    delta waves appear
    Occasional 'sleep spindles'

Sleep spindles

D { Moderately deep sleep
    Delta waves
    K complexes

K complexes   K complexes

E { Deep sleep
    Large, slow delta waves

100μV[

1s

**CLASSIFICATION OF STAGES OF HUMAN
SLEEP BASED ON EEG**

103. CONSCIOUSNESS AND SLEEP—PART II

II. States of consciousness (cont.)

B. SLEEP (CONT.)

3. The brain, as a whole, does not rest during sleep, and there is no general inhibition of activity of cerebral neurons. In actuality, there is a considerable amount of neuronal activity during slow-wave sleep, and many areas of the brain are more active during REM sleep than during wakefulness. In addition, blood flow and oxygen consumption by the brain do not decrease in sleep

4. During sleep there is a change in distribution or reorganization of neuronal activity, some individual neurons being less active than during waking, although the brain as a whole remains relatively active

5. Sleep may not be a period of generalized rest but may represent a period of rest for certain specific elements, during which time they can replenish substrates important for generation of potentials

   a. It has been suggested that the importance of sleep lies not in short-term recovery but in the relatively long-term chemical and structural changes that the brain must undergo to make learning and memory possible

   b. We do not know for certain what the functional significance of sleep is

C. RETICULAR FORMATION AND CONSCIOUSNESS

1. Consciousness is the result of interplay between three neuronal systems, one causing arousal and the other two sleep, and all three are parts of the reticular formation which lies in the central core of the brainstem in the center of the neural pathways ascending and descending between the brain and spinal cord

2. Neural structures in the reticular formation have been found to be essential for the maintenance of a waking EEG since stimulation causes EEG arousal, and destruction results in coma and an EEG characteristic of the sleeping state

D. Other areas of the brain, the cortex and certain hypothalamic areas, also play a role in wakefulness and alert state

E. Two neuronal systems control sleep by opposing the tonic activity of the reticular activating system

1. One is in the center core of the brainstem

2. The other is the sleep center in the pons, whose activity induces paradoxical sleep

3. The two centers interact with each other in a true feedback phenomenon, so that the cycling of sleep and wakefulness is due, in part, to the slow accumulation and dissipation of chemical transmitters

4. Of the two basic components of the sleep-wake cycle, the waking mechanisms seem to be more easily activated than those causing sleep

III. Disturbances of sleep: precisely how normal sleep and waking occur is uncertain

A. INSOMNIA (WAKEFULNESS: INABILITY TO SLEEP): rarely a sign of organic disease

B. HYPERSOMNIA (INCREASED SLEEPINESS): common in fatigue and while taking sedatives

C. NARCOLEPSY: condition usually affecting adults, in which the person is suddenly overcome by irresistible sleep even in the middle of a meal or under monotonous conditions

1. Narcoleptics also suffer from sleep paralysis (the mind wakes up before the body, and for a time after waking, patient is unable to move limbs; dream hallucinations (patient wakes, but dream continues); sleepwalking (body wakes up before mind); and cataplexy (with intense emotion, such as fright, a sudden outburst of crying or laughter occurs)

NEURAL STRUCTURES & CONNECTIONS
IMPORTANT IN MAINTAINING
SLEEP-WAKING CYCLE

(sleep-inducing connections
are in dotted lines)

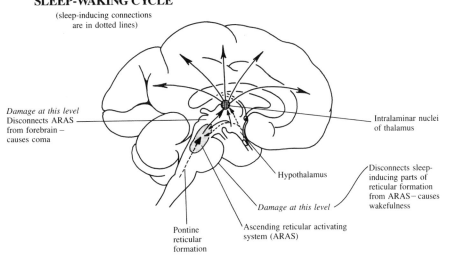

Damage at this level
Disconnects ARAS
from forebrain –
causes coma

Intralaminar nuclei
of thalamus

Hypothalamus

Disconnects sleep-
inducing parts of
reticular formation
from ARAS – causes
wakefulness

Damage at this level

Pontine
reticular
formation

Ascending reticular activating
system (ARAS)

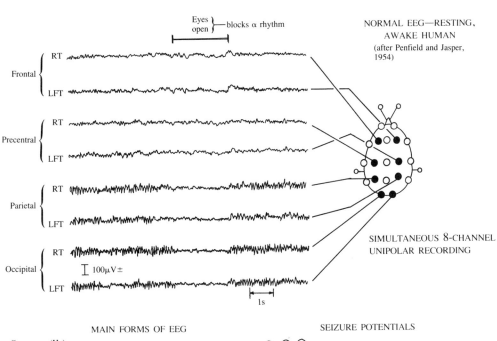

Eyes open — blocks α rhythm

NORMAL EEG—RESTING,
AWAKE HUMAN
(after Penfield and Jasper,
1954)

Frontal { RT / LFT

Precentral { RT / LFT

Parietal { RT / LFT

Occipital { RT / LFT

100μV±

1s

SIMULTANEOUS 8-CHANNEL
UNIPOLAR RECORDING

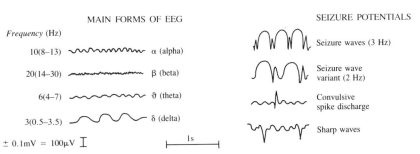

MAIN FORMS OF EEG

Frequency (Hz)

10(8–13) α (alpha)

20(14–30) β (beta)

6(4–7) ϑ (theta)

3(0.5–3.5) δ (delta)

± 0.1mV = 100μV

1s

SEIZURE POTENTIALS

Seizure waves (3 Hz)

Seizure wave
variant (2 Hz)

Convulsive
spike discharge

Sharp waves

104. MIGRAINE

I. Introduction: many people who have symptoms related to the nervous system have functional, not structural, problems, e.g., symptoms of headache and giddiness. Explanations of their cause are hard to find and may have a psychological basis, but there is no doubt that they are real. The skull and brain are incapable of feeling pain. It is from the scalp, its muscles and blood vessels, and from the dura mater and its venous sinuses, as well as blood vessels at the base of the brain, that pain can arise. Headaches most commonly originate in the muscles, blood vessels, and dura mater, and are produced by tension in, or stretching of, these structures

II. Migraine is due to a disturbance in the behavior of cranial blood vessels. What causes an attack is unknown, but it seems certain that chemical changes in and around the blood vessel walls play an important role
   A. SYMPTOMS are produced by spasms of vessels inside the skull and dilation of vessels outside of the skull (theory); the spasm produces the warning of the attack aura, and the dilation produces the headache
   B. MIGRAINE ATTACKS sometimes bear a relationship to anxiety, menstruation, excitement, and some foods (e.g., chocolate) but may also come "out of the blue"
   C. ATTACKS usually begin in youth, even in childhood, and affect women more than men
   D. MIGRAINE AND ALLERGIC DISORDERS (asthma, eczema) tend to be familial and hereditary
   E. THE AURA: symptoms depend on the blood vessel that goes into spasm
      1. Not all attacks have a clear aura. Many begin with flashing, scintillating lights zigzagging across the visual fields, sometimes forming colored patterns
      2. Other visual auras include homonymous hemianopia, bitemporal hemianopia, or complete loss of vision in both eyes or patches of loss (scotoma)
         a. Above symptoms are due to spasms affecting the vessels supplying the optic nerves, the chiasma, or the occipital visual cortex
      3. If the middle cerebral artery is involved, the patient develops numbness of one hand and arm, one half of the face and tongue, and even dysphagia
      4. If the basilar artery is involved, there may be bilateral blindness, vertigo, tinnitus, tingling in both hands and legs, and dysarthria
      5. Symptoms last for periods ranging from a few minutes to as long as an hour, die away, and are replaced by the headache
   F. THE HEADACHE is one sided but often affects the entire head or occipital region
      1. Due to dilatation of superficial arteries, the head throbs violently
      2. The headache increases to a maximum when vomiting occurs; then there is improvement and the patient goes to sleep, probably to awake free from pain
      3. Occasionally it lasts 24–48 hours; vomiting is sometimes prolonged and constant

III. Specific types of migraine
   A. OPHTHALMOPLEGIC MIGRAINE: due, in most cases, not to migraine, but to pressure on the 3rd cranial nerve by an aneurysm on the circle of Willis
   B. FAMILIAL HEMIPLEGIC MIGRAINE: can be diagnosed only if other family members have identical attacks. See frank paralysis of one arm and one leg (hours or days)
   C. COUGH HEADACHES: due to vascular distention during coughing
   D. PERIODIC MIGRANOUS NEURALGIA (histamine cephalgia or cluster headaches)
      1. Migraine rarely disturbs sleep; this condition always does
      2. Attacks occur every night for weeks, followed by months or years of remission
      3. Last $\frac{1}{2}$ to 2 hours and is very severe
   E. MUSCLE TENSION HEADACHES: commonest form of chronic headaches affecting muscles of the scalp, resulting in prolonged headaches lasting for days or weeks
      1. Described as "pressure" or a "tight band" around head and down neck
      2. There is no vomiting, aura, or sleep disturbance

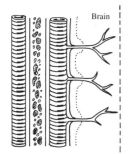
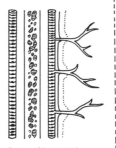
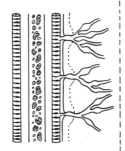
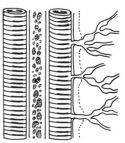

Normal cerebral vasculature Vessels respond to autonomic nerves except parenchymal aa. (in brain substance), which have no nerve supply—respond to local metabolic needs of brain tissue

Spasm of innervated aa. Reduced cerebral flow systemic release of serotonin contributes to local vasoconstriction Localized ischemia results in *aura*

Parenchymal aa. dilate in response to anoxia, acidosis, and other effects on brain due to diminished blood supply

Local vasodilatation of intraparenchymal vsls. Decrease in peripheral resistance causes marked local dilation of ipsilateral innervated intra- and extracranial aa. Serotonin levels drop

HEADACHE PHASE OF MIGRAINE

(after Diamond, S. & Medina, J.L. Clinical Symposia, Ciba. Vol. 33:2. 1981)

MIGRAINE

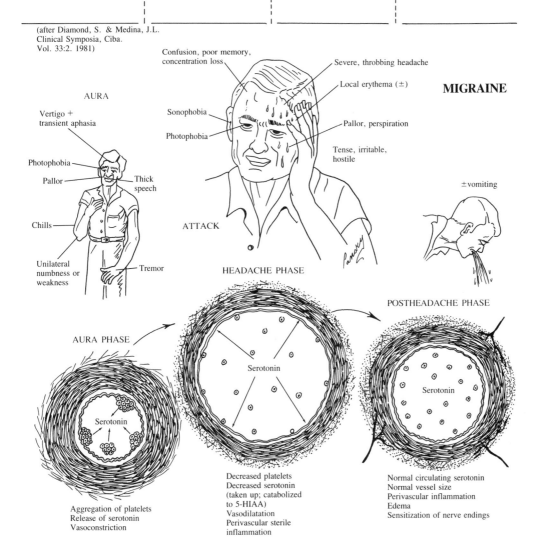

AURA

Vertigo + transient aphasia

Photophobia

Pallor

Chills

Unilateral numbness or weakness

Thick speech

Tremor

Confusion, poor memory, concentration loss

Sonophobia

Photophobia

Severe, throbbing headache

Local erythema (±)

Pallor, perspiration

Tense, irritable, hostile

ATTACK

±vomiting

HEADACHE PHASE

POSTHEADACHE PHASE

AURA PHASE

Serotonin

Serotonin

Serotonin

Aggregation of platelets Release of serotonin Vasoconstriction

Decreased platelets Decreased serotonin (taken up; catabolized to 5-HIAA) Vasodilatation Perivascular sterile inflammation

Normal circulating serotonin Normal vessel size Perivascular inflammation Edema Sensitization of nerve endings

Unit Seven

AUTONOMIC NERVOUS
SYSTEM

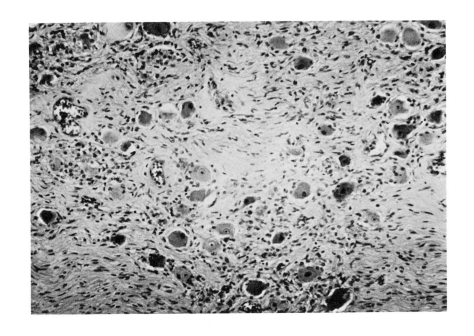

105. AUTONOMIC NERVOUS SYSTEM: BASIC PLAN

I. **Introduction:** the peripheral nervous system consists of 2 main divisions, the autonomic nervous system (ANS) and the craniospinal nervous system, which are highly integrated both anatomically and functionally

A. BY DEFINITION, the ANS or involuntary system is entirely a motor system and comprises all efferent (motor) nerves not otherwise classified as somatic motor

B. AUTONOMIC RESPONSES usually occur without conscious control or awareness

C. AUTONOMIC NEURONS are distributed to smooth muscle and cardiac muscle, arrector pili muscle (hair follicle), and myoepithelial cells (glands)

D. THE ANS IS ANATOMICALLY SUBDIVIDED according to location of its preganglionic cell bodies, into sympathetic (thoracolumbar) and parasympathetic (craniosacral) components.

II. **Structural organization of the ANS** consists of numerous paravertebral, prevertebral (collateral), cranial, and terminal ganglia associated with either cranial or spinal nerves

A. PARAVERTEBRAL AND PREVERTEBRAL GANGLIA belong to the sympathetic division
   1. The paravertebral ganglia form 2 sympathetic chains extending from the base of the skull to the coccyx
   2. The prevertebral ganglia are located in perivascular plexuses, the abdominal aorta being the most prominent site for these plexuses
   3. Paravertebral and prevertebral ganglia are connected to spinal cord segments T1 through L2 or L3 by means of spinal nerves from these same segments

B. CRANIAL AND TERMINAL GANGLIA are part of the parasympathetic division. Cranial ganglia are associated with cranial nerves III, VII, and IX; terminal ganglia are found within or near the various internal organs and are associated with cranial nerve X and spinal cord segments S2, S3, S4

C. THE ANS IS A 2-NEURON CHAIN SYSTEM
   1. The preganglionic neuron has its cell body within the CNS, and its axon carries the impulse from the CNS to a ganglion where it synapses
   2. The postganglionic neuron has its cell body in the ganglion, and its axon leaves the ganglion to innervate smooth muscle, cardiac muscle, and glands. Postganglionic neurons outnumber preganglionics by a ratio of about 32:1; thus, a single preganglionic may serve to stimulate a number of postganglionic fibers

III. **Morphologic and functional distinction between autonomic and somatic motor outflow**

A. SENSORY AND INTERNUNCIAL NEURONS to both autonomic and somatic efferent neurons are structurally and functionally similar (same afferent pathway feeds both)

B. THE SOMATIC SYSTEM has its cell bodies located exclusively within the CNS anterior gray horn; the ANS has its first-order preganglionic neuron cell bodies within the CNS, but the second-order postganglionic neuron cell bodies are located outside the CNS in a ganglion

C. IN SOMATIC REFLEX ARCS, inhibition is exerted by one neuron upon another, but never by a nerve cell upon an effector (muscle cell—central inhibition)

D. IN AUTONOMIC REFLEX ARCS, presynaptic fibers may inhibit preganglionic neurons. In addition, some autonomic postganglionic fibers inhibit the act of the effector organs they innervate

E. UNLIKE SOMATIC CONTROL, STRUCTURES INNERVATED BY THE ANS can contract, secrete, and function normally in the absence of the nervous system, e.g., the small intestine, completely isolated from the CNS, continues to show peristalsis and responds to stretch and other stimuli—autoregulation

F. UNLIKE THE WELL-DEVELOPED END-PLATE REGION OF THE SKELETAL MUSCLE FIBERS, THE ANS lacks a well-defined neuromuscular or neuroglandular junction. Instead, many varicosities (swellings) are found along the entire length of the terminal nerves

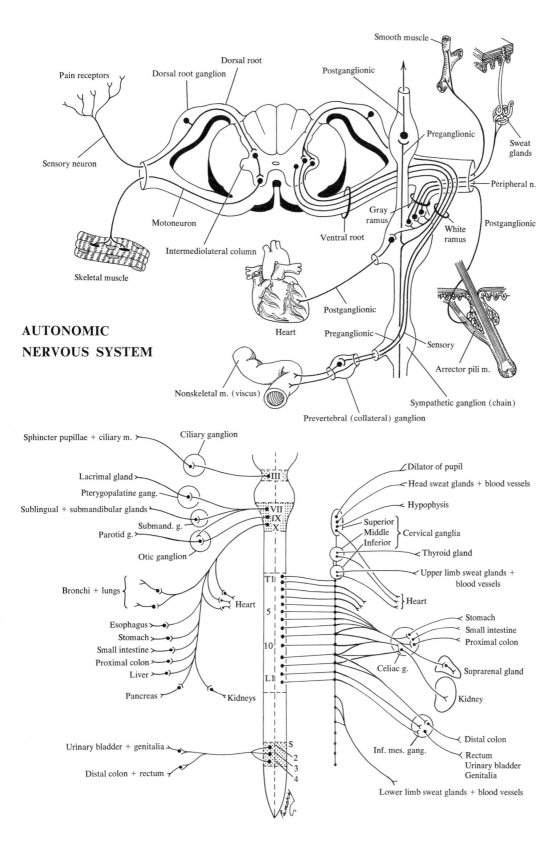

**AUTONOMIC
NERVOUS SYSTEM**

Pain receptors

Dorsal root ganglion

Dorsal root

Postganglionic

Smooth muscle

Preganglionic

Sweat glands

Sensory neuron

Peripheral n.

Motoneuron

Gray ramus

White ramus

Postganglionic

Ventral root

Intermediolateral column

Skeletal muscle

Postganglionic

Heart

Preganglionic

Sensory

Arrector pili m.

Nonskeletal m. (viscus)

Sympathetic ganglion (chain)

Prevertebral (collateral) ganglion

Sphincter pupillae + ciliary m.

Ciliary ganglion

Dilator of pupil

Head sweat glands + blood vessels

Lacrimal gland

III

Hypophysis

Pterygopalatine gang.

VII

Sublingual + submandibular glands

IX

Superior
Middle
Inferior

Cervical ganglia

Submand. g.

X

Parotid g.

Thyroid gland

Otic ganglion

Upper limb sweat glands + blood vessels

Bronchi + lungs

T1

Heart

Heart

5

Stomach

Esophagus

Small intestine

Stomach

Proximal colon

Small intestine

10

Proximal colon

Celiac g.

Suprarenal gland

Liver

L1

Pancreas

Kidneys

Kidney

Distal colon

Urinary bladder + genitalia

S

Inf. mes. gang.

Rectum
Urinary bladder
Genitalia

2
3
4

Distal colon + rectum

Lower limb sweat glands + blood vessels

-227-

106. SYMPATHETIC DIVISION OF THE AUTONOMIC NERVOUS SYSTEM—PART I

A. **Introduction:** most sympathetic ganglia are linked together in chainlike structures called the sympathetic chain or trunk (bilateral) extending from the cranium to the coccyx

A. THE GANGLIA OF EACH CHAIN are arranged segmentally except in the cervical region where they fuse into 3 large ganglia: superior, middle, and inferior cervical

II. **Preganglionic sympathetic axons** originate from cell bodies in the lateral gray column of the spinal cord at levels T1 through L2 or L3

A. THESE FIBERS exit the spinal cord in the ventral roots (along with somatic motor axons), but separate from the spinal nerves to form white rami communicantes
   1. These fibers enter the sympathetic chain, synapse with postganglionic cells in the chain ganglia, or pass through to synapse in prevertebral ganglia
   2. Preganglionic fibers do not necessarily synapse with a ganglion cell in the adjacent area, but may enter the chain and ascend or descend for several segments before synapsing. Connecting strands between ganglia of these chains are partly explained by these ascending-descending preganglionic fibers

B. PREGANGLIONIC WHITE RAMI are myelinated (B-type fibers), have a glistening white appearance in fresh specimens (not evident in cadaver), and their cell bodies are located at levels T1 to L2 or L3

III. **Postganglionic sympathetic axons** originate from cell bodies in the chain ganglia and return to the spinal nerves via the gray rami communicantes or go directly to various viscera

A. POSTGANGLIONIC NEURONS arise from all levels of the sympathetic chains and run in close association with blood vessels. Although their origin is limited, they ultimately fan out to cover the entire body, including the head, neck, and pelvic contents
   1. Postganglionic axons that emerge from chain ganglia as gray rami are mostly nonmyelinated (C-type fibers)

B. EACH SPINAL NERVE receives a gray ramus which is distributed with the nerve to smooth muscle of blood vessels, sweat glands, and arrector pili muscles
   1. Gray rami, associated with all pairs of spinal nerves, outnumber white rami

C. COLLATERAL OR PREVERTEBRAL GANGLIA are other collections of postganglionic cell bodies outside the sympathetic trunks belonging to the sympathetic division
   1. Chief among the named prevertebral/collateral ganglia are the celiac, superior mesenteric, and inferior mesenteric ganglia
      a. These ganglia are located at origins of major arteries having the same names
      b. Collections of preganglionic axons called thoracic splanchnic nerves pass through the sympathetic chains without synapsing and pass then through the diaphragm to synapse on postganglionic cells of these named ganglia
      c. These splanchnic nerves are examples of an exception to the rule that the sympathetic division has very short preganglionic axons
   2. Other less well-defined collateral ganglia (ovarian, hypogastric, etc.) are present in association with extensive autonomic nerve plexuses

SYMPATHETIC TRUNK AND NERVES

Superior cervical ganglion
Gray ramus
C5
Middle cervical ganglion
C6
Ansa subclavia
Vertebral n.
Subclavian a.
C7
C8
Spinal n.
Cervicothoracic ganglion
T1
Cervical splanchnic n.
T2
White ramus
2nd thoracic ganglion
Post. mediastinal
Splanchnic nn.
Cardioaortic
Bronchial
Esophageal
Greater splanchnic n.
Lesser splanchnic n.
Least splanchnic n.
10th ganglion
Rami communicantes
Intermediomedial column
Intermediolateral column
Post. vagal trunk
Greater splanchnic n.
Lesser splanchnic n.
Celiac gang.
Celiac a.
Aorticorenal gang.
Sup. mesenteric gang. + plexus
Renal a.
Sup. mesenteric a.
Inf. mesenteric plexus
Lumbar splanchnics
Lumbar trunk + ganglia
Bladder
Sup. hypogastric plexus (presacral n.)
Smooth m. sphincter
S2
3
4
Pudendal n.
MALE
Inf. hypogastric (pelvic) plexus + ganglia
Uterus
Rectum
FEMALE
Bladder
Vagina
Sphincter ani internus

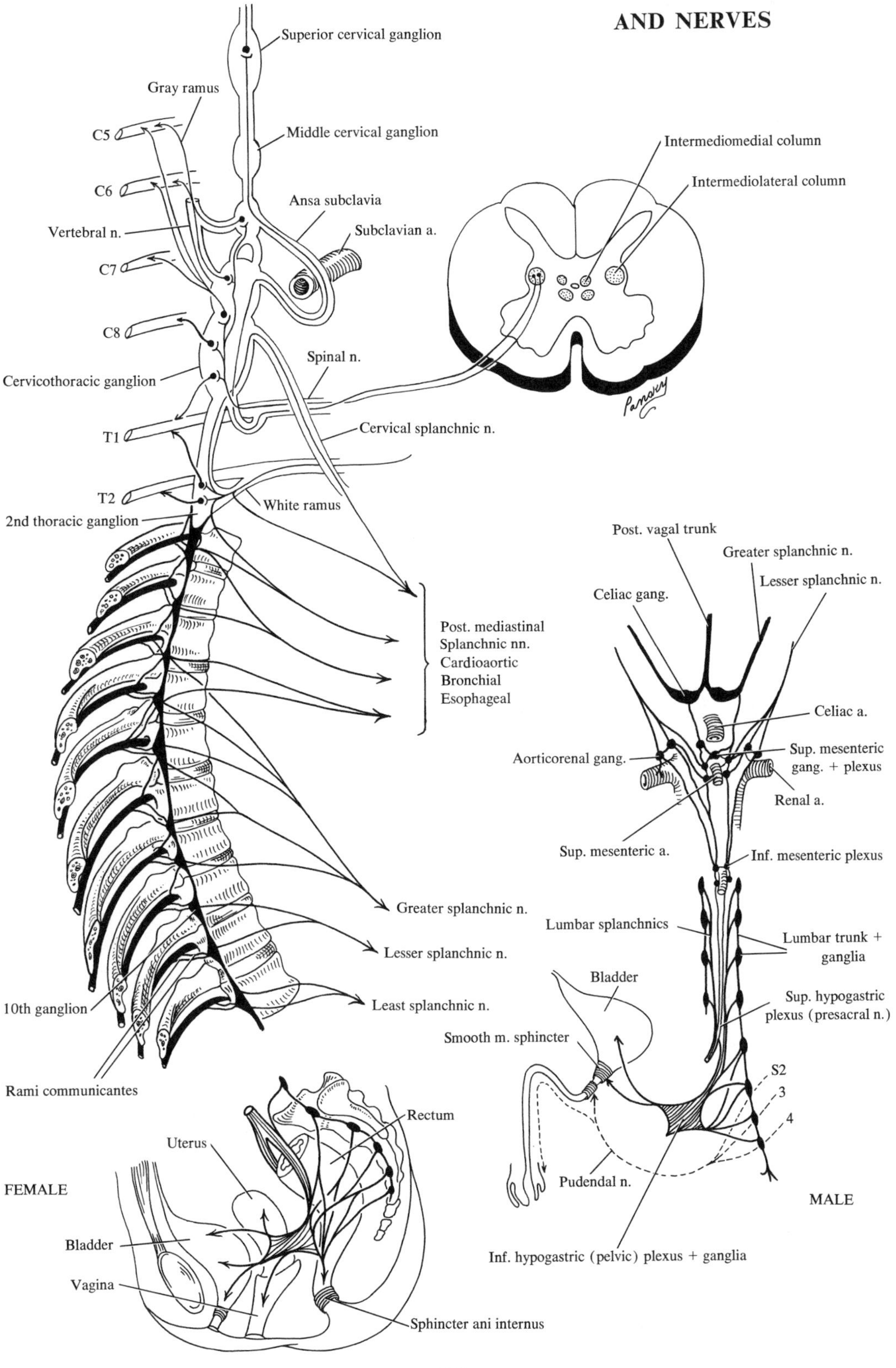

107. SYMPATHETIC DIVISION OF THE AUTONOMIC NERVOUS SYSTEM—PART II

I. Summary of the general pattern for the sympathetic division

A. ALL PREGANGLIONICS originate from spinal levels T1 through L2 or L3. Thus, innervation of the viscera, head, and extremities must originate from the thoracolumbar outflow regardless of the point at which they finally emerge from the chain

B. CERTAIN DISSECTABLE NERVES emerging from the chain contain mostly preganglionic fibers (thoracic splanchnics), while others contain mostly postganglionic fibers (cardiac sympathetics and most sympathetic fibers to the head, neck, and extremities)

C. GRAY RAMI COMMUNICANTES emerge from the chain and join all spinal nerves, being distributed in a segmental pattern to the body wall and musculature in accordance with the segmental distribution of spinal nerves

D. SYMPATHETIC FIBERS emerging from the superior cervical ganglion are distributed to the head, following the arterial supply of that region

E. THORACIC VISCERA receive their sympathetic innervation from fibers emerging from the chain at levels C1 through T5

F. MOST OF THE ABDOMINAL VISCERA are supplied by the thoracic splanchnics which arise from the sympathetic chain at levels T5 through T12; the remainder of the abdominal and pelvic viscera are supplied by fibers emerging from the sympathetic chain at levels L1 through S3.

G. THE SYMPATHETIC SYSTEM stimulates activities that are accompanied by an expenditure of energy stores, including acceleration of heart rate, increased force of heartbeat, rise of arterial pressure, elevation of blood sugar levels, and direction of blood flow to skeletal muscles (at the expense of visceral and cutaneous circulation)

H. THE SYMPATHETIC SYSTEM, unlike the discrete parasympathetic system, functions as a total unit throughout the body, since each sympathetic preganglionic neuron synapses with up to 30 or more postganglionic neurons and each of the latter ends on many effector cells. In addition, norepinephrine (secreted at postganglionic terminals) as well as epinephrine and norepinephrine from the adrenal medulla are denatured slowly

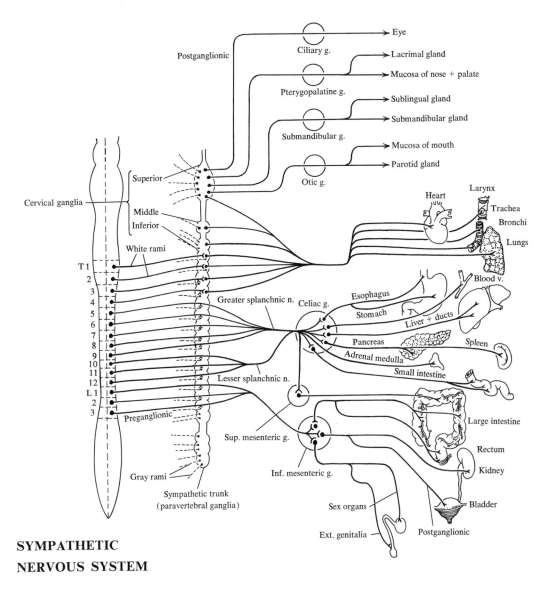

Postganglionic

Eye — Ciliary g.

Lacrimal gland — Pterygopalatine g.

Mucosa of nose + palate

Sublingual gland

Submandibular gland — Submandibular g.

Mucosa of mouth

Parotid gland — Otic g.

Cervical ganglia

Superior
Middle
Inferior

White rami

Heart
Larynx
Trachea
Bronchi
Lungs
Blood v.

T 1
2
3
4
5
6
7
8
9
10
11
12
L 1
2
3

Greater splanchnic n.

Celiac g.

Esophagus
Stomach
Liver + ducts
Pancreas
Spleen
Adrenal medulla
Small intestine

Lesser splanchnic n.

Preganglionic

Sup. mesenteric g.

Large intestine

Rectum
Kidney

Gray rami

Inf. mesenteric g.

Sympathetic trunk
(paravertebral ganglia)

Sex organs

Bladder

Ext. genitalia

Postganglionic

SYMPATHETIC
NERVOUS SYSTEM

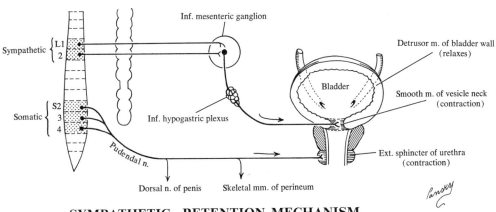

Inf. mesenteric ganglion

Sympathetic { L1
2

Detrusor m. of bladder wall
(relaxes)

Bladder

Somatic { S2
3
4

Smooth m. of vesicle neck
(contraction)

Inf. hypogastric plexus

Pudendal n.

Ext. sphincter of urethra
(contraction)

Dorsal n. of penis   Skeletal mm. of perineum

Pansky

SYMPATHETIC—RETENTION MECHANISM

108. PARASYMPATHETIC DIVISION OF THE AUTONOMIC NERVOUS SYSTEM

I. **Introduction:** preganglionic fibers of the craniosacral outflow originate from the brainstem and from sacral lateral gray horns of levels S2, S3, and S4 of the spinal cord. Fibers from the cranial portion are distributed through cranial nerves III (oculomotor), VII (facial), IX (glossopharyngeal), and X (vagus)

A. THE GANGLIA OF THE PARASYMPATHETIC DIVISION are widely scattered and irregularly spaced when compared with the sympathetic division

   1. Parasympathetic ganglia lie close to the organ innervated or, in some cases, within the wall of the organ supplied, accounting for the characteristically short postganglionic fibers of this division

   2. Most preganglionic neurons run without interruption from their CNS origin either to the wall of the viscus they supply or where they synapse with terminal ganglion cells associated with the pelvic plexuses

B. PARASYMPATHETIC FIBERS OF CRANIAL NERVES III, VII, AND IX supply ganglia of the head, whereas cranial nerve X (vagus, wanderer) supplies fibers to the heart, lungs, and most of the abdominal viscera via the prevertebral plexuses

C. THE SACRAL PORTION supplies fibers to the remainder of the abdominal contents not supplied by the vagus and to most of the pelvic viscera. The pelvic nerve (nervus erigens) supplies parasympathetic fibers to most of the large intestine and to the pelvic viscera and genitalia via the hypogastric plexus

II. **The craniosacral subdivision of the autonomic nervous system:** preganglionic fibers of the cranial and sacral subdivisions form synaptic relations with postganglionic neurons in cranial autonomic or terminal ganglia respectively

A. CRANIAL PARASYMPATHETICS: in the cranial region, 4 parasympathetic ganglia are related topographically to the major divisions of the trigeminal (V) nerve

   1. Ciliary ganglion: located along the lateral surface of the optic nerve

   2. Pterygopalatine ganglion: lies in the pterygopalatine fossa near the maxillary nerve

   3. Submandibular ganglion: positioned over the submandibular gland near the lingual nerve

   4. Otic ganglion: located medial to the mandibular nerve as it exits the foramen ovale

B. SACRAL PARASYMPATHETICS

   1. Preganglionic fibers exit the spinal cord at levels S2, S3, and S4 to form the pelvic nerve (nervus erigens) and pass to the terminal ganglia of the pelvic plexuses, myenteric and submucosal plexuses of the descending colon and rectum

   2. Postganglionic fibers from terminal ganglia supply the urinary bladder, the descending colon, the rectum, and accessory reproductive organs

   3. Sacral autonomic fibers innervate all viscera not supplied by the vagus nerve

C. LARGEST SOURCE OF PREGANGLIONIC PARASYMPATHETICS is the dorsal motor nucleus of the vagus nerve, which supplies most thoracic and abdominal viscera (except those in the pelvis)

   1. In the thorax: preganglionic vagal fibers enter the pulmonary, esophageal, and cardiac plexuses, to be distributed to the terminal (intrinsic) ganglia of the heart and bronchial musculature

   2. In the abdomen: preganglionic vagal fibers go to the stomach and pass through the celiac and subsidiary plexuses, to end in the terminal ganglia of the intestine, liver, pancreas, and kidneys

      a. These terminal ganglia form the extensive ganglionated plexuses of Auerbach (myenteric) and of Meissner (submucosal), which extend the entire length of the digestive tube and are composed of many small aggregates of ganglion cells interconnected by delicate fiber bundles

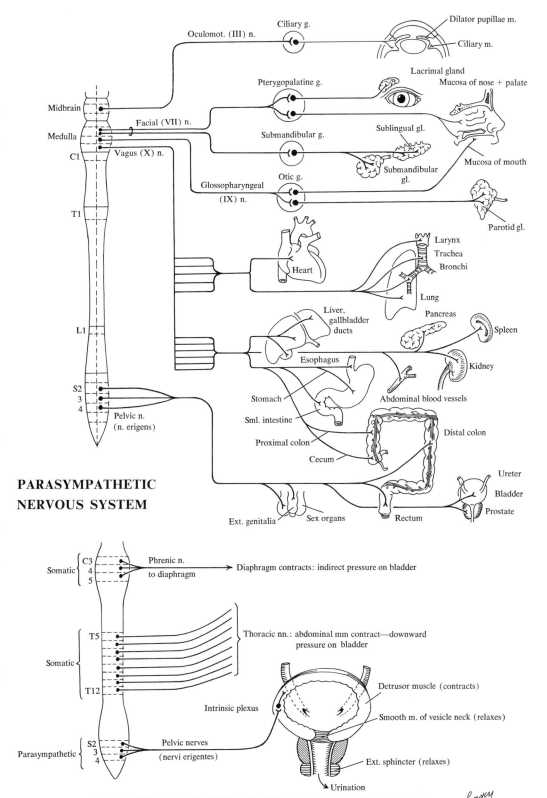

Oculomot. (III) n.

Ciliary g.

Dilator pupillae m.

Ciliary m.

Pterygopalatine g.

Lacrimal gland

Mucosa of nose + palate

Midbrain

Facial (VII) n.

Medulla

Vagus (X) n.

C1

Submandibular g.

Sublingual gl.

Submandibular gl.

Mucosa of mouth

T1

Glossopharyngeal (IX) n.

Otic g.

Parotid gl.

Larynx

Trachea

Bronchi

Heart

Lung

Liver, gallbladder ducts

Pancreas

Spleen

L1

Esophagus

Kidney

Abdominal blood vessels

Stomach

Sml. intestine

Proximal colon

Cecum

Distal colon

S2

3

4

Pelvic n. (n. erigens)

PARASYMPATHETIC NERVOUS SYSTEM

Ureter

Bladder

Prostate

Ext. genitalia

Sex organs

Rectum

Somatic { C3 4 5

Phrenic n. to diaphragm

Diaphragm contracts: indirect pressure on bladder

Somatic { T5 ... T12

Thoracic nn.: abdominal mm contract—downward pressure on bladder

Intrinsic plexus

Detrusor muscle (contracts)

Smooth m. of vesicle neck (relaxes)

Parasympathetic { S2 3 4

Pelvic nerves (nervi erigentes)

Ext. sphincter (relaxes)

Urination

Pansky

PARASYMPATHETIC—EXPULSION MECHANISM

-233-

109. AUTONOMICS OF THE HEAD

I. **Preganglionic fibers from white rami of the upper thoracic nerves** go mainly to the superior cervical ganglion

II. **Postganglionic fibers from the superior cervical ganglion** are distributed to the lower 4 cranial nerves, the upper 3 or 4 cervical nerves, the pharynx, the superior cervical cardiac nerve, and the external and internal carotid arteries

III. **Nerve plexuses** are formed about the carotid arteries by postganglionic fibers from the superior cervical ganglion. These postganglionics pass through cranial autonomic ganglia (without synapsing) to join cranial nerves for distribution to face and scalp

IV. **Dual sympathetic-parasympathetic innervation** is also provided to the salivary glands, intrinsic muscles of the eye, and mucous membranes of the pharynx and nose
A. FOUR PAIRS OF CRANIAL AUTONOMIC GANGLIA, each of which receives a sympathetic, parasympathetic, and sensory root, mediate this dual autonomic supply (only parasympathetic nerves synapse here; others pass through without synapse)
   1. The ciliary ganglion is located between the optic nerve and the lateral rectus muscle and is comprised of
      a. A parasympathetic root from cells of the Edinger-Westphal nucleus and the medial portion of the CN IIIrd nucleus
      b. A sympathetic root of postsynaptic fibers from the superior cervical ganglion via the carotid plexus
      c. A sensory root from the nasociliary branch of the ophthalmic (V) nerve
      d. Short ciliary nerves leave ganglion with postganglionic (III) fibers to sphincter pupillae and ciliary muscles, postganglionic sympathetics to the dilator pupillae muscle, and sensory fibers from the eyeball
   2. The pterygopalatine ganglion is located in the pterygopalatine fossa, closely associated with the maxillary (V) nerve, and is composed of
      a. A parasympathetic preganglionic root from cells of the superior salivatory nucleus via the glossopalatine nerve, great petrosal nerve, and Vidian nerve (VII)
      b. A sympathetic postganglionic root from the internal carotid plexus via the deep petrosal nerve, which joins the great petrosal nerve to form the Vidian nerve
      c. A sensory root from the maxillary (V) nerve, but some fibers arise in cranial nerves VII and IX via the tympanic plexus and Vidian nerve
      d. Postganglionic parasympathetic fibers carry VII secretory fibers to lacrimal gland
   3. The otic ganglion is located medial to the mandibular nerve just below the foramen ovale in the infratemporal fossa and is composed of
      a. A parasympathetic preganglionic root from the inferior salivatory nucleus (medulla) which is distributed via the glossopharyngeal (IX) nerve (tympanic branch), tympanic plexus, and lesser petrosal nerve to the ganglion
      b. A sympathetic postganglionic root from the superior cervical ganglion follows the plexus on the middle meningeal artery to the ganglion
      c. A sensory root receives fibers from the IXth nerve and from the geniculate ganglion of the VIIth nerve via the tympanic plexus and the lesser petrosal nerve
      d. Postganglionics of IX follow the auriculotemporal (V) nerve to the parotid gland
   4. The submandibular ganglion is located on medial side of mandible between lingual nerve and mandibular duct, lying over submandibular gland; composed of
      a. A parasympathetic preganglionic root from the superior salivatory nucleus via the glossopalatine, chorda tympani, and lingual nerves
      b. A sympathetic postganglionic root from the plexus along the maxillary artery
      c. A sensory root from the geniculate ganglion which is distributed via the glossopalatine, chorda tympani, and lingual nerves
      d. Postganglionic (VII) fibers to submandibular and sublingual glands (secretory)

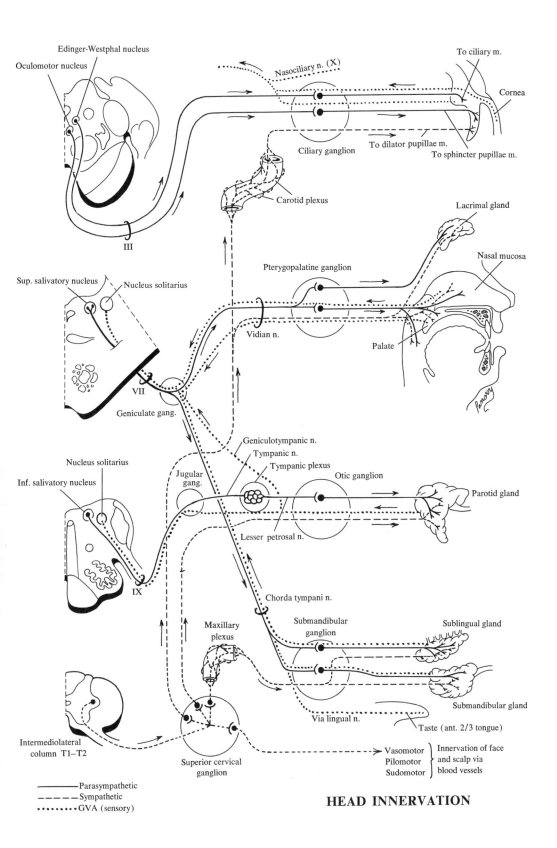

Edinger-Westphal nucleus

Oculomotor nucleus

To ciliary m.

Nasociliary n. (X)

Cornea

Ciliary ganglion

To dilator pupillae m.

To sphincter pupillae m.

Carotid plexus

III

Lacrimal gland

Nasal mucosa

Sup. salivatory nucleus

Nucleus solitarius

Pterygopalatine ganglion

Vidian n.

Palate

VII

Geniculate gang.

Geniculotympanic n.

Tympanic n.

Tympanic plexus

Jugular gang.

Otic ganglion

Parotid gland

Nucleus solitarius

Inf. salivatory nucleus

Lesser petrosal n.

IX

Chorda tympani n.

Maxillary plexus

Submandibular ganglion

Sublingual gland

Submandibular gland

Via lingual n.

Taste (ant. 2/3 tongue)

Intermediolateral column T1–T2

Superior cervical ganglion

Vasomotor
Pilomotor
Sudomotor

Innervation of face and scalp via blood vessels

———— Parasympathetic

– – – – Sympathetic

•••••• GVA (sensory)

HEAD INNERVATION

110. FUNCTIONAL ANATOMY OF THE AUTONOMIC NERVOUS SYSTEM

I. Transmitter chemicals of the ANS

A. IN BOTH DIVISIONS OF THE ANS, the chemical transmitter for the ganglionic synapse between pre- and postganglionic fibers is acetylcholine

1. The transmitter at the junction between parasympathetic postganglionic fibers and effector organs is also acetylcholine
2. The transmitter between the sympathetic postganglionic fiber and its effector organ is norepinephrine (a member of the catecholamine family)
   a. An exception is sympathetic postganglionic fibers to sweat glands which release acetylcholine
   b. One ganglion, the adrenal medulla, in the sympathetic system never develops long postganglionic fibers, but on activation of its preganglionic nerves, the cells discharge their transmitter into the bloodstream
      i. Release a mixture of 80% epinephrine and 20% norepinephrine
      ii. Are really hormones rather than neurotransmitters and are transmitted via blood and intestinal fluid to receptor sites on effector cells sensitive to them

B. FIBERS THAT RELEASE ACETYLCHOLINE are called cholinergic fibers; those that release norepinephrine are called adrenergic fibers

C. MANY DRUGS STIMULATE OR INHIBIT the synaptic functions of the ANS

1. Those whose actions mimic the actions of the sympathetic system are called sympathomimetic drugs (amphetamine and phenylephrine)
2. Several choline compounds and the mushroom poison muscarine mimic parasympathetic actions and are called parasympathomimetic drugs

D. DRUGS THAT BLOCK THE ACTIONS of the autonomic system are sympatholytic and parasympatholytic. The plant belladonna (deadly nightshade) yields atropine, which is a parasympatholytic drug

II. Actions and dual innervation of the ANS: the actions of the ANS depend not only on the chemical released by the postganglionic cell but also on the effector cell's receptor sites and intracellular machinery

A. AN IMPORTANT CHARACTERISTIC of autonomic nerves is dual innervation of most effector organs, often with opposite effects. Skeletal muscle cells, in contrast, do not receive dual innervation since somatic nerves are excitatory

1. Action potentials over the sympathetic nerves to the heart increase heart rate; those over the parasympathetic fibers decrease heart rate
2. In the intestine, activation of sympathetic fibers reduces smooth muscle contraction in the intestinal wall; parasympathetics increase contraction

B. TO PREVENT THE TWO SYSTEMS FROM CONFLICTING, they are usually activated reciprocally; i.e., as activity of one system is enhanced, that of the other is depressed

C. BECAUSE SMOOTH MUSCLES, HEART MUSCLE, AND GLANDS PARTICIPATE AS EFFECTORS in all homeostatic control of our internal environment, there is a wide array of effects

1. Sympathetic system helps the body to cope with challenges from the outside environment and involves stress and emotions (fear and rage)
2. Parasympathetic system is more responsible for internal maintenance, such as digestion, defecation, and urination, and is more active during recovery or rest

D. AUTONOMIC RESPONSES usually occur without conscious control or awareness, but

1. Discrete visceral and glandular responses can be learned; animals have learned to control their own heart rates or manipulate vessel dilatation or contraction
2. Small segments of the autonomic response can be regulated independently
3. Autonomic responses, made up of many components, are variable but rather than being gross, undiscerning discharges are finely tailored to specific bodily demands

CHEMICAL TRANSMITTERS

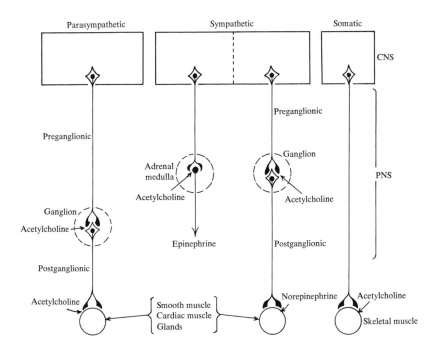

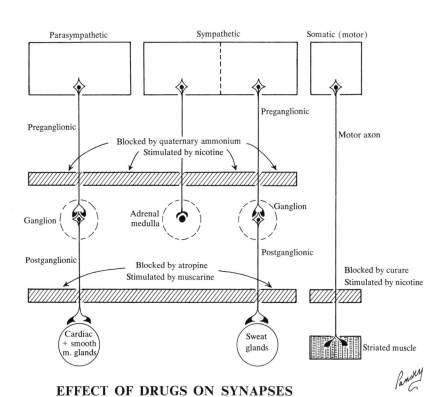

EFFECT OF DRUGS ON SYNAPSES

Unit Eight

THE CEREBELLUM

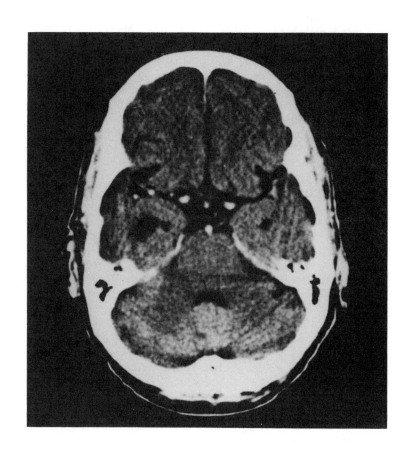

111. CEREBELLUM: GROSS ANATOMY

I. General information: the cerebellum occupies most of the posterior cranial fossa, lies dorsal to the brainstem, and is attached to the pons, medulla, and mesencephalon by 3 pairs of thick fiber bundles, the cerebellar peduncles, found on the ventral aspect of the cerebellum. They include

A. SUPERIOR CEREBELLAR PEDUNCLE (BRACHIUM CONJUNCTIVUM): the bridge between the midbrain and the cerebellum, with fibers largely projecting from the cerebellum toward the midbrain and thalamus

B. MIDDLE CEREBELLAR PEDUNCLE (BRACHIUM PONTIS): the bridge between the pons and the cerebellum, with fibers projecting to the cerebellum

C. INFERIOR CEREBELLAR PEDUNCLE (RESTIFORM BODY): between the medulla and the cerebellum, with fibers projecting both to and from the cerebellum

II. Other relationships

A. THE FOURTH VENTRICLE separates the most medial portion of the ventral surface of the cerebellum from the pons and medulla

B. THE TENTORIUM CEREBELLI (TRANSVERSE FOLD OF DURA) separates the cerebellum from the occipital lobe of the cerebrum

III. The cerebellum is composed of 2 large lateral masses, the cerebellar hemispheres and a midline portion known as the vermis (wormlike)

A. ITS SURFACE is comprised of numerous parallel folia, with each primary folium being subdivided under the surface into secondary and tertiary folia

B. IN MIDSAGITTAL SECTION (THROUGH THE VERMIS), the cerebellum has the appearance of a highly branched tree (arbor vitae)

   1. The cerebellar cortex is a layer of gray matter which covers the surface

   2. The medullary center is an internal core of white matter consisting of fibers from the cortex to cerebellar nuclei, incoming fibers which terminate on neurons in the gray matter, and fibers connecting different cortical areas

C. EMBEDDED IN THE MEDULLARY CORE are gray substance, the cerebellar nuclei: the dentate nucleus (largest, most lateral, irregular), the emboliform nucleus (vertical plate of cells), the globose nuclei (one or more rounded gray areas), and the fastigial nuclei (near the midline, in the core of the vermis). These nuclei are present in each of the cerebellar hemispheres

IV. Lobes of cerebellum

A. ANTERIOR LOBE (PALEOCEREBELLUM) lies anterior to the primary fissure and consists of most of the vermis and of the anterior (superior) aspect of the cerebellar hemispheres. This phylogenetically old lobe is primarily associated with proprioceptive (spinocerebellar) and exteroceptive input from the head and body, including some from the vestibular system. It has a significant role in the regulation of muscle tone

B. POSTERIOR LOBE (NEOCEREBELLUM): the largest part of the cerebellum, positioned between the other two lobes. It consists of the main bulk of the cerebellar hemispheres and part of the vermis. This phylogenetically new lobe (essentially a mammalian structure) receives connections of the cerebrum and plays an essential role in the muscular coordination of phasic movements

C. FLOCCULONODULAR LOBE (ARCHICEREBELLUM) consists of the paired flocculi of the hemispheres (small appendages in the posterior inferior region) and the unpaired nodulus, which is the inferior part of the vermis. The archicerebellum represents the cerebellar portion of the vestibular system and phylogenetically is its oldest part. It is actually a specialized portion of the somatic afferent column that is dominated by direct, indirect, and feedback connections of the vestibular system. It plays a significant role in regulating muscle tone, equilibrium, and posture through its influences on the trunk muscles

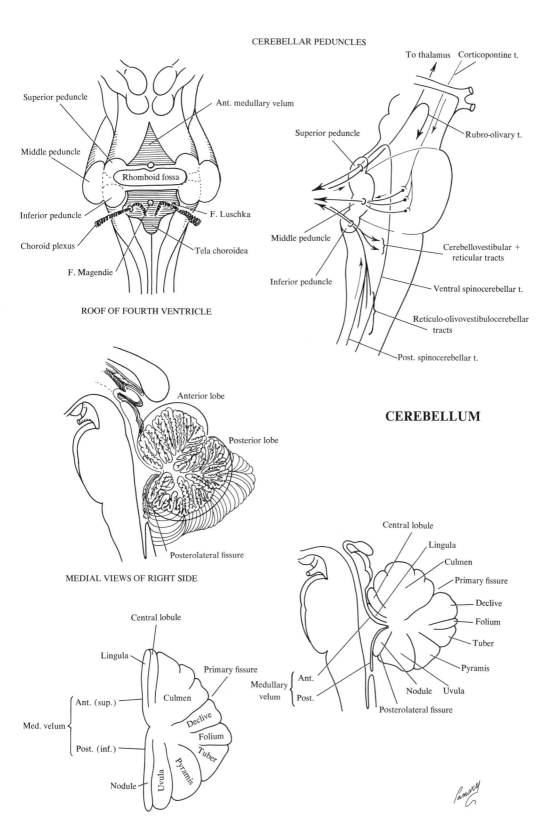

CEREBELLAR PEDUNCLES

Superior peduncle

Ant. medullary velum

Middle peduncle

Rhomboid fossa

Inferior peduncle

F. Luschka

Choroid plexus

Tela choroidea

F. Magendie

ROOF OF FOURTH VENTRICLE

To thalamus   Corticopontine t.

Superior peduncle

Rubro-olivary t.

Middle peduncle

Inferior peduncle

Cerebellovestibular + reticular tracts

Ventral spinocerebellar t.

Reticulo-olivovestibulocerebellar tracts

Post. spinocerebellar t.

Anterior lobe

Posterior lobe

CEREBELLUM

Posterolateral fissure

MEDIAL VIEWS OF RIGHT SIDE

Central lobule

Lingula

Culmen

Primary fissure

Declive

Folium

Tuber

Pyramis

Nodule   Uvula

Posterolateral fissure

Medullary velum { Ant. Post.

Central lobule

Lingula

Primary fissure

Culmen

Ant. (sup.)

Med. velum {

Post. (inf.)

Declive

Folium

Tuber

Pyramis

Uvula

Nodule

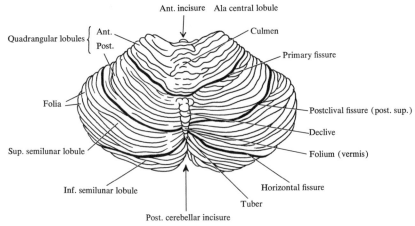

Ant. incisure   Ala central lobule

Quadrangular lobules { Ant.
                       Post.

Culmen

Primary fissure

Folia

Postclival fissure (post. sup.)

Declive

Sup. semilunar lobule

Folium (vermis)

Inf. semilunar lobule

Horizontal fissure

Tuber

Post. cerebellar incisure

CEREBELLUM

VIEW FROM ABOVE AND BEHIND (NATURAL SIZE)

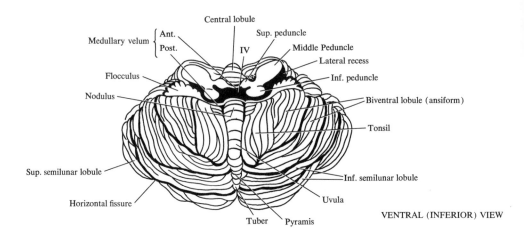

Central lobule

Medullary velum { Ant.
                  Post.

Sup. peduncle

Middle Peduncle

IV

Lateral recess

Flocculus

Inf. peduncle

Nodulus

Biventral lobule (ansiform)

Tonsil

Sup. semilunar lobule

Inf. semilunar lobule

Horizontal fissure

Uvula

Tuber   Pyramis

VENTRAL (INFERIOR) VIEW

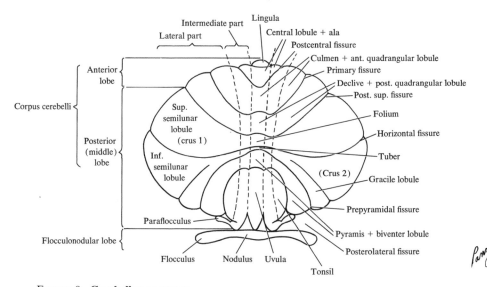

Lingula

Intermediate part

Central lobule + ala

Lateral part

Postcentral fissure

Culmen + ant. quadrangular lobule

Anterior lobe

Primary fissure

Declive + post. quadrangular lobule

Corpus cerebelli {

Post. sup. fissure

Sup. semilunar lobule (crus 1)

Folium

Horizontal fissure

Posterior (middle) lobe

Inf. semilunar lobule

Tuber

(Crus 2)

Gracile lobule

Paraflocculus

Prepyramidal fissure

Pyramis + biventer lobule

Flocculonodular lobe {

Posterolateral fissure

Flocculus   Nodulus   Uvula

Tonsil

FIGURE 9. **Cerebellar anatomy.**

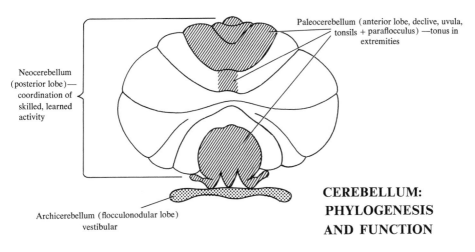

Neocerebellum (posterior lobe)— coordination of skilled, learned activity

Paleocerebellum (anterior lobe, declive, uvula, tonsils + paraflocculus) —tonus in extremities

Archicerebellum (flocculonodular lobe) vestibular

CEREBELLUM: PHYLOGENESIS AND FUNCTION

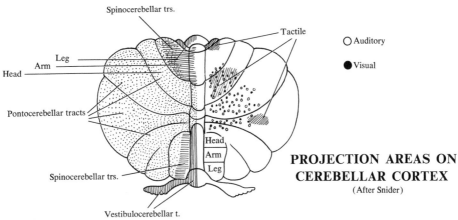

Spinocerebellar trs.

Tactile

Leg
Arm
Head

Pontocerebellar tracts

Spinocerebellar trs.

Head
Arm
Leg

Vestibulocerebellar t.

○ Auditory
● Visual

PROJECTION AREAS ON CEREBELLAR CORTEX
(After Snider)

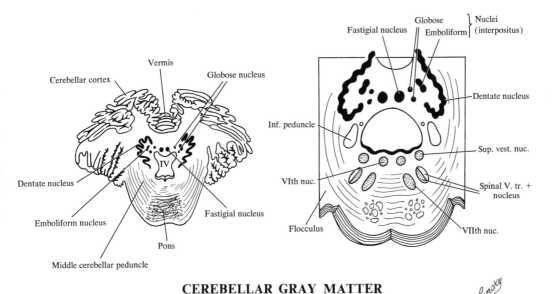

Vermis

Cerebellar cortex

Globose nucleus

Dentate nucleus

Emboliform nucleus

Pons

Middle cerebellar peduncle

Fastigial nucleus

Fastigial nucleus

Globose
Emboliform } Nuclei (interpositus)

Inf. peduncle

Dentate nucleus

Sup. vest. nuc.

VIth nuc.

Spinal V. tr. + nucleus

Flocculus

VIIth nuc.

CEREBELLAR GRAY MATTER

Panoky

FIGURE 10. **Cerebellar projections and nuclei.**

112. CEREBELLAR CORTEX: GENERAL LAYERS

I. **General information:** cortical surface area is greatly increased by numerous long gyri known as folia. Its total surface area is 75% of that of the cerebral cortex. Some 85% of its free surface is hidden from surface view and lies deep within the sulci of the folia. All cerebellar folia have the same microscopic structure and consist of 3 layers of neurons

II. **Internal structure of the cerebellum** consists of a highly convoluted cortex which is superficial to the central core of white matter. It is composed of 5 types of neurons arranged in 3 distinct layers

A. MOLECULAR LAYER: the outer layer which is the synaptic integrative layer. It receives its input mainly from the granular layer and deep cerebellar nuclei and partly from outside the cerebellum. It consists mainly of nerve fibers (axons and dendrites) and some basket and stellate cells.

B. PURKINJE CELL LAYER: the middle cell layer. It is the output layer of the cerebellar cortex which discharges to the intrinsic cerebellar nuclei and contains the Purkinje cell bodies

C. GRANULAR LAYER: the innermost layer immediately adjacent to the white medullary core and the main receptive layer which receives input from outside the cerebellum. It contains the granule and Golgi (inner stellate) cell bodies

D. TWO TYPES OF AFFERENT FIBERS are seen in the cortex. Mossy fibers terminate in synaptic contact with granule cells of the innermost layer, and climbing fibers enter the molecular layer and wind among the Purkinje cell dendrites. The only fibers leaving the cortex are Purkinje cell axons, which terminate in the central nuclei of the cerebellum. An exception are the fibers from the flocculonodular lobe cortex, which end in brainstem nuclei

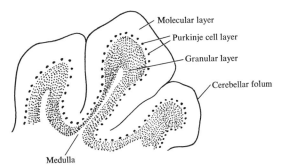

Molecular layer
Purkinje cell layer
Granular layer
Cerebellar folum
Medulla

TRANSVERSE SECTION OF
CEREBELLAR CONVOLUTION
(After Cajal)

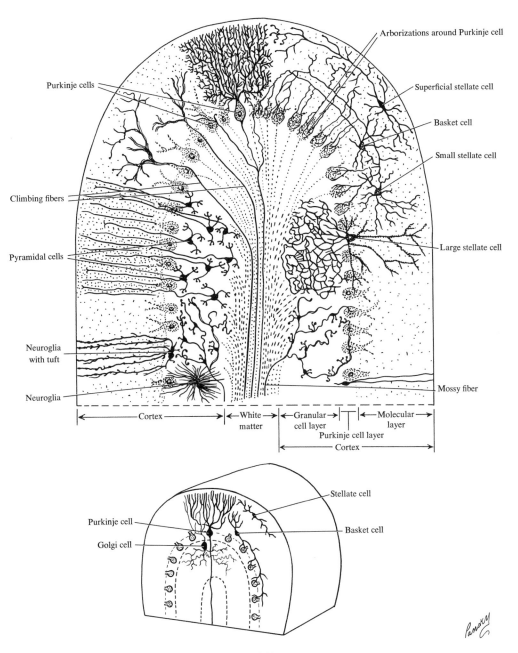

Purkinje cells

Climbing fibers

Pyramidal cells

Neuroglia
with tuft

Neuroglia

Arborizations around Purkinje cell

Superficial stellate cell

Basket cell

Small stellate cell

Large stellate cell

Mossy fiber

Cortex — White matter — Granular cell layer — Molecular layer
Purkinje cell layer
Cortex

Purkinje cell
Golgi cell

Stellate cell
Basket cell

113. CEREBELLAR CORTEX: CELL AND NERVE TYPES AND THEIR ROLES

I. Granule cells are small multipolar neurons whose nerve cell bodies are densely packed in the granular layer

A. AXONS ascend into molecular layer and bifurcate as a T to form parallel fibers that run parallel to the long axis of the folium

B. PARALLEL FIBERS synapse with dendritic branches of Purkinje cells which also ascend into the molecular layer. As many as 40,000 parallel fibers may synapse (all excitatory) on each Purkinje cell

C. DENDRITES are synaptically related to mossy fibers which are terminal branches of axons of cells that lie outside the cerebellum

II. Purkinje cells are arranged side by side in a single row between the molecular layer and the inner granular layer

A. DENDRITES ascend into the molecular layer, where they branch profusely

B. AXONS run through the granular layer into the white matter, where recurrent collateral branches are given off to other Purkinje cells. The axons end in cerebellar nuclei (or Deiter's cells)

III. Basket cells: their cell bodies are in the lower part of the molecular layer

A. DENDRITES are also in the molecular layer and are excited by parallel fibers

B. AXONS run perpendicular to parallel fibers and give collaterals to many Purkinje cells. They end in a rete of terminals (a basket appearance) about the cell bodies of the Purkinje cells

C. FUNCTION: inhibit the Purkinje cells

IV. Stellate cells are similar to basket cells and are often found in a small clump of mossy fibers. A volley of impulses in the stellate cells produces a line of excited Purkinje cells (with adjacent parallel bands of Purkinje cells being inhibited by the basket cells)

V. Golgi cells: cell bodies are found in the upper granular cell layer

A. DENDRITES branch extensively within the molecular layer and make contact with parallel fibers of the Purkinje cells

B. AXONS arborize extensively in the granular layer to eventually synapse on dendrites of the granule cells

C. FUNCTION: inhibit the mossy fiber, granule cell synapse

D. EXCITED BY parallel fibers and climbing fibers collaterals; inhibited by Purkinje cell axon collaterals

VI. Mossy fibers are axons that provide the primary cerebellar inputs from pontine nuclei and spinocerebellar pathways

A. CELL BODIES lie outside the cerebellum

B. EACH MOSSY FIBER ENDS in 3–5 mosslike rosettes that synapse with several granule cells. Each granule cell may receive input from as many as six mossy fibers

C. FUNCTION: to excite the granule cell

VII. Climbing fibers are axons of neurons whose cell bodies are located in the inferior olivary nucleus

A. A CLIMBING FIBER MAKES CONTACT with a single Purkinje cell and branches progressively as it "climbs" its dendrites

B. A CLIMBING FIBER SYNAPSES with primary and secondary dendritic branches of Purkinje cells at many sites (100–300)

C. EACH ACTION POTENTIAL in a climbing fiber causes a giant excitatory postsynaptic potential (EPSP) which, in turn, produces a burst of repetitive Purkinje cell firing

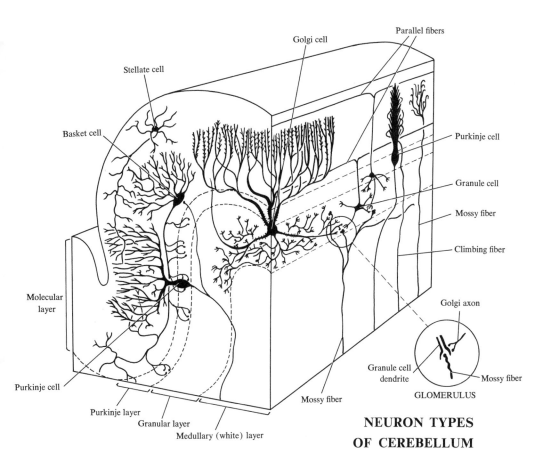

Stellate cell

Golgi cell

Parallel fibers

Basket cell

Purkinje cell

Granule cell

Mossy fiber

Climbing fiber

Molecular
layer

Golgi axon

Granule cell
dendrite

Mossy fiber

GLOMERULUS

Purkinje cell

Purkinje layer

Granular layer

Medullary (white) layer

Mossy fiber

**NEURON TYPES
OF CEREBELLUM**

CEREBELLAR "WIRING DIAGRAM"

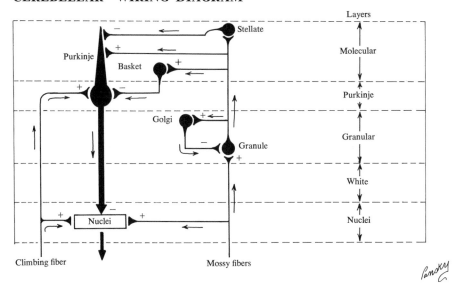

Layers

Stellate

Molecular

Purkinje

Basket

Purkinje

Golgi

Granular

Granule

White

Nuclei

Nuclei

Climbing fiber

Mossy fibers

114. SUBCORTICAL CEREBELLAR NUCLEI

I. General location: four pairs of nuclei are buried deep within the white medullary core of the cerebellum and are positioned dorsal and dorsolateral to the fourth ventricle

II. Name and relationship

A. DENTATE NUCLEI are the largest and most lateral of these deep nuclei. Each resembles a crumpled paper bag with its opening directed medially when cut in the horizontal plane

B. EMBOLIFORM NUCLEI* lie just medial to the dentate and usually appear somewhat plug (embolus)-shaped in their position, close to what appears to represent the opening or hilus of the dentate nucleus

C. GLOBOSE NUCLEI* are usually represented as 2 or more small ovoid nuclear masses just medial to the emboliform

D. FASTIGIAL NUCLEI are the most medial and are located just lateral to the fastigium (roof) of the fourth ventricle

III. Relationship of cerebellar cortex and subcortical nuclei: in general, the cerebellar nuclei receive mainly the axons from ipsilateral Purkinje cells

A. LATERAL AREAS OF THE CORTEX project to laterally positioned dentate nuclei

B. PARAMEDIAN AREAS OF THE CORTEX project to the emboliform and globose nuclei

C. MEDIAN AREA (VERMIS) OF THE CORTEX projects to the globose and fastigial nuclei. The latter nucleus also receives some fibers from the vestibular nuclei

D. PURKINJE CELL AXONS terminate in the central nuclei. Olivocerebellar fibers also pass near the nuclei on their way to the cortex and may send collaterals there. The fastigial nucleus also receives afferents from the vestibular nerve and nuclei. A few rubrocerebellar fibers end in the globose and emboliform nuclei

   1. Purkinje cells have an inhibitory role with respect to neurons of the central nuclei

   2. Excitation by afferents other than Purkinje cells (mainly olivocerebellar fibers) predominates over Purkinje cell inhibition to maintain a tonic discharge of impulses from the central nuclei to the thalamus and brainstem

---

*The emboliform and globose nuclei are collectively called the nucleus interpositus.

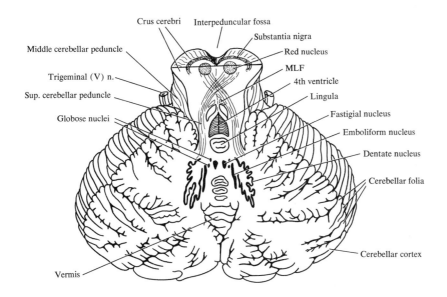

SECTION THROUGH CEREBELLUM AND MESENCEPHALON—
LEVEL OF SUPERIOR CEREBELLAR PEDUNCLE

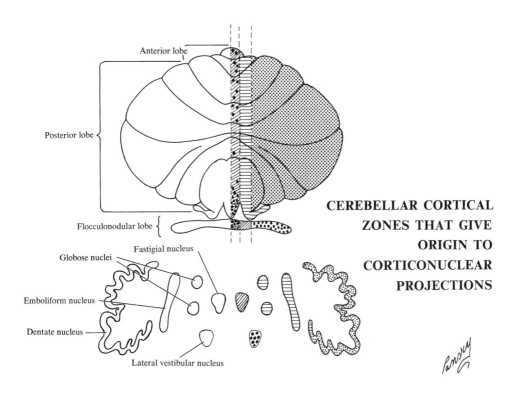

CEREBELLAR CORTICAL
ZONES THAT GIVE
ORIGIN TO
CORTICONUCLEAR
PROJECTIONS

115. SUMMARY OF CEREBELLAR CONNECTIONS

I. General: all fibers carrying information to and from the cerebellum do so by way of the cerebellar peduncles

A. INCOMING (AFFERENT) FIBERS

1. Superior cerebellar peduncle (brachium conjunctivum) is the main efferent pathway of the cerebellum. Its afferent connections are the ventral (anterior) spinocerebellar tract to the paleocerebellum; the tectocerebellar tract from the midbrain tectum to the ipsilateral neocerebellum, which represents the optic and auditory input; and the trigeminocerebellar tract, which conveys general afferent impulses from the head

2. Middle cerebellar peduncle (brachium pontis) is entirely afferent. It conveys fibers from contralateral pontine nuclei which receive innervation from the cerebral cortex (corticopontocerebellar tract). Each cerebellar hemisphere monitors the activity of the opposite cerebral hemisphere

3. Inferior cerebellar peduncle (restiform body) is predominantly an afferent pathway. Conveys "unconscious" exteroceptive and proprioceptive fibers from the spinal cord and vestibular system

   a. Posterior (dorsal) spinocerebellar tract projects directly to paleocerebellum

   b. Accessory (lateral) reticular nuclei, gracile and cuneate nuclei, and other medullary reticular nuclei which receive input from the spinal cord, project to the ipsilateral paleo- and neocerebellum

   c. Accessory inferior olivary nuclei (paleo-olivary) receive cord input and project to the paleocerebellum

   d. Main inferior olivary nucleus (neo-olivary) receives major input from the cerebral cortex, caudate nucleus, globus pallidus, red nucleus, and brainstem reticular formation and project crossed fibers to all parts of the neocerebellum, dentate nucleus, and nucleus interpositus

   e. Direct vestibulocerebellar fibers project to the archicerebellum of the same and opposite sides

B. OUTGOING (EFFERENT) FIBERS: no direct cerebellospinal pathways exist

1. Superior cerebellar peduncle consists primarily of the efferent fibers from the dentate, emboliform, and globose nuclei, and the entire outflow completely crosses over in the lower midbrain as the decussation of the brachium conjunctivum

   a. The dentate nucleus designates efferent tracts via the dentatorubral, dentatothalamic, and dentatoreticular fibers

   b. Most fibers from the dentate nucleus project rostrally to the thalamus (ventral lateral and thalamic reticular nuclei), with some fibers also to the red nucleus, inferior olive, and brainstem reticular formation

   c. While the dentate projects to the thalamus, the globose and emboliform nuclei project mainly to the red nucleus, inferior olive, and reticular formation

   d. This results in two additional loops which permit the cerebellum to monitor other areas of the nervous system.
      i. Cortico/ponto/cerebello/dentato/rubro/thalamo/cortical pathway
      ii. Cerebello/dentato/rubro/olivo/cerebellar pathway

2. Middle cerebellar peduncle has no efferent fibers

3. Inferior cerebellar peduncle: outflow is through the juxtarestiform body (on the medial aspect of the peduncle) and includes the following

   a. Fastigiobulbar tract: from the fastigial nuclei to the vestibular nuclei (primarily the lateral nucleus) and reticular nuclei of the pons and medulla
      i. The uncinate fasciculus of Russell (hooked bundle) formed by fibers from the nuclei fastigii of both sides, hooks around the rostral border of the superior cerebellar peduncle before passing through the restiform body

   b. Some direct fibers from the flocculonodular cortex pass to the vestibular nuclei

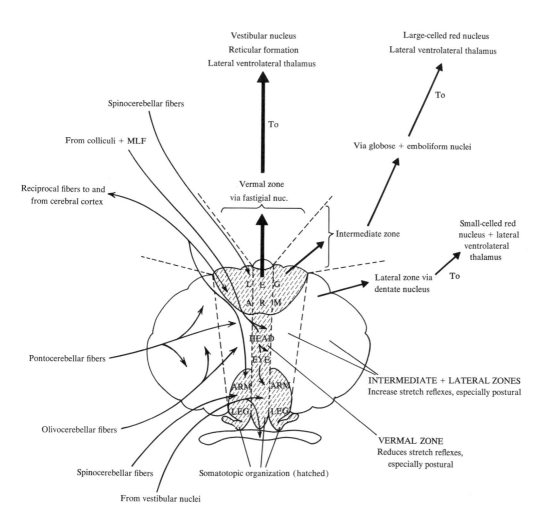

Vestibular nucleus
Reticular formation
Lateral ventrolateral thalamus

Large-celled red nucleus
Lateral ventrolateral thalamus

Spinocerebellar fibers

From colliculi + MLF

To

To

Via globose + emboliform nuclei

Reciprocal fibers to and
from cerebral cortex

Vermal zone
via fastigial nuc.

Intermediate zone

Small-celled red
nucleus + lateral
ventrolateral
thalamus

LEG

ARM

Lateral zone via
dentate nucleus

To

Pontocerebellar fibers

HEAD
EYE

ARM   ARM

LEG   LEG

INTERMEDIATE + LATERAL ZONES
Increase stretch reflexes, especially postural

Olivocerebellar fibers

VERMAL ZONE
Reduces stretch reflexes,
especially postural

Spinocerebellar fibers

Somatotopic organization (hatched)

From vestibular nuclei

SCHEMATIC OF FLATTENED CEREBELLAR CORTEX
PARASAGITTAL ZONES + MAJOR CONNECTIONS

(After Lewis)

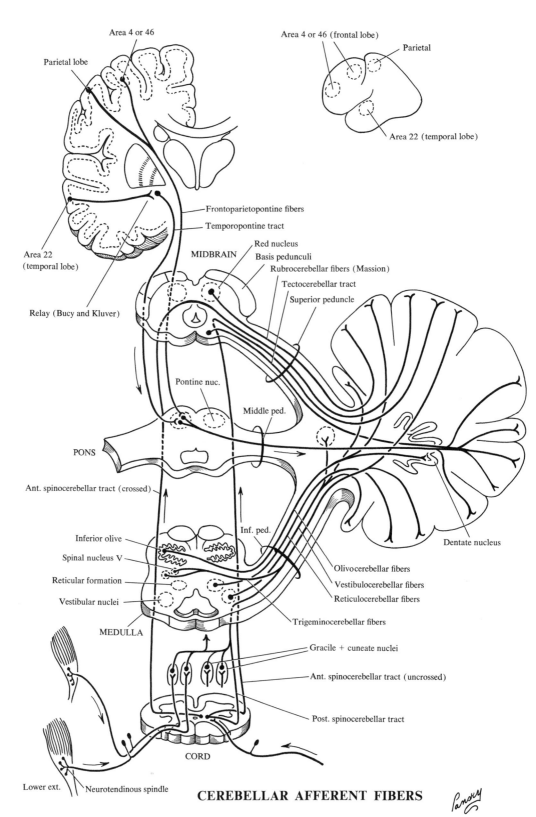

Area 4 or 46

Parietal lobe

Area 4 or 46 (frontal lobe)

Parietal

Area 22 (temporal lobe)

Frontoparietopontine fibers

Temporopontine tract

Red nucleus

Basis pedunculi

Rubrocerebellar fibers (Massion)

Tectocerebellar tract

Superior peduncle

MIDBRAIN

Area 22 (temporal lobe)

Relay (Bucy and Kluver)

Pontine nuc.

Middle ped.

PONS

Ant. spinocerebellar tract (crossed)

Inf. ped.

Inferior olive

Spinal nucleus V

Reticular formation

Vestibular nuclei

MEDULLA

Dentate nucleus

Olivocerebellar fibers

Vestibulocerebellar fibers

Reticulocerebellar fibers

Trigeminocerebellar fibers

Gracile + cuneate nuclei

Ant. spinocerebellar tract (uncrossed)

Post. spinocerebellar tract

CORD

Lower ext.

Neurotendinous spindle

CEREBELLAR AFFERENT FIBERS

FIGURE 11.

-252-

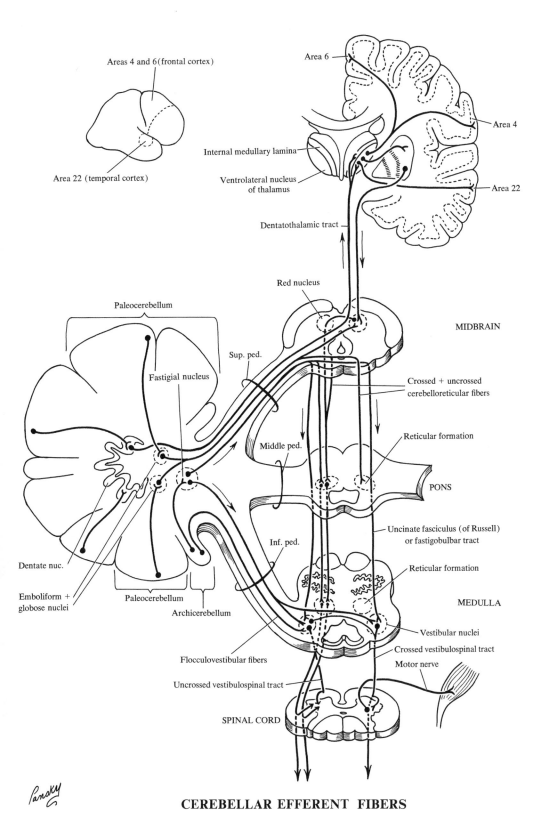

Areas 4 and 6 (frontal cortex)

Area 22 (temporal cortex)

Area 6

Area 4

Internal medullary lamina

Ventrolateral nucleus
of thalamus

Area 22

Dentatothalamic tract

Red nucleus

MIDBRAIN

Paleocerebellum

Sup. ped.

Fastigial nucleus

Crossed + uncrossed
cerebelloreticular fibers

Reticular formation

Middle ped.

PONS

Uncinate fasciculus (of Russell)
or fastigobulbar tract

Inf. ped.

Reticular formation

Dentate nuc.

MEDULLA

Emboliform +
globose nuclei

Paleocerebellum

Vestibular nuclei

Archicerebellum

Crossed vestibulospinal tract

Flocculovestibular fibers

Motor nerve

Uncrossed vestibulospinal tract

SPINAL CORD

Pansky

CEREBELLAR EFFERENT FIBERS

FIGURE 12.

116. SUMMARY OF FUNCTIONAL CEREBELLAR ANATOMY

I. Normal cerebellar function

A. THE CEREBELLUM IS RESPONSIBLE FOR MUSCLE SYNERGY (coordination) throughout the body

B. THE CEREBELLUM COORDINATES the action of muscle groups and times their contractions so that movements are performed smoothly and accurately

C. THE CEREBELLUM ACTS AS A MONITOR of other centers of the brain. It is not the initiator of muscle contraction or movement in general, but monitors centers that do initiate such motion

D. VOLUNTARY MOVEMENTS can proceed without its assistance, but such movements are clumsy and disorganized. Lack of motor skills as a result of cerebellar dysfunction is called asynergia or cerebellar ataxia

E. THE CEREBELLUM IS THE MODULATOR (REGULATOR) of motor activities and is integrated with the vestibular system for the maintenance of muscle tone and equilibrium, with general proprioceptive and exteroceptive receptors, and with the auditory and visual systems

F. THE CEREBELLUM IS BASICALLY A SOMATIC AFFERENT ORGAN (head ganglion of proprioception)
  1. Although the cerebellum receives large numbers of afferent fibers, conscious perception does not occur in it, nor do its efferent fibers give rise to conscious sensations elsewhere in the brain
  2. The cerebellum receives and processes unconscious afferent stimuli from various afferent (sensory) centers and passes the information on to the motor systems, to be utilized by them to make the necessary corrections in specific muscles or muscle groups (via the cerebellar-cerebral cortical connections or interconnections)
  3. As far as can be determined, the cerebellum plays no role in the appreciation of conscious sensations or in intelligence

II. Cerebellar dysfunction

A. ATAXIA is caused by cerebellar damage and is manifested in the following ways
  1. Disturbances of posture and gait may be very pronounced. Lesions of the midline region of the cerebellum cause difficulty in maintaining an upright stance. The gait is staggering, like that seen in drunkenness
    a. The loss of equilibrium is due to the lack of muscle synergy and not to a defect in the pathway of conscious proprioception
  2. Decomposition of movement: an action that requires the cooperative movement of several joints is not properly coordinated, but is broken down into its component parts
  3. Dysmetria is shown by the inability to stop a movement at the desired point. In reaching the hand toward an object, the patient either overshoots the goal or stops before it is reached
  4. Dysdiadochokinesis is the inability to stop one movement and follow it immediately by the directly opposite action, as in the case of rapid alternating movements of pronation-supination of the hands
  5. Scanning speech is due to asynergy of the muscles used in speaking. The spacing of sounds is irregular, with pauses in the wrong places

B. HYPOTONIA: muscle tone is decreased and may be ascertained by palpation

C. ASTHENIA: muscles affected by cerebellar lesions are weaker and tire more easily than normal muscles

D. TREMOR: of cerebellar dysfunction is usually an intention tremor. It is evident during purposeful movements, but absent or diminished with rest

E. NYSTAGMUS is present, with cerebellar lesions being accounted for by irritation of vestibular fibers in the cerebellum. It may also be due to the effect of pressure on the vestibular nuclei of the brainstem ventral to the cerebellum

CEREBELLAR CONNECTIONS–
SUMMARY

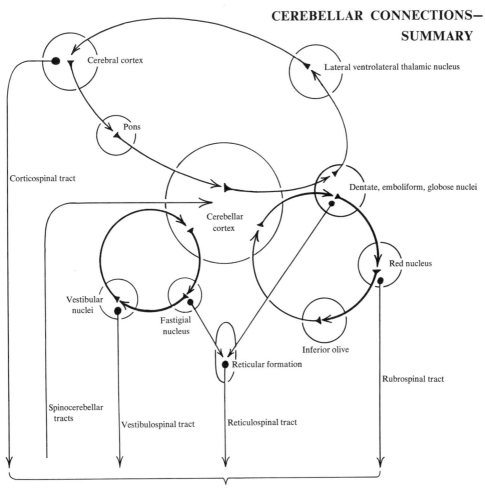

Cerebral cortex

Lateral ventrolateral thalamic nucleus

Pons

Corticospinal tract

Dentate, emboliform, globose nuclei

Cerebellar
cortex

Red nucleus

Vestibular
nuclei

Fastigial
nucleus

Inferior olive

Reticular formation

Rubrospinal tract

Spinocerebellar
tracts

Vestibulospinal tract

Reticulospinal tract

Connecting to motor systems

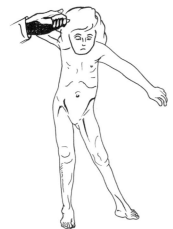

POSITION AS RESULT OF
CEREBELLAR ATAXIA

CONTORTIONIST MOVEMENT
AS RESULT OF CEREBELLAR
HYPOTONIA

Pansky

THE RETICULAR
FORMATION; LIMBIC
AND OLFACTORY SYSTEMS

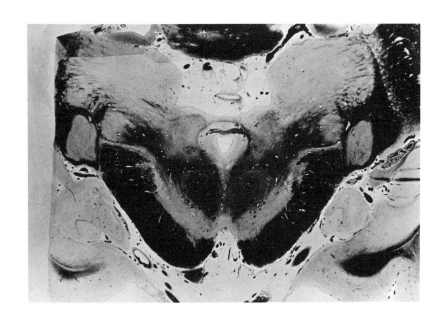

117. RETICULAR FORMATION:
GENERAL CONSIDERATIONS

I. **Definition:** an area in the tegmentum of the midbrain and pons and its continuation into the medulla and upper spinal cord, consisting of small groups of nerve cells interspersed among horizontally and vertically running nerve fibers, and excluding the cranial nerve nuclei and roots of the long fiber tracts

II. **Anatomy of the reticular formation:** general considerations
   A. THE RETICULAR FORMATION (RF) BEGINS JUST ROSTRAL TO THE PYRAMIDAL DECUSSATION and extends rostrally through the mesencephalon
   B. THE RF IS FOUND ONLY IN THE MEDULLA, PONS, AND MIDBRAIN and is a phylogenetically old part of the brain (constitutes a prominent part of brainstem in lower forms)
   C. THE RF COMPRISES THOSE BRAINSTEM AREAS that are made up of diffuse aggregations of cells of different types and sizes, separated by fibers extending in all directions
   D. NERVE ROOTS AND BUNDLES OF MYELINATED AND UNMYELINATED FIBERS derived from phylogenetically more recent structures intersect the RF
   E. CIRCUMSCRIBED CELL GROUPS such as the red nucleus, olivary nucleus, and cranial nerve nuclei are not included in the RF
   F. THE RF IS SEPARATED from the ependyma of the 4th ventricle and aqueduct by a zone of subependymal (periaqueductal) central gray matter
   G. IN THE LOWER MEDULLA, the RF lies ventral to the central canal, appears to be continuous with the gracilis and cuneate nuclei dorsally, and extends laterally almost to the spinal nuclei of the Vth (trigeminal) cranial nerve
   H. IN THE UPPER MEDULLA, the RF lies between the central gray matter (dorsally) and the inferior and accessory olives (ventrally) and extends laterally to the minor tract area and trigeminal spinal nucleus
   I. IN THE PONTINE TEGMENTUM, the RF lies between the subependymal gray matter and the lemnisci
   J. IN THE MIDBRAIN, the RF reaches its largest extent and extends from the periaqueductal gray matter to reach the substantia nigra and completely surrounds the red nucleus
   K. THE RF ends in the diencephalon as the reticular nucleus

III. **Specific reticular nuclei**
   A. IN THE MEDULLA, the 5 major nuclei will be considered
      1. Lateral reticular nucleus is located near the surface of the medulla and extends from the level of the caudal border of the inferior nucleus to the midolivary level; it projects to the cerebellum
      2. Paramedian nucleus lies adjacent to midline of medulla and projects to cerebellum
      3. Ventral reticular nucleus is comprised of cells scattered in the medial ⅔ of the area bounded by the inferior olive (ventrally), gracilis and cuneate nuclei (dorsally), and the nucleus of the trigeminal spinal tract (laterally)
      4. Magnocellular nucleus lies just rostral to the ventral reticular nucleus; comprised of large neurons intermingled with some smaller ones
      5. Parvicellular nucleus is located medial to the spinal nucleus of V and ventral to the vestibular nuclei. It extends from midolivary level to rostral limit of the medulla. Considered sensory because collaterals from sensory system terminate here
   B. IN THE TEGMENTUM OF THE PONS, 3 separate nuclei are mentioned, but 2 of these are rostral extensions of the medullary reticular nuclei
      1. Caudal pontine reticular nucleus is a forward extension of the magnocellular nucleus of the medulla and is comprised of large cells
      2. Rostral pontine reticular nucleus is located in the medial portion of the tegmentum and is comprised only of small nerve cell bodies
      3. Parvicellular reticular nucleus is located in the lateral tegmentum of the pons; it is a forward extension of the parvicellular nucleus

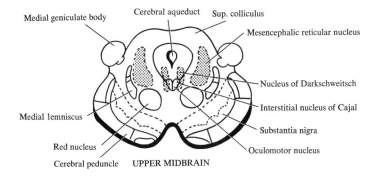

Medial geniculate body   Cerebral aqueduct   Sup. colliculus

Mesencephalic reticular nucleus

Nucleus of Darkschweitsch

Interstitial nucleus of Cajal

Medial lemniscus

Substantia nigra

Red nucleus

Oculomotor nucleus

Cerebral peduncle   **UPPER MIDBRAIN**

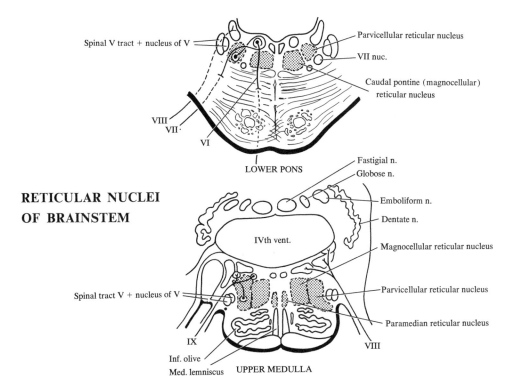

Spinal V tract + nucleus of V

Parvicellular reticular nucleus

VII nuc.

Caudal pontine (magnocellular) reticular nucleus

VIII
VII·
VI

LOWER PONS

RETICULAR NUCLEI
OF BRAINSTEM

Fastigial n.
Globose n.

Emboliform n.

Dentate n.

IVth vent.

Magnocellular reticular nucleus

Spinal tract V + nucleus of V

Parvicellular reticular nucleus

Paramedian reticular nucleus

IX

VIII

Inf. olive
Med. lemniscus   **UPPER MEDULLA**

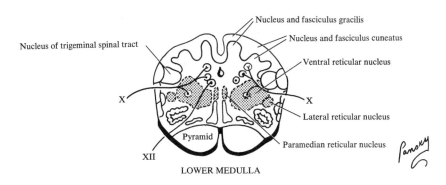

Nucleus and fasciculus gracilis

Nucleus and fasciculus cuneatus

Nucleus of trigeminal spinal tract

Ventral reticular nucleus

X

X

Lateral reticular nucleus

Pyramid

Paramedian reticular nucleus

XII

LOWER MEDULLA

Pansky

118. RETICULAR FORMATION: SPECIFIC RETICULAR NUCLEI AND RELATED INFORMATION

III. Specific reticular nuclei (cont.)

C. IN THE MIDBRAIN, the RF is a forward extension of the pontine RF, and the mesencephalic reticular nucleus is the major nucleus found in this part of the brainstem
   1. Mesencephalic reticular nucleus is surrounded by the red nucleus, tectum, and ascending lemnisci
   2. From the midbrain, the RF extends into the diencephalon as the zona incerta of the subthalamus
   3. Pallidotegmental and corticofugal fibers from the precentral gyrus end in this part of the RF, which suggests a "motor" function
   4. Mesencephalic RF does not send fibers as far caudally as the spinal cord

IV. Some points of importance relative to reticular nuclei and formation

A. LARGE CELLS OF THE RF are restricted to its medial part (medial $\frac{2}{3}$), where they are intermingled with both small and large cells
B. THE LATERAL ONE THIRD OF THE RF has only small cells
C. THE LARGEST CELLS are found in more caudal locations of the RF (i.e., caudal pons and medulla)
D. NO TYPICAL GOLGI TYPE II CELLS (cells with short axons) in the RF
E. ALL AXONS OF THE RF CELLS appear to project at least some distance caudal and/or rostral
F. DENDRITES OF RF CELLS are long and radiating. Most of the dendrites are spread out in a plane perpendicular to the long axis of the brainstem
G. EACH HALF OF THE RF BORDERS (peripherally) on the long ascending and descending fiber bundles traversing the brainstem (MLF, medial lemniscus, spinothalamic tract, etc.) and on particular nuclei, which in certain places intrude somewhat into its territory
H. INDISTINCT BOUNDARIES exist between different reticular nuclei. However, rather obvious architectonic differences between relatively small areas of the RF do exist. There are also differences between areas with regard to fiber connections and functions

V. Reticular activating system (ascending activating system or activating system)

A. WHEN ONE GOES FROM A DROWSY OR RELAXED STATE to one of attention or alertness, the EEG shows low-voltage fast activity. This is referred to as desynchronization, activation, or arousal reaction
   1. This reaction can be elicited by stimulating a great variety of structures within the CNS
   2. Lesions placed in the central parts of the brainstem result in the manifestation of behavioral somnolence and electrocortical synchrony
B. THE RF DOES NOT CORRESPOND to the reticular activating system. The RF is too restrictive and is a morphologic designation
C. THE RETICULAR ACTIVATING SYSTEM does include part of the RF, as well as the nonspecific thalamic nuclei (midline, intralaminar, and reticular)
   1. These nonspecific nuclei project to hypothalamic and other specific thalamic nuclei
   2. These nonspecific nuclei receive fibers from the RF and the spinal cord
   3. The thalamic nuclei project to the cerebral cortex to influence its level of activity
D. RAPHE NUCLEI receive fibers from and give off fibers to the same part of the nervous system as does the RF, with only minor differences
E. NEURONS OF THE RAPHE NUCLEI are rich in serotonin, whereas some components of reticular nuclei of the tegmentum of the pontomesencephalic brainstem are rich in norepinephrine and dopamine

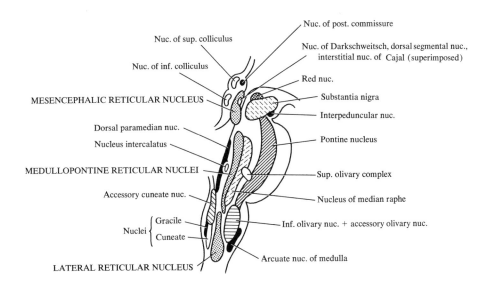

Nuc. of sup. colliculus

Nuc. of inf. colliculus

MESENCEPHALIC RETICULAR NUCLEUS

Dorsal paramedian nuc.

Nucleus intercalatus

MEDULLOPONTINE RETICULAR NUCLEI

Accessory cuneate nuc.

Nuclei { Gracile / Cuneate

LATERAL RETICULAR NUCLEUS

Nuc. of post. commissure

Nuc. of Darkschweitsch, dorsal segmental nuc., interstitial nuc. of Cajal (superimposed)

Red nuc.

Substantia nigra

Interpeduncular nuc.

Pontine nucleus

Sup. olivary complex

Nucleus of median raphe

Inf. olivary nuc. + accessory olivary nuc.

Arcuate nuc. of medulla

BRAINSTEM FORMATIONS (UNSEGMENTED)

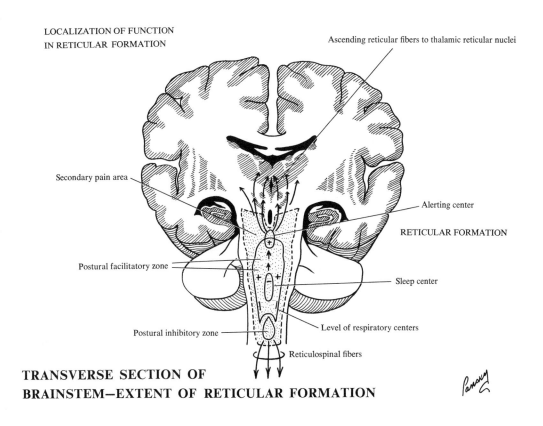

LOCALIZATION OF FUNCTION
IN RETICULAR FORMATION

Ascending reticular fibers to thalamic reticular nuclei

Secondary pain area

Postural facilitatory zone

Postural inhibitory zone

Alerting center

RETICULAR FORMATION

Sleep center

Level of respiratory centers

Reticulospinal fibers

TRANSVERSE SECTION OF
BRAINSTEM—EXTENT OF RETICULAR FORMATION

119. RETICULAR FIBER CONNECTIONS

I. Afferent fiber connections to the RF come from the following sources:
A. SPINORETICULAR FIBERS
   1. A massive influx of these direct fibers ascend in the spinal cord in the anterolateral funiculus. They appear to reach all parts of the RF bilaterally, but most are distributed to the medial $\frac{2}{3}$ of the RF and to the magnocellular nucleus of the medulla and pontine reticular nuclei
   2. RF located in pons-mesencephalon transitional zone of brainstem also receives a large influx of spinoreticular fibers.
B. SPINOTHALAMIC TRACT FIBERS and collaterals terminate on lateral reticular nucleus, which in turn projects to the cerebellum
C. CRANIAL NERVE FIBERS listed below send many sensory collaterals to RF
   1. Trigeminal sensory nucleus fibers
   2. Vestibular pathway fibers
   3. Ascending auditory pathway fibers
   4. Optic fibers from superior colliculus enter RF via tectoreticular fibers
D. CEREBELLUM
   1. Fastigial nuclei via the uncinate fasciculus are distributed throughout the medial $\frac{2}{3}$ of the medullary reticular formation
   2. Dentate nuclei via the superior cerebellar peduncle
E. HYPOTHALAMUS sends fibers from the lateral nuclear group to the mesencephalic reticular formation
F. GLOBUS PALLIDUS (medial portion) provides small input to the mesencephalic tegmentum RF
G. CEREBRAL CORTEX fibers to the RF are prominent (mainly from sensorimotor area)
   1. Corticoreticular fibers supply the region of the magnocellular and pontine tegmental nuclei
   2. Fibers from cerebral cortex also project directly to mesencephalic RF

II. Efferent fibers from the RF pass mainly to spinal cord, thalamus, and other nuclei within and above the brainstem
A. RETICULOSPINAL FIBERS (DESCENDING TO THE SPINAL CORD)
   1. From medial $\frac{2}{3}$ of pontine RF to end in laminae VII and VIII
   2. From medial $\frac{2}{3}$ of medullary RF, fibers descend (crossed and uncrossed) to end in laminae VII and VIII of spinal cord
   3. The reticulospinal fibers terminate on interneurons of the spinal cord
   4. No somatotopic pattern is noted in the origin of fibers
   5. Many RF fibers contain biogenic amines and serotonin

III. Ascending fibers
A. GIANT CELLS in caudal portion of RF do not have ascending projections
B. ASCENDING FIBERS originate mainly from pontine and medullary nuclei
C. ASCENDING RF FIBERS are mostly uncrossed, but some do decussate or cross
D. MANY RF CELLS FROM PONTINE AND MEDULLARY REGIONS have axons that project beyond the mesencephalon
E. AN IMPORTANT GROUP OF EFFERENT FIBERS of the mesencephalic reticular formation passes rostrally via mamillary peduncle, dorsal longitudinal fasciculus, and medial forebrain bundle to mamillary bodies, lateral hypothalamic areas, and preoptic and septal areas, respectively

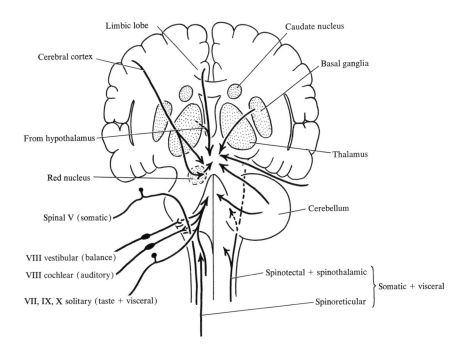

INPUT TO BRAINSTEM RETICULAR FORMATION

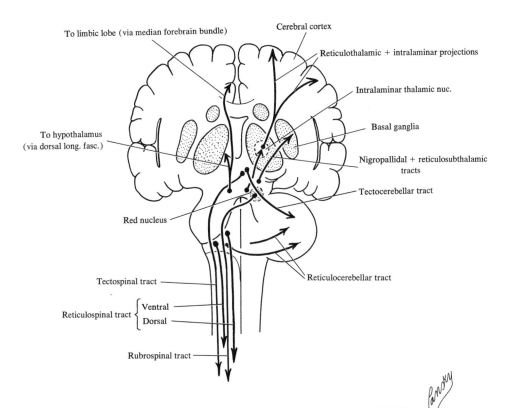

OUTPUT FROM BRAINSTEM RETICULAR FORMATION

120. LIMBIC AND OLFACTORY SYSTEMS

I. **Introduction:** the term "limbic system" was derived from the concept of a "limbic lobe" presented in 1787 by the French anatomist Broca. The word "limbic" means to border, fringe, or ring, and the term "limbic system" was first used to describe a series of structures that form a border or ring about the medial wall of the cerebrum. Various approaches have been used to classify limbic structures and to relate them to each other based on a common principle such as (1) evolutionary development, (2) role in olfaction, and (3) fiber connections and interconnections

II. **Limbic lobe** designates brain tissue that surrounds the rostral brainstem and interhemispheric commissures; it underlies the neocortical mantle. It appears early phylogenetically and demonstrates a certain consistent morphology both grossly and microscopically. Grossly it includes the following structures
   A. SUBCALLOSAL AND CINGULATE GYRI
   B. CINGULUM, a long association pathway, which connects the orbital frontal cortex and hippocampal cortex
   C. PARAHIPPOCAMPAL GYRUS AND UNDERLYING HIPPOCAMPAL FORMATION
   D. ISTHMUS AND VARIOUS GYRI SURROUNDING THE OLFACTORY FIBERS

III. **Structures in the limbic system** form a ring or loop positioned deep on the medial surface of the cerebrum and include the following gray areas and fiber bundles
   A. LIMBIC LOBE: ring-shaped convolution on the medial cerebrum
   B. AMYGDALA: amygdaloid nuclear complex
   C. HYPOTHALAMUS: specifically the mamillary bodies
   D. CERTAIN MIDBRAIN NUCLEI that have reciprocal relations with the hypothalamus
   E. THALAMUS: anterior and dorsomedial nuclei
   F. FORNIX, INDUSIUM GRISEUM (SUPRACALLOSAL GYRUS), AND MAMILLOTHALAMIC TRACT
   G. STRIA TERMINALIS, STRIA MEDULLARIS, AND HABENULA NUCLEI
   H. SEPTAL AREA AND NUCLEI
   I. ORBITOFRONTAL CORTEX, PREPYRIFORM CORTEX, AND OLFACTORY TUBERCLE

IV. **Primary roles of the limbic system:** numerous neuroanatomic substrates underlying human behavioral and emotional expression reside in the limbic system. This system is functionally associated with
   A. EMOTIONAL ACTIVITIES essential for the self-preservation of the individual (e.g., feeding, fight, and flight)
   B. ACTIVITIES ESSENTIAL FOR THE PRESERVATION OF THE SPECIES (e.g., mating, procreation, and care of the young)
   C. VISCERAL ACTIVITIES associated with both of the above
   D. MECHANISMS FOR MEMORY
   E. NUMEROUS OTHER ACTIVITIES OF THE HYPOTHALAMUS (e.g., modulation and regulation)

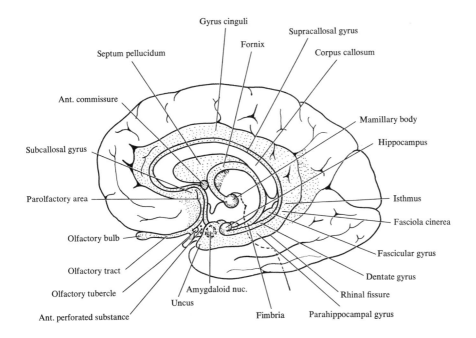

Gyrus cinguli

Septum pellucidum

Fornix

Supracallosal gyrus

Corpus callosum

Ant. commissure

Mamillary body

Hippocampus

Subcallosal gyrus

Parolfactory area

Isthmus

Fasciola cinerea

Olfactory bulb

Fascicular gyrus

Olfactory tract

Dentate gyrus

Olfactory tubercle

Rhinal fissure

Amygdaloid nuc.

Uncus

Ant. perforated substance

Fimbria

Parahippocampal gyrus

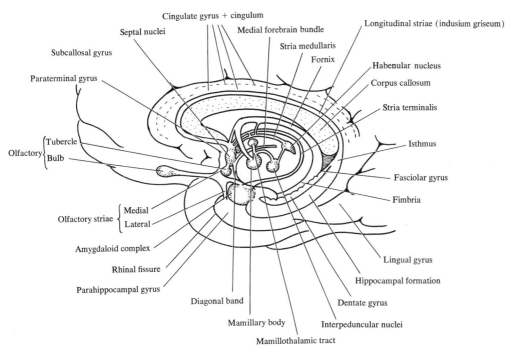

Cingulate gyrus + cingulum

Septal nuclei

Medial forebrain bundle

Stria medullaris

Longitudinal striae (indusium griseum)

Subcallosal gyrus

Fornix

Paraterminal gyrus

Habenular nucleus

Corpus callosum

Stria terminalis

Olfactory {Tubercle / Bulb}

Isthmus

Fasciolar gyrus

Fimbria

Olfactory striae {Medial / Lateral}

Amygdaloid complex

Lingual gyrus

Rhinal fissure

Hippocampal formation

Parahippocampal gyrus

Diagonal band

Dentate gyrus

Mamillary body

Interpeduncular nuclei

Mamillothalamic tract

LIMBIC SYSTEM

121. HIPPOCAMPAL FORMATION

I. **The hippocampal formation** consists of the hippocampus (Ammon's horn), dentate gyrus, and subicular complex (part of the parahippocampal gyrus which continues with the hippocampus)

A. IT DEVELOPS EMBRYOLOGICALLY on the medial wall of the temporal lobe along the hippocampal fissure

1. The hippocampal fissure parallels the choroid fissure, the point of invagination of the choroid plexus into the temporal horn of the lateral ventricle
2. Both fissures arch anteriorly and inferiorly from the interventricular foramina to the tip of the temporal horn of the lateral ventricle
3. The hippocampal fissure deepens, and the invaginated part extends into the temporal horn, as the hippocampus or ram's horn as it is often called
4. The dentate and parahippocampal gyri form from the lips of this fissure

B. THE ALVEUS, a thin layer of white matter comprised of axons from the cells of the hippocampus, lines the whole ventricular surface of the hippocampal formation and constitutes the main efferent pathway

1. Pyramidal cell is the principal cell type of the hippocampus
2. Basket cells, a second cell type, lie proximal to the pyramidal cells

C. THE HIPPOCAMPUS has the following 3 major cellular layers

1. Plexiform layer is the external layer located adjacent to the ventricular surface (alveus), which contains axons of pyramidal cells and afferent neurons
2. Pyramidal layer is comprised of pyramidal cells
3. Polymorphic cell layer
4. Secondary lamina are formed (from alvear surface outward): stratum oriens (composed of basal dendrites and basket cells), pyramidal layer, stratum radiatum, stratum lacunosum, and stratum moleculare. The last 3 are the molecular layer

D. THE FIMBRIA is formed by axons in the alveus which course over the hippocampus to its medial border. These fibers make up the beginning of the fornix system. Its thin free border is continuous and contiguous with the epithelium of the choroid fissure

E. PARAHIPPOCAMPAL GYRUS (ENTORHINAL CORTEX, AREA 28) extends from the hippocampal fissure to the collateral sulcus laterally. Its superior portion is called the subiculum

F. MOST OF THE CORTEX of the parahippocampal gyrus has a neocortex (isocortex) structure of 6 layers. The subiculum, hippocampal formation, and the dentate gyrus have an allocortex structure of 3 layers

G. THE INDUSIUM GRISEUM (LONGITUDINAL STRIAE OR SUPRACALLOSAL GYRUS) is a thin gray vestigial convolution of the hippocampal formation

1. It remains positioned above the developing fibers of the corpus callosum
2. It contains 2 thin bands of myelinated fibers (white matter) which appear as the medial and lateral longitudinal ridges or striae on the superior surface of the corpus callosum in each cerebral hemisphere

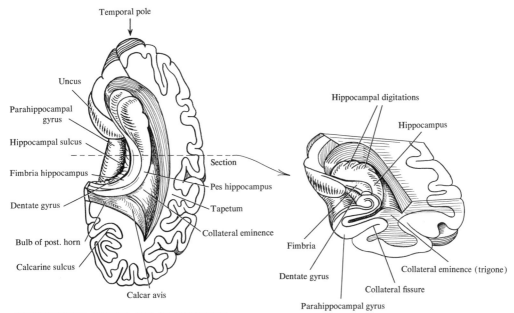

**SUPERIOR VIEW OF INFERIOR
HORN OF LATERAL VENTRICLE**

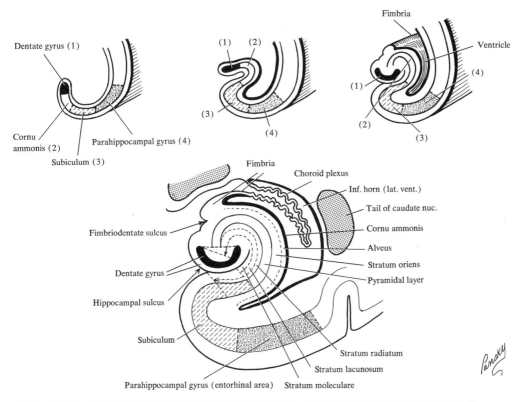

CORONAL SECTION OF INFERIOR HORN OF LATERAL VENTRICLE

122. FIBER CONNECTIONS OF THE HIPPOCAMPAL FORMATION: AFFERENTS

I. Afferent fiber connections

A. ENTORHINAL CORTEX (AREA 28) occupies the anterior end of the parahippocampal gyrus and is one primary site where afferent fibers arise
   1. This cortical area is a part of the piriform lobe (olfactory association cortex)
   2. No direct nerve fibers connect this lobe with the olfactory area
   3. Fibers from area 28 give rise to two major pathways
      a. Medial portion sends fibers via the "alvear path" to the hippocampus
      b. Lateral portion sends fibers via the "perforant path," which traverses the subiculum on its way to the hippocampus and dentate gyrus
   4. Axons making up these two pathways establish synaptic contacts with dendrites of pyramidal cells in the hippocampus. They do not make direct contact with granule cells in the dentate gyrus
B. DIAGONAL BAND FIBER COMPONENTS from the medial septal nucleus enter the hippocampal formation via the fimbria of the fornix and are distributed in a diffuse fashion to cells in the hippocampal formation
C. THE CINGULUM projects fibers to the entorhinal cortex, which in turn relays impulses on to the hippocampus
D. THE INDUSIUM GRISEUM may send some fibers to the hippocampal formation
E. CORTICAL AREAS such as the piriform and prefrontal cortices contribute some fibers to the hippocampus
F. ANTERIOR THALAMIC NUCLEUS projects to the presubicular portion of the hippocampal formation
G. OLFACTORY, VISUAL, AUDITORY, GUSTATORY, AND SOMATOSENSORY CORTICAL AREAS indirectly supply the hippocampal formation via synaptic contacts with area 28 (entorhinal cortex). None of the direct afferent pathways to the hippocampal formation appears to transmit olfactory impulses

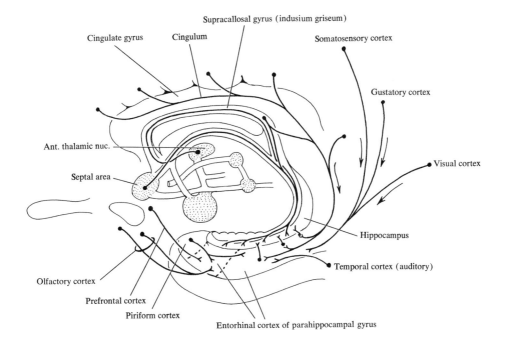

Supracallosal gyrus (indusium griseum)

Cingulate gyrus

Cingulum

Somatosensory cortex

Gustatory cortex

Ant. thalamic nuc.

Septal area

Visual cortex

Hippocampus

Temporal cortex (auditory)

Olfactory cortex

Prefrontal cortex

Piriform cortex

Entorhinal cortex of parahippocampal gyrus

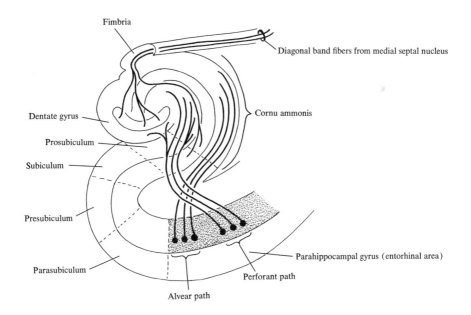

Fimbria

Diagonal band fibers from medial septal nucleus

Dentate gyrus

Cornu ammonis

Prosubiculum

Subiculum

Presubiculum

Parahippocampal gyrus (entorhinal area)

Parasubiculum

Perforant path

Alvear path

FIBER CONNECTIONS OF HIPPOCAMPAL FORMATION— AFFERENTS

123. FIBER CONNECTIONS OF THE HIPPOCAMPAL FORMATION: EFFERENTS

I. Efferent fiber connections

A. THE FORNIX comprises the main efferent fiber system of the hippocampal formation

   1. Myelinated axons of the large pyramidal cells of the hippocampus form the thin surface lining of the lateral ventricle (temporal horn) known as the alveus. These fibers converge to form the fimbria of the fornix

   2. The ''crura'' of the fornix are terms applied to the fimbriae at the level of the posterior end of the hippocampus as they bend backward and arch under the splenium of the corpus callosum

   3. Hippocampal or fornical commissure is made up of a thin sheet of crossing fibers from the ventral aspect of the fornix that serve to connect the 2 hippocampi. This commissural connection is usually poorly developed in man

   4. Body of the fornix is formed as the 2 crura join and run forward inferior to the corpus callosum to the level of the rostral margin of the thalamus

   5. Anterior columns of the fornix are formed as the body splits into 2 separate fiber bundles that bend and continue inferiorly between the anterior commissure (ventrally) and the interventricular foramina (dorsally)

      a. The columns of the fornix extend into the hypothalamus, where they divide into medial and lateral parts

      b. Most of the fibers in the columns end in the mamillary bodies

      c. Some fibers exit the columns to enter the anterior thalamic nuclei

   6. Precommissural and postcommissural fornices are formed by fibers within the columns from the fimbria which regroup and separate off from the columns as components rostral and caudal to the anterior commissure

      a. Precommissural fibers pass rostral to the anterior commissure and (1) supply the septal area, (2) fibers located near the anterior pole of the hippocampal formation extend to the most lateral region of the dorsal septal area, and (3) fibers located more caudally in the hippocampal formation extend to the more medial parts of the dorsal septal area

      b. Postcommissural fibers pass just caudal to the anterior commissure and

         i. Descend through the hypothalamus to supply the mamillary bodies, adjacent parts of the hypothalamus, anterior thalamic nucleus, and perhaps areas of the mesencephalon

         ii. Fibers take origin from subicular complex cells instead of the pyramidal cells of the hippocampus

B. MAMILLOTHALAMIC TRACT is comprised of fibers that connect the mamillary body and the anterior thalamic nucleus of the same side

   1. The anterior thalamic nucleus has reciprocal connections with the cingulate gyrus (cortex) via fibers that run through the medullary core

   2. These and other connections show the complex pathways through which impulses from the hippocampal formation may be projected to different parts of the neuroaxis

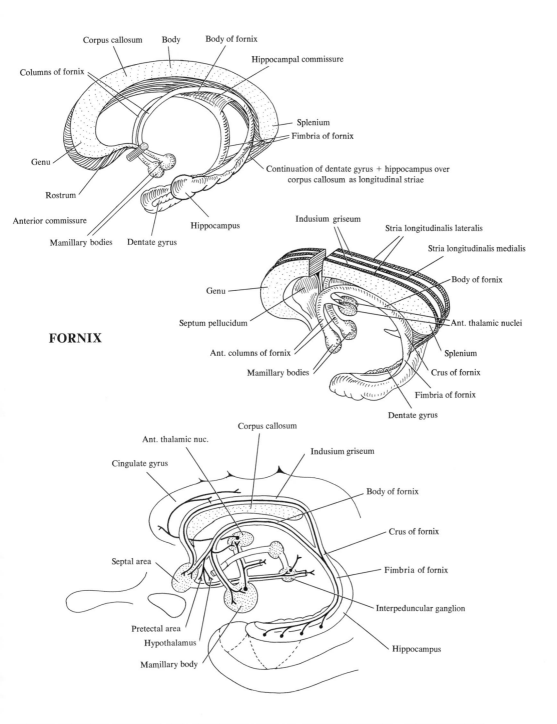

FORNIX

FIBER CONNECTIONS OF HIPPOCAMPAL FORMATION— EFFERENTS

124. FUNCTIONS OF THE HIPPOCAMPAL FORMATION

I. **Introduction:** both morphologic and functional evidence supports the thesis that the hippocampal formation takes part in a wide variety of activities such as endocrine, learning, recent memory, expression of emotion, hypothalamic, and autonomic function

A. ENDOCRINE FUNCTIONS are regulated or modulated by the hippocampal formation
   1. In the ventral region of this structure, estradiol-concentrating neurons are numerous and closely packed. Corticosterone is also localized in heavy concentrations in the ventral aspect of this formation
   2. Corticosterone injections cause a general inhibitory effect on the neurons in the hippocampus
   3. Hippocampal lesions (or section of the fornix) usually result in the disruption of the diurnal rhythm of ACTH release
   4. Stimulation of the hippocampal formation caused the inhibition of ovulation in spontaneously ovulating rats

B. LEARNING AND RECENT MEMORY ACTIVITIES, as revealed in experimental animal studies and human clinical investigations, strongly indicate a significant role for the hippocampal formation. EEG and ablation experiments support the thesis that the formation is involved in learning processes
   1. In humans with lesions of the hippocampus, learning and memory disabilities are common. The Korsakoff syndrome is such a disturbance, which demonstrates both retrograde and antegrade amnesia. The patient has difficulty in remembering newly acquired information and in recalling events in the recent past
      a. Results from toxic effects of alcohol or from vitamin B deficiency, causing damage to either the hippocampus or its target neurons
   2. Patients subjected to temporal lobectomies (for the treatment of psychomotor epilepsy) exhibit a disorder referred to as short-term memory defect, which is characterized by not being able to learn and remember new simple facts, such as names, but by being able to recall events that occurred in the distant past and those just prior to surgery with little or no difficulty

C. EXPRESSION OF EMOTION, HYPOTHALAMIC ACTIVITIES, AND AUTONOMIC FUNCTIONS
   The hippocampal formation plays a role in controlling aggressive behavior, but its effect upon aggression are not uniform, as observed when
      a. Stimulation of areas in the vicinity of the amygdala appears to facilitate the onset of aggressive behavior
      b. Stimulation of areas close to the septal area suppresses such reactions
      c. Hippocampal formation may exert its effects on hypothalamic aggression via its differential synaptic connections in the septal area since both the amygdala and areas closer to the septal area project their axons to different areas of the septum. The septal area may serve as a relay nucleus for impulses from the hippocampal formation to the hypothalamus
   2. In humans, aggressive behavior has been associated with lesions, tumors, and seizure activity involving the hippocampal formation and temporal lobe
      a. Seizures of the temporal lobe involving the hippocampus (partial or temporal lobe seizures) may result in confusion and cause hallucinations involving gustatory and auditory sensations
      b. An attempt to direct or move a patient with temporal lobe seizures may initiate aggressive behavior
   3. Hippocampectomized animals appear more active, more exploratory, and exhibit an increased startle reaction in response to sudden loud noises

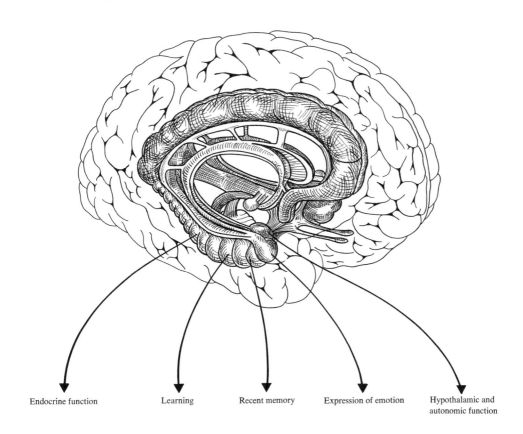

Endocrine function   Learning   Recent memory   Expression of emotion   Hypothalamic and autonomic function

FUNCTIONS OF THE HIPPOCAMPAL FORMATION

125. AMYGDALA (AMYGDALOID BODY OR NUCLEAR COMPLEX): ANATOMY AND CONNECTIONS

I. **Introduction:** the term "amygdala" means almond-shaped and is used to describe this series of nuclear masses (nerve cell bodies) and associated nerve fibers that are located in the dorsomedial portion of the temporal lobe of the cerebrum and form the medial, anterior, and superior portions of the ventral tip of the temporal horn of the lateral ventricle. It is continuous with the uncus and closely related to the hippocampus. It fuses with the tail of the caudate nucleus, and the stria terminalis originates from its more caudal portion. In addition to subnuclear groups, it consists of a cortical mantle which includes the piriform lobe. It is also intimately related to the adjacent cortex known as the prepiriform area (rostral levels) and the periamygdaloid cortex (caudal levels)

II. **The amygdaloid nuclear complex** is divided into 2 main groups of nuclei
A. CORTICOMEDIAL AMYGDALOID is continuous with the anterior perforated substance, diagonal band, putamen, caudate nucleus, and cortical areas surrounding the uncus
   1. This portion of the complex is small in man and includes the anterior area (poorly differentiated), lateral olfactory tract nucleus (poorly developed), medial amygdaloid nucleus, and cortical amygdaloid nucleus
B. Basolateral amygdaloid is the largest and best-differentiated part of the nuclear complex. It is continuous with the claustrum and cortex of the parahippocampal gyrus and includes the lateral, basal, and accessory basal nuclei

III. **Fiber connections of the amygdaloid complex** are incompletely established in the human brain. Information on nerve fiber connections has been obtained from electrophysiologic and nerve degeneration experiments in cats and dogs
A. AFFERENT CONNECTIONS are closely linked with cortical tissue and include
   1. Olfactory bulb and anterior olfactory nucleus fibers enter the corticomedial nuclear group via the lateral olfactory stria
   2. Prepiriform and piriform cortices send indirect olfactory connections to the basolateral group of nuclei
   3. Orbitofrontal cortex has reciprocal connections with the basolateral group
   4. Diagonal band nuclei send direct fiber connections to the basolateral group
   5. Inferior temporal gyrus sends direct connections to the lateral nucleus
   6. Medial thalamus sends fibers to the anterior amygdala
   7. Ventromedial hypothalamus: to the amygdala (diffusely) via stria terminalis
   8. Brainstem reticular formation (dopaminergic neurons) to central amygdala
B. EFFERENT CONNECTIONS
   1. Stria terminalis is the most completely established efferent pathway
      a. It arises mainly from corticomedial group of nuclei and adjacent piriform cortex
      b. It conveys fibers to the septal areas, medial preoptic area, anterior medial hypothalamus, ventromedial hypothalamic nucleus, habenular nucleus (via stria medullaris), and anterior perforated substance
   2. Ventral amygdalofugal pathways take origin in the basolateral amygdala, piriform and periamygdala cortices
      a. They extend to the anterior perforated substance, preoptic area, septal area, dorsomedial nucleus of the thalamus, hypothalamic nuclei, parahippocampal gyrus, and brainstem reticular formation via the medial forebrain bundle

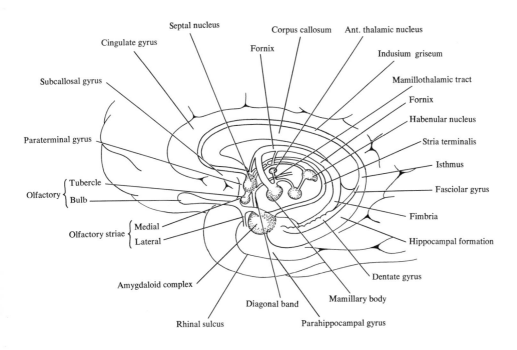

Septal nucleus

Cingulate gyrus

Corpus callosum

Ant. thalamic nucleus

Fornix

Indusium griseum

Subcallosal gyrus

Mamillothalamic tract

Fornix

Paraterminal gyrus

Habenular nucleus

Stria terminalis

Isthmus

Olfactory { Tubercle / Bulb

Fasciolar gyrus

Olfactory striae { Medial / Lateral

Fimbria

Hippocampal formation

Amygdaloid complex

Dentate gyrus

Diagonal band

Mamillary body

Rhinal sulcus

Parahippocampal gyrus

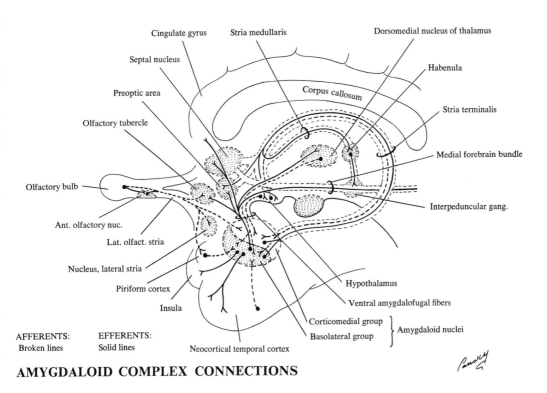

Cingulate gyrus

Stria medullaris

Dorsomedial nucleus of thalamus

Septal nucleus

Habenula

Preoptic area

Corpus callosum

Stria terminalis

Olfactory tubercle

Medial forebrain bundle

Olfactory bulb

Interpeduncular gang.

Ant. olfactory nuc.

Lat. olfact. stria

Nucleus, lateral stria

Piriform cortex

Hypothalamus

Insula

Ventral amygdalofugal fibers

Corticomedial group

AFFERENTS: Broken lines

EFFERENTS: Solid lines

Basolateral group

} Amygdaloid nuclei

Neocortical temporal cortex

AMYGDALOID COMPLEX CONNECTIONS

126. FUNCTIONS OF THE AMYGDALA

I. Olfaction does not seem to be closely related to the amygdaloid complex since bilateral lesions or destruction does not impair olfactory discrimination

II. Emotional behavior is regulated by the amygdala through its effects upon the hypothalamus. These effects upon the hypothalamic rage mechanism may be inhibitory or facilitatory, though the former is often more pronounced
A. LESIONS of this complex in rats, cats, and monkeys result in "taming effects"
B. HYPEREMOTIONAL STATE may also result from lesions in this same area
C. ELECTRICAL STIMULATION OF THE CORTICOMEDIAL GROUP OF NUCLEI results in the inhibition of aggressive behavior and stimulation of the lateral nucleus facilitates aggressive activities
D. STIMULATION OF THE AMYGDALOID REGION in humans causes feelings of confusion, fear, and amnesia of events that occurred during the stimulation

III. Endocrine function is influenced by the amygdaloid complex as indicated
A. NEURONS IN THE AMYGDALA have the ability to concentrate estrogen
B. OVULATION IS INDUCED by the stimulation of the corticomedial group of nuclei
C. OVULATION IS PREVENTED by transection of the stria terminalis
D. ACTH IS RELEASED when the nasolateral nucleus is stimulated or when a lesion is made in the corticomedial group
E. GROWTH HORMONE IS RELEASED when the basolateral nucleus is stimulated, whereas the corticomedial group serves to inhibit its release

IV. Somatic responses, observed when the amygdala is stimulated, include
A. TURNING OF THE HEAD AND EYES to the side opposite that stimulated
B. EXHIBITING COMPLEX RHYTHMIC MOVEMENTS related to chewing and swallowing

V. Autonomic responses, observed when the amygdala is stimulated, include
A. CHANGES IN RESPIRATORY RATE, RHYTHM, AND AMPLITUDE, as well as inhibition of respiration
B. INCREASES AND DECREASES IN ARTERIAL BLOOD PRESSURE and changes in heart rate
C. INCREASED SECRETION OF DIGESTIVE JUICES, which may cause peptic ulcers

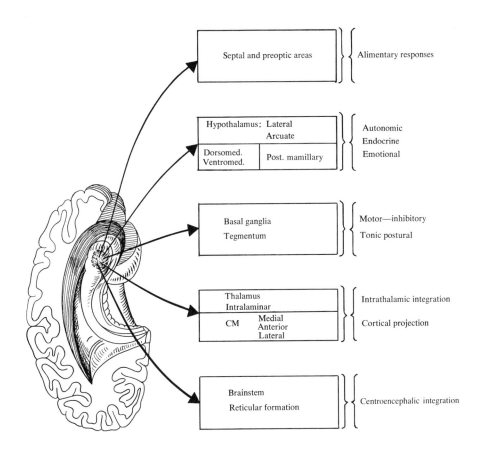

Septal and preoptic areas	Alimentary responses
Hypothalamus: Lateral / Arcuate / Dorsomed. Ventromed. / Post. mamillary	Autonomic / Endocrine / Emotional
Basal ganglia / Tegmentum	Motor—inhibitory / Tonic postural
Thalamus / Intralaminar / CM / Medial / Anterior / Lateral	Intrathalamic integration / Cortical projection
Brainstem / Reticular formation	Centroencephalic integration

GENERAL FUNCTIONS OF THE AMYGDALA OLFACTORY SYSTEM

127. OLFACTORY SYSTEM: FUNCTIONAL ANATOMY

I. **The rhinencephalon (nose-brain)** consists of the olfactory nerves, bulbs, tracts, striae, paraolfactory (subcallosal) areas, anterior perforated substance, prepiriform cortex, amygdaloid complex, and the anterior olfactory nuclei. Vertebrates more dependent on olfactory senses (macrosomatic) than on vision (microsomatic) have more prominent rhinencephalic structures. Man is an example of the latter

A. PERIPHERAL OLFACTORY RECEPTORS are located in small, specialized areas of the nasal mucosa referred to as the olfactory epithelium

   1. It is usually pseudostratified columnar epithelium in type

   2. It is located specifically on the superior concha, roof of the nasal cavity, and on the superior portion off the nasal septum. The mucosa lies above the path of the major air currents entering the nose, so that odorous molecules must diffuse up to the cells or be drawn up by respiratory changes

   3. The receptor cells are modified neurons originally from the CNS itself. They have 2 processes (bipolar): one passes to the brain to form the olfactory nerve; the other has fine cilia and extends out to the surface of the olfactory mucosa

     a. There are about 100 million of these cells in the olfactory epithelium interspersed among sustentacular cells

     b. The mucosal end of the olfactory cell forms a knob called the olfactory vesicle, from which large numbers of olfactory hairs or cilia (0.3 μm in length) project into the mucus that coats the inner surface of the nasal cavities. These projecting olfactory hairs are believed to react to odors in the air and then to stimulate the olfactory cells

     c. Spaced among the olfactory cells in the olfactory membrane are many small glands of Bowman that secrete mucus onto the surface of the membrane

   4. To be detected, an odorous substance must release molecules that diffuse into the air and pass into the nose to the region of the olfactory mucosa. Molecules then dissolve in the layer of mucus over the receptors and form a relation with the receptor that depolarizes the membrane to initiate an action potential in the afferent nerve fiber

II. **Physiologic basis for discrimination of the thousands of different odor qualities** is speculative

A. THERE ARE NO APPARENT MICROSCOPIC DIFFERENCES in the receptor cells, yet, meaningful differences are thought to exist at the molecular level

B. ODOR DISCRIMINATION may depend on a large number of different types of interactions between the odor-substance molecules and receptor sites. Assuming the receptor cells have 20–30 different types of receptor sites, each may be able to interact with many different odor molecules but responds best to a specific molecule

C. OLFACTORY DISCRIMINATION not only depends on action potential patterns generated in the different afferent neurons but also varies with attentiveness. The state of the olfactory mucosa is affected by congestion, hunger (sensitivity is greater when hungry), sex (women have a keener sensitivity) and smoking (decreases sensitivity)

D. KNOWLEDGE OF THE ODOR of a substance is aided by the stimulation of other receptors accounting for odors of pungence, acridity, coolness, and irritation

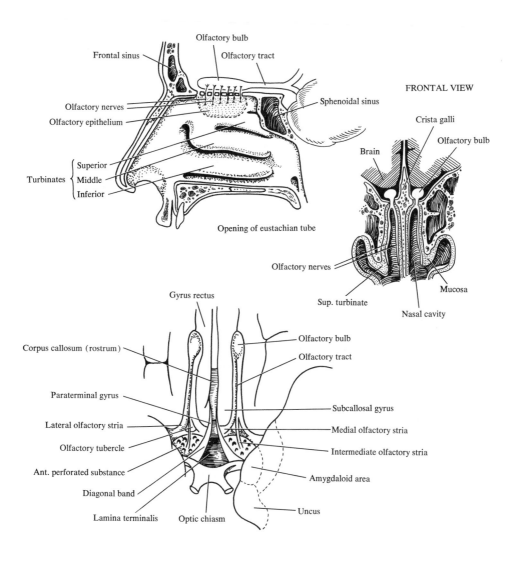

Olfactory bulb

Frontal sinus

Olfactory tract

FRONTAL VIEW

Olfactory nerves

Sphenoidal sinus

Olfactory epithelium

Crista galli

Olfactory bulb

Brain

Turbinates { Superior Middle Inferior }

Opening of eustachian tube

Olfactory nerves

Mucosa

Sup. turbinate

Nasal cavity

Gyrus rectus

Corpus callosum (rostrum)

Olfactory bulb

Olfactory tract

Par(A)terminal gyrus

Subcallosal gyrus

Lateral olfactory stria

Medial olfactory stria

Olfactory tubercle

Intermediate olfactory stria

Ant. perforated substance

Amygdaloid area

Diagonal band

Lamina terminalis

Optic chiasm

Uncus

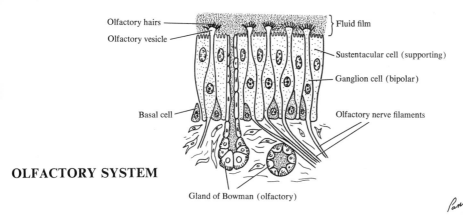

Olfactory hairs

Fluid film

Olfactory vesicle

Sustentacular cell (supporting)

Ganglion cell (bipolar)

Basal cell

Olfactory nerve filaments

OLFACTORY SYSTEM

Gland of Bowman (olfactory)

128. OLFACTORY NERVES, BULBS, AND STRIAE

I. **The olfactory nerves, bulbs, and striae** comprise the most peripheral portion of the olfactory apparatus and possess the following relationships

A. RECEPTOR CELLS possess elongated proximal processes (grouped as filaments) which pass through the openings of the cribriform plate of the ethmoid bone as the olfactory nerves, which then pierce the surface of the olfactory bulb to synapse with 2nd-order neurons in the rostral part of the bulb

   1. Mitral cells are large, triangular-shaped 2nd-order neurons that synapse with olfactory fila, tufted cells, or granule cells

   2. Tufted cells are more peripheral in position and smaller than mitral cells. They synapse with olfactory fila in a bushy synapse known as the olfactory glomerulus

   3. Granule cells are the smallest 2nd-order neurons of the bulb and serve an inhibitory function

B. THE ANTERIOR OLFACTORY NUCLEUS is located in the caudal portion of the bulb and consists of clusters of neurons. The cells of origin of the olfactory part of the anterior commissure are located here

C. THE PAIRED OLFACTORY TRACTS are narrow caudal projections of the bulbs which lie in the olfactory sulcus of the cerebrum. Each tract bifurcates into lateral, medial, and intermediate striae which take the following routes

   1. Most of the fibers of the medial olfactory stria enter the subcallosal area, then the rostral portion of the anterior commissure, and finally return to the contralateral olfactory bulb

   2. Fibers that end in the olfactory trigone within the anterior perforated substance appear to be mitral cell processes

   3. Fibers of the lateral stria are mainly mitral cell processes; they run along the lateral margin of the anterior perforated substance to end in the prepiriform cortex in the vicinity of the uncus. This area (prepiriform cortex) is designated as the primary olfactory cortex. Area 28, the parahippocampal gyrus, is considered the secondary olfactory cortex. Take special note that the olfactory system, unlike the other sensory systems, does not utilize the thalamus as a relay center

II. **Clinical significance**

A. ANOSMIA: total or partial loss of smell may be caused by the following factors

   1. The common cold is perhaps the most frequent cause of the loss of smell

   2. Other causes include nasal irritants, ethmoid bone fractures, meningitis, drug intoxications, psychoses, and uncinate fits

B. HYPEROSMIA has been associated with hysterias and drug addiction

C. THE SUBARACHNOID SPACE, which is in close physical proximity to the olfactory mucosa, may also become involved in certain infections involving the olfactory system

D. THE PIRIFORM AREA has been directly identified as the functional olfactory cortex in humans, since electrical stimulation causes olfactory sensations

   1. Epileptic seizures originating in the vicinity of the uncus may begin with an illusion of smell or taste (often unpleasant)

   2. The seizures may include motor phenomena (e.g., chewing movements and lip smacking), as well as changes of consciousness. The seizures are known as uncinate fits

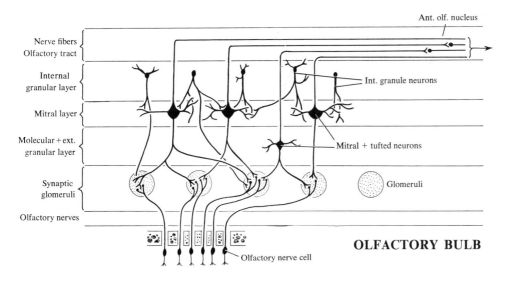

Ant. olf. nucleus

Nerve fibers
Olfactory tract

Internal
granular layer

Int. granule neurons

Mitral layer

Molecular + ext.
granular layer

Mitral + tufted neurons

Synaptic
glomeruli

Glomeruli

Olfactory nerves

Olfactory nerve cell

OLFACTORY BULB

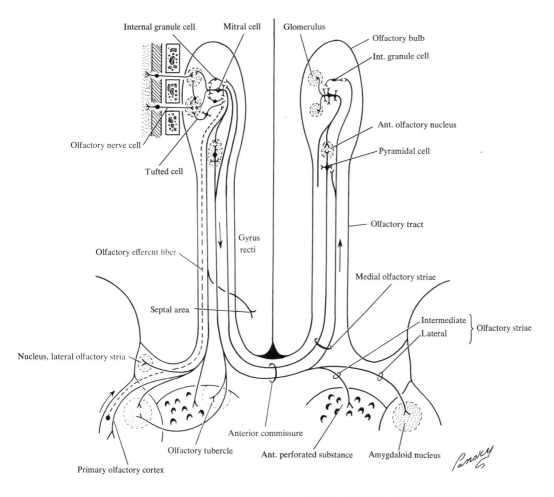

Internal granule cell   Mitral cell   Glomerulus

Olfactory bulb

Int. granule cell

Olfactory nerve cell

Tufted cell

Ant. olfactory nucleus

Pyramidal cell

Olfactory tract

Olfactory efferent fiber

Gyrus
recti

Septal area

Medial olfactory striae

Intermediate
Lateral

Olfactory striae

Nucleus, lateral olfactory stria

Anterior commissure

Olfactory tubercle

Ant. perforated substance   Amygdaloid nucleus

Primary olfactory cortex

OLFACTORY PATHWAYS

– 281 –

Unit Ten

ORGANS OF
SPECIAL SENSATION

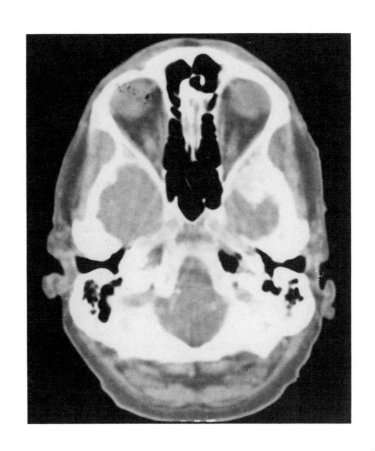

129. ANATOMY OF TASTE

I. Introduction

A. The receptors for the sense of taste are chemoreceptors that respond to certain chemicals in the external environment near the receptors

B. In man, the sense of taste is less important than the receptors for vision or hearing (not true for most other animals)

C. Taste and smell affect a person's appetite, initiate digestion, and aid in the avoidance of harmful substances, but do not exert strong or essential influences

D. It is normally impossible to perceive pure taste without sensing an overlay of smell, because true taste is a much cruder and less sensitive sensation than smell, and taste thresholds are significantly higher than thresholds of smell

II. The anatomy of taste

A. The specialized receptor organs for the sense of taste consist of about 10,000 taste buds located on the tongue, roof of the mouth (palate), tonsillar pillars, pharynx, and larynx (the latter receptors are not activated by substances in the mouth but fire during swallowing)

B. The taste buds on the anterior $\frac{2}{3}$ of the tongue are located on the free surface of the fungiform papillae, with each papilla containing 1–8 taste buds. These papillae, which are scattered singly, are especially numerous near the tip and margins of the tongue

   1. In the region of the V-shaped sulcus terminalis is a linear series of 3–14 large circumvallate papillae which cover a portion of the posterior $\frac{1}{3}$ of the tongue. The numerous taste buds on these papillae are found along the sides of the papillae, facing the troughs that surround the circumvallate papillae. The taste buds of the vallate papillae atrophy in old age, but apparently there is little to no decrease in taste sensitivity with age

   2. Moderate numbers of taste buds are found on the foliate papillae, which are located along the posterolateral surface of the tongue

C. The taste buds

   1. No structural difference in the taste buds or taste cells of the young adult has been shown to exist

   2. Each taste bud contains upward of 25–40 neuroepithelial taste cells and other less differentiated cells, the sustentacular cells, which act as reserve cells to replenish the taste cells when they die out

      a. The receptor taste cells are continuously turning over and each mature taste cell is replaced every 200–300 hours

   3. The outer tips of the taste cells are arranged around a minute taste pore

   4. From the tips of the taste cells, several microvilli (taste hairs), about 2–3 μm long and 0.2 μm wide, protrude outward through the taste pore to approach the cavity of the mouth, are bathed by mouth fluids, and provide the receptor surface for taste

      a. The receptor sites are probably located on the surface of the microvilli

      b. Not only does each taste cell have many different types of receptor cells, but the proportion of different types varies from taste cell to taste cell; thus, each taste cell can respond to several stimuli, and there is no evidence that a specific taste cell responds to only one group of stimuli

      c. Another regulatory factory for the types of stimuli is that the proteins on the cell membrane may act to open or close "gates" of the membrane to different types of potential taste stimuli

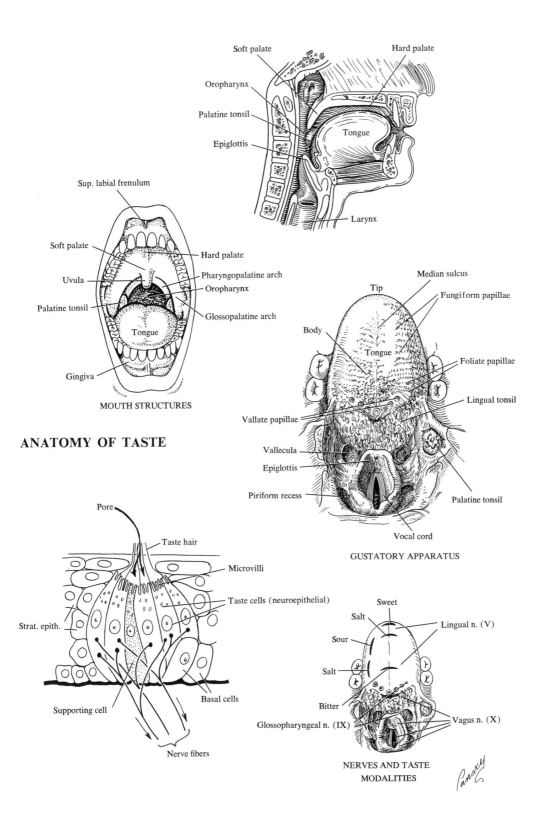

Soft palate

Hard palate

Oropharynx

Palatine tonsil

Epiglottis

Tongue

Larynx

Sup. labial frenulum

Soft palate

Hard palate

Uvula

Pharyngopalatine arch

Oropharynx

Palatine tonsil

Glossopalatine arch

Tongue

Gingiva

MOUTH STRUCTURES

ANATOMY OF TASTE

Median sulcus

Tip

Fungiform papillae

Body

Tongue

Foliate papillae

Lingual tonsil

Vallate papillae

Vallecula

Epiglottis

Piriform recess

Palatine tonsil

Vocal cord

GUSTATORY APPARATUS

Pore

Taste hair

Microvilli

Taste cells (neuroepithelial)

Strat. epith.

Supporting cell

Basal cells

Nerve fibers

Sweet

Salt

Sour

Salt

Bitter

Lingual n. (V)

Glossopharyngeal n. (IX)

Vagus n. (X)

**NERVES AND TASTE
MODALITIES**

-285-

130. SENSATION OF TASTE

I. Taste nerve fibers: interwoven among the taste cells is a branching terminal network of several taste nerve fibers that are stimulated by the taste cells. These fibers invaginate deeply into folds of the taste cell membranes, so that there is an extremely intimate contact between the taste cell and the nerves

A. ONE NERVE FIBER may innervate several receptor cells and one neuroepithelial receptor cell may be innervated by the nerve fibers of several nerve cells

B. THERE IS NO ONE-TO-ONE RELATIONSHIP between nerve and cell; thus the receptor cells lack specificity in the kind of chemical they respond to, as well as in the way they connect to the brain (each cell and each neuron respond to several taste qualities)

C. THE AFFERENT FIBERS, however, show different firing patterns in response to different substances; e.g., a fiber may fire faster when stimulated by salt, but only sporadically when stimulated by sugar (others may respond just the opposite). Thus, quality is achieved, and awareness depends on the number of neurons stimulated rather than the specific ones

　　1. The variations in intensity of taste are transmitted by the differences in the frequency of firing of nerve impulses

　　2. Adaptation is rapid (in seconds or minutes), and frequency signals quantity as well

　　3. In addition, information is increased by temperature, texture, and odor

II. Mechanism of stimulation: the mechanism by which taste receptors are stimulated and action potentials generated is not known, but there are many theories

A. IT HAS BEEN SUGGESTED THAT STEP 1 IS A LOOSE BINDING of ions or individual dissolved molecules that enter the taste pores with specific sites on the receptor cell membrane (substances must be dissolved to be tasted)

B. THE CHEMICAL-SUBSTANCE-RECEPTOR SITE COMBINATION alters the cell membrane, forming pores through which ions move to change the membrane potential of the cell

　　1. It is known that the membrane of the receptor cells depolarizes when the cell is chemically stimulated

　　2. The fact that a single receptor will respond to more than one quality of taste is explained by the single receptor cell having different sites, each able to bind with different types of molecules

III. Taste sensations are divided into 4 basic modalities or groups: sweet, sour, salt, and bitter

A. DIFFERENT TYPES OF TASTE BUDS OR RECEPTOR CELLS, to support the specificity, have not been identified. In fact, a single receptor cell can respond to varying degrees to many different substances, but not equally to each sensation

B. SALT AND SWEET, however, are perceived more acutely on the tongue; bitter and sour are perceived most acutely on the palate

　　1. The tip of the tongue is sensitive to all 4 stimuli, but especially to sweet and salty substances; the sides of the tongue, to sour substances; and the base, to bitter substances

C. THE MECHANISM by which individual taste qualities are derived from the mixed sensory signals of the taste pathways is still unknown

MODALITIES OF TASTE
(MOST SENSITIVE AREAS)

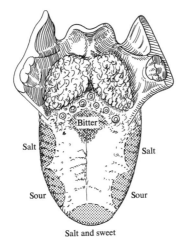

Bitter

Salt   Salt

Sour   Sour

Salt and sweet

SENSATION OF TASTE

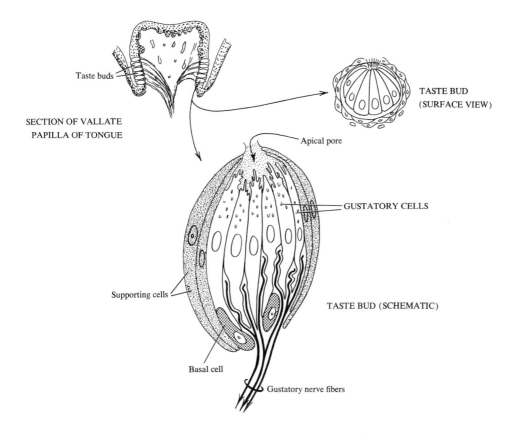

Taste buds

SECTION OF VALLATE
PAPILLA OF TONGUE

TASTE BUD
(SURFACE VIEW)

Apical pore

GUSTATORY CELLS

Supporting cells

TASTE BUD (SCHEMATIC)

Basal cell

Gustatory nerve fibers

131. PATHWAYS OF TASTE

I. **First-order fibers:** taste fibers of the facial (via the chorda tympani), glossopharyngeal, and vagus nerves pass into the medulla and end in the rostral portions of the nucleus solitarius (this portion being called the nucleus gustatorius)
A. TASTE IMPULSES are conveyed from the anterior $\frac{2}{3}$ of the tongue by the fibers of the chorda tympani (VII) and from the posterior $\frac{1}{3}$ of the tongue, palate, and pharynx by fibers in the branches of the glossopharyngeal (IX) and vagus (X) nerves
B. THE SENSORY NERVES terminate either in the taste buds or as free nerve endings in the connective tissues and epithelium other than the taste buds. Many of the free nerve endings are of terminals of general sensory fibers of the trigeminal (V) nerve and must not be confused with those for taste

II. **Second-order fibers** have axons which probably ascend as uncrossed fibers in association with the medial lemniscus and reticular formation to their termination in the most medial part of the ventral posteromedial (VPM) nucleus of the thalamus
A. SOME FIBERS OF THE GUSTATORY PATHWAYS from the nucleus solitarius to the thalamus may cross (not proven) and ascend in association with the contralateral medial lemniscus

III. **Third-order fibers** from the VPM nucleus pass through the posterior limb of the internal capsule and terminate in the lower end of the postcentral gyrus (opercular region of the parietal lobe) and possibly in the cortex of the insula and superior temporal gyrus

IV. **The reflex-type brainstem connections** of the taste pathway nuclei with the autonomic nuclei for salivation (superior and inferior salivatory nuclei) form the basis for the salivation reflexes that accompany taste responses to food stimuli on the tongue
A. OTHER AUTONOMIC BRAINSTEM CONNECTIONS are responsible for "gustation sweating," in which localized facial and forehead sweating develops in response to spicy or peppery food (can be pathologic if excessive)
B. TASTE-SENSITIVE NEURONS in the VPM thalamic nucleus also appear to respond to thermal (although not to mechanical) stimulation of the tongue

V. **The nuances of true taste** are the result of transmission via many patterns of afferent impulses in the various nerve fibers and of processing in the nuclei of the taste pathways (comparable to the transmission of sound qualities)
A. A DECREASE IN TASTE SENSITIVITY WITH AGE is paralleled by a decrease in the total number of taste buds after the age of 40. The center of the tongue, with only a few taste buds, is, in any event, relatively taste blind.
B. THE ABILITY TO TASTE CERTAIN CHEMICALS has a hereditary basis. Some people have a taste blindness to certain substances that others can taste
C. SOME TASTE ABNORMALITIES are known to be associated with alterations of body metabolism as a consequence of the action of certain chemical substances
D. TASTE MAY HAVE A ROLE IN THE PRESERVATION OF LIFE ITSELF in some animals in that it enables some animals to select foods that contain nutrients or ions that they need to survive; i.e., the rat needs calcium salts to survive and selects foods high in this substance

VI. **Lesions of the VIIth nerve,** proximal to emergence of the chorda tympani but distal to the geniculate ganglion, result in ipsilateral loss of taste, but not in a loss of general sensation on anterior $\frac{2}{3}$ of tongue. IXth nerve lesions cause a loss of both general sensation and taste on the posterior $\frac{1}{3}$ of tongue

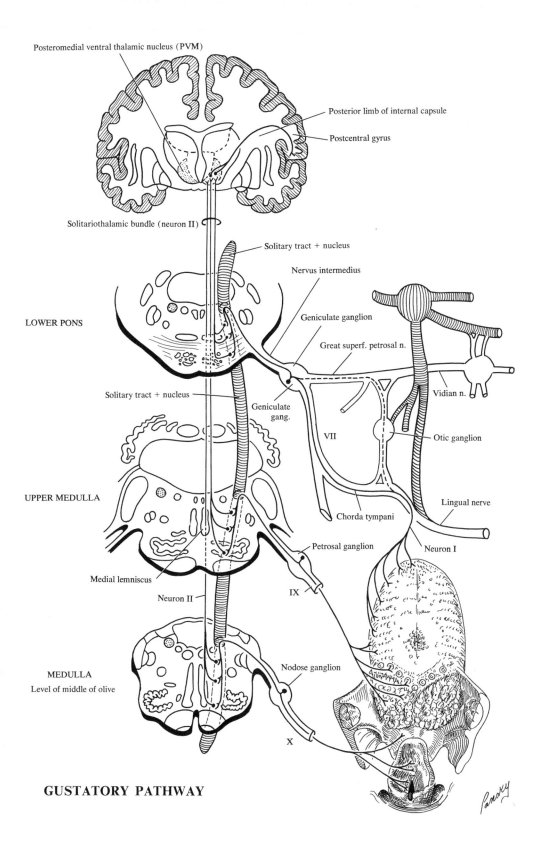

Posteromedial ventral thalamic nucleus (PVM)

Posterior limb of internal capsule

Postcentral gyrus

Solitariothalamic bundle (neuron II)

Solitary tract + nucleus

Nervus intermedius

Geniculate ganglion

Great superf. petrosal n.

LOWER PONS

Solitary tract + nucleus

Geniculate gang.

Vidian n.

VII

Otic ganglion

UPPER MEDULLA

Lingual nerve

Medial lemniscus

Chorda tympani

Neuron I

Petrosal ganglion

Neuron II

IX

MEDULLA
Level of middle of olive

Nodose ganglion

X

GUSTATORY PATHWAY

132. ANATOMY OF THE EAR

I. The ear is usually considered to be composed of 3 portions: outer, middle, and inner ears

A. THE OUTER EAR tends to gather and direct sound waves toward the structures of hearing
   1. Pinna: fleshy, cartilaginous flap attached to the skull
   2. External auditory meatus: tube lined with thin skin and lubricated by the brownish waxy cerumen, produced by ceruminous glands in the wall

B. THE MIDDLE EAR: air-filled cavity which contains the eardrum or tympanum, 3 small ear bones or ossicles, the tendons of 2 small muscles (tensor tympani and stapedius); it communicates posteriorly with the mastoid air cells and anteriorly with the nasopharynx via the auditory or eustachian tube
   1. The eardrum (tympanic membrane) forms the lateral wall of the middle ear cavity and is composed of skin (laterally), mucous membrane (medial); it is set into vibration by sound waves
   2. The ear ossicles: malleus (lateral), incus (intermediate), and stapes (medial)
      a. Act as a system of levers to transmit eardrum vibrations to the fluid-filled cavities of the inner ear (stapes to vestibule via the foramen ovale)
      b. The bones decrease the amplitude of the eardrum oscillation by a factor of 10 but increase the pressure at the inner ear by a factor of 13
   3. Tensor tympani tendon attaches to the malleus; stapedius tendon to the stapes
      a. Contraction limits movement of ossicles, prevents damage with loud sounds
   4. The auditory (eustachian) tube allows equalization of air pressure on the 2 sides of the eardrum, preventing rupture

C. THE INNER EAR consists of a series of cavities and channels (osseous labyrinth) hollowed out in the petrous temporal bone, which contains a closed membranous labyrinth of living tissues. The osseous labyrinth is filled with perilymph; the membranous labyrinth with endolymph
   1. Bony labyrinth consists of semicircular canals, a vestibule, and a cochlea, and opens into the subarachnoid space via the perilymphatic duct
   2. Membranous labyrinth
      a. The utriculus and sacculus lie in the bony vestibule and contain the maculae, which are structures for sensing static posture. The essential mechanism of operation of the maculae is the force of gravity acting on the otoliths and the bending of hair cells of the maculae. Nerve fibers from the latter pass to the cerebellum, over cranial nerve VIII, where appropriate muscular responses are initiated to correct head positions
      b. The membranous semicircular canals lie in their bony counterparts, are 3 in number, and are placed in 3 mutually perpendicular planes (superior, lateral, and posterior). Head movement, or centrifugal force, causes the fluid in the canals to move toward or away from the sense organs located in the ampullae (the cristae) near the terminal ends of the canals
         i. The cristae bend and impulses pass over the VIIIth nerve to the cerebellum, where muscular responses correct body position. Thus, the maculae signal a held (static) position; the cristae are triggered by motion (kinetic)
      c. The cochlear duct is the membranous structure of the cochlea and divides the latter into 3 compartments or scalae: scala tympani and vestibuli, filled with perilymph, and the duct itself (scala media) filled with endolymph. The duct contains the spiral organ of Corti (of hearing), which rests on the basilar membrane, a tapered structure composed of 25,000 strands of connective tissue. The organ of Corti is an organized complex of supporting cells and hair cells (neuroepithelial sensory end organs: 3500 inner and 20,000 outer hair cells). The specialized cells also contain microvilli or stereocilia

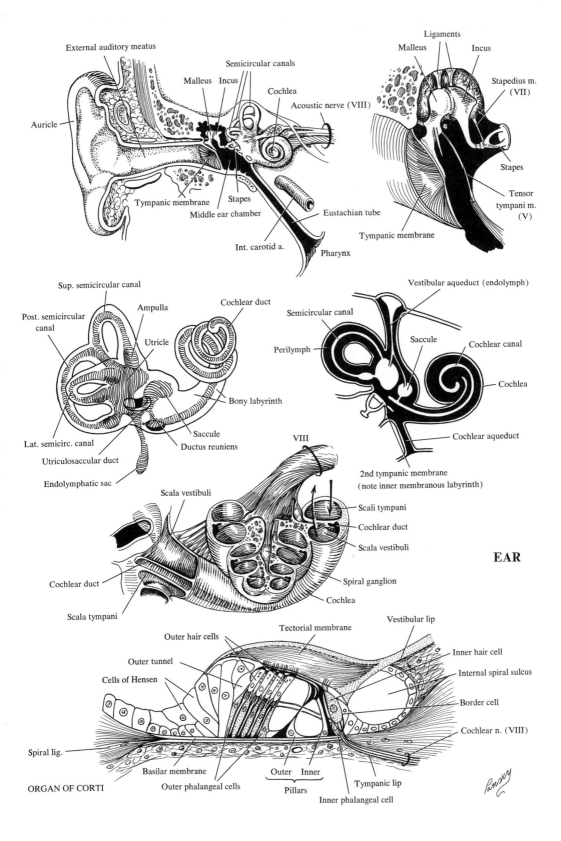

External auditory meatus

Semicircular canals

Malleus Incus

Cochlea

Acoustic nerve (VIII)

Auricle

Tympanic membrane

Stapes

Middle ear chamber

Int. carotid a.

Pharynx

Eustachian tube

Ligaments

Malleus

Incus

Stapedius m. (VII)

Stapes

Tensor tympani m. (V)

Tympanic membrane

Sup. semicircular canal

Ampulla

Cochlear duct

Post. semicircular canal

Utricle

Lat. semicirc. canal

Saccule

Ductus reuniens

Utriculosaccular duct

Endolymphatic sac

Bony labyrinth

Vestibular aqueduct (endolymph)

Semicircular canal

Saccule

Cochlear canal

Perilymph

Cochlea

Cochlear aqueduct

2nd tympanic membrane (note inner membranous labyrinth)

VIII

Scala vestibuli

Scali tympani

Cochlear duct

Scala vestibuli

Spiral ganglion

Cochlea

Cochlear duct

Scala tympani

EAR

Vestibular lip

Tectorial membrane

Outer hair cells

Inner hair cell

Outer tunnel

Internal spiral sulcus

Cells of Hensen

Border cell

Cochlear n. (VIII)

Spiral lig.

Basilar membrane

Outer Inner

Tympanic lip

ORGAN OF CORTI

Outer phalangeal cells

Pillars

Inner phalangeal cell

Pansky

-291-

133. SENSATION OF HEARING: GENERAL INFORMATION

I. Hearing
A. GENERAL INFORMATION
1. Sound energy is transmitted through air as a disturbance of air molecules; since there are no air molecules in a vacuum, there is no sound in a vacuum
2. The disturbances of air molecules that make up a sound wave consist of
   a. Regions of compression, in which air molecules are close together and pressure is high, alternating with areas of rarefaction, where the molecules are farther apart and the pressure is lower
3. The individual molecules travel only short distances, but the disturbances passed from one molecule to another can travel many miles. It is in these disturbances or sound waves that sound energy is transmitted
4. The sound dies out only when enough of the original sound energy has been dissipated so that a sound wave can no longer disturb the air molecules around it
5. When the waves of rarefaction and compression are regularly spaced, the tone is said to be pure (tuning fork)
   a. Waves of speech and other sounds are not regularly spaced but are complex waves of many frequencies of vibration
6. Sounds keenly heard by human ears are from sources vibrating at frequencies of 1000 to 4000 Hz, but the entire range of frequencies that are audible extends from 20 to 20,000 Hz
7. The frequency of vibration of the sound source (wave) is related to the pitch we hear (analyzed in the cochlea, due to selective vibration of different areas of the organ of Corti for different pitches)
   a. The faster the vibration, the higher the pitch
8. We also detect loudness (intensity), analyzed partially by the cochlea, due to the amplitude of the bounce of the basilar membrane; and tonal quality or timbre of sound, which appears to be a function of the brain
   a. The difference between the packing or pressure of air molecules in a zone of compression and a zone of rarefaction gives us amplitude and is related to the loudness of the sound
9. The degree of purity of the sound wave (number of sound frequencies in addition to the fundamental tone) is related to the quality or timbre of the sound
10. We can distinguish about 400,000 different sounds
B. TESTS OF HEARING
1. A rough estimate can be obtained by recording the distance at which a patient can hear a whisper or ticking watch with one ear while the other is covered
2. The Rinne test differentiates between conductive and sensorineural deafness. The stem of a tuning fork (256 or 512 Hz in frequency) is placed on the mastoid process (bone conduction) until the vibration is no longer heard by the patient and then the fork is held at the external auditory canal (air conduction). Air conduction should be greater than bone conduction, so the sound should be heard about twice as long if air conduction is intact (normal Rinne test)
3. Audiometers can measure frequencies and intensities through the entire hearing range

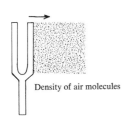

Density of air molecules

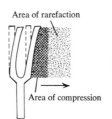

Area of rarefaction

Area of compression

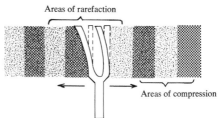

Areas of rarefaction

Areas of compression

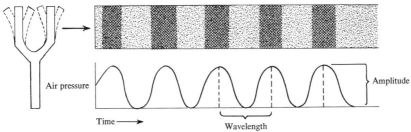

Air pressure

Time ⟶

Amplitude

Wavelength

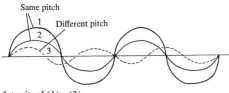

Same pitch

1

2

Different pitch

3

Intensity of (1) > (2)
(3) is least intense

3 SOUND WAVES

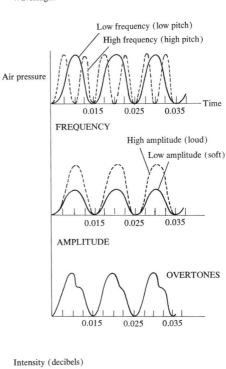

Air pressure

Low frequency (low pitch)

High frequency (high pitch)

0.015   0.025   0.035

Time

FREQUENCY

High amplitude (loud)

Low amplitude (soft)

0.015   0.025   0.035

AMPLITUDE

OVERTONES

0.015   0.025   0.035

DECIBEL RATINGS OF SOUNDS	
Sound	**Rating Decibels**
Silence	0
Ticking watch	20
Residential street	40
Moving stream	50
Auto at 30 ft	60
Conversation at 3 ft	70
Loud radio	80
Truck—15 ft	90
Car horn—15 ft	100
Pneumatic drill—3 ft	120
Amplified rock music	130
Airplane prop.—15 ft	130
Jet at takeoff	150

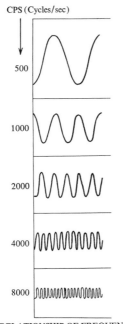

CPS (Cycles/sec)

500

1000

2000

4000

8000

RELATIONSHIP OF FREQUENCY
OF SOUND WAVE TO PITCH

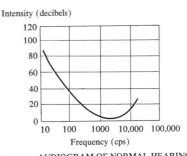

Intensity (decibels)

120
100
80
60
40
20
0

10   100   1000   10,000   100,000

Frequency (cps)

AUDIOGRAM OF NORMAL HEARING
(20–20,000 cps)

134. ANATOMY OF HEARING: SOUND TRANSMISSION

I. Anatomy of hearing: the transmission of sound

A. THE 1ST STEP IN HEARING is usually the entrance of the pressure wave into the external auditory meatus

1. The waves reverberate from the side and end of the canal to fill it with the continuous vibrations of pressure waves

2. The air molecules, under slightly higher pressure during a wave of compression, push against the tympanic membrane, causing it to bow inward

   a. The distance the membrane moves, although small, is a function of the force and velocity with which the air molecules hit it and is related to the loudness of the sound

3. During the following wave of rarefaction, the membrane moves back to its original position

4. The tympanic membrane is very highly sensitive and responds to all varying pressures of the sound waves, vibrating slowly for slow-frequency sounds and rapidly for high tones

   a. The membrane separates the external ear canal from the middle ear cavity, and the pressures in the two air-filled chambers are usually equal at atmospheric pressure. However, differences are possible as a result of atmospheric changes, and this may distort the membrane and cause pain

   b. The eustachian tube has a slitlike ending and is normally closed, but can be opened by yawning, swallowing, or sneezing to equalize the pressure

B. THE 2ND STEP IN HEARING consists of transmission of the sound energy from the tympanic membrane, through the middle ear, to the receptor cells in the inner ear, which are surrounded by fluid

1. The middle ear serves to transfer movements of air in the outer ear to the fluid-filled chambers of the inner ear

2. The tympanic membrane is coupled by a chain of 3 small ossicles, the malleus, incus, and stapes, to a membrane-covered opening, the oval window (fenestra ovale)

   a. The total force on the oval window is the same as on the membrane, but due to its smaller size, the force per unit area (pressure) is multiplied 15–20 times

   b. Additional advantage is reached through the lever action of the 3 ossicles

   c. Thus, the tiny amounts of energy involved are transferred to the inner ear with little loss

3. In addition, the amount of energy transmitted to the inner ear is further modified by contraction of 2 small muscles in the middle ear that alter tension on the tympanic membrane (tensor tympani muscle) and the position of the stapes (stapedius muscle) on the oval window

   a. The muscles protect the receptor apparatus from intense sound stimuli and aid intent listening over certain frequencies

EXTERNAL EAR

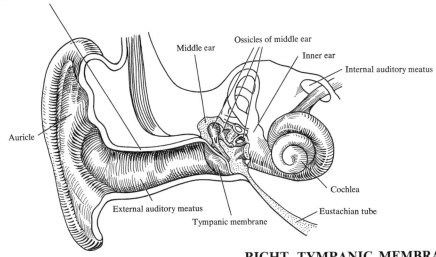

Middle ear

Ossicles of middle ear

Inner ear

Internal auditory meatus

Auricle

Cochlea

External auditory meatus

Eustachian tube

Tympanic membrane

RIGHT TYMPANIC MEMBRANE

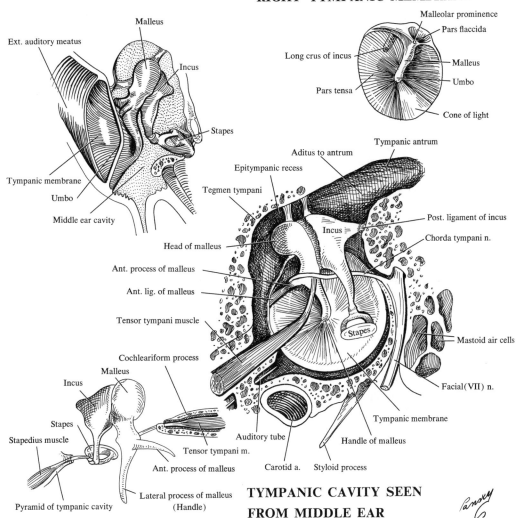

Malleus

Ext. auditory meatus

Incus

Stapes

Tympanic membrane

Umbo

Middle ear cavity

Malleolar prominence

Pars flaccida

Long crus of incus

Malleus

Umbo

Pars tensa

Cone of light

Aditus to antrum

Tympanic antrum

Epitympanic recess

Tegmen tympani

Post. ligament of incus

Incus

Chorda tympani n.

Head of malleus

Ant. process of malleus

Ant. lig. of malleus

Tensor tympani muscle

Stapes

Mastoid air cells

Cochleariform process

Facial (VII) n.

Malleus

Incus

Tympanic membrane

Stapes

Stapedius muscle

Auditory tube

Handle of malleus

Tensor tympani m.

Pyramid of tympanic cavity

Ant. process of malleus

Carotid a.

Styloid process

Lateral process of malleus
(Handle)

TYMPANIC CAVITY SEEN
FROM MIDDLE EAR

Panoky

135. ANATOMY OF HEARING: INNER EAR RECEPTORS

I. **The pressure wave** pushes in on the tympanic membrane, the chain of ossicles rocks the footplate of the stapes against the oval window, and it bows into the vestibule of the inner ear, creating a pressure wave in the inner ear scala vestibuli, which takes 2 pathways to dissipate

A. THE WAVE MAY PASS TO THE END OF THE SCALA VESTIBULI, pass around the end of the basilar membrane via the helicotrema, and return via a second chamber, the scala tympani, where it finally ends at still another membrane-covered window, the round window (fenestra cochleae) which bows out into the middle ear cavity

B. MOST OF THE PRESSURE WAVE, HOWEVER, IS TRANSMITTED TO THE BASILAR MEMBRANE, which is deflected into the scala tympani
   1. The basilar membrane is narrow and relatively stiff in the cochlea closest to the middle ear cavity, but becomes more elastic as it extends the length of the cochlear spiral. The stiff proximal part vibrates immediately in response to pressure changes transmitted in the scala vestibuli, but the distal portion responds more slowly
   2. With each change in pressure in the inner ear, the wave of vibrations is made to travel down the basilar membrane and frequencies of the incoming wave are sorted out by the membrane. The region of maximal displacement varies with frequency of vibration of the sound source
      a. Membrane properties are such that the cochlea nearest the oval window and middle ear resonates best with high-frequency tones and has its greatest amplitude of vibration when high-pitched tones and short waves are heard. The traveling wave dies out once it is past this region
      b. The basilar membrane acts as a filter and low-frequency waves progress farther along the membrane. Thus, lower tones (low pitch or frequency and long waves) travel out along the membrane for a greater distance and are heard at the apex
   3. Where the basilar membrane is maximally displaced, the stimulation of the hair cells (receptors), that ride on it, is the greatest
      a. Fine hairs on the receptor cells are in contact with the overhanging tectorial membrane, an acellular protein structure similar to epidermal keratin, with specialized parts such as Hardesty's membrane and Hensen's stripe, into which the hair cells are embedded
      b. As the basilar membrane is displaced by pressure waves in the scala vestibuli, the hair cells move in relation to the tectorial membrane and are displaced. Incoming sound energy is transformed from vibrating molecules of pressure waves to electric events in the hair cells. Hair cell movement causes depolarization of the hair cells. However, the cells are easily damaged by high-intensity waves

C. THE GENERATOR POTENTIALS formed by the activated hair cells lead to action potentials in the peripheral endings of the afferent nerves and are transmitted to the CNS
   1. The greater the loudness (energy of the sound wave), the greater the basilar membrane movement. Intensity discrimination depends on the length of the membrane set into motion
   2. The greater the amplitude of the generator potential, the greater the frequency of action potentials in the afferent nerve
   3. Musical sounds, chords, and harmonics are the result of several frequencies vibrating in simple periodic oscillations (rhythms). Noises are the result of frequencies not in periodic oscillation

D. SOUND IS LOCALIZED to the side where it is loudest. Comparison of its onset and intensities at each ear are clues in localizing its source

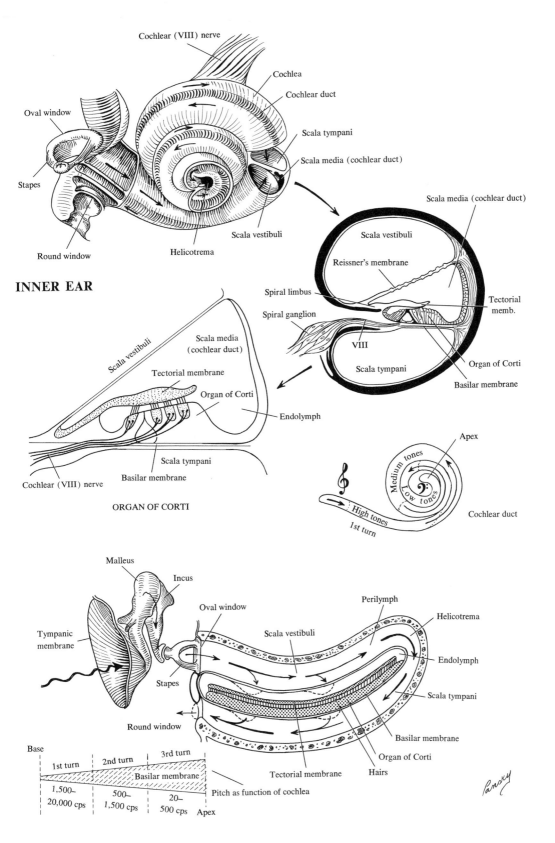

INNER EAR

Cochlear (VIII) nerve

Cochlea

Cochlear duct

Scala tympani

Scala media (cochlear duct)

Oval window

Stapes

Round window

Helicotrema

Scala vestibuli

Scala media (cochlear duct)

Scala vestibuli

Reissner's membrane

Spiral limbus

Spiral ganglion

Tectorial memb.

VIII

Organ of Corti

Scala tympani

Basilar membrane

Scala vestibuli

Scala media (cochlear duct)

Tectorial membrane

Organ of Corti

Endolymph

Scala tympani

Basilar membrane

Cochlear (VIII) nerve

ORGAN OF CORTI

Apex

Medium tones

Low tones

High tones

1st turn

Cochlear duct

Malleus

Incus

Oval window

Perilymph

Helicotrema

Tympanic membrane

Scala vestibuli

Endolymph

Stapes

Scala tympani

Round window

Basilar membrane

Organ of Corti

Tectorial membrane

Hairs

Base

1st turn

2nd turn

3rd turn

Basilar membrane

Pitch as function of cochlea

1,500–20,000 cps

500–1,500 cps

20–500 cps

Apex

Pansky

-297-

136. AUDITORY PATHWAYS

I. The ascending pathway of the auditory system is composed of sequences of neurons, arranged in series and in parallel. A precise designation of 2nd-, 3rd-, and 4th-order neurons is difficult. The basic sequence is

A. NEURONS OF THE 1ST ORDER: from cell bodies in the spiral ganglion of the cochlear nerve, extending from the spiral organ of Corti, terminate centrally in the dorsal and ventral cochlear nuclei

B. NEURONS OF THE 2ND ORDER: from cell bodies in the cochlear nuclei, axons ascend as crossed fibers in the lateral lemnisci to the inferior colliculus
1. Fibers from the dorsal cochlear nucleus decussate through the posterior tegmentum as the dorsal acoustic stria, ascend in the contralateral lemniscus, and end in the nucleus of the lateral lemniscus and inferior colliculus
2. Fibers from the dorsal part of the ventral cochlear nucleus pass dorsal to the inferior cerebellar peduncle, decussate through the intermediate tegmentum as the intermediate acoustic stria, ascend in the contralateral lemniscus, and end in the nucleus of the lateral lemniscus and inferior colliculus
3. Fibers from the ventral cochlear nucleus pass through the anterior tegmentum as the large ventral acoustic stria (part of the trapezoid body). These fibers (1) end in the ipsilateral and contralateral reticular formation, superior olivary nuclei, and nuclei of the trapezoid body, and (2) ascend in the contralateral lateral lemniscus and end in the contralateral nucleus of the lateral lemniscus and inferior colliculus
   a. Decussating fibers of the ventral acoustic stria and those from the superior olivary and trapezoid nuclei are collectively called the trapezoid body

C. NEURONS OF THE 3RD ORDER: from cell bodies in the inferior colliculus, axons pass through the brachium of the inferior colliculus to the medial geniculate body (parvocellular or dorsal portion) of the thalamus

D. NEURONS OF THE 4TH ORDER: from cell bodies in the medial geniculate body, axons pass via the auditory (geniculocortical) radiations to the primary auditory cortex (transverse temporal gyri of Heschl, areas 41 and 42)

E. ALTHOUGH MEANINGFUL CONSCIOUS HEARING requires the neural activity of the cerebral cortex, the latter is not essential for auditory reflexes. "Listening" and reacting to cochlear input can occur with only the lower brain stations intact. The nucleus of the inferior colliculus and brainstem reticular formation may be the integrative centers for this reflex

II. Descending (centrifugal) efferent fibers in the auditory pathways run parallel to the ascending fibers and are integrated in the feedback control of auditory input and are involved with processing and sharpening the ascending auditory influences by channeling the essential neural information (signal) and inhibiting and suppressing unwanted neural activity or noise

A. CORTICOGENICULATE FIBERS: temporal cortex to medial geniculate body

B. CORTICOCOLLICULAR FIBERS: temporal cortex to inferior colliculus

C. GENICULOCOLLICULAR FIBERS: from medial geniculate body to inferior colliculus

D. COLLICULAR EFFERENTS: from inferior colliculus to nuclei of the superior olivary complex, lateral lemniscus, and dorsal and ventral cochlear nuclei

E. THE EFFERENT COCHLEAR BUNDLE OF FIBERS: from the superior olivary complex to the hair cells of the spiral organ of Corti. Consists of many crossing fibers together termed the olivocochlear bundle and of fewer uncrossed fibers, the peduncle of the olive. Conveys inhibitory influences directly to the hair cells

III. Deafness—3 types: transmission, a defect of the eardrum or ossicle activity (e.g., osteosclerosis); cochlear, a defect in the organ of Corti; or central, injury to the auditory cortex or auditory nerve

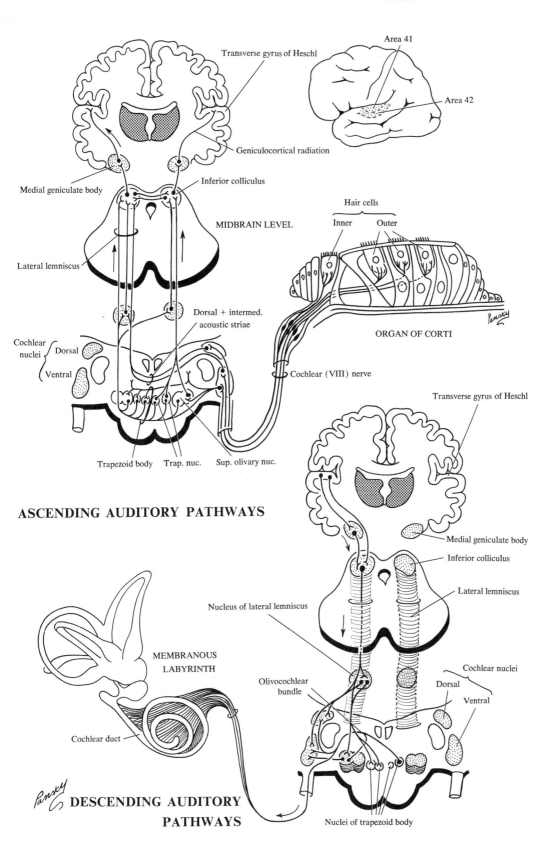

Transverse gyrus of Heschl

Area 41

Area 42

Geniculocortical radiation

Medial geniculate body

Inferior colliculus

MIDBRAIN LEVEL

Hair cells

Inner   Outer

Lateral lemniscus

Dorsal + intermed.
acoustic striae

ORGAN OF CORTI

Cochlear
nuclei { Dorsal

Ventral

Cochlear (VIII) nerve

Trapezoid body   Trap. nuc.   Sup. olivary nuc.

ASCENDING AUDITORY PATHWAYS

Transverse gyrus of Heschl

Medial geniculate body

Inferior colliculus

Lateral lemniscus

Nucleus of lateral lemniscus

MEMBRANOUS
LABYRINTH

Cochlear nuclei

Olivocochlear
bundle

Dorsal

Ventral

Cochlear duct

**DESCENDING AUDITORY
PATHWAYS**

Nuclei of trapezoid body

137. ANATOMY OF THE VESTIBULAR SYSTEM—PART I

I. **The vestibular system** contains mechanoreceptors specialized to detect changes in both motion and position of the head

A. THE RECEPTORS are part of the vestibular apparatus which is housed in the bony channels or labyrinths of the inner ear, bilaterally

　　1. The vestibular apparatus is a membranous labyrinth which forms 3 semicircular canals with their enlarged ends or ampullae. The latter contain the cristae ampullaris, which are the sensory receptors composed of hair cells and supporting cells. In addition, there are the utricle and saccule, which lie in the bony vestibule and contain a ridge or macula composed of hair cells and supporting cells. Endolymph fills the membranous labyrinth

　　　　a. The 3 canals, on each side, are arranged at right angles to each other

　　　　b. The actual receptors of the canals are hair cells that lie at the ends of the afferent neurons

　　　　c. The sensory hairs are ensheathed by a gelatinous mass which blocks the channel of the semicircular canals at that point

　　　　d. The hair cells are secondary receptor cells of the vestibular sense organs and act as transducers. Two types of sensory hairs are seen in the hair cells

　　　　　　i. Stereocilia (modified microvilli) and kinocilia (modified cilia)

　　　　　　ii. Each hair cell has about 70 stereocilia (sensory hairs) and 1 kinocilium

　　　　e. The vestibular sensory receptors consist of 2 types of sensory cells and supporting cells

　　　　　　i. Sensory cell type I is flask-shaped, with a round base and constricted neck enclosed by a chalice-shaped afferent nerve ending (associated with vestibular afferent fiber synapses)

　　　　　　ii. Sensory cell type II is a cylindric columnar cell with a basically located synaptic region. Several afferent and efferent nerve terminals synapse with these hair cells

II. **Receptor system of canals**

A. WHEN THE HEAD IS MOVED, the bony labyrinth, membranous labyrinth, and attached bodies of the hair cells turn with it, and the canals thus signal the rate of change of motion of the head

B. THE ENDOLYMPH of the membranous labyrinth is neither attached to bone nor automatically pulled with it, but due to inertia, the fluid tends to keep its original position

C. THE HAIRS are pulled against the relatively stationary column of endolymph and are bent as the bodies of the hair cells move with the skull

D. SPEED AND MAGNITUDE OF HEAD MOVEMENT determine the degree to which the hairs bend and the hair cells are stimulated

E. THE HAIRS SLOWLY RETURN TO NORMAL POSITION as inertia is overcome

　　1. The hair cells are stimulated during changes in rate of motion (during acceleration of the head) but, during motion at a constant speed, stimulation of the hair cells ceases

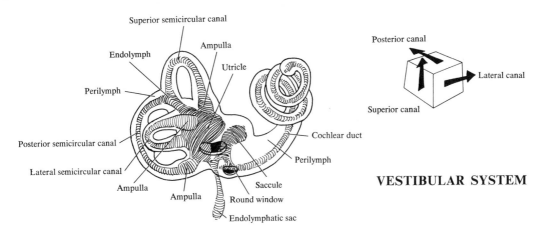

Superior semicircular canal
Ampulla
Endolymph
Utricle
Perilymph
Posterior semicircular canal
Lateral semicircular canal
Ampulla
Ampulla
Saccule
Round window
Endolymphatic sac
Cochlear duct
Perilymph

Posterior canal
Lateral canal
Superior canal

VESTIBULAR SYSTEM

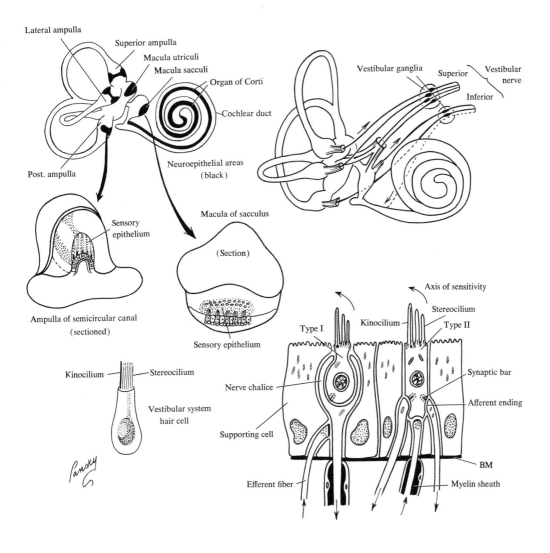

Lateral ampulla
Superior ampulla
Macula utriculi
Macula sacculi
Organ of Corti
Cochlear duct
Post. ampulla
Neuroepithelial areas
(black)

Vestibular ganglia
Superior
Vestibular nerve
Inferior

Sensory epithelium
Ampulla of semicircular canal
(sectioned)

Macula of sacculus
(Section)
Sensory epithelium

Kinocilium
Stereocilium
Vestibular system
hair cell

Nerve chalice
Supporting cell
Efferent fiber

Axis of sensitivity
Stereocilium
Kinocilium
Type I
Type II
Synaptic bar
Afferent ending
BM
Myelin sheath

138. ANATOMY OF THE VESTIBULAR SYSTEM—PART II

I. Receptor system of canals (cont.)

A. THE MECHANISM OF SYNAPTIC TRANSMISSION or the nature of the transmitter substance is not known even though the junction between the hair cells and their afferent nerve fibers has the features of a chemically mediated synapse

B. THE AFFERENT NERVE FIBERS are activated at a relatively slow resting frequency even when the head is motionless

C. THE SHEARING FORCE that bends the hairs on the receptor cells is related to the frequency of action potentials in the afferent nerve. When the hairs are bent one way, the rate of firing speeds up; when they are bent in the opposite direction, the firing frequency slows down

D. IF THE HEAD IS TURNED LEFT, the hairs of the left canal move in that direction, causing a higher rate of firing action potentials in the left vestibular nerve
  1. In the right canal, the hair movement is in the opposite direction, and the rate of firing in the right nerve drops below resting level
  2. Thus, there is an imbalance of input to the CNS that is very significant

II. The utricle and saccule contain the receptors of the vestibular system that provide information about the position of the head relative to the direction of the forces of gravity. The macula utriculi is maximally stimulated when the head is bent forward or backward and minimally when the head is erect. The macula sacculi is maximally stimulated when the head is bent to the side

A. THE RECEPTOR CELLS OF THE UTRICLE AND SACCULE are mechanoreceptors sensitive to the movement of their projecting cilia or hairs
  1. The hair cells of the utricle and saccule are collected into groups from which the hairs protrude into a gelatinous substance
  2. Tiny calcium carbonate stones or otoliths are embedded in the gelatinous covering of the hair cells, making the substance heavier than the endolymph fluid that surrounds it
  3. When the head is bent, the gelatinous-otolith material changes position, pulled by the gravitational forces, to the lowest point in the utricle and saccule
    a. The shearing forces of the gelatinous-otolith substance against the hair cells bend the hairs and stimulate the receptor cells

B. INFORMATION FROM THE VESTIBULAR APPARATUS serves 2 purposes
  1. It controls the muscles that move the eyes so that in spite of changes in head position, the eyes remain fixed on the same point
  2. The information is used in reflex mechanism to maintain the upright posture
    a. In cats, dogs, etc., the vestibular apparatus plays an important role in postural fixation of the head, orientation in space, and reflexes accompanying locomotion
    b. In man, very few postural reflexes are known to depend primarily on vestibular input, even though the vestibular apparatus is referred to as the sense organs of balance

STRUCTURE OF CRISTA OF AMPULLA OF SEMICIRCULAR CANAL

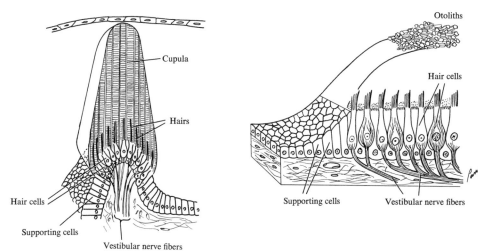

Cupula

Hairs

Hair cells

Supporting cells

Vestibular nerve fibers

Otoliths

Hair cells

Supporting cells

Vestibular nerve fibers

STRUCTURE OF MACULA
OF UTRICLE AND SACCULE

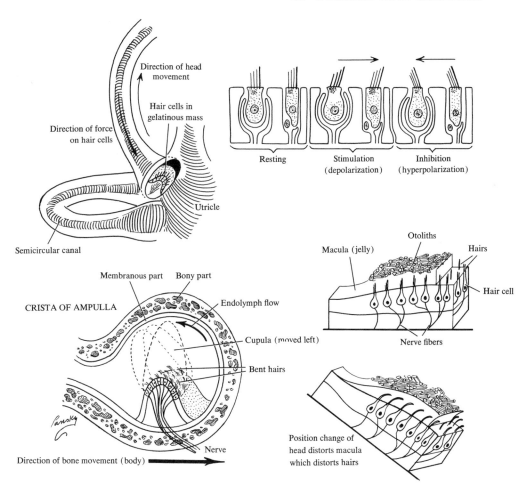

Direction of head movement

Hair cells in gelatinous mass

Direction of force on hair cells

Utricle

Semicircular canal

Resting

Stimulation (depolarization)

Inhibition (hyperpolarization)

Otoliths

Macula (jelly)

Hairs

Hair cell

Nerve fibers

Membranous part   Bony part

CRISTA OF AMPULLA

Endolymph flow

Cupula (moved left)

Bent hairs

Nerve

Direction of bone movement (body)

Position change of head distorts macula which distorts hairs

139. VESTIBULAR PATHWAYS

I. The vestibular pathways consist of suprasegmental pathways distributed to the brainstem, spinal cord, and cerebellum

A. THE CELL BODIES of the 19,000 nerve fibers of each vestibular nerve lie in the vestibular (Scarpa's) ganglion, which is divided into a pars superior and pars inferior

1. Nerves from receptors of the anterior and lateral semicircular canals and the utricle have their cell bodies in the pars superior
2. Nerves from the posterior semicircular canal and saccule have their cell bodies in the pars inferior

B. THE PRIMARY VESTIBULAR FIBERS pass into the upper medulla deep to the inferior cerebellar peduncle, bifurcate into short ascending and descending branches, and terminate in specific regions in each of the 4 vestibular nuclei (lateral, medial, superior, and inferior) in the upper lateral medulla, and in the nodule of the archicerebellar cortex (flocculonodular lobe)

1. The primary vestibular fibers from the cristae (semicircular canals) are distributed mainly to parts of the superior and medial nuclei; those from the maculae (utricle and saccule) terminate in portions of the medial and inferior vestibular nuclei
2. Vestibular nuclei receive input primarily from the labyrinth and archicerebellum
3. Influence from the fastigial nuclei to the vestibular nuclei are facilitatory; those from the cerebellar cortex (vermis and flocculonodular lobe) are inhibitory

C. THE VESTIBULAR NUCLEI project secondary vestibular fibers to

1. Upper brainstem (pons and midbrain) via the medial longitudinal fasciculus (MLF)
2. Cervical spinal segments via the medial vestibulospinal tract
3. The full length of the spinal cord via the lateral vestibulospinal tract
4. The archicerebellum and fastigial nucleus of the cerebellum via the juxtarestiform body (medial part of the inferior cerebellar peduncle)
5. Brainstem reticular formation
6. Hair cells of the macula and crista of the membranous labyrinth via the vestibular nerve as the vestibular efferent fibers

D. ALL CONTRALATERAL PROJECTIONS from the vestibular nuclei to the MLF decussate at the level of the vestibular nuclei

E. THE MLF IS COMPOSED of many ascending and descending fibers from the vestibular nuclei and is important because it helps to integrate the vestibular system with movements of the eyes, neck, and head

II. Eye movements and the MLF: functional interaction of the vestibular receptors resulting in eye movements is mediated via ascending fibers that project from the vestibular nuclei to the MLF of both sides and ascend in the brainstem

A. THE SUPERIOR VESTIBULAR NUCLEUS PROJECTS ascending uncrossed fibers; the inferior and medial nuclei project both crossed and uncrossed fibers; and the lateral nucleus projects largely crossed fibers via the MLF

1. The pathway is called the vestibulomesencephalic pathway and connects with the nuclei of cranial nerves III, IV, and VI to form the basis of many coordinated (conjugate) eye movements, synchronizing movements of both eyes to the side, upward, and downward (i.e., stimulation of the nerve innervating the crista ampullaris of the lateral canal on one side results in lateral conjugate deviation of the eyes to the contralateral side as a result of contractions of the lateral rectus muscle of the right eye and the medial rectus muscle of the left eye)

B. VESTIBULAR PROJECTIONS to the nuclei of the extraocular muscles are very precise

1. The most potent vestibular projections are primarily to horizontal eye movements (lateral and medial recti mm.). Effects of vestibular influences on elevation and depression of the eyes are minimal

PRIMARY VESTIBULAR FIBERS AND NUCLEI

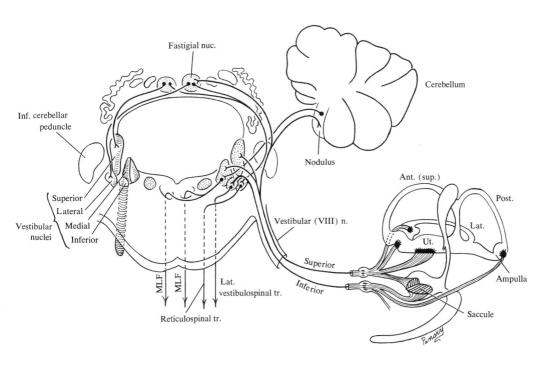

Fastigial nuc.

Cerebellum

Inf. cerebellar peduncle

Nodulus

Ant. (sup.)

Post.

Vestibular (VIII) n.

Lat.

Ut.

{ Superior
Lateral
Medial
Inferior }

Vestibular nuclei

Superior

Ampulla

Inferior

MLF MLF

Lat. vestibulospinal tr.

Saccule

Reticulospinal tr.

Pansky

VESTIBULAR CONNECTIONS TO EYE MOVEMENTS

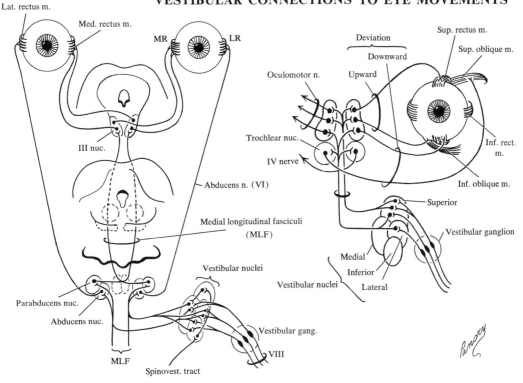

Lat. rectus m.

Med. rectus m.

MR

LR

Sup. rectus m.

Deviation

Sup. oblique m.

Downward

Oculomotor n.

Upward

III nuc.

Trochlear nuc.

IV nerve

Inf. rect. m.

Abducens n. (VI)

Inf. oblique m.

Medial longitudinal fasciculi (MLF)

Vestibular nuclei

Superior

Vestibular ganglion

Parabducens nuc.

Medial
Inferior
Lateral

Abducens nuc.

Vestibular nuclei

MLF

Vestibular gang.

VIII

Spinovest. tract

Pansky

-305-

140. VESTIBULAR SYSTEM: EYE, HEAD, AND BODY MOVEMENTS

I. Eye movements and the MLF (cont.)

A. THE INTERSTITIAL NUCLEUS OF CAJAL AND THE NUCLEUS OF DARKSCHEWITSCH are midbrain nuclei in the ventrolateral periaqueductal gray at the lower level of the IIIrd nucleus

1. Fibers from the interstitial nucleus descend (some cross in the posterior commissure) bilaterally in the MLF and terminate in the IIIrd, IVth, and medial vestibular nuclei of the brainstem and in laminae VII and VIII of the cord

a. Lesions of these nuclei result in impaired vertical eye movement

2. The role of the nucleus of Darkschewitsch and its fibers is unknown. It does not project through the MLF

B. NYSTAGMUS: as the head and body pivot and circle, the eyes attempt to fix on an object in space (slow component); as the head and body continue to circle, the eyes snap quickly in the direction in which the head is circling (fast or quick component or compensating component)

1. Nystagmus is named by the direction of the fast component

2. Following a sudden stop after spinning, the direction of nystagmus is reversed (postrotatory nystagmus) because the semicircular canals stop when the head stops, but the endolymph, with its momentum, continues to move and thus pushes the cupula and hairs of the hair cells in the opposite direction from that of the spin

3. Postrotating nystagmus is associated with a fast component and sensation of vertigo or dizziness, which are in the opposite direction, and a slow component, tendency to fall, pastpointing, and endolymph movement, all in the same direction

4. Pastpointing is the phenomenon of missing an object when reaching to touch it

5. The plane of the head in the spin determines the nystagmus: horizontal, oblique, vertical, rotatory, etc.

II. The vestibular system and head and body movements

A. THE VESTIBULAR NUCLEI project crossed and uncrossed fibers (largely from the medial and inferior vestibular nuclei) that descend in the MLF as the vestibulospinal pathway, which terminates in cervical segments of the spinal cord and conveys influences of inhibition to extensor tonus

B. THE LATERAL VESTIBULAR NUCLEUS projects uncrossed fibers to all levels of the cord via the lateral vestibulospinal tract, which is somatotopically organized; the rostroventral part goes to cervical spinal levels; the dorsocaudal part of the nucleus projects to lumbosacral spinal levels

1. The archicerebellum has some influence via the lateral vestibular nucleus on the lateral vestibulospinal tract and to lower motor neurons in the cord. The lateral vestibular nucleus can be thought of as a displaced cerebellar nucleus since it is the only nucleus receiving direct input from the Purkinje cells of the cerebellar cortex

C. THE SOMATIC VESTIBULOSPINAL PATHWAYS play a role in muscular activities of the body and extremities associated with postural movements and balance

1. The vestibulospinal pathways exert facilitatory influences on extensor tonus via the alpha and gamma motor neurons of the spinal nerves

2. In addition, spinovestibular fibers project ascending influences to the vestibular nuclei from the cord

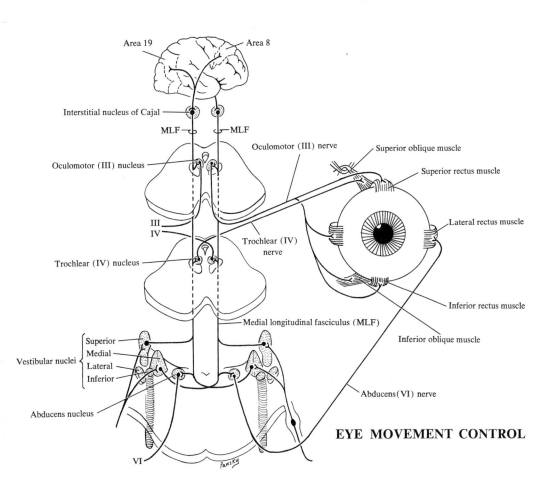

Area 19

Area 8

Interstitial nucleus of Cajal

MLF — MLF

Oculomotor (III) nerve

Superior oblique muscle

Superior rectus muscle

Oculomotor (III) nucleus

Lateral rectus muscle

III
IV

Trochlear (IV) nerve

Trochlear (IV) nucleus

Inferior rectus muscle

Medial longitudinal fasciculus (MLF)

Inferior oblique muscle

Vestibular nuclei { Superior
Medial
Lateral
Inferior

Abducens nucleus

Abducens (VI) nerve

VI

Pansky

EYE MOVEMENT CONTROL

L SLOW PHASE ← **NYSTAGMUS** → R FAST PHASE

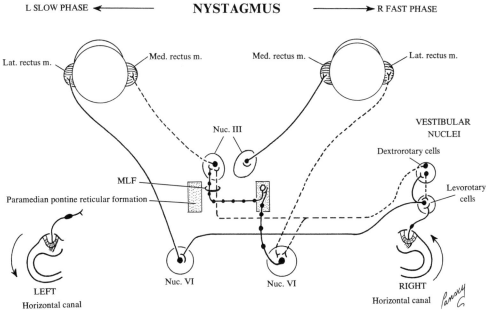

Lat. rectus m.

Med. rectus m.

Med. rectus m.

Lat. rectus m.

Nuc. III

VESTIBULAR
NUCLEI

Dextrorotary cells

MLF

Levorotary
cells

Paramedian pontine reticular formation

Nuc. VI

Nuc. VI

LEFT
Horizontal canal

RIGHT
Horizontal canal

Pansky

141. VESTIBULAR SYSTEM: PATHWAYS AND SPECIAL FEATURES

I. Vestibular connection with cerebellum and reticular formation

A. PRIMARY FIBERS of the vestibular nerve and secondary fibers from the inferior and medial vestibular nuclei pass through the juxtarestiform body and course as mossy fibers to the ipsilateral flocculonodular lobe and uvula (the flocculonodular lobe and adjacent cortex are often called the vestibulocerebellum)

    1. In addition, fibers from these vestibular nuclei project to the fastigial nuclei

B. THE MACULAE OF THE UTRICLE AND SACCULE are the major sources of vestibular input to the cerebellum

C. THE PURKINJE CELLS OF THE VESTIBULOCEREBELLUM AND VERMIS relay inhibitory influence to the fastigial nuclei and lateral vestibular nuclei

    1. The fastigial nucleus projects fibers conveying excitatory influences to part of each vestibular nucleus and to the reticular formation (lateral reticular nucleus and nucleus reticularis pontis caudalis)

    2. The vestibular nuclei probably receive most of their input from the cerebellum

D. THE VESTIBULOCEREBELLUM exerts its influence on the coordination of the axial muscles of the neck and vertebral column in balance and posture through

    1. The lateral vestibular nucleus and lateral vestibulospinal tract

    2. The medial vestibular nucleus and the medial vestibulospinal tract

    3. The caudal pontine reticular nucleus and the pontine reticulospinal tract

    4. Fibers from the vestibular pathway to the pontine and medullary reticular formation are integrated in the reticular system and are involved in producing nausea, vomiting, and sweating after vestibular stimulation

II. Vestibular efferent pathway to the membranous labyrinth

A. EFFERENT FIBERS from the brainstem pass through the vestibular nuclei and end in the vestibular receptors of the membranous labyrinth (the nuclei of origin and the role of these fibers are controversial). They probably exert inhibitory influences and ameliorate the effects of motion sickness and its aftersensation nystagmus

III. Functional features of the vestibular system: the labyrinth plays a significant role in maintenance of head and neck position and body posture

A. VERTIGO: the dizziness or sensation associated with lack of equilibrium involves the subjective sense of rotation, either of the surroundings or of the subject

    1. It is a cardinal sign of labyrinthine and vestibular dysfunction, but may occur in normal subjects with excessive stimulation of the semicircular canals

    2. In vestibular disease, nystagmus occurs and is a consequence of influences conveyed from the vestibular system via the MLF to the extraocular muscles

B. DIZZINESS, HEADACHE, NAUSEA, AND VOMITING are symptoms of motion sickness, are primarily due to the excessive stimulation of the utricle and saccule, and involve the vestibulocerebellar projections

    1. Drugs like Dramamine raise the threshold of vestibular stimulation and help prevent motion sickness*

C. MÉNIÈRE'S DISEASE is a disorder probably resulting from abnormal circulation of the vestibular endolymph. Its symptoms are tinnitus (ringing in the ear), attacks of vertigo, pallor, vomiting, nausea, and increased respiration. The cristae ampullaris are normal

D. LOSS OF BOTH LABYRINTHS is not followed by vertigo or nystagmus. Normal locomotion under these circumstances requires visual cues

E. MAN IS NOT AS DEPENDENT ON HIS VESTIBULAR SYSTEM as on his visual and general proprioceptive systems. Injury to the labyrinths, vestibular nuclei, or vestibular pathways may lead to nystagmus, tendency to fall to one side, pastpointing, and some erect posture problems, but these attenuate in time as other proprioceptive cues are used

---

*Because the labyrinth is not functional during the 1st year of life, infants do not get motion sickness.

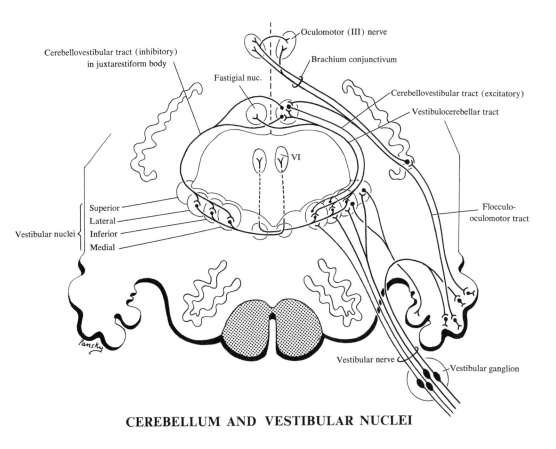

CEREBELLUM AND VESTIBULAR NUCLEI

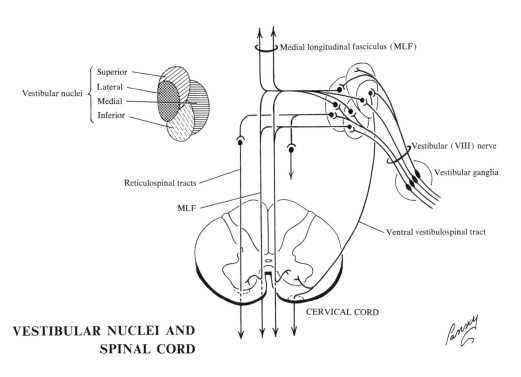

VESTIBULAR NUCLEI AND
SPINAL CORD

142. ANATOMY OF THE EYE

I. **Anatomy of the eye:** the eye is an organ capable of transducing light to visual impulses and consists of three layers of tissue

A. THE OUTER FIBROUS TUNIC is composed of the sclera and cornea

  1. The sclera or white of the eye: tough, fibrous protective structure affording attachment for the eye muscles that rotate the eyeball
  2. The cornea: avascular, anterior, transparent portion which serves as a fixed lens that initially bends the light rays to a near focus on the retina

B. THE MIDDLE VASCULAR TUNIC OR UVEAL TRACT (UVEA) is composed of

  1. The choroid: highly vascularized, richly black-pigmented posterior layer which absorbs stray light rays and is coexistent with the photoreceptive retina, which it supplies with nutriment and oxygen
  2. The ciliary body contains the ciliary muscles for fine focusing of the lens, and the iris, a thin, muscular diaphragm (circulating and radiating fibers) that regulates the opening (pupil) which controls light entering the eye
  3. The lens and supporting zonular fibers (zonules): biconvex, elliptical in shape

C. THE INNER TUNIC OR RETINA: a 10-layered structure containing the photosensitive or visual receptors, the rods and cones, and neurons leading to the optic nerve

  1. Is an extension of the forebrain and divided into a photoreceptive layer (neuroretina) and pigment epithelium to the ora serrata, an unpigmented layer (ciliary retina) over the ciliary body, and a pigmented epithelium (iridic retina) over the posterior iris
  2. Its posterior portion is modified into 3 concentric regions: the central 5–6 mm of retina, the macula lutea (3 mm) with a pit in its center, the fovea centralis (0.4 mm), and the blind spot or optic disk just medial to the posterior pole (site of emergence of the optic nerve)
  3. The functional retina terminates anteriorly at the ora serrata

D. ADDITIONAL ESSENTIAL STRUCTURES

  1. The eye is protected by the eyelids and lubricated by the lacrimal apparatus
  2. The aqueous humor (anterior) chamber between the cornea and the lens is filled with an alkaline, watery solution secreted by the ciliary processes
     a. Glaucoma: leading cause of blindness in the United States; develops commonly when the aqueous humor fails to drain properly; pressure builds up in the eyeball and can, if untreated, lead to optic nerve damage
  3. The large vitreous humor chamber lies behind the lens, and is a viscous and gelatin-filled chamber
  4. Iridocorneal filtration angle: lateral circular border of the anterior chamber
  5. Canal of Schlemm (aqueous vein) drains anterior chamber into venous system

II. **Visual sensation:** the receptors of the eye are sensitive to only that small part of the spectrum of electromagnetic radiation called light

A. ELECTROMAGNETIC RADIATION has both wavelength and particlelike properties

  1. The radiant energy is propagated as small, discrete packets called photons
  2. Radiant energy, however, is described in terms of wavelengths and frequencies
     a. Wavelength is the distance between 2 successive wave peaks; varies from fractions of a millimeter (bottom of spectrum) to several miles (top of spectrum)
     b. Frequency is the number of wave peaks (cycles) that pass a point in a given time and is expressed in cycles per second or herz (Hz)
        i. Photons of visible light oscillate from 4×10^{14} to 7×10^{14} times per second
     c. Light of different wavelengths is associated with different color sensations: a wavelength of 540 nm looks green; one of 565 nm looks red
     d. Wavelength times frequency equals velocity: velocity of light in free space is a fundamental constant of nature, thus as wavelength decreases, frequency increases

EYE STRUCTURE

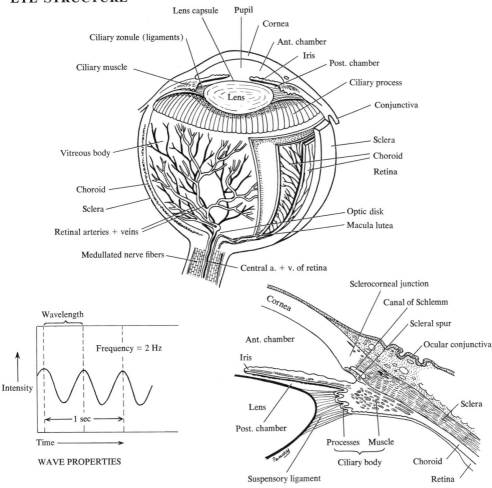

Lens capsule
Pupil
Cornea
Ant. chamber
Iris
Post. chamber
Ciliary process
Conjunctiva
Ciliary zonule (ligaments)
Ciliary muscle
Lens
Sclera
Choroid
Retina
Vitreous body
Choroid
Sclera
Optic disk
Macula lutea
Retinal arteries + veins
Medullated nerve fibers
Central a. + v. of retina

Wavelength
Frequency = 2 Hz
Intensity
1 sec
Time
WAVE PROPERTIES

Cornea
Sclerocorneal junction
Canal of Schlemm
Scleral spur
Ocular conjunctiva
Ant. chamber
Iris
Lens
Post. chamber
Sclera
Processes Muscle
Choroid
Retina
Suspensory ligament
Ciliary body

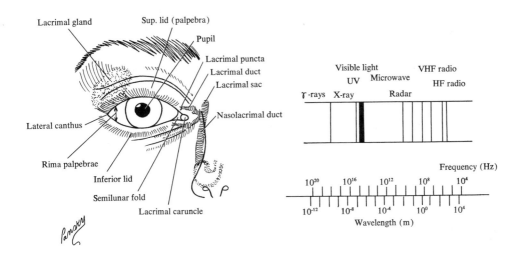

Lacrimal gland
Sup. lid (palpebra)
Pupil
Lacrimal puncta
Lacrimal duct
Lacrimal sac
Nasolacrimal duct
Lateral canthus
Rima palpebrae
Inferior lid
Semilunar fold
Lacrimal caruncle

Visible light   VHF radio
UV   Microwave   HF radio
γ-rays X-ray   Radar

Frequency (Hz)
10^{20} 10^{16} 10^{12} 10^{8} 10^{4}
10^{-12} 10^{-8} 10^{-4} 10^{0} 10^{4}
Wavelength (m)

143. VISUAL OPTICS

I. **The optics of vision:** the image of the object being viewed must be focused on the retina of the eye, a thin layer of neural tissue lining the back of the eyeball, where light-sensitive receptor cells of the eye are found. When a light wave hits a boundary between 2 substances, like the cornea of the eye, the rays are bent and travel in a new direction. The degree of bending depends on the frequency of the light and the angle at which it enters the 2nd medium

A. THE LENS AND CORNEA are the optical systems that focus the image of the object on the retina. The shape of both, as well as the lengh of the eyeball, determines the point where the light rays converge on the retina

   1. The cornea plays a greater role than the lens in focusing, because light rays are bent more in passing from air into it than pass into and out of the lens

      a. The corneal surface is curved so that light rays from a single point source hit it at different angles and are bent different amounts. However, all are directed to a point after leaving the lens

      b. When the object has more than one dimension, the image to the retina is upside down to the original light source, and is also reversed left to right

   2. The lens makes all adjustments for distance by changing its shape (accommodation), although the cornea does most of the focusing on the retina

      a. The shape of the lens is controlled by a muscle (ciliary muscle) that flattens the lens when distant objects are to be focused, and allows it to become more spherical to provide for more bending of light for near objects

      b. Cells are added to the outer surface of the lens throughout life. Cells at the center are the oldest and farthest from nutrient fluids that bathe the lens exterior. They age and die first, making the lens more stiff and less able to accommodate

      c. Decrease in refractive power of lens (reduced elasticity with age) until it is almost nonaccommodating is called presbyopia (corrected with convex lenses)

      d. Cells of the lens can become opaque and impair detailed vision, forming a cataract (defective lens can be removed and vision restored by compensating lenses)

B. VISUAL DEFECTS occur when the lens or cornea does not have a smooth spherical surface, resulting in astigmatism (improper eyeball, corneal, or lens irregularities can be compensated for by corrective lenses)

C. THE EYEBALL, when normal, focuses sharp images on the retina (emmetropic)

   1. Defects in vision occur if the eyeball is too long in relation to lens size; thus, images of far objects are focused in front of the retina, resulting in nearsightedness or myopia (inability to see distant objects clearly)

   2. If the eyeball is too short for the lens, distant objects focus well but near objects are focused behind the retina, resulting in farsightedness or hypermetropia (hyperopia)

D. THE IRIS controls the amount of light entering the eye by means of its ringlike sphincter and dilator pupillae muscles. In addition, it contains connective tissue and pigmented cells which give it its color

   1. The pupil (hole in the center of the iris) allows light to enter the eyeball

   2. The iris sphincter muscle reflexly contracts in bright light, decreasing the pupil diameter, not only reducing the amount of light entering the eye but also directing it to the central and most optically accurate part of the lens

      a. The sphincter pupillae muscle, conversely, relaxes in dim light and, with help of dilator pupillae muscle, allows for maximum sensitivity and opening of the pupil

II. **Glaucoma** is a dangerous disorder of the aqueous system of the anterior chamber of eye

A. FLUID may be secreted excessively by ciliary body (open-angle glaucoma)

B. DILATED IRIS may fold into area of trabeculae, blocking drainage into canal of Schlemm (closed-angle glaucoma)

C. PRESSURE IN EYE becomes excessive, altering circulation in retina, and results in permanent blindness, unless treated

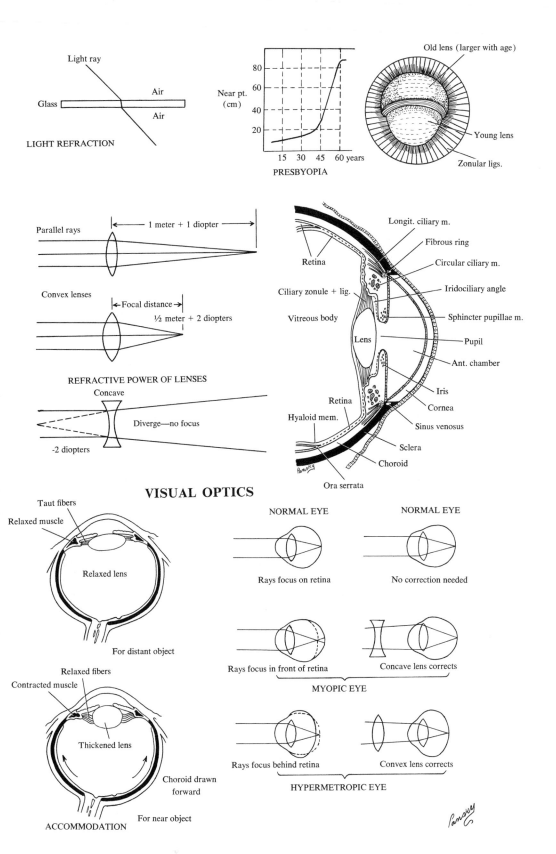

LIGHT REFRACTION

PRESBYOPIA

Old lens (larger with age)

Young lens

Zonular ligs.

Parallel rays

1 meter + 1 diopter

Convex lenses

Focal distance

½ meter + 2 diopters

REFRACTIVE POWER OF LENSES

Concave

Diverge—no focus

-2 diopters

Longit. ciliary m.

Fibrous ring

Circular ciliary m.

Iridociliary angle

Sphincter pupillae m.

Pupil

Ant. chamber

Iris

Cornea

Sinus venosus

Sclera

Choroid

Ora serrata

Retina

Ciliary zonule + lig.

Vitreous body

Lens

Retina

Hyaloid mem.

VISUAL OPTICS

Taut fibers

Relaxed muscle

Relaxed lens

For distant object

Relaxed fibers

Contracted muscle

Thickened lens

Choroid drawn forward

ACCOMMODATION   For near object

NORMAL EYE

Rays focus on retina

NORMAL EYE

No correction needed

Rays focus in front of retina   Concave lens corrects

MYOPIC EYE

Rays focus behind retina   Convex lens corrects

HYPERMETROPIC EYE

144. RETINA

I. The retina contains the receptor cells, called either rods or cones because of their microscopic appearance. Only the rods and cones respond directly to light; all other components of the pathway are influenced only by synaptic input to them. Each retina contains about 120 million rods and 7 million cones, but only about 1 million optic nerve fibers leading from it

A. BOTH CELL TYPES contain light-sensitive molecules called photopigments, whose major function is to absorb light

    1. Light energy causes the photopigments to change their molecular configuration, which then alters the properties of the receptor-cell membranes in which they are located

        a. Unlike other receptor cells, in response to stimulation, the membrane decreases its permeability to sodium ions, which hyperpolarizes the receptor-cell membrane. This seemingly releases other neurons in the visual pathway from inhibition

B. FOUR KINDS OF PHOTOPIGMENTS

    1. Rhodopsin is very sensitive to low levels of illumination and is found in the rods, in man, with a maximum scotopic sensitivy of 500 nm

    2. Erythrolabe, chlorolabe, and cyanolabe are found in the cones and are sensitive to light wavelengths of the 3 primary colors: red, green, and blue, respectively

        a. Color blindness: genetic, sex-linked trait in which cones that respond to red or green light are missing and the person is blind to these colors

C. COMPOSITION OF PHOTOPIGMENTS: all four are made up of a protein (opsin) bound to a chromatophore molecule (retinal or retinaldehyde or retinene or vitamin A aldehyde)

    1. Opsins are hydrophobic proteins with one retinal group per molecule, a carbohydrate side chain, no phospholipid, and a molecular weight of about 27,000. They are buried in the membrane of the discs of the photoreceptor outer segments

    2. All 4 photopigments possess the 11-*cis* retinaldehyde as the chromophore but are united to 4 different opsins, which confers the specific light sensitivities on the photopigment, i.e., whether it responds to all light or selectively to red, blue, or green

    3. Light acts on the chromatophore and it splits away from the opsin, changing its molecular configuration and generating a nerve impulse. Retinal eventually breaks down into vitamin A. Most of the latter is transformed back to retinal, which combines in the dark with opsin to regenerate rhodopsin

        a. After the breakdown of the photopigment in the presence of light, the chromatophore molecule is rearranged and rejoined to opsin to restore the photopigment

        b. Thus, the only action of light in vision is to change the chromatophore; everything else in the sequence leading to vision (chemical, physiologic, or psychologic) is a "dark" consequence of the light reaction

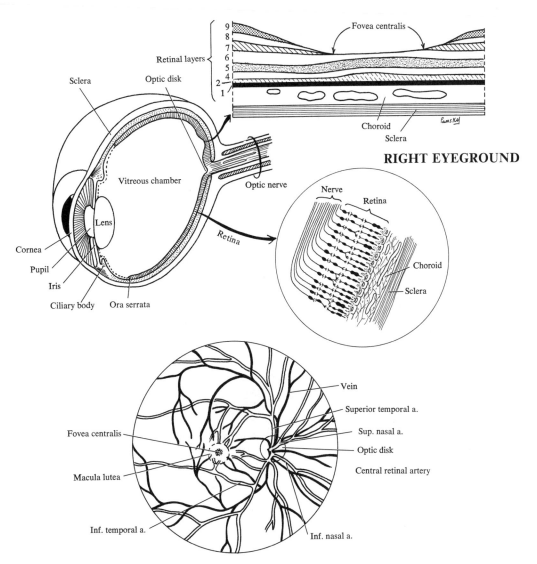

RIGHT EYEGROUND

RHODOPSIN CYCLE

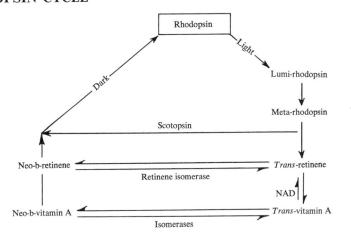

145. RODS AND CONES

I. **The rod receptor cell** contains rhodopsin, a pigment called visual purple, which is very sensitive and can detect small amounts of light (scotopic vision). The rods act as photoreceptors during conditions of poor lighting and for night vision

  A. THE ROD RESPONSES do not indicate color, show only shades of gray, do not indicate brightness, and their acuity is very poor

  B. THE RODS ARE MOST NUMEROUS in the peripheral retina, nearest the lens, and are absent from the retinal center (the fovea)

    1. There is greater sensitivity of the peripheral retina to weak light; as many as 600 rods can converge on the same optic nerve fiber, with marked summation of the peripheral ganglion cell

II. **The cones** are of 3 types, each with one of the 3 photopigments for color vision. They operate only at high illumination levels (photopic vision), are the photoreceptors for day vision, and have a very high visual acuity

  A. THE CONES are concentrated in the center of the retina (the fovea), which we use for finely detailed vision

    1. Near the fovea, fewer and fewer rods and cones converge on each fiber and both become more slender, which progressively increases the acuity of vision near the center of the retina

III. **Synapses:** each receptor cell in the retina (rod or cone) synapses on a 2nd neuron, a bipolar cell, which in turn synapses on a ganglion cell. The axons of the latter form the optic nerve, which passes out of the eye directly to the brain

  A. CONE RECEPTOR CELLS have relatively direct lines to the brain because each bipolar cell receives synaptic input from few cones, and each ganglion cell receives synaptic input from few bipolar cells. In the fovea, where only cones exist, there is little convergence; instead, the cones are represented by about equal numbers of bipolar and ganglion cells. The number of optic nerve fibers leading from the fovea is almost equal to the number of cones there

    1. The above results in a lack of convergence, which provides precise information about the retinal areas stimulated but little chance for summation of subthreshold events to fire the ganglion cell

  B. MANY ROD CELLS, conversely, converge on bipolar and ganglion cells so that spatial and temporal summation are very good, although acuity is poor. Thus, a low intensity of light stimulus that would cause only a subthreshold response in a cone ganglion can cause an action potential in a rod ganglion cell, explaining why objects in a darkened room are indistinct and appear only in shades of gray

  C. SENSITIVITY OF THE EYE improves after being in the dark for some time, as a result of dark adaptation

    1. The theory states that the excitability of the rod vision pathways depends on the number of intact rhodopsin molecules in the rods. In bright light, so many rod rhodopsin molecules break down that they are ineffective and vision is due to cone activation. If one moves to a darkened room from a lighted one, there are relatively few intact rhodopsin molecules initially available, but as the rhodopsin gradually regenerates in the dark, visual sensitivity improves

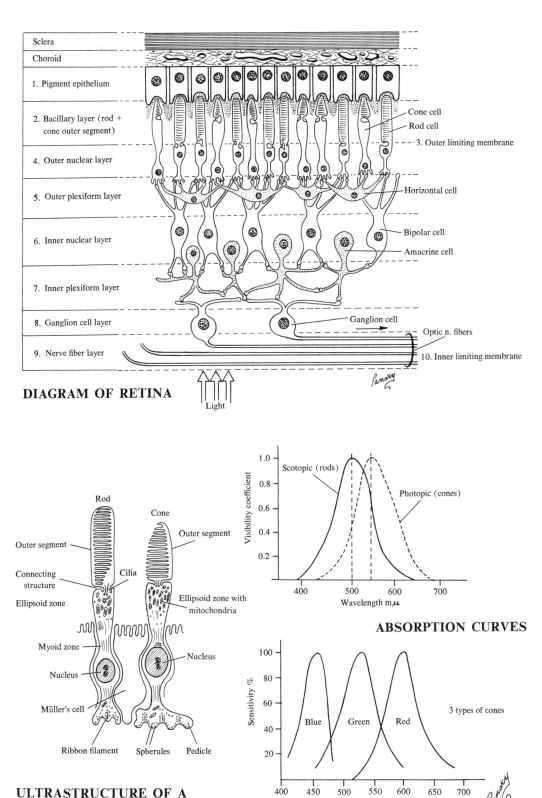

Sclera	
Choroid	
1. Pigment epithelium	
2. Bacillary layer (rod + cone outer segment)	Cone cell / Rod cell / 3. Outer limiting membrane
4. Outer nuclear layer	
5. Outer plexiform layer	Horizontal cell
6. Inner nuclear layer	Bipolar cell / Amacrine cell
7. Inner plexiform layer	
8. Ganglion cell layer	Ganglion cell
9. Nerve fiber layer	Optic n. fibers / 10. Inner limiting membrane

DIAGRAM OF RETINA

Light

Rod
Cone

Outer segment
Outer segment
Connecting structure
Cilia
Ellipsoid zone
Ellipsoid zone with mitochondria
Myoid zone
Nucleus
Nucleus
Nucleus
Müller's cell
Ribbon filament   Spherules   Pedicle

ULTRASTRUCTURE OF A ROD AND CONE

Visibility coefficient

Scotopic (rods)
Photopic (cones)

Wavelength mμ

ABSORPTION CURVES

Sensitivity %

Blue   Green   Red

3 types of cones

Wavelength mμ

-317-

146. VISUAL (RETINOGENICULOSTRIATE) PATHWAYS— PART I

I. General summary of visual pathways: impulses from the retinae pass backward through the optic nerves. At the optic chiasma, all fibers from the opposite nasal halves of the retinae cross and join fibers from the temporal retinae of the same side to form the optic tracts. The fibers of each tract synapse in the lateral geniculate body, and from here, the geniculocalcarine fibers pass through the optic radiation or geniculocalcarine tract to the optic or visual cortex in the calcarine area of the occipital lobe. Visual fibers also pass to the brain, hypothalamus, superior colliculi, and pretectal nuclei

A. THE BOUNDED SPACE OF THE ENVIRONMENT SEEN BY ONE EYE (monocular vision) while that eye is fixed on a stationary or fixed point is called the field of vision or monocular field of that eye. Normally we see with both eyes (binocular vision) and have a binocular field of vision formed by the almost complete overlapping of the monocular fields
   1. The temporal hemiretina of one eye sees the same half of the visual field as the nasal hemiretina of the other eye
   2. When the 2 hemiretinas do not see precisely the same part of the visual fields, the condition is diplopia or double vision

II. Stimulation of rods and cones with light results in photochemical decomposition and acts on the membrane of the receptor to cause a receptor potential that lasts as long as the light continues. It is a hyperpolarization signal rather than a depolarization signal

A. RECEPTOR POTENTIALS are transmitted unchanged through the bodies of the rods and cones since neither generate action potentials. The receptor potentials themselves, however, acting at the synaptic bodies, induce signals in the successive neurons by electronic current flow across the synaptic gaps

III. Stimulation of the bipolar, amacrine, and horizontal cells: these are all cells seen in the inner nuclear layer of the retina

A. THE SYNAPTIC BODIES OF THE RODS AND CONES make intimate contact with the dendrites of both bipolar and horizontal cells and the hyperpolarization is transferred directly to both, with the excitation signal summating in both.

B. THE BIPOLAR CELLS are the main transmitting link for the visual signal from the rods and cones to the ganglion cells and are entirely excitatory

C. THE HORIZONTAL CELLS spread horizontally and transmit signals laterally as far as several hundred microns. They transmit signals mainly to bipolar cells in the areas lateral to the excited rods and cones. They are inhibitory, enhance contrasts in the visual scene, and are important in helping differentiate colors

D. THE AMACRINE CELLS inhibit the ganglion cells, but response is transient, not steady

E. THE INTERPLEXIFORM CELLS are interspersed among cell bodies of bipolar cells and are postsynaptic to the amacrine and presynaptic to the horizontal and bipolar cells. They constitute a feedback loop from inner to outer layers of retinal synapses

V. Neuroglial cells, similar to those of the brain's gray matter, are found in the retinal inner layers. In addition, there are large numbers of modified neuroglial cells called Müller cells, which extend from the interface between the innermost retinal fiber layer and vitreous body to the junction of the rod and rod fiber, and cone and cone fiber, forming the inner and outer limiting membranes. They extend through the retina with lateral processes given off which intervene between retinal neural elements, giving these cells a supporting role. They have a metabolic, a neuroendocrine, and possibly a nerve transmission modulator function, among other possibilities

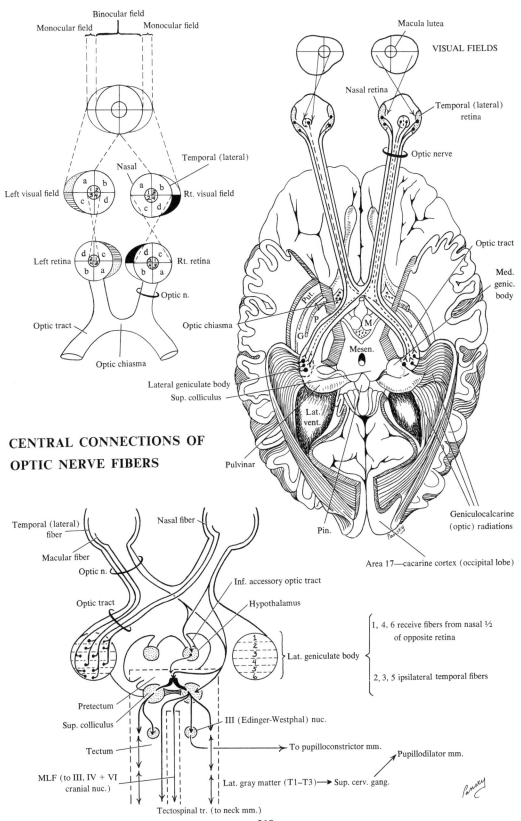

Binocular field

Monocular field   Monocular field

Macula lutea

VISUAL FIELDS

Nasal retina

Temporal (lateral) retina

Optic nerve

Nasal   Temporal (lateral)

Left visual field

Rt. visual field

Optic tract

Med. genic. body

Left retina   Rt. retina

Optic n.

Optic chiasma

Lateral geniculate body

Optic tract

Sup. colliculus

Optic chiasma

Pulvinar

CENTRAL CONNECTIONS OF
OPTIC NERVE FIBERS

Geniculocalcarine (optic) radiations

Pin.

Area 17—cacarine cortex (occipital lobe)

Temporal (lateral) fiber   Nasal fiber

Macular fiber

Optic n.

Inf. accessory optic tract

Optic tract

Hypothalamus

1, 4, 6 receive fibers from nasal ½ of opposite retina

Lat. geniculate body

2, 3, 5 ipsilateral temporal fibers

Pretectum

Sup. colliculus

III (Edinger-Westphal) nuc.

Tectum

To pupilloconstrictor mm.

Pupillodilator mm.

MLF (to III, IV + VI cranial nuc.)

Lat. gray matter (T1–T3) ⟶ Sup. cerv. gang.

Tectospinal tr. (to neck mm.)

147. VISUAL (RETINOGENICULOSTRIATE) PATHWAYS— PART II

I. **Stimulation of the ganglion cells:** bipolar, horizontal, and amacrine cells all stimulate the ganglion cells

A. THE GANGLION CELLS TRANSMIT THEIR SIGNALS via the optic nerve fibers in the form of action potentials, and they transmit continuous nerve impulses (5 per second) even when not stimulated. Thus, the visual signal is superimposed on this basic level of stimulation and can be either excitatory or inhibitory

B. THE GANGLION CELLS TRANSMIT luminosity signals (intensity), contrasts in the visual scene, changes in intensity, and color signals

   1. Thus color analysis begins in the retina and is not entirely a brain function

C. FOUR TYPES OF GANGLION CELLS are present according to the manner in which their discharge is initiated: classified as belonging to an on-system (discharge when light is on), an off-system (discharge at end of light stimulation), an on-off system (discharge at the beginning and end of light), and a steady-background system (discharge spontaneously with no light needed*

II. **Function of the lateral geniculate bodies:** each of which is composed of 6 nuclear layers

A. LAYERS 1, 4, AND 6 receive signals from the nasal retina of the opposite eye; layers 2, 3, and 5 receive signals from the temporal part of the ipsilateral retina

B. ALL LAYERS OF THE LATERAL GENICULATE BODY relay visual information to the visual cortex through the geniculocalcarine tract (optic radiations)

C. THE PAIRING OF LAYERS FROM THE 2 EYES plays a major role in the fusion of vision because corresponding retinal fields in the 2 eyes connect with respective neurons that are approximately superimposed over each other in the successive layers

D. THE SIGNALS RECORDED IN THE RELAY NEURONS OF THE GENICULATE BODY are similar to those recorded in the ganglion cells of the retina; a few transmit luminosity but the majority transmit signals of contrasting borders in the visual image, and many are responsive to movement of objects across the visual scene

   1. The signals in the geniculate body differ from the retina in that a greater number of complex interactions are found as a result of convergence of excitatory and inhibitory signals from 2 or more ganglion cells on the relay neurons of the geniculate body

III. **Function of the primary visual striate (area 17) cortex:** the ability to detect spatial organization of the visual scene, namely, forms of objects, brightness, shading, etc. It is dependent on the primary visual cortex (mainly in the calcarine fissure) located bilaterally on the medial aspect of each occipital cortex

A. SPECIFIC POINTS OF THE RETINA connect with specific points of the visual cortex; i.e., the right halves of the 2 respective retinae connect with the right visual cortex and the left with the left

   1. The macula is represented at the occipital pole, and the peripheral regions of the retina are represented in concentric circles farther and farther forward from the occipital pole

   2. The upper part of the retina is seen superiorly in the cortex, the lower part inferiorly

B. COLOR ANALYSIS: there are specific cells in the primary cortex that are stimulated by color intensity or by contrasts of the opponent colors. These effects are identical to those seen in the lateral geniculate body except that the proportion of cells excited by opponent color contrasts is less here

   1. Since neuronal excitation by color contrasts is a means of deciphering color, it appears that the cortex is concerned with an even higher order of detection of color than simply deciphering color itself, a process mainly completed by the time the signals pass through the geniculate bodies

*The on-sets, off-sets, and on-off sets can also be identified in the lateral geniculate body and the primary visual cortex of the optic pathways.

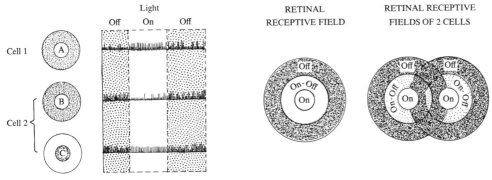

Light
Off   On   Off

Cell 1

Cell 2

RETINAL
RECEPTIVE FIELD

RETINAL RECEPTIVE
FIELDS OF 2 CELLS

A = On center gang. cell   C = Off center gang. cell with light in periphery
B = Off center gang. cell

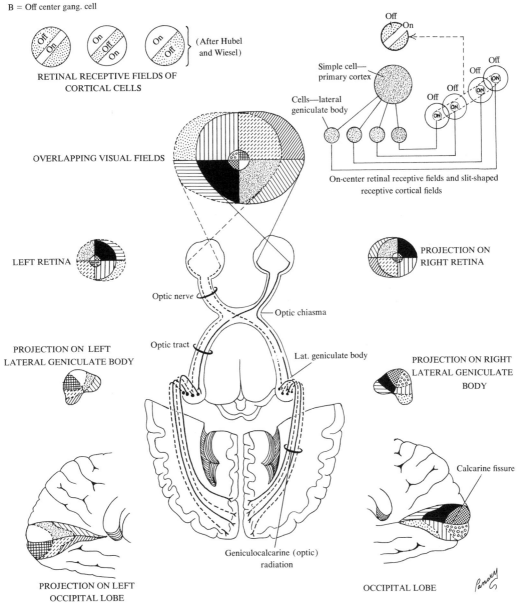

RETINAL RECEPTIVE FIELDS OF
CORTICAL CELLS

(After Hubel
and Wiesel)

Simple cell—
primary cortex

Cells—lateral
geniculate body

On-center retinal receptive fields and slit-shaped
receptive cortical fields

OVERLAPPING VISUAL FIELDS

LEFT RETINA

PROJECTION ON
RIGHT RETINA

Optic nerve

Optic chiasma

Optic tract

Lat. geniculate body

PROJECTION ON LEFT
LATERAL GENICULATE BODY

PROJECTION ON RIGHT
LATERAL GENICULATE
BODY

Calcarine fissure

Geniculocalcarine (optic)
radiation

PROJECTION ON LEFT
OCCIPITAL LOBE

OCCIPITAL LOBE

148. OPTIC RADIATIONS AND SUPERIOR COLLICULUS

I. **The optic radiation** is also called the geniculocalcarine tract. Fibers from the lateral geniculate nucleus (body) project through the retrolentiform and sublentiform parts of the internal capsule and curve around the lateral wall of the lateral ventricle to end in the cortex adjacent to the calcarine fissure

A. ALL FIBERS OF THE RADIATION do not pass directly backward to the occipital lobe. Instead they form a broad sheet covering the posterior and inferior horns of the ventricle.

1. Fibers representing superior visual quadrants (inferior retinal quadrants) loop out into the temporal lobe as Meyer's loop before turning posteriorly. Temporal lobe damage may thus produce a visual defect

2. Fibers representing inferior visual quadrants (superior retinal quadrants) swing back posteriorly and superiorly around the body and posterior horn of the ventricle

B. A RETINOTOPIC ORGANIZATION is maintained in the optic radiations

1. Fibers from inferior visual fields are more superior, while those representing superior visual fields loop farthest out into the temporal lobe

2. Macular fibers occupy a broad middle area

3. The visual pathway in the radiation is more dispersed than elsewhere and individual fibers carry information from only one eye. Thus, damage here often results in deficits that are slightly different for the 2 eyes

C. THE VISUAL PATHWAY ends retinotopically in the cortex around the calcarine fissure (area 17 of Brodmann)

1. Inferior visual fields project to cortex above the calcarine fissure; superior visual fields to cortex below the fissure

2. The maculae are represented more posteriorly

3. Peripheral fields are represented more anteriorly

4. Many myelinated fibers ramify in the visual cortex in a discrete layer that is seen as a thin white stripe (line of Gennari) by the naked eye and accounts for its name—striate cortex

D. VISUAL OR STRIATE CORTEX parallels the calcarine fissure. It is surrounded by area 18, which in turn is surrounded by area 19. Areas 18 and 19 are both referred to as the visual association cortex and are interconnected with area 17

II. **Superior colliculus:** major function in humans is not known, but it probably serves in the reflex control and regulation of many eye and head movements and certain aspects of vision, particularly in attention and perception associated with vision. The major inputs (in primates) are visual, arising in the retina and striate cortex

A. RETINAL INPUT FIBERS bypass the lateral geniculate nucleus, pass in the brachium of the superior colliculus, and end retinotopically in the colliculus. Many fibers are collaterals of axons that end in the geniculate nucleus

B. CORTICAL INPUT from area 17 projects to the colliculi via the brachium and ends in a retinotopic pattern

C. IN ADDITION TO VISUAL INPUTS, the superior colliculus receives somatosensory (spinotectal) and auditory inputs from the inferior colliculus, and inputs from widespread cortical areas

D. EFFERENT CONNECTIONS include projections to the reticular formation, inferior colliculus, cervical spine (tectospinal tract), and posterior thalamus (pulvinar and lateral geniculate nuclei)

RETINOGENICULATE & GENICULOCALCARINE PATHWAYS

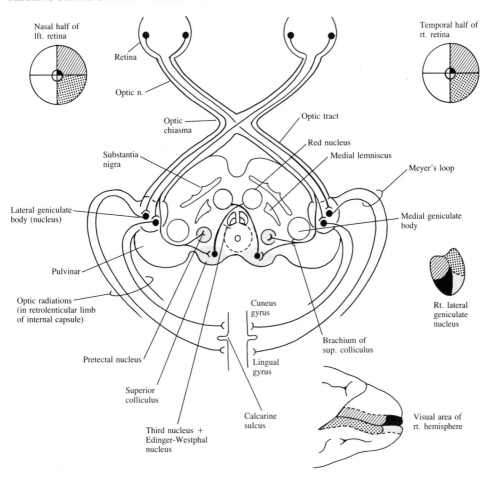

Nasal half of
lft. retina

Retina

Optic n.

Optic
chiasma

Substantia
nigra

Lateral geniculate
body (nucleus)

Pulvinar

Optic radiations
(in retrolenticular limb
of internal capsule)

Pretectal nucleus

Superior
colliculus

Third nucleus +
Edinger-Westphal
nucleus

Temporal half of
rt. retina

Optic tract

Red nucleus

Medial lemniscus

Meyer's loop

Medial geniculate
body

Rt. lateral
geniculate
nucleus

Cuneus
gyrus

Brachium of
sup. colliculus

Lingual
gyrus

Calcarine
sulcus

Visual area of
rt. hemisphere

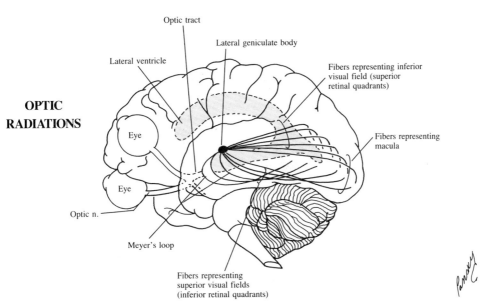

**OPTIC
RADIATIONS**

Optic tract

Lateral geniculate body

Lateral ventricle

Fibers representing inferior
visual field (superior
retinal quadrants)

Fibers representing
macula

Eye

Eye

Optic n.

Meyer's loop

Fibers representing
superior visual fields
(inferior retinal quadrants)

149. OTHER VISUAL PATHWAYS

I. Signals from the primary cortex project laterally in the occipital cortex into visual association areas (areas 18, 19) for additional processing of visual information

A. HERE THE NEURONS RESPOND TO MORE COMPLEX PATTERNS such as geometrical design, curving borders, angles, etc.

B. DESTRUCTIVE LESIONS OF THE VISUAL ASSOCIATION AREAS result in difficulty with certain types of visual perception and learning such as dyslexia or word blindness (difficulty in understanding the meaning of words one sees)

II. The tectal system or secondary visual system is difficult to define and comprises a sequence of retina to the superior colliculus, to pulvinar of the thalamus, and to the inferotemporal cortex

III. Other pathways from the retina involve various reflexes and the pineal gland

A. SUPERIOR COLLICULUS AND PRETECTUM are important nuclear structures intercalated in the pupillary reflex, accommodation, and conjugate eye movements

B. THE ACCESSORY OPTIC SYSTEM involves several small nuclei in the midbrain tegmentum and has a significant role in the function of the pineal gland

IV. Retinohypothalamic fibers exist in a variety of mammals (including monkeys), which end in the suprachiasmatic nucleus, a small hypothalamic nucleus above the optic chiasma. Similar connections are assumed possible in humans

V. Accommodation (near reflex) is essential for visual acuity. Three things occur reflexly when attention is directed to nearby objects: (1) convergence of the 2 eyes, so that image of object falls on both fovea, (2) contraction of the ciliary muscle and thickening of lens (accommodation) so that image is in focus on retina, and (3) pupillary constriction, which reduces any light aberration and increases depth of field

A. IS MEDIATED BY THE CEREBRAL CORTEX, unlike pupillary light reflex

1. One can voluntarily focus on a nearby object

2. It is regulated by a negative feedback mechanism that automatically adjusts the focal power of the lens

3. The pathway includes the optic pathways from the eye to the cerebral cortex; in turn, a projection back from cortical area 19 (and from frontal eye field area 8) via the optic radiation through the brachium of the superior colliculus terminates in the nucleus of the superior colliculus and/or pretectal area

   a. Through a chain of interneurons, the collicular nuclei are connected to the nucleus of cranial nerve III, stimulating medial rectus motor neurons and preganglionic motor neurons of the Edinger-Westphal nucleus. The latter project via the IIIrd nerve and synapse in the ciliary ganglion, with postganglionic fibers innervating the ciliary muscles in the ciliary body. Thus, pupillary constriction and convergence of the eye accompany accommodation to near vision

VI. Depth perception: the most precise and accurate means for localizing objects in space utilizes the images seen simultaneously by both eyes, called stereopsis or solid vision

A. BASED ON THE FACT that the visual cortex on one side receives input from the contralateral nasal hemiretina and the ipsilateral temporal hemiretina, and the presence of an interhemispheric link through the corpus callosum

B. THE SOLE BASIS FOR STEREOPSIS is the horizontal disparity between the 2 hemiretinal images; the horizontal disparities between the 2 hemiretinas viewing the same visual field obtain the cues of depth and, in turn, the processing in the neurons of the visual pathways leads to depth perception

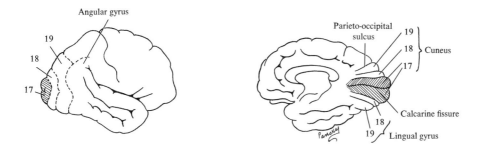

ACCOMMODATION REFLEX

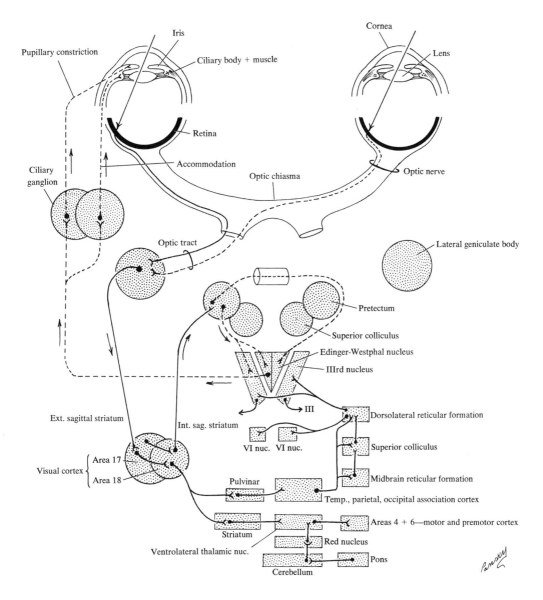

150. VISUAL REFLEXES

I. Pupillary constriction (direct light) and consensual reflexes

A. THE PUPILS CONTRACT when the eyes are exposed to bright light and reduce the intensity of light reaching the retina. This involuntary light reflex is in response to direct light stimulation and its pathway is

1. Retinal ganglion cell axons via the optic nerve, optic chiasma (some fibers decussate), optic tract, and brachium of superior colliculus (bypassing the lateral geniculate body) terminate in the pretectum. Interneurons project to the contralateral pretectum via the posterior commissure. Interneurons interact with other interneurons which, in turn, synapse with preganglionic neurons of the oculomotor nerve (arising in the Edinger-Westphal nucleus). These preganglionic neurons project via the IIIrd nerve and synapse in the ciliary ganglion from which postganglionic neurons innervate the constrictor muscle of the iris

2. When only one eye is exposed to bright light, the pupils of both eyes constrict (consensual reflex). This is a result of crossing of some fibers in the optic chiasma and interconnections across the midline of the bilateral pretectum

II. Pupillary dilatation

A. THE PUPILS DILATE when the eyes are exposed to dim light. Optic pathways act through the reticular formation and descending sympathetic pathways (in the dorsolateral tegmentum of the brainstem and in the anterior half of the cervical spinal cord) to stimulate preganglionic neurons of the sympathetic intermediolateral cell column at C8 and T1 levels. The preganglionic fibers ascend in the sympathetic chain and synapse in the superior cervical ganglion, and postganglionic neurons follow adjacent blood vessels to innervate the dilator muscle of the iris. Interruption of these fibers may result in Horner's syndrome

III. Corneal reflex: touching the cornea causes the eyelids to close reflexly. The afferent fibers are in the ophthalmic nerve (V), and the efferent fibers of the reflex arc are in the facial nerve (VII) to the orbicularis oculi muscle

IV. Argyll-Robertson pupil may occur in syphilis of the CNS. In this syndrome, the pupil is small in dim light and does not constrict further when the eye is exposed to bright light. This same pupil will respond by constricting further during accommodation-convergence reaction (pupils do accommodate)

A. THE USUAL STATED CAUSE for this dissociation of pupillary reflex is a lesion in the pretectum and may be due to diabetes, tertiary syphilis, or encephalitis. Some feel that the cause is a local disease of the myoepithelial cells of the dilator pupillae, because the patient's pupil dilates sluggishly, or not at all, when the sphincter is paralyzed by atropine. The contraction of the sphincter in accommodation is stronger than that in response to light (sufficient to overcome the inertia of a diseased dilator)

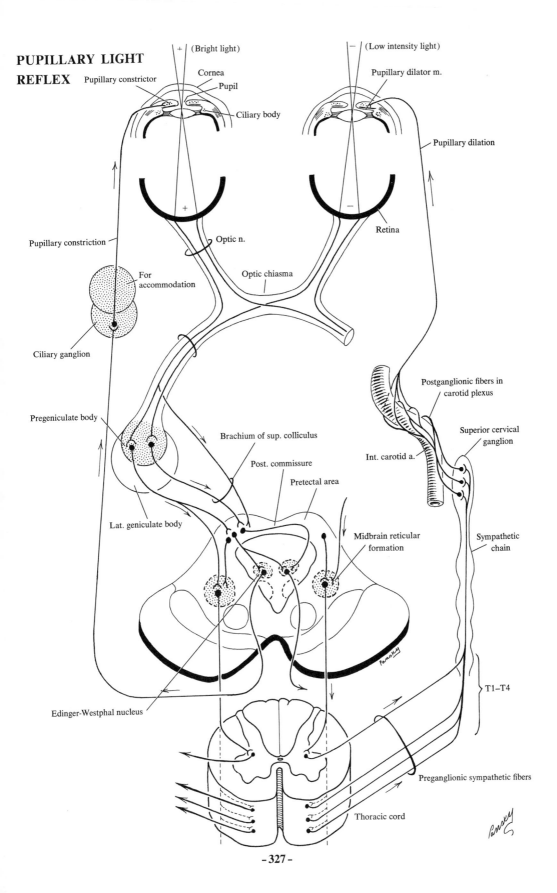

PUPILLARY LIGHT REFLEX

+ | (Bright light)

− (Low intensity light)

Cornea

Pupillary constrictor

Pupil

Ciliary body

Pupillary dilator m.

Pupillary dilation

+

−

Pupillary constriction

Optic n.

Retina

For accommodation

Optic chiasma

Ciliary ganglion

Pregeniculate body

Postganglionic fibers in carotid plexus

Superior cervical ganglion

Int. carotid a.

Brachium of sup. colliculus

Post. commissure

Pretectal area

Lat. geniculate body

Midbrain reticular formation

Sympathetic chain

Edinger-Westphal nucleus

T1–T4

Preganglionic sympathetic fibers

Thoracic cord

151. EYE MOVEMENTS AND THEIR CONTROL

I. **Binocular vision:** the same points in the field of vision common to the two eyes must be focused on corresponding loci in the two retinas. The simultaneous movement of both eyes in the same direction is called conjugate movement (convergence is a disconjugate movement, when the eyes look at a close-up vision together)

A. THE OCULOMOTOR SYSTEM regulates these movements and comprises several central pathways and the lower motor neurons of cranial nerves III, IV, and VI, innervating the extraocular muscles

1. Each set of muscles (medial and lateral recti, superior and inferior recti, and the superior and inferior oblique) is reciprocally innervated so that the contraction of one muscle of each pair is synergistically synchronized with the relaxation of the other muscle in order to direct the gaze to any position

2. The oculomotor system comprises four subdivisions

   a. The saccadic system (search for visual targets): the movement is a small rapid jerk called a saccade. Saccades search the visual field, move the visual image over the receptors and prevent adaptation. Saccades are the fastest movement in the body. They occur during sleep with the eyes closed. If stopped, all color detail fades in seconds.

   b. The smooth pursuit system (tracking of visual objects): smooth movements cause the eye to follow objects in the visual fields. They require continual feedback of visual information about the moving object

   c. The vestibular system (compensation for head movements): if a stationary visual object is focused on the fovea and the head is moved, the eyes must be moved an equal distance for the object to remain focused on the fovea (i.e., if the head moves down, the eyes must move up)

   d. The vergence system (convergence): used to track a visual object in depth through the visual field by turning the eyes inward as the object gets closer and outward as it moves away

3. The four movements appear to be controlled by separate neurologic systems, yet they cooperate in all movements

   a. The control system depends on information from the retina

   b. When the eyes and head are fixed, the image of a moving object moves across the receptors and gives rise to retinal signals

   c. When the eyes follow a moving object, however, the image remains stationary on the retina and the retina cannot signal movement, but we see movement because the rotation of the eye in the head can apparently give rise to perception of movement, even estimate velocity

   d. In movement, one must decide what is moving and what is stationary with regards to a reference frame. If the information about movement is by vision alone (and not due to body movement), we generally assume that the larger objects in the visual field are stationary

B. VOLITIONAL MOVEMENTS OF THE EYE: the frontal eye field in area 8 is the cortical center which influences voluntary eye movements mediated through cranial nerves III, IV, and VI and is called the voluntary eye field

C. AUTONOMIC EYE MOVEMENTS AND EYE FIXATION: our eyes can scan a line or follow a moving object without any volitional effort

1. The neuronal center and pathways involved with locking the eyes on an object once it has been located comprise the involuntary fixation mechanism and include reflex pathways from retina to primary visual cortex to association visual cortex and then via corticotectal fibers to the superior colliculus, from which connections are made with cranial nerve nuclei of III, IV, and VI. Area 19 may play a role in this mechanism

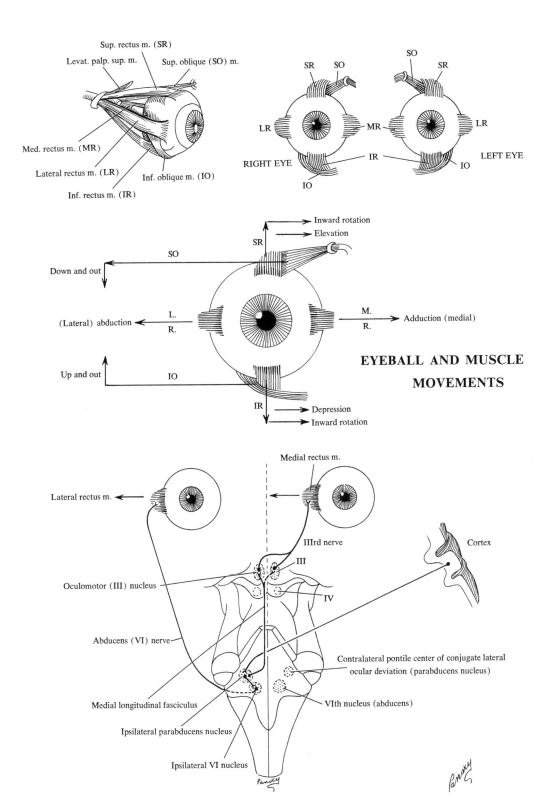

Sup. rectus m. (SR)

Levat. palp. sup. m.

Sup. oblique (SO) m.

Med. rectus m. (MR)

Lateral rectus m. (LR)

Inf. oblique m. (IO)

Inf. rectus m. (IR)

SR   SO

SO   SR

LR

MR

LR

RIGHT EYE

IR

LEFT EYE

IO

IO

Inward rotation

Elevation

SR

SO

Down and out

(Lateral) abduction

L.
R.

M.
R.

Adduction (medial)

Up and out

IO

IR

Depression

Inward rotation

**EYEBALL AND MUSCLE
MOVEMENTS**

Medial rectus m.

Lateral rectus m.

IIIrd nerve

III

Cortex

Oculomotor (III) nucleus

IV

Abducens (VI) nerve

Contralateral pontile center of conjugate lateral
ocular deviation (parabducens nucleus)

Medial longitudinal fasciculus

VIth nucleus (abducens)

Ipsilateral parabducens nucleus

Ipsilateral VI nucleus

CONJUGATE LATERAL DEVIATION OF EYES

152. LESIONS OF THE OPTIC PATHWAYS
AND RELATED PHENOMENA

I. Impairment of a small area of the retina results in scotoma (blind spot) in the field viewed by the eye

II. Complete interruption of the optic nerve results in permanent blindness; in one eye this is called monocular blindness
A. THE BLIND EYE can still accommodate and show a consensual light reflex because the normal eye activates the intact reflex arcs through the pretectum and superior colliculus and the optic nerve to the blind eye

III. A fixed dilated pupil may be a sign of pressure on the oculomotor nerve

IV. A midline lesion of the optic chiasma may interrupt the decussating retinofugal fibers, resulting in a loss of reception in the nasal hemiretina of each eye, accompanied by loss of vision in the temporal half of the visual field of each eye, resulting in bitemporal hemianopia (hemianopsia)
A. SUCH A DEFECT IS HETERONYMOUS: portions of both retinas that view different areas of the fields of vision are involved
B. INTERRUPTION OF THE NONDECUSSATING FIBERS on both sides of the optic chiasma produces a binasal hemianopia: loss of reception in the temporal hemiretina of each eye accompanied by loss of vision in the nasal half of the visual field of each eye

V. Complete interruption of the optic tract or lateral geniculate body or the entire primary visual cortex on one side (e.g., right) results in a left contralateral homonymous hemianopia or blindness in the field of vision on the side opposite the lesion
A. HOMONYMOUS refers to corresponding regions of both retinas, hence to a single visual field

VI. Partial lesions of the visual pathways produce partial defects in the visual fields
A. A LOWER QUADRANTIC HOMONYMOUS ANOPSIA results from a contralateral lesion in the upper (superior) half of the optic radiation or the entire primary visual cortex (area 17) above the calcarine fissure
   1. This field defect occurs because the pathways from the upper temporal quadrant of the ipsilateral retina and upper nasal quadrant of the contralateral retina are interrupted
B. A LESION OF THE LOOP OF MEYER will produce a contralateral upper quadrantic anopsia because pathways from the lower temporal quadrants of the ipsilateral retina and the lower nasal quadrant of the contralateral retina are interrupted

VII. Lesions of a portion of the primary visual cortex may result in a homonymous field defect in which the field defect of one eye is exactly superimposed on that of the other eye; the loss is said to be congruous
A. A LESION IN THE OPTIC TRACT OR RADIATION may result in a field defect in which the fields from each eye are not exactly superimposed; the loss is said to be incongruous

VIII. A blow or injury that damages the choroid layer of the eye may result in sympathetic ophthalmia; not only may sight be impaired in the injured eye but the other eye also gradually loses its sight. This happens because the body responds to certain substances released from the injured choroid layer, resulting in an immunologic reaction that destroys both eyes

IX. Absence of pigment: true albinos shun light and have difficulty seeing because pigment is absent in the iris, choroid, and retinal pigment layers. Visual acuity is rarely better than 20/100 to 20/200

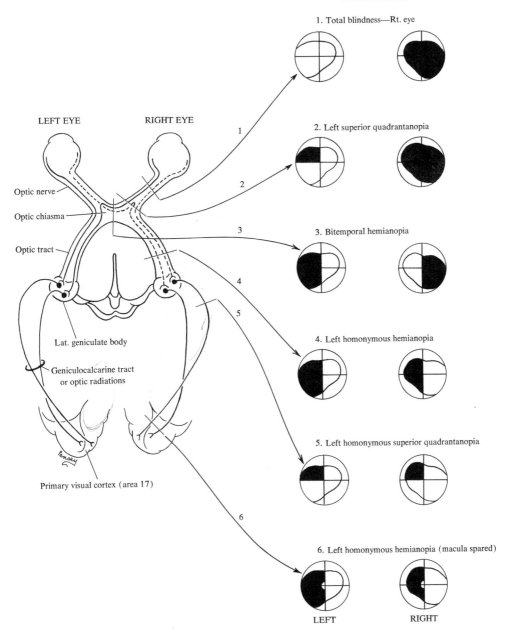

VISUAL FIELDS

1. Total blindness—Rt. eye

2. Left superior quadrantanopia

3. Bitemporal hemianopia

4. Left homonymous hemianopia

5. Left homonymous superior quadrantanopia

6. Left homonymous hemianopia (macula spared)

LEFT EYE   RIGHT EYE

Optic nerve

Optic chiasma

Optic tract

Lat. geniculate body

Geniculocalcarine tract
or optic radiations

Primary visual cortex (area 17)

LEFT   RIGHT

**DEFECTS IN VISUAL FIELDS
PRODUCED BY LESIONS IN
REGIONS OF THE VISUAL PATHWAYS**

PART II

Physiologic Mechanisms of Neuroscience

Unit Eleven

NEUROPHYSIOLOGY

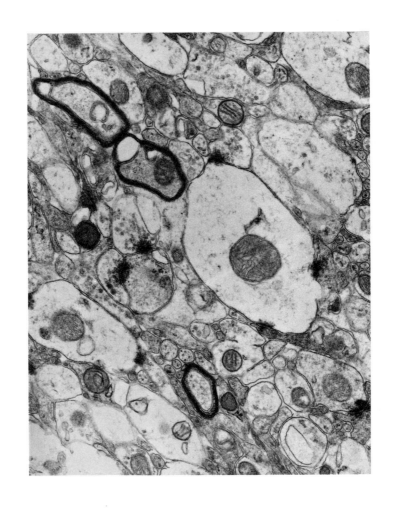

153. PHYSIOLOGIC ACTIVITIES OF NEURONS: FUNCTIONAL ANATOMY OF THE NEURON

I. **The human nervous system** consists of neurons and glial cells which make up the brain and spinal cord. It also includes the many nerve processes that run between them, as well as the receptors, muscles, and glands they innervate

A. THE BRAIN AND SPINAL CORD form the central nervous system (CNS) and are protected by the skull and vertebral column, respectively

B. ALL NERVE CELLS (or parts of them) that lie outside of the skull and vertebral column are part of the peripheral nervous system (PNS)

II. **The neuron, the basic unit of the nervous system:** the human brain has about 25 billion cells

A. EACH NEURON is bounded by a cell membrane that encloses the cell contents

  1. Only about 10% of the cells in the nervous system are neurons

    a. The remaining cells are glial cells which probably support the neurons physically and sustain them metabolically

  2. The neuron consists of 3 parts, each associated with a particular function

    a. The dendrites (usually several) and cell body or soma

      i. Form a series of branched cell outgrowths connected to the cell body and serve as an extension of the cell membrane of the body

      ii. There are great variations in dendritic formations: profusely branched; ratio of soma surface to dendritic surface may be larger in others; some neurons have no dendrites; and dendrites can be several hundred micrometers long

      iii. The dendrites and soma are the site of the specialized junctions with other neurons through which signals are passed to the cell

      iv. The soma contains the nucleus and other organelles involved in metabolic processes and maintains the metabolism of the neuron, including its growth and repair (diameters of neuronal cell bodies are in the order of 5–100 mm)

    b. The axon (neurite or nerve fiber) is a long single process extending from the cell body (usually longer than the dendrite)

      i. Its first portion plus the part of the cell body where it is joined to the cell is called the initial segment

      ii. Can give off collaterals (branches) along its course, and near its end undergoes branching into many axon terminals, the last part of which is enlarged and is responsible for transmitting a signal from the neuron to the cell contacted by the terminal. Axons and collaterals vary in length from a few micrometers to well over a meter

      iii. Every axon in a peripheral nerve lies in a "tube" formed by special glial cells, the Schwann cell. Together, the axon and Schwann sheath (neurolemma) around it are called the nerve fiber

        (a) The number of nerve fibers bundled together to form a nerve varies from dozens or hundreds to tens of thousands

        (b) About $\frac{1}{3}$ of all nerve fibers are encased, during growth, by the Schwann cell winding around the axon so that between the axon and the Schwann-cell body, a sheath is formed consisting of a lipid-protein mixture called myelin; nerve fibers "insulated" in this manner are called myelinated nerve fibers, A or B fibers

        (c) The myelin sheath is discontinuous, being interrupted at regular intervals which, under the microscope, look like constrictions or nodes and are called the nodes of Ranvier (occur about every 1–2 mm)

      iv. Nerve fibers without the myelin sheath are called unmyelinated or C fibers

        (a) The unmyelinated fibers are also unsheathed by Schwann cells; often a Schwann cell may enclose several unmyelinated axons

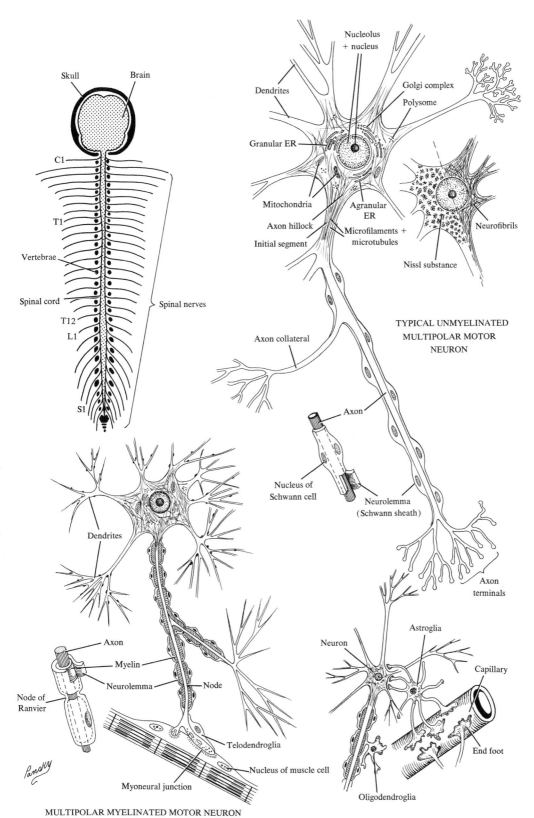

Skull   Brain

C1

T1

Vertebrae

Spinal cord

T12

L1

S1

Spinal nerves

Nucleolus
+ nucleus

Dendrites

Golgi complex

Polysome

Granular ER

Mitochondria

Agranular
ER

Axon hillock

Microfilaments +
microtubules

Initial segment

Neurofibrils

Nissl substance

TYPICAL UNMYELINATED
MULTIPOLAR MOTOR
NEURON

Axon collateral

Axon

Nucleus of
Schwann cell

Neurolemma
(Schwann sheath)

Axon
terminals

Dendrites

Astroglia

Neuron

Capillary

Axon

Myelin

Neurolemma

Node

Node of
Ranvier

End foot

Telodendroglia

Nucleus of muscle cell

Myoneural junction

Oligodendroglia

MULTIPOLAR MYELINATED MOTOR NEURON

– 337 –

154. MEMBRANE POTENTIALS: PRINCIPLES OF ELECTRICITY—CHEMICAL REACTIONS AND ELECTRIC CHARGES

I. **Introduction:** a difference in potential, the membrane potential, usually exists between the inside of a cell and the extracellular fluid around it

A. THE FUNCTION OF A CELL, in many cells (muscle cells, glandular cells, etc.), can be controlled by the size of the membrane potential

B. THE SPECIALIZED ROLE OF THE NERVOUS SYSTEM is to propagate changes in membrane potential in its cells and to transmit them to other cells

II. **Chemical reactions** are basically electric in nature since they involve exchanging or sharing of negatively charged electrons between atoms to form ions or bonds

A. MOST CHEMICAL REACTIONS RESULT IN NEUTRAL MOLECULES, containing equal numbers of electrons and protons, but in some cases the ions formed have a net electric charge
   1. Organic molecules: negative carboxyl group, $RCOO^-$, or positive amino group, RNH_3^+
   2. Inorganic ions: Na^+, K^+, and Cl^-

B. ELECTRICAL PHENOMENA: the cell environment contains many charged particles and the interaction of these plays an important role in cell function
   1. Except for water, the major chemical components of the extracellular fluid are Na^+ and Cl^- ions; the intercellular fluid contains high concentrations of K^+ ions and organic molecules with ionized groups, particularly proteins and phosphate compounds (RPO_4^{2-})

C. ALL PHYSICAL AND CHEMICAL PHENOMENA can be described using the fundamental units of measurements: length, time, mass, temperature, and electric charge
   1. Each of these units is independent in that none can be defined in terms of a combination of the others; e.g., electric charge is a fundamental unit of measurement and cannot be defined in terms of length, time, mass, or temperature

III. **Electric charges:** two types of charge that behave symmetrically are positive $(+)$ charges and negative $(-)$ charges

A. LIKE CHARGES repel each other; unlike charges attract each other
   1. When electrons were discovered, they behaved like the charged particles which we called negative. Protons, on the other hand, behaved like positive charges. Thus, electrons are negative and protons are positive
      a. All electrical phenomena result from the interaction of negative electrons and positive protons
      b. The total number of $(+)$ and $(-)$ particles in the universe are said to be equal, thus, the universe is electrically neutral, as are atoms, which contain an equal number of electrons and protons

B. AN ELECTRICAL FORCE draws the opposite charges together. When $(+)$ and $(-)$ charges are separated, the force is measured and the relation between the amount of force, quantity of charge, and the distance separating the charges can be observed
   1. The amount of force acting between electric charges increases when the charged particles are moving closer together and with increasing quantity of charge
   2. Energy (E) is the ability to do work (W) and work is the product of force (F) and distance (D): $W = FD$
      a. If opposite charges come together as a result of attracting forces between them, work is done and energy released by the moving particles, since a force is exerted over a distance between them opposite to their electric attracting force. Thus, energy must be added to the system, in the form of work, in order to separate the charges

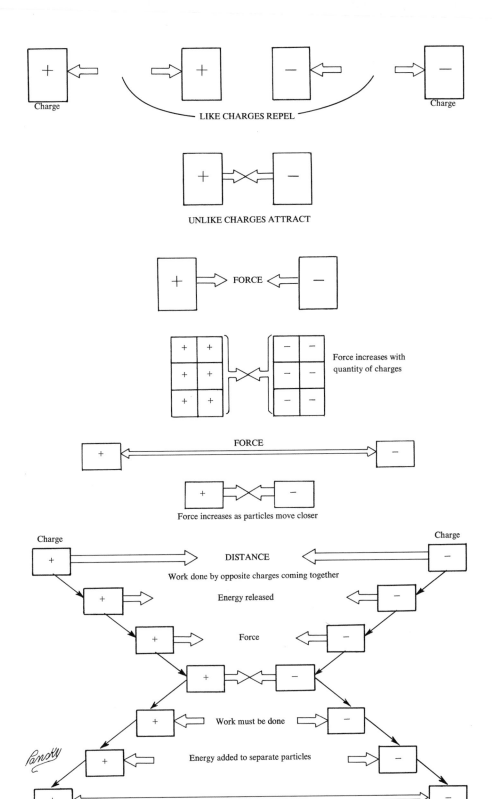

Charge

LIKE CHARGES REPEL

Charge

UNLIKE CHARGES ATTRACT

FORCE

Force increases with quantity of charges

FORCE

Force increases as particles move closer

Charge

Charge

DISTANCE

Work done by opposite charges coming together

Energy released

Force

Work must be done

Energy added to separate particles

Panofy

155. MEMBRANE POTENTIALS: PRINCIPLES OF ELECTRICITY—VOLTAGE, CURRENT, RESISTANCE

I. Electric charges (cont.)

A. AN ELECTRIC FIELD is produced when charges tend to separate because of diffusion, which opposes further separation
   1. The electric field is a vector which has both magnitude and direction
   2. The potential difference is the sum of all the electric field vectors
B. WHEN ELECTRIC CHARGES ARE SEPARATED, they have the potential for doing work if they are allowed to come together again
   1. Voltage is a measure of the potential of a separated electric charge to do work and is defined as the amount of work done by an electric charge when moving from one point in a system to another
      a. Voltage is always measured with respect to 2 points in a system, so one refers to the potential difference between the 2 points, and the units of measurements are known as volts (the terms "voltage" and "potential difference" are synonymous)
   2. The total amount of charge that can be separated in most biological systems is very small and thus the potential difference is very small
      a. The potential across the nerve cell membrane is about −70 millivolts (mV), negative inside the membrane

II. Current is the movement of electric charge per unit of time

A. IF AN ELECTRIC CHARGE IS SEPARATED BETWEEN 2 POINTS, there is a potential difference between these points, and the electric force of attraction between the opposite charges tends to make the charges flow, producing a current
   1. The amount of charge that does move, the current, depends on the nature of the material lying between the separated charges, which may be either a conductor or an insulator
      a. The space may be occupied by a copper wire, a solution of water and ions, glass, or rubber, or it may even be empty (a vacuum)
   2. The ability of an electric charge to move through these media varies
      a. The flow depends on the number of charged particles in the material that are able to move and carry current
      b. The amount of current also depends on the interactions between the moving charges and the materials
   3. The ability of a material to store charges is determined by its dielectric constant
B. RESISTANCE is the hindrance of the movement of electric charge through a particular material
   1. The higher the resistance, the lower the amount of current flow for any given voltage
      a. Glass and rubber have high electric resistance and are used as insulators
      b. Materials of low resistance, such as copper wire, are used as conductors
   2. Pure water is a relatively poor conductor because it contains very few charged particles, but when sodium chloride (NaCl) is added, the solution becomes a relatively good conductor with a low resistance since sodium (Na) and chloride (Cl) ions provide charges that can carry the current
C. THE WATER COMPARTMENTS inside and outside the cells in the body contain numerous charged particles (ions) which are capable of moving between areas of charge separation
   1. Lipid contains very few charged particles and thus has a high electric resistance
      a. The lipid component of the cell membrane provides a region of high electric resistance separating two water compartments (inside and outside of the cell) of low resistance

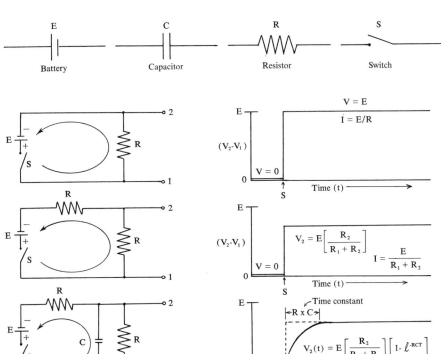

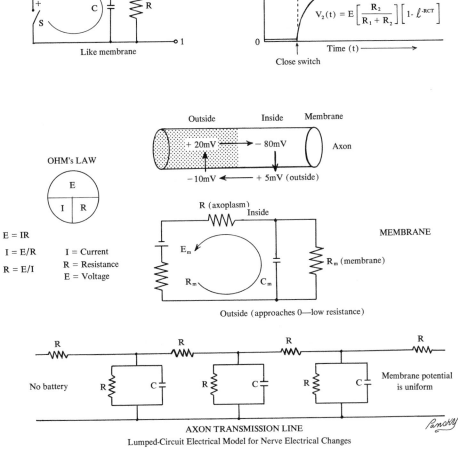

AXON TRANSMISSION LINE
Lumped-Circuit Electrical Model for Nerve Electrical Changes

156. ELECTRICAL AND CHEMICAL PROPERTIES OF A RESTING CELL: DIFFUSION POTENTIAL

I. **Measuring the membrane potential:** the intracellular and extracellular spaces of a cell are filled with aqueous salt solutions in which most of the molecules dissociate into (+) or (−) charged ions (atoms or molecules)

A. CATIONS are positively (+) charged atoms; anions are negatively (−) charged atoms
B. AN ELECTRIC POTENTIAL DIFFERENCE (VOLTAGE) across a cell membrane can be determined by inserting a fine electrode into the cell and another into the extracellular fluid around the cell, and connecting the two by means of a voltmeter
   1. All cells of the body exhibit a membrane potential (true for nerve and muscle cells *only* when they are not stimulated). The inside of the cell is (−) charged with respect to the outside, which is relatively (+) charged
      a. The nonstimulated potential is called the resting potential and varies from 5 to 100 mV, depending on the cell type and its chemical environment

II. **Diffusion potential** is generated across a membrane largely due to the tendency for K^+ to move down its concentration chemical gradient* so that the inside of the cell is negative with respect to the outside of the cell (outward K^+ and inward Cl^- diffusion)

A. INTRODUCTION: the different ionic concentrations on either side of the cell would soon cancel each other out by simple diffusion of the mobile particles, were it not prevented or compensated for by
   1. K^+ ions move through the membrane in both directions, but because of higher concentrations inside the cell, the ions pass through the membrane pores from the inside $30\times$ more often than from the outside, resulting in a net efflux of K^+ ions. The motive force is the higher osmotic pressure of K^+ inside the cell
   2. The osmotic pressure difference for K^+ would soon equalize except for an opposing force applied by an electric field which is the membrane potential
      a. When a K^+ ion leaves the cell, since it is (+) charged, an excess positive charge appears on the outer surface of the membrane (capacitor) which corresponds to an excess of (−) charges now present on the membrane's inside. Thus, the membrane potential serves to oppose the efflux of more ions and the efflux of (+) charges in itself builds up an electric potential that impedes the efflux of more (+) charges from the cell
B. PERMEABLE MEMBRANE: If we have 2 solutions of different ionic composition and ignore the nature of the membrane between them, they would react as follows
   1. With solution 1 of dilute NaCl at a concentration of 0.1 M and that of solution 2 also of dilute NaCl but at a concentration of 0.01 M, and if the membrane between is permeable to all ion species
      a. Both Na^+ and Cl^- on side 1 are more concentrated, and they will diffuse down their concentration gradients and move from side 1 to side 2
      b. Mobility and movement of Cl^-, however, are 50% greater than those of Na^+, so Cl^- will move to side 2 faster and side 2 will transiently become slightly (−) charged with respect to side 1. This electric gradient, due to differential diffusion of charged particles in solution, is called the diffusion potential and disappears in time as the concentrations on both sides equalize
   2. If we change solutions to resemble extracellular fluid (0.15 M NaCl), and the intracellular fluid is 0.15 M KCl, and again ignore the separation membrane
      a. Cl^- concentrations on the 2 sides are equal, but Na^+ and K^+ are not
      b. The mobility of K^+, like Cl^-, is about 50% greater than Na^+, and thus K^+ will diffuse down its concentration gradient faster than Na^+; positive charge will initially leave side 2 faster than it enters, and side 2 will become electronegative with respect to side 1. Again, in time, equilibrium will prevail

---

*The magnitude of the diffusing tendency from one area to another is proportionate to the difference in concentration of the substance in 2 areas.

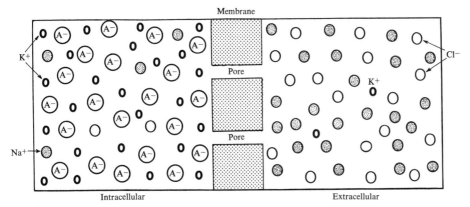

ION DISTRIBUTION

IONIC CONCENTRATIONS INSIDE AND OUTSIDE A NERVE CELL
(mmol/liter of water)

Ion	Extracellular	Intracellular
Na^+	150	15
K^+	5	150
Cl^-	125	10
HCO_3^-	27	8
A^- (Protein)	——	155
Resting potential: -70 to -90mv		

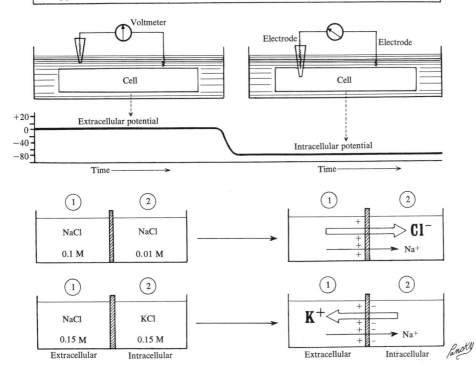

DIFFUSION POTENTIALS THROUGH A COMPLETELY PERMEABLE MEMBRANE

-343-

157. ELECTRICAL AND CHEMICAL PROPERTIES OF A RESTING CELL: EQUILIBRIUM POTENTIAL

I. Diffusion potential (cont.)

A. EQUILIBRIUM POTENTIAL deals with intra- and extracellular concentrations, but now involves the addition of a selectively permeable membrane between compartments, so that K^+ can pass through but not Na^+ or Cl^-

1. All the Na^+ will remain on side 1 and some of the K^+, diffusing down its gradient, will be added to it; side 1 will become relatively positive
2. This diffusion potential, however, will not disappear or equalize in time, but will be of greater magnitude, depending on the concentration gradient of K^+
3. The concentration gradient, which causes net diffusion of particles from a region of higher to a region of lower concentrations, is called the concentration force
   a. As this force moves K^+ from side 2 to side 1 and the latter becomes more (+), the electric potential difference begins to influence the movement of the (+)-charged K particles [they are attracted by the relatively negative charge of side 2 and repulsed by the (+) charge of side 1]
   b. The electrical potential difference that tends to return K^+ to side 2 (repulsion of like charges and attraction of opposite charges) is called the electric force
   c. As long as the concentration force driving K^+ from side 2 to 1 is greater than the electric force driving the opposite direction, there will be net movement of K^+ from side 2 to side 1 and the potential difference will increase
   d. Side 1 will become more and more positive until the electric force opposing the entry of K^+ equals the concentration force favoring its entry. The membrane potential at which the latter occurs is called the equilibrium potential, and, at that time, there is no net movement of the ion because the forces acting on it are exactly balanced
      i. The value of the equilibrium potential for any ion depends on the concentration gradient for that ion across the membrane (if the concentrations on the 2 sides are equal, the concentration force is zero and the electric potential needed to oppose it is also zero)
      ii. The K^+ concentrations of neurons and extracellular fluid typically show the equilibrium potential for K^+ to be about 90 mV, the inside of the cell being (−) with respect to the outside
      iii. For neurons, the Na^+ equilibrium potential is 60 mV, the inside being positive

B. THE NERNST EQUATION can be used to calculate the equilibrium potential for K^+ ions:

$$Ek = \frac{RT}{ZF} \ln \frac{[K^+]_o}{[K^+]_i} = -61 \log \frac{[K^+]_i}{[K^+]_o}$$

Ek = equilibrium potential (Nernst potential) for K^+, R = gas constant, T = temperature in °Kelvin, Z = valence for K^+, F = Faraday constant, $[K^+]_o$ and $[K^+]_i$ are the K^+ concentrations outside and inside the cell. Calculating for $[K^+]_o$ = 4 mM/L and $[K^+]_i$ = 155 mM/L, Ek = −97 mV

C. OTHER IONS, including chloride (Cl^-) and sodium (Na^+), contribute to the membrane potentials of living cells. The permeability to Cl^- varies among different cell types, and there is always some inward passive diffusion of Na^+ ions. The influence of these ions and of K^+ on the membrane potential is included in an extension of the Nernst equation called the Goldman constant field equation:

$$Em = \frac{RT}{F} \ln \frac{P_K[K^+]_o = P_{Na}[Na^+]_o + P_{Cl}[Cl^-]_i}{P_K[K^+]_i + P_{Na}[Na^+]_i + P_{Cl}[Cl^-]_o}$$

P_K, P_{Na}, and P_{Cl} = permeabilities of K^+, Na^+, and Cl^-, respectively

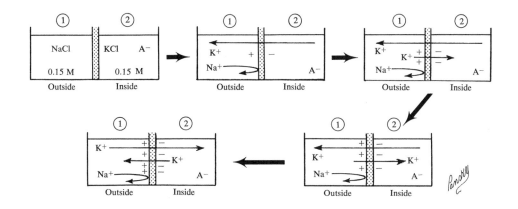

Cell permeable only to K+—diffusion down concentration gradient leaves inside of cell negatively charged

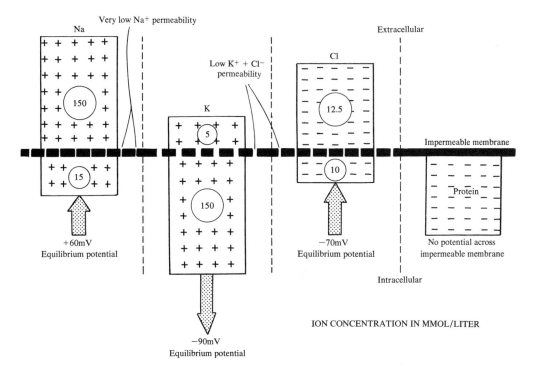

ION CONCENTRATION IN MMOL/LITER

Arrows are the magnitude of transmembrane (equilibrium) potentials needed to maintain internal + external ion concentration across a permeable membrane (Arrow direction: effect of equilibrium potentials needed to maintain with regard to ion exchange)
(After Lewis)

158. ELECTRICAL AND CHEMICAL PROPERTIES OF A RESTING CELL: RESTING MEMBRANE POTENTIAL

I. Diffusion potential (cont.)

A. THE RESTING CELL MEMBRANE POTENTIAL: why is there a negative resting potential in a nerve or muscle cell?

1. The source of the resting potential is the unequal distribution of diffusible ions, particularly K^+, inside and outside the cell; i.e., K^+ are about $20-100\times$ greater in concentration inside a muscle cell than outside it

2. The distribution of Cl^- ions is the reverse of that of the K^+ ions; generally, about $20-100\times$ lower inside the cell than outside

3. The majority of intracellular anions are large protein anions (A^-), not Cl^-; Na^+ concentration is about $5-15\times$ lower inside the cell than outside

4. Extracellular salt solution (salty like blood) is essentially an NaCl solution with an NaCl content of about 9 g/L, equal to a physiologic solution

5. In a nerve cell at rest, the K^+ ion concentration is greater inside the cell than outside, and the Na^+ concentration is the opposite, with the cell membrane being about $50-75\times$ more permeable to K^+ than to Na^+

6. Resting membrane potential (about -80 mV) is close to K^+ equilibrium potential because the membrane is not perfectly impermeable to Na^+, which continually diffuses down its electric and concentration gradients, adding a small amount to the inside of the cell. Measured membrane potential of a neuron (cell body) is closer to -70 mV than to the K^+ equilibrium potential of -90 mV, resulting in a membrane not at the K^+ equilibrium potential and a continual net diffusion of K^+ out of the cell

B. THE SODIUM-POTASSIUM PUMP: membrane active transport mechanisms use energy from cellular metabolism to pump Na^+ back out of the cell and K^+ back in

1. For each 3 Na^+ ions pumped out of the cell, 2 K^+ ions are pumped into it

2. The pump makes a small direct contribution to the membrane potential because of the net outward pumping of cations

3. The pump makes an essential indirect contribution to membrane potential by maintaining the concentration gradients down which the ions diffuse. The membrane potential is due primarily to the diffusion of these ions

C. SUMMARY

1. Some K^+ ions diffuse out of the cell down their concentration gradient and some move in down the electric gradient, because the membrane potential is not as negative as the K^+ equilibrium potential (-70 mV rather -90 mV). Less K^+ enters the cell passively than leaves it

2. The difference is relatively small and is made up by active transport via the membrane pump such that the total K^+ entering equals that leaving—no gain or loss

3. Na^+ is driven passively into the cell by both electric and concentration forces, but because the membrane permeability of a resting cell to Na^+ is so low, the amount entering is low

4. There is no passive force to remove Na^+ from the cell. That which enters must be actively transported out by the membrane pump. The amount of Na^+ pumped out equals, in most cases, the amount of K^+ pumped in

5. When more than 1 ion species can diffuse across the membrane, the membrane permeability properties, as well as the concentration gradient of each species, must be considered when evaluating membrane potential

   a. If the membrane is impermeable to a given ion species, no ion of that species can cross it and contribute to diffusion potential, regardless of the electric and concentration gradients that are present

   b. The greater the membrane permeability to an ion species, the greater the influence of that ion on the diffusion potential

6. Since the resting membrane is more permeable to K^+ than Na^+ (and the concentration differences are about the same), the resting membrane potential is much closer to the K^+ equilibrium potential than to that of Na^+

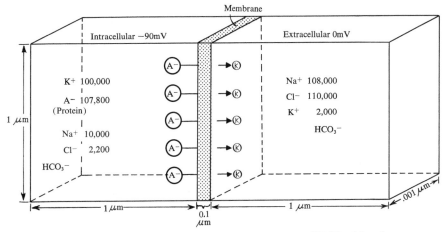

MEMBRANE CHARGE DURING RESTING POTENTIAL (K+ diffused through
membrane but stays fixed at its outside due to A⁻ charge inside)

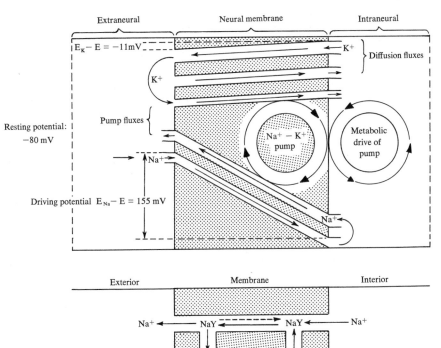

Na⁺ − K⁺ TRANSPORT THROUGH MEMBRANE VIA CARRIER MOLECULES (after Glynn)

159. ACTION POTENTIAL: INTRODUCTION AND IONIC BASIS

I. **Introduction:** resting potential is needed for nerve cells and muscle fibers to fulfill their body functions. The nerve cell, however, must pick up information and transmit it in order to coordinate and integrate body function. When the cells work and are active, brief (+) charges in the membrane potential occur and are called action potentials, nerve impulses, or spikes. The latter are initiated in 3 ways: by activation of receptors; by synaptic input from other neurons; or spontaneously by intrinsic changes in the cell membrane (pacemaker neurons)

A. THE MEMBRANE POTENTIAL UNDERGOES RAPID CHANGE when nerve and muscle cells seem physiologically active, suddenly changing from -90 mV to $+30$ mV and then rapidly returning to its original value. This rapid change of membrane potential or sequence of depolarization and repolarization of the membrane that occurs spontaneously when the membrane is depolarized beyond the threshold potential is called an action potential and may only last about $\frac{1}{1000}$ second. Excitable membranes can transmit action potentials along their surfaces, transmitting a signal from one portion of the nerve or muscle to another

B. PHASES OF THE ACTION POTENTIAL
1. The action potential begins with a rapid (+) change in potential, the upstroke or rising phase, which lasts about 0.2–5 milliseconds in nerves and muscles
    a. During this rising phase, the cell loses its negative resting charge of polarization and this phase is also called the depolarization phase
       i. The membrane is said to be depolarized or stimulated when its potential is less negative than the resting membrane potential (closer to zero)
2. The depolarization phase sets processes in motion to restore the resting membrane charge. The excitation-induced depolarization phase is followed by a spontaneous repolarization (moving from zero back to the resting potential)
3. When the potential moves beyond its resting level, it is hyperpolarizing

C. THE ACTION POTENTIAL AND EXCITATION ARE TRIGGERED when the membrane, starting from the resting potential, is depolarized to about -50 mV. The potential from which it starts is called the threshold. The membrane charge is unstable at threshold
1. The charge dissipates rapidly and automatically and usually even reverses polarity, resulting in a rapid rise of the action potential that goes beyond the zero potential and is called the overshoot phase
2. The condition of a spontaneous progressive discharge of the membrane, triggered at threshold, is called excitation and lasts only a brief time, less than 1 millisecond

II. **Ionic basis**

A. HIGH INTRACELLULAR K^+ CONCENTRATION is a prerequisite for resting potential; high extracellular Na^+ concentration is necessary for action potential. Excitability depends on low intracellular Na^+ concentration so that Na^+ can flow into the cell

B. THE ACTION POTENTIAL MUST RESULT FROM A TRANSIENT CHANGE in either the concentration gradients or the membrane permeabilities, the latter being the case
1. In the resting state, the membrane is 50–75$\times$ more permeable to K^+ than Na^+ ions; thus, the magnitude and polarity of the resting potential are due almost entirely to the movement of K^+ ions out of the cell

C. DURING THE ACTION POTENTIAL membrane permeability to Na^+ and K^+ is altered
1. In the rising phase, membrane permeability to Na^+ increases 600-fold, due to brief opening of "sodium gates," and Na^+ ions rush into the cell. Tetrodotoxin can block this process. At this stage there is little change in K^+ ion permeability of the membrane. More (+) charge enters the cell in the form of Na^+ ions than leaves it in the form of K^+ ions; the membrane potential decreases and eventually reverses polarity, becoming (+) inside and (−) outside the membrane

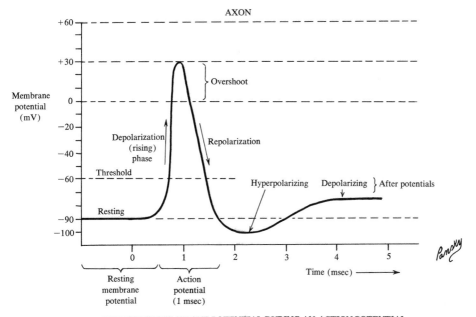

CHANGES IN MEMBRANE POTENTIAL DURING AN ACTION POTENTIAL

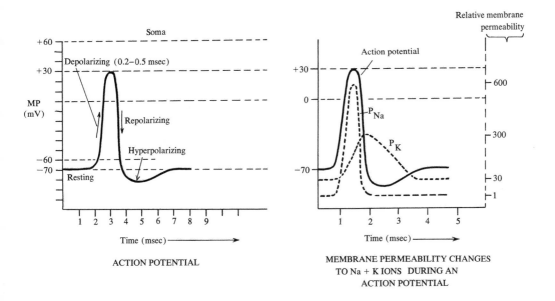

ACTION POTENTIAL

MEMBRANE PERMEABILITY CHANGES
TO Na + K IONS DURING AN
ACTION POTENTIAL

160. ACTION POTENTIAL: IONIC BASIS
AND SUMMARY EVENTS

I. Ionic basis (cont.)

A. ACTION POTENTIAL IN NEURONS lasts about 1 millisecond (0.001 second) and the membrane quickly returns to resting level because

1. Permeability to Na^+ is altered when the membrane potential is changed; e.g., hyperpolarization causes a decrease in Na^+ permeability, whereas depolarization results in an increase in Na^+ permeability

2. Increased Na^+ permeability (activation) is rapidly turned off (Na^+ inactivation)

3. The membrane permeability to K^+ increases over its resting stage due to opening of "potassium gates." Tetraethylammonium (TEA) selectively blocks this process

4. As the membrane becomes more (+) inside and the drive for Na^+ entry is reduced, Na^+ permeability decreases toward its resting value and Na^+ entry quickly decreases (this alone would restore the potential to its resting level)

5. The entire process is further speeded up by a simultaneous increase in K^+ permeability, causing more K^+ to move out of the cell down its concentration gradient, allowing K^+ diffusion to regain predominance over Na^+ diffusion and the membrane to return to its resting level

6. Massive movements are not necessary; e.g., only 1 of every 100,000 K^+ ions in the cell need diffuse out to change the membrane to its resting value, and very few Na^+ ions need enter the cell to cause depolarization during an action potential, resulting in virtually no change in the concentration gradients during that action potential

B. SUMMARY OF IONIC MOVEMENTS DURING ACTION POTENTIAL

1. As a result of depolarization beyond the threshold, the Na^+ conductance increases rapidly (increased permeability to Na^+)

2. The K^+ conductance also increases, but with a time lag

3. Therefore, initially, Na^+ ions flow rapidly into the cell, creating further depolarization of the membrane, and the membrane potential moves in the direction of the Na^+ equilibrium potential at $+60$ mV

4. Then K^+ ions flow out of the cell, restoring the resting potential charge and repolarizing the membrane to its resting potential

5. Despite large changes in the conductance of the membrane during the action potential, the ionic shifts through the membrane are small in relation to the number of ions surrounding the membrane

6. The Na^+ ions that flowed into the cell with the action potential are expelled again in the course of time by the Na^+ pump

   a. The active Na^+ transport compensates not only for the resting Na^+ influx but also for the Na^+ influx during excitation

   b. The active Na^+ transport is of no importance for the individual action potential; if the pump is blocked (poisoned with dinitrophenol), despite active transport elimination, thousands of action potentials can occur before intracellular Na^+ concentration becomes too high to render the cell inexcitable

   c. Energy-consuming processes such as the Na^+ pump are only necessary to maintain the concentration gradients

   d. We have no explanation for the physiochemistry of how Na^+ permeability of the membrane is altered when the level of the membrane potential changes

   e. Whatever the responsible mechanisms, they are found only in the cell membranes of nerve and muscle cells; changing the membrane potential in other types of cells does not result in a change in membrane permeability

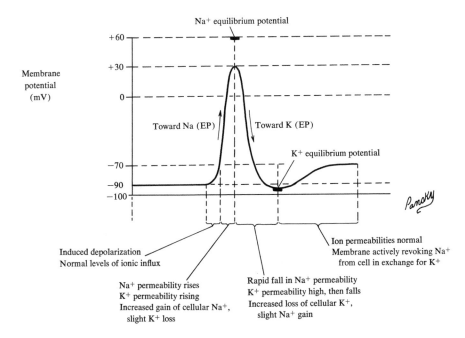

MEMBRANE ION PERMEABILITY CHANGES DURING AN ACTION POTENTIAL

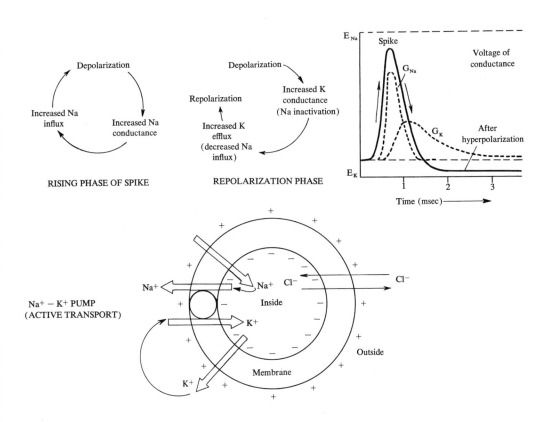

161. ACTION POTENTIAL: THRESHOLD AND REFRACTORY PERIOD

I. **Threshold:** partially depolarizing an excitable membrane initiates an action potential only when the strength of the stimulus is sufficient to depolarize the membrane potential to a critical level, known as the threshold potential. This is a threshold stimulus

A. SUBTHRESHOLD STIMULI (WEAKER STIMULI) do not initiate an action potential. Stronger stimuli, suprathreshold stimuli, elicit action potentials, but the response is no different from that following a threshold stimulus

B. THRESHOLD POTENTIAL FOR MOST EXCITABLE MEMBRANES is 5–15 mV more depolarized than the resting membrane's potential (if the resting potential of a neuron is −70 mV, the threshold's potential may be −60 mV). To initiate an action potential in such a membrane, the potential must be decreased by at least 10 mV

C. EXPLANATION OF THRESHOLD: at any potential between resting and threshold, Na^+ movement into the cell is less than K^+ ion movement out of the cell despite the increased Na^+ permeability which prevents further depolarization beyond that induced directly by the stimulus and drives the potential rapidly back to resting levels as soon as the stimulus is removed

   1. Once threshold is reached, the process is no longer dependent on stimulus strength, and the depolarization continues, to become an action potential solely because the membrane permeability changes allow Na^+ ions to diffuse down the electric and concentration gradients which exist across the membrane

   2. Action potentials triggered either by stimuli just strong enough to depolarize the membrane a bit above threshold or by very strong stimuli are identical

   3. Action potentials occur either maximally or not at all (all-or-none law). A single action potential cannot convey any information about the magnitude of the stimulus that initiated it

   4. The ability to distinguish between levels of stimuli, e.g., a loud sound or whisper, depends on the number of action potentials transmitted per unit of time (depends on frequency of action potentials and not on their size)

   5. Most nerve cells respond at frequencies between 0 and 100 action potentials per second (some even higher for brief periods of time)

II. **Refractory periods:** If one applies a threshold strength stimulus to a membrane and then stimulates the membrane with a 2nd threshold stimulus, the membrane does not always respond to the 2nd stimulus. The membrane appears unresponsive for a period of time even though identical stimuli are applied. The membrane, during this period, is said to be refractory to the 2nd stimulus

A. TWO SEPARATE REFRACTORY PERIODS ASSOCIATED WITH AN ACTION POTENTIAL: if the 2nd stimulus is increased to suprathreshold, levels can be distinguished

   1. During the 1-millisecond period of the action potential spike, a 2nd stimulus will not produce a 2nd action potential response no matter how strong it is; the membrane is said to be in its absolute refractory period

   2. Following the above, there is an interval during which a 2nd action potential response can be produced, only if the stimulus strength is greater than threshold—the relative refractory period (can last 10–15 milliseconds or longer)

B. REFRACTORY PERIOD MECHANISMS are related to the membrane mechanisms that alter the Na^+ and K^+ permeability

   1. The absolute refractory period corresponds to the period of inactivation of Na^+ ion channels

   2. The relative refractory period corresponds to the period of recovery of Na^+ channels. K^+ permeability becomes high during repolarization and is continuous throughout the negative and positive afterpotentials

C. AFTER AN ACTION POTENTIAL, time is needed to return the membrane structure to its original resting state. Refractory periods limit the number of action potentials an excitable membrane can produce in a given period of time

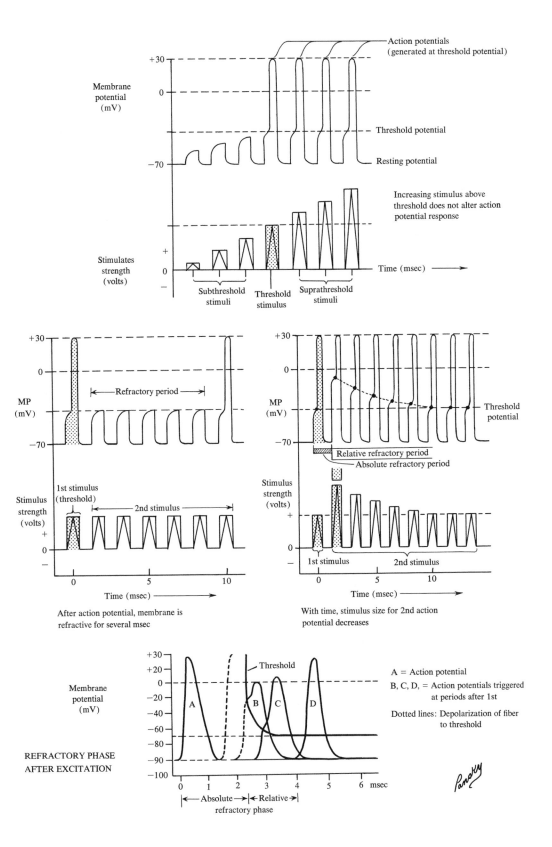

Action potentials
(generated at threshold potential)

Membrane
potential
(mV)

+30

0

−70

Threshold potential

Resting potential

Increasing stimulus above
threshold does not alter action
potential response

Stimulates
strength
(volts)

+

0

−

Time (msec)

Subthreshold
stimuli

Threshold
stimulus

Suprathreshold
stimuli

MP
(mV)

+30

0

−70

Refractory period

1st stimulus
(threshold)

2nd stimulus

Stimulus
strength
(volts)

+

0

−

0   5   10

Time (msec)

After action potential, membrane is
refractive for several msec

MP
(mV)

+30

0

−70

Threshold
potential

Relative refractory period
Absolute refractory period

Stimulus
strength
(volts)

+

0

−

1st stimulus   2nd stimulus

0   5   10

Time (msec)

With time, stimulus size for 2nd action
potential decreases

Membrane
potential
(mV)

+30
+20
0
−20
−40
−60
−80
−90
−100

Threshold

A

B   C   D

A = Action potential

B, C, D, = Action potentials triggered
at periods after 1st

Dotted lines: Depolarization of fiber
to threshold

REFRACTORY PHASE
AFTER EXCITATION

0   1   2   3   4   5   6   msec

|← Absolute →|← Relative →|
refractory phase

Panaky

−353−

162. PROPAGATION OF AXON DEPOLARIZATION

I. **Introduction:** If the action potential of a nerve fiber is measured at 2 points and the nerve is stimulated at one end, an action potential 1st appears at the measuring point closest to the site of stimulation and a little later at the 2nd measuring point. Thus, the action potential is propagated or conducted from the site of stimulation past the 2nd electrode

A. VELOCITY OF CONDUCTION is determined by dividing the distance between the 2 points that the action potential passes by the time it takes to travel that distance
   1. Conduction velocities in nerve fibers range from 1 m/s to greater than 100 m/s; they depend on characteristics of fiber and are typical for each type

B. ONE ACTION POTENTIAL does not itself travel along the membrane, but each action potential triggers, by local current flow, a new one at an adjacent area of the membrane
   1. The old action potential provides the electric stimulus that depolarizes the new membrane site to just past its threshold potential
   2. When this happens, the Na^+ activation cycle at the new membrane site takes over and an action potential occurs there
   3. Once the new site is depolarized to threshold, the action potential generated there is solely dependent on the electrochemical gradients and membrane permeability properties at the new site
   4. Since the above factors are the same as those involved in the generation of the old one, the new action potential is identical to the old. No distortion takes place as the signal passes along the membrane and the signal at the end of the membrane is the same as that at the beginning

C. PROCESS OF PROPAGATION
   1. When an action potential is initiated at a site, Na^+ permeability changes so that Na^+ rushes in across the membrane, making the inside of the cell relatively more ($+$) and leaving the outside more ($-$). At this time, the remainder of the cell membrane is at its normal resting potential
   2. Like charges attract and unlike ones repel, and thus current (the flow of positive charges) flows away from the activated membrane region through the cytoplasm and toward the activated region through the extracellular fluid
   3. The addition of ($+$) charge to the inside of the cell and removal of ($+$) charge from the outside decrease the potential difference across the membrane; this initial depolarization, due to local current flow, acts as a stimulus to trigger an action potential
   4. Meanwhile, at the original membrane site, Na^+ inactivation is occurring and K^+ permeability is increasing so that the membrane is repolarizing
   5. These processes repeat themselves until the end of the membrane is reached
   6. By convention, the direction of movement of the ($+$) ions is labeled the direction of current flow (negatively charged particles, however, can and do move in the opposite direction)
   7. The change in potential or electric charge moves from one point to another much faster than the ions themselves move between the points because of the influence of the molecules themselves. However, local current flow is decremental—its amplitude decreases with increasing distance
   8. Even though local current flow is decremental because membranes depend on local current flow, it is over a short distance and doesn't matter because only the amount of local current flow which is necessary to depolarize action potential occurs at the new site and is transmitted without a decrement, since the new action potentials are continually generated along the membrane

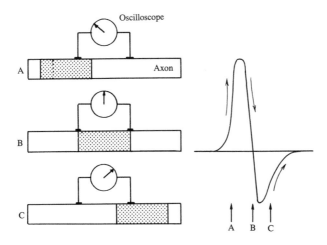

DIPHASIC RECORDING OF ACTION POTENTIAL WITH EXTERNAL ELECTRODES

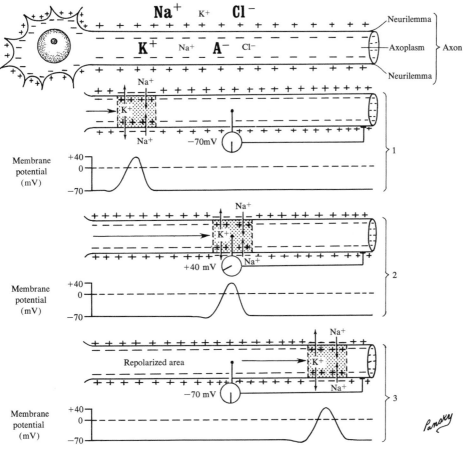

PROPAGATION OF A NERVE IMPULSE

163. PROPAGATION OF ACTION POTENTIAL: DIRECTION AND VELOCITY

I. Propagation of the action potential (cont.)

A. DIRECTION OF ACTION POTENTIAL PROPAGATION
1. The only direction of action potential propagation is away from the stimulation site
2. The action potentials in skeletal muscle membrane are initiated near the middle of the cell and propagate from this region toward the 2 ends, but in most nerve cell membranes, action potentials are initiated at one end of the cell; this unidirectional propagation is determined by the stimulus location and not by the inability to conduct in the other direction

B. VELOCITY OF ACTION POTENTIAL PROPAGATION
1. Velocity depends on the fiber diameter and whether or not the fiber is myelinated
   a. The larger the fiber diameter, the faster the action potential propagation because a large fiber offers less resistance to local current flow and adjacent regions are brought to threshold faster
   b. Myelin electrically insulates the membrane, making it more difficult for current to flow between intra- and extracellular fluid compartments
   c. Action potentials do not occur along the sections of membrane protected by myelin but occur only where the myelin coating is interrupted at the nodes of Ranvier where the membrane is exposed to the extracellular fluid. Thus, the action potential appears to jump from node to node as it propagates along the myelinated fiber; this is called saltatory conduction
   d. The membrane of nodes adjacent to the active node is brought to threshold faster and undergoes an action potential sooner than if no myelin were present
2. The velocity of action-potential propagation in large myelinated fibers can exceed 250 miles per hour

Fiber Type	Fiber Diameter (μm)	Conduction Velocity (m/sec)	Spike Duration (msec)	Absolute Refractory Period (msec)	Functions
A (α)	12–22	70–120	0.4–0.5	0.2–1.0	Somatic motor, primary muscle-spindle afferents Propioceptors
A (β)	5–13	30–70	0.4–0.5	0.2–1.0	Touch, kinesthesia (muscle sense), pressure
A (γ)	3–8	15–40	0.4–0.7	0.2–1.0	Motor (to mm spindle) Touch and pressure
A (δ)	1–5	12–30	0.4–1.0	0.2–1.0	Pain and temperature (heat and cold) Pressure
B	1–3	3.0–5.0	1.2	0.6–1.2	Automatic preganglionic fibers
C (dorsal root)	0.2–1.2	0.2–2.0	2	2	Pain, reflexes
C (sympathetic nerves)	0.3–1.3	0.7–2.3	2	2	Postganglionic sympathetic

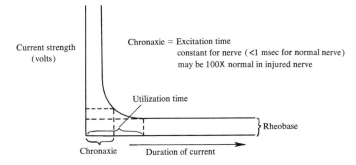

Chronaxie = Excitation time
    constant for nerve (<1 msec for normal nerve)
    may be 100X normal in injured nerve

STRENGTH—DURATION RELATIONSHIPS (EXCITATION TIME)

(Rheobase: Least strength which an applied current could have and still excite the membrane)

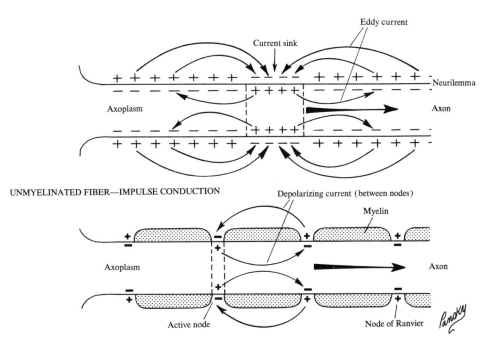

UNMYELINATED FIBER—IMPULSE CONDUCTION

MYELINATED FIBER—IMPULSE CONDUCTION
(SALATORY CONDUCTION)

164. CLASSIFICATION OF NEURONS AND RECEPTORS

I. Classification of neurons: regardless of shape they can be placed into 3 classes

A. AFFERENT NEURONS lie largely outside of the CNS

   1. At their peripheral endings have receptors, which, in response to physical or chemical environmental changes, cause action potentials to be generated in the afferent neuron
   2. These neurons carry information from receptors into the brain or spinal cord. After transmission to the CNS, some of the afferent information may be perceived as a conscious sensation
   3. Afferent nerve fibers from the viscera are called visceral afferents; all others are called somatic afferents

B. EFFERENT NEURONS also lie largely outside of the CNS and transmit the final integrated information from the CNS out to the effector organs (muscles or glands)

   1. Efferent neurons to skeletal muscles are called motor neurons; all the rest belong to the autonomic nervous system and are called autonomic efferents

C. THE INTERNEURONS lie within the CNS, where they originate and terminate

   1. 99% of all nerve cells belong to this group
   2. The interneurons and their connections in large part account for thoughts, feelings, learning, language, etc.
   3. The number of interneurons in the pathway between the afferent and efferent neurons varies with the complexity of the action
   4. One type of reflex, the stretch reflex, has no interneurons, whereas stimuli invoking memory or language may involve thousands of interneurons

II. Receptors: different energy forms (sensory elements) relay information about our external and internal environment: pressure, temperature, pain, light, etc.

A. TELERECEPTORS: for our distant environment via our eyes and ears

B. EXTEROCEPTORS: for our immediate environment via skin receptors

C. PROPRIOCEPTORS: about the attitude and position of the body in space, with receptors in the labyrinth and those of muscles, tendons, and joints

D. INTEROCEPTORS AND VISCEROCEPTORS: for information on events in the viscera

E. ALONE can deal with the variety of energy forms; the rest of the system can use only action potentials or small subthreshold changes in membrane potentials. Thus, all energy forms must be translated into the language of action potential

   1. The receptors are either specialized peripheral endings of neurons or separate cells intimately connected to them
      a. Not all neurons have specialized receptor regions—seen only on afferent neurons
   2. There are many types of receptors, each of which is specific, namely, responds to one form of energy or stimulus more than another, although virtually all receptors can be activated by several different forms of energy
      a. The receptors of the eye normally respond to light (electromagnetic radiation with wavelengths between 400 and 800 nm, violet to blue) but can be activated by intense mechanical stimuli such as a ''poke'' in the eye and we see ''stars''
      b. Receptors of the ear, on the other hand, are specialized to transduce sound waves with frequencies from 20 to 20,000 Hz (cycles per second)—sensory modalities (include vision, audition, etc.)
      c. Usually much more energy is needed to excite a receptor by energy forms to which it is not specific, the receptor being highly sensitive to its own special energy form, which is an adequate stimulus

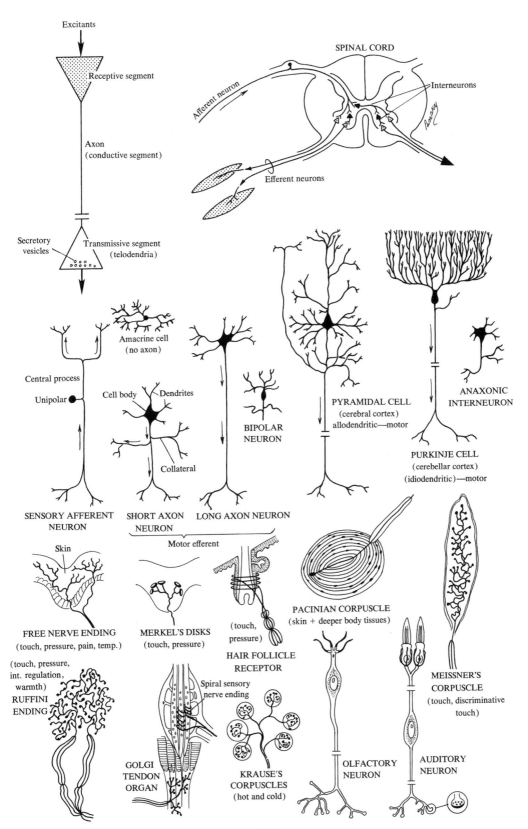

Excitants

Receptive segment

Axon
(conductive segment)

Secretory
vesicles

Transmissive segment
(telodendria)

SPINAL CORD

Afferent neuron

Interneurons

Efferent neurons

Amacrine cell
(no axon)

Central process

Unipolar

Cell body — Dendrites

Collateral

BIPOLAR
NEURON

PYRAMIDAL CELL
(cerebral cortex)
allodendritic—motor

ANAXONIC
INTERNEURON

PURKINJE CELL
(cerebellar cortex)
(idiodendritic)—motor

SENSORY AFFERENT
NEURON

SHORT AXON
NEURON

LONG AXON NEURON

Motor efferent

Skin

FREE NERVE ENDING
(touch, pressure, pain, temp.)

MERKEL'S DISKS
(touch, pressure)

(touch,
pressure)

HAIR FOLLICLE
RECEPTOR

PACINIAN CORPUSCLE
(skin + deeper body tissues)

(touch, pressure,
int. regulation,
warmth)
RUFFINI
ENDING

Spiral sensory
nerve ending

MEISSNER'S
CORPUSCLE
(touch, discriminative
touch)

GOLGI
TENDON
ORGAN

KRAUSE'S
CORPUSCLES
(hot and cold)

OLFACTORY
NEURON

AUDITORY
NEURON

165. GENERATOR POTENTIAL:
RECEPTOR ACTIVATION

I. Receptors (cont.)

A. RECEPTOR ACTIVATION OR THE GENERATOR POTENTIAL

1. The following refers only to the general mechanisms for receptor activation in which the receptor is the peripheral ending of the afferent neuron and responds to pressure or mechanical changes, namely, the transformation of mechanical energy into electrochemical energy of action potentials which take place at the very end of the afferent neuron (in the part lacking a myelin sheath)

   a. Mechanical stimuli (e.g., pressure) bend, stretch, or press on the receptor membrane and somehow, perhaps by opening pores in the membrane, increase membrane permeability

   b. With increased permeability, ions move across the membrane down their electric and concentration gradients (the permeability change is to Na^+ and one or more other ions)

   c. Since the intracellular fluid of all nerve cells has a higher concentration of K^+ and a lower concentration of Na^+ than the extracellular fluid, and the inside of the reacting neuron is about 70 mV negative with respect to the outside, change in membrane permeability at the receptor results in a net outward diffusion of a small number of K^+ ions and a simultaneous movement of a larger number of Na^+ ions inward

   d. The result is a net movement of (+) charge into the cell leading to a decrease in membrane potential (depolarization), with the movement of Na^+ into the nerve fiber playing the major role in depolarizing the nerve ending (K^+ and other ions such as Cl^- have less effect). The initial depolarization of the receptor is called the generator potential (g.p.)

   e. The g.p. occurs at the unmyelinated nerve ending where the cell membrane is specialized for excitation by an adequate or appropriate stimulus

   f. The g.p. cannot by itself produce action potentials, but the depolarization is conducted by local current flow a short distance from the nerve ending to the first node in the myelin sheath. If the amount of depolarization that reaches the first node is large enough to bring the membrane there to threshold, an action potential is initiated; it is the action potential and not the generator potential that propagates along the nerve fiber

   g. If, after one action potential is fired, local current flow through the first node remains sufficient to depolarize the node again to threshold, another action potential occurs, and as long as the first node is depolarized to threshold, action potentials continue to fire and propagate along the membrane of the afferent nerve

   h. The action potential is all-or-none and its amplitude is always the same, regardless of the size of the stimulus. The generator potentials, however, are not all-or-none; their amplitude and duration vary with the stimulus and other factors which might influence action potential frequency in the afferent neuron

   i. Summary: generator potential is a local response whose only function is to trigger an action potential; the latter traveling along the nerve fiber to its terminations

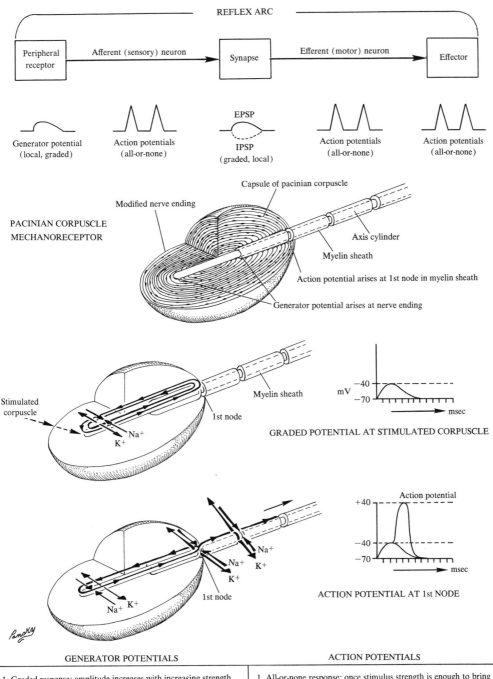

REFLEX ARC

| Peripheral receptor | Afferent (sensory) neuron → | Synapse | Efferent (motor) neuron → | Effector |

Generator potential
(local, graded)

Action potentials
(all-or-none)

EPSP

IPSP
(graded, local)

Action potentials
(all-or-none)

Action potentials
(all-or-none)

PACINIAN CORPUSCLE
MECHANORECEPTOR

Capsule of pacinian corpuscle
Modified nerve ending
Axis cylinder
Myelin sheath
Action potential arises at 1st node in myelin sheath
Generator potential arises at nerve ending

Stimulated corpuscle
Myelin sheath
1st node
Na^+
K^+

-40
-70
mV
msec

GRADED POTENTIAL AT STIMULATED CORPUSCLE

Na^+
Na^+
K^+
K^+
Na^+
K^+
1st node

Action potential
$+40$
-40
-70
msec

ACTION POTENTIAL AT 1st NODE

GENERATOR POTENTIALS	ACTION POTENTIALS
1. Graded response; amplitude increases with increasing strength or velocity of stimulus	1. All-or-none response; once stimulus strength is enough to bring membrane to threshold, further stimulus strength has no effect on amplitude
2. Can be added together. If a 2nd stimulus arrives before the GP of the 1st is over, the GP is added to the depolarization from 1st	2. Cannot be added together
3. Has no refractory period	3. Has a refractory period (1 msec)
4. Conducted passively; decreases in magnitude with increasing distance along nerve fiber	4. Propagated without loss of amplitude along nerve fiber
5. Duration: >1–2 msec; varies	5. Duration: 1–2 msec

166. RECEPTORS: AMPLITUDES AND INTENSITY CODING

I. Amplitude and duration of the generator potential (g.p.)

A. AMPLITUDE determines action potential frequency (number of action potentials fired per unit of time); it does not determine action potential size

B. AMPLITUDE VARIATIONS are due to the stimulus intensity, rate of change of the stimulus application, summation of successive g.p.s, and adaptation

   1. G.p.s become larger with greater intensity of the stimulus because the permeability changes increase and the transmembrane ion movements are greater (possible increase in number or size of pores allowing transmembrane ion flow)

   2. Amplitude of the g.p. also rises with a greater rate of change of stimulus application, as well as rate of removal of stimulus (so-called off-response, which further depolarizes the membrane, leading to a burst of action potentials)

   3. Summation of generator potentials

      a. Amplitude can be varied by adding 2 or more g.p.s—possible because they are a graded phenomenon and last 5–10 milliseconds; thus, if the nerve ending is stimulated again before the g.p. from a preceding stimulus has died, the 2 summate and make a large single g.p.

   4. Adaptation is a decrease in frequency of action potentials in the afferent neuron despite a constant stimulus energy and is not due to nerve fatigue (which does not occur in nerve membrane). It occurs because

      a. Stimulus energy can be dissipated in the tissues as it passes through them to reach the receptor

      b. Receptor membrane responsiveness can decrease with time and the g.p. amplitude drops even though the energy reaching the receptor stays the same

      c. Even if the g.p. amplitude remains unchanged, there can be a decreased frequency of action potentials in response to a constant stimulus because the membrane changes at the 1st node

C. SUMMARY: the magnitude of a given g.p. varies with stimulus intensity, rate of change of stimulus application, summation, adaptation, and cessation of the stimulus. Amplitude of the g.p. does not determine the amplitude of the action potentials in the afferent neuron but determines how many, if any, action potentials take place

II. Intensity coding: there are different stimulus intensities

A. INCREASED STIMULUS STRENGTH means a larger g.p. and higher frequency of firing action potentials

   1. There is an upper limit to the positive correlation between stimulus intensity and action potential frequency

      a. When stimulus strength becomes great, the g.p. reaches a maximum; further increase in rate of firing of action potentials by that receptor cannot take place

      b. Though a certain receptor cannot generate a higher frequency of action potentials, receptors at other branches of the same neuron can be stimulated

   2. Most afferent neurons have many branches, each with a receptor at its ending

   3. Receptors at different branches do not respond with equal ease to a given stimulus

      a. As stimulus strength increases, more and more receptors begin to respond and action potentials generated by these receptors propagate along the branch to the main afferent nerve fiber and, if the membrane is not refractory, they increase the frequency of action potential there

   4. In addition to the increased frequency of firing in a single neuron, similar receptors on the nerve endings of other afferent neurons are also activated as stimulus strength increases because a larger area is affected. This calling in of receptors on additional nerve cells is called recruitment

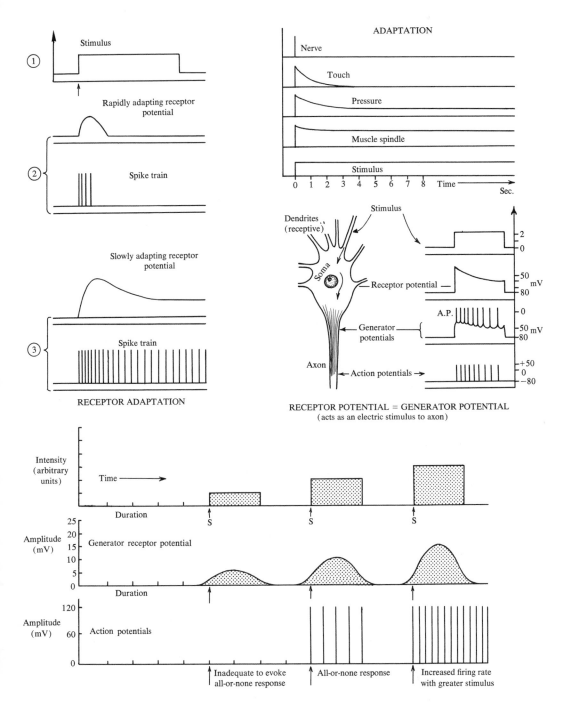

ADAPTATION

Nerve

Touch

Pressure

Muscle spindle

Stimulus

0 1 2 3 4 5 6 7 8   Time ——→
                              Sec.

Stimulus

Dendrites (receptive)

Soma

Receptor potential —

2
0

50
80 mV

Generator potentials

A.P.

0
50 mV
80

Axon   ← Action potentials →

+50
0
−80

① Stimulus

② Rapidly adapting receptor potential

Spike train

③ Slowly adapting receptor potential

Spike train

RECEPTOR ADAPTATION

RECEPTOR POTENTIAL = GENERATOR POTENTIAL
(acts as an electric stimulus to axon)

Intensity (arbitrary units)

Time ——→

Duration

S   S   S

Amplitude (mV)

25
20
15
10
5
0

Generator receptor potential

Duration

Amplitude (mV)

120

60

0

Action potentials

Inadequate to evoke all-or-none response

All-or-none response

Increased firing rate with greater stimulus

RELATIONSHIP OF STIMULUS INTENSITY, AMPLITUDE
OF RECEPTOR POTENTIAL, AND FREQUENCY OF
DISCHARGE IN A NERVE CONTINUOUS WITH THE RECEPTOR

167. SYNAPSES: ANATOMY AND DEFINITIONS

I. Introduction: A synapse is an anatomically specialized junction between 2 neurons (usually axonal endings on another cell) where the electric activity in 1 neuron influences the excitability of the 2nd

A. MOST SYNAPSES OCCUR between the axon terminals of 1 neuron and the cell body or dendrites of a 2nd—the axodendritic synapse; there may, however, be axosomatic, axoaxonal somatosomatic, somatoaxonic, and dendrodendritic synapses as well

B. THE NEURONS CONDUCTING TOWARD THE SYNAPSES are called presynaptic and those conducting information away are called postsynaptic
   1. Every postsynaptic neuron has thousands of synaptic junctions on the surface of its dendrites or cell body so that information from hundreds of presynaptic nerve cells converges on them

C. EACH ACTIVATED SYNAPSE produces either excitatory or inhibitory potentials in the postsynaptic cell. The excitability of a postsynaptic neuron at any time depends on the number of active excitatory and inhibitory synapses converging upon it
   1. If the postsynaptic neuron reaches threshold and generates a response, action potentials are transmitted along its axon to the terminal branches, which diverge to influence the excitability of many other cells
   2. Postsynaptic neurons thus function as neural integrators. Their output reflects the sum of incoming pieces of information arriving as excitatory and inhibitory inputs

D. THERE ARE 2 KINDS of synapse. Most are chemical synapses which release a chemical neurotransmitter from the presynaptic nerve ending. Molecules of the neurotransmitter diffuse across a synaptic cleft to the postsynaptic membrane. The postsynaptic cell is depolarized by attachment of the transmitter to specific receptors. Electrical synapses (electrotonic synapses) occur between some neurons in the CNS. They consist of gap junctions (bridged junctions) at which there are intercellular communication channels. They allow ions and small molecules to pass rapidly from cell to cell. Neurons interconnected by electrical synapses can fire synchronously.

E. IF AN AXON ENDS ON A SKELETAL MUSCLE FIBER, this synapse is called a neuromuscular junction (NMJ). Synapses also occur on smooth muscle and glandular secretory cells.

II. Functional anatomy of a chemical synapse

A. THE PRESYNAPTIC AXON TERMINAL (synaptic knob) is a slight swelling at the end of the axon

B. THE SYNAPTIC CLEFT separates the pre- and postsynaptic neurons and prevents direct propagation of the action potential from the pre- to the postsynaptic cell

C. MEMBRANE-ENCLOSED VESICLES found in the synaptic knob transmit information across the cleft by means of a chemical agent stored in the small vesicles
   1. When the action potential in the presynaptic neuron reaches the axon terminal and depolarizes the knob, small amounts of the chemical transmitter are released from the knob into the cleft. Once released from the vesicles, the transmitter diffuses across the cleft and combines with receptor sites (a membrane molecular configuration with which chemical transmitters combine) on the part of the postsynaptic cell (the subsynaptic membrane) lying under the knob

D. THE COMBINATION OF THE TRANSMITTER with the receptor sites causes changes in the permeability properties of the postsynaptic membrane and in the membrane potential of the postsynaptic cell
   1. There is a synaptic delay (less than 0.001 second) between excitation of the nerve terminal of the presynaptic neuron and membrane potential changes in the postsynaptic cell, which is a function of the release mechanism that frees transmitter substance from the knob
   2. Synaptic activity is terminated when the transmitter is chemically transformed into an ineffective substance and simply diffuses away from the receptor sites or is taken back up by the synaptic knob

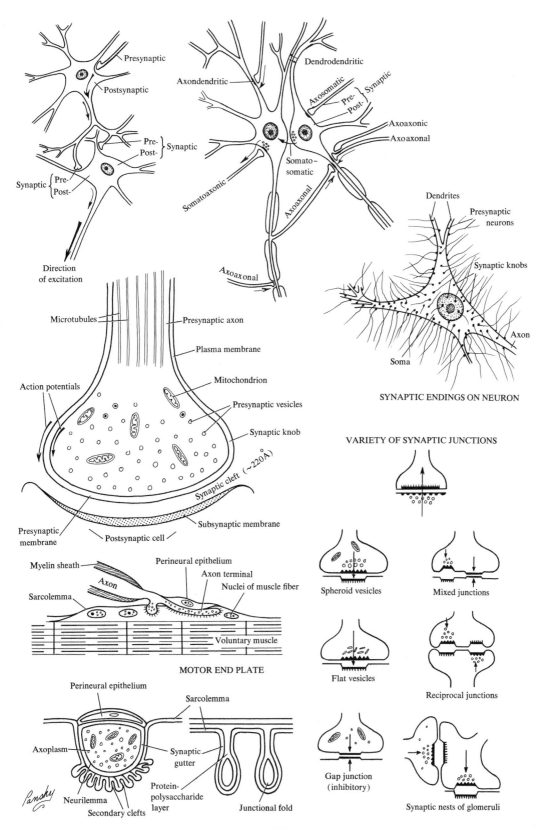

Presynaptic

Postsynaptic

Pre- } Synaptic
Post-

Synaptic { Pre-
Post-

Direction
of excitation

Axondendritic

Dendrodendritic

Axosomatic

Pre- } Synaptic
Post-

Axoaxonic
Axoaxonal

Somato–
somatic

Somatoaxonic

Axoaxonal

Axoaxonal

Dendrites

Presynaptic
neurons

Synaptic knobs

Axon

Soma

SYNAPTIC ENDINGS ON NEURON

Microtubules

Presynaptic axon

Plasma membrane

Mitochondrion

Action potentials

Presynaptic vesicles

Synaptic knob

Synaptic cleft (~220Å)

Presynaptic
membrane

Subsynaptic membrane

Postsynaptic cell

VARIETY OF SYNAPTIC JUNCTIONS

Myelin sheath

Perineural epithelium

Axon terminal

Axon

Nuclei of muscle fiber

Sarcolemma

Voluntary muscle

MOTOR END PLATE

Spheroid vesicles

Mixed junctions

Flat vesicles

Reciprocal junctions

Perineural epithelium

Sarcolemma

Axoplasm

Synaptic
gutter

Neurilemma

Protein-
polysaccharide
layer

Secondary clefts

Junctional fold

Gap junction
(inhibitory)

Synaptic nests of glomeruli

Panshy

-365-

168. SYNAPSES: EXCITATORY AND INHIBITORY

I. Functional anatomy of a synapse (cont.)

A. EXCITATORY SYNAPSES
1. When an excitatory synapse is activated, it increases the likelihood that the membrane potential of the postsynaptic cell will reach threshold and that the cell will undergo an action potential
2. The chemical transmitter receptor site combination increases the permeability of the subsynaptic membrane to (+) charged ions so that they are free to move according to the electric and chemical forces acting on them
3. At the synaptic membrane of excitatory synapses, there is a simultaneous movement of a relatively small number of K^+ out of the cell and a larger number of Na^+ into it. The net effect is the movement of (+) ions into the neuron which slightly depolarizes the postsynaptic cell; referred to as the excitatory postsynaptic potential (EPSP). Its only function is to help trigger an action potential

B. INHIBITORY SYNAPSES
1. When an inhibitory synapse is activated by presynaptic release of γ-aminobutyric acid (GABA), changes are produced that lessen the likelihood that the cell will undergo an action potential
2. At inhibitory synapses, the combination of the chemical transmitter with the receptor sites on the subsynaptic membrane also changes the permeability of the membrane, usually to Cl^- but sometimes instead to K^+ ions
   a. The greater permeability to Cl^- or K^+ is responsible for the changes in membrane potential associated with activity of an inhibitory synapse
   b. The resting membrane potential approximates the Cl^- equilibrium potential
3. The increased negativity or hyperpolarization is called an inhibitory postsynaptic potential (IPSP)
   a. When a neuron is acted on by an inhibitory synapse, its membrane potential is moved farther from the threshold level (hyperpolarization)

C. CHEMICALLY GATED CHANNELS
1. The activation of an excitatory postsynaptic cell results in simultaneous movement of Na^+ and K^+ through large chemically gated channels in the postsynaptic membrane. This is unlike the action potential in which Na^+ and K^+ ions move sequentially through separate small voltage-gated channels in the membrane
2. Chemically gated channels open in response to the concentration of neurotransmitter (acetylcholine at the NMJ and some CNS synapses). This results in depolarization of the postsynaptic membrane, and is followed by activation of voltage-gated channels and initiation of a postsynaptic action potential

D. CHEMICAL TRANSMITTERS. The neurotransmitter at skeletal neuromuscular junctions is acetylcholine. In the CNS, in addition to acetylcholine, the neurotransmitters include norepinephrine (noradrenaline) in hippocampus and cerebral cortex, glutamate and aspartate (synapses on motor neurons), glycine (interneurons activated by Ia afferent fibers from stretch receptors), and GABA (inhibitory neurotransmitter in brain and spinal cord)

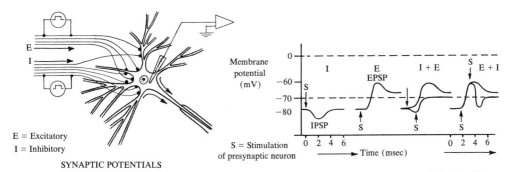

E = Excitatory
I = Inhibitory

SYNAPTIC POTENTIALS

Membrane potential (mV)

S = Stimulation of presynaptic neuron

Time (msec)

SUM OF I + E DETERMINES IF IMPULSE IS
INITIATED AT INITIAL AXON SEGMENT
(After Curtis and Eccles)

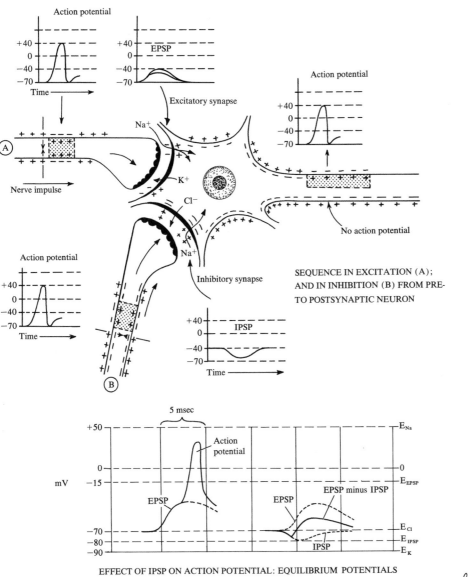

Action potential

EPSP

Time

Excitatory synapse

Na^+

K^+

Cl^-

Na^+

Nerve impulse

Action potential

No action potential

Inhibitory synapse

Action potential

Time

IPSP

Time

SEQUENCE IN EXCITATION (A);
AND IN INHIBITION (B) FROM PRE-
TO POSTSYNAPTIC NEURON

5 msec

Action potential

EPSP

EPSP

EPSP minus IPSP

IPSP

mV

E_{Na}

E_{EPSP}

E_{Cl}

E_{IPSP}

E_K

EFFECT OF IPSP ON ACTION POTENTIAL: EQUILIBRIUM POTENTIALS
OF Na^+, K^+, Cl^-, EPSP, AND IPSP ARE SEEN

Panossy

– 367 –

169. SYNAPSES: ACTIVATION OF POSTSYNAPTIC CELL

I. Functional anatomy of a synapse (cont.)

A. ACTIVATION OF THE POSTSYNAPTIC CELL: In most neurons, postsynaptic integration is possible because one excitatory synaptic event is not sufficient by itself to change the membrane potential of the postsynaptic neuron from resting to threshold level; an EPSP produced by a single presynaptic fiber is smaller than 1 mV

    1. Whereas changes of up to 25 mV are needed to depolarize the membrane from resting to threshold

    2. It requires the combined effect of many fibers to bring the postsynaptic membrane to its threshold level

B. THERE IS A GENERAL DEPOLARIZATION OF THE MEMBRANE toward threshold when excitatory synaptic activity predominates (known as facilitation) and a hyperpolarization when inhibition predominates

C. WHEN 2 SYNAPTIC POTENTIALS SUMMATE as a result of successive stimulation of the same presynaptic fiber, it is called temporal summation

D. IF 2 EPSP's SUMMATE BUT ORIGINATE AT DIFFERENT PLACES on the postsynaptic neuron, this is called spatial summation

E. INHIBITORY POTENTIALS can show both temporal and spatial summation

F. DIFFERENT PARTS OF THE NEURON have different thresholds

    1. The neuronal cell body and larger dendritic branches reach threshold when their membrane is depolarized at about 25 mV from the resting level

    2. In many cells, the initial segment of the neuron has a threshold that is less than half of the body

G. THE SUBSYNAPTIC MEMBRANE is depolarized at an activated excitatory synapse and hyperpolarized at an activated inhibitory synapse

    1. By the mechanism of local current flow, current flows through the cytoplasm from an excitatory synapse and toward the inhibitory one

    2. Thus, the entire body, including the initial segment, becomes slightly depolarized during activation of an excitatory synapse and slightly hyperpolarized during activation of an inhibitory synapse

H. SYNAPTIC EVENTS last more than 10 × as long as action potentials

I. THE GREATER THE DEPOLARIZATION due to synaptic events, the greater the number of action potentials fired (up to the limit imposed by the duration of the absolute refractory period)

J. NEURONAL RESPONSES are almost always in the form of so-called bursts or trains of action potentials

K. THE BIAS OF THE CELL'S ACTIVITY by synapse grouping is greatly lessened by the fact that the initial segment acts to average all the synaptic input as a result of its lower threshold

    1. However, those synapses nearer the initial segment have a greater influence on cell activity than those at the ends of the dendrites, and this synaptic placement provides a mechanism for giving different inputs a greater or lesser influence on the postsynaptic cell's output. In motoneurons, inhibitory synapses are predominantly on the soma and therefore to the initial segment than are excitatory synapses, which are predominantly on the dendrites

L. IN SOME CELLS, action potentials can be initiated in regions other than the initial segment

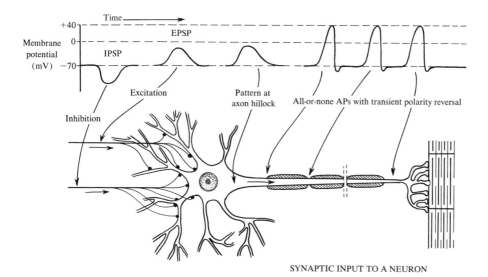

SYNAPTIC INPUT TO A NEURON

Membrane potential (mV)
+40
0
-70
Time
EPSP
IPSP
Inhibition
Excitation
Pattern at axon hillock
All-or-none APs with transient polarity reversal

Excitatory synapse
Na⁺
Inhibitory synapse
Cl⁻
K⁺

Threshold
Excitatory
-70
mV Membrane potential
Time (msec)
Threshold
Inhibitory
-70

Inhibitory synapse
3
Excitatory synapses
2
1

Time
0
Trans-membrane potential (mV)
-70
Stimulate
① ① ① ① ② ① ③ ① ③
+ +
②
Temporal summation due to successive stimulation of same neuron
Spatial summations: summate 2 different presynaptic neurons

Independent firing A + B stimulate 1,2,5,6
A   B
Fired together stimulate all 6
1 2 3 4 5 6
Subthreshold

A   B
Spatial facilitation
Summation
1 2 3 4 5 6
Convergence

A   B
1 2 3 4 5 6
4   4
6
Occlusion: when fired together (8 < 6)

Synaptic knob
Presynaptic
Post.
Divergence
Pre.
Post.
Pre.
Post.

-369-

170. SYNAPSES: CHEMICAL TRANSMITTERS AND PRESYNAPTIC INHIBITION

I. Chemical transmitters

A. PRESYNAPTIC NEURONS influence postsynaptic neurons only by means of chemical transmitters

B. THERE ARE MANY DIFFERENT SYNAPTIC TRANSMITTERS, but all synaptic endings from a single presynaptic cell probably liberate the same transmitter

C. SUBSTANCES THOUGHT TO BE TRANSMITTERS are listed on the opposite page

D. SYNAPSES OPERATE IN ONLY ONE DIRECTION because all the transmitter is stored on 1 side of the synaptic cleft. In contrast, action potentials travel along a nerve fiber in either direction but, due to 1-way conduction across synapses, action potentials pass through the nervous system in only 1 direction

E. THE CHEMICAL TRANSMITTER continues to combine with the receptor sites on the subsynaptic membrane until it is inactivated by combining chemically with another substance and is then taken up again by the presynaptic endings or just diffuses away from the synaptic region
   1. If the transmitter substance is not removed quickly from the receptor sites, a single volley of impulses in the presynaptic fibers may cause a prolongation of firing in the postsynaptic neuron

F. SYNAPSES ARE VULNERABLE to many drugs and toxins that can modify the synthesis, storage, release, inactivation, or uptake of the transmitter substance, or block the receptor sites on the subsynaptic membrane to prevent combination with the transmitter
   1. The tetanus bacillus toxin acts at the inhibitory synapses on the motor neuron by blocking the receptor sites. This eliminates inhibitory input to the motor neurons and allows the unchecked excitatory inputs to have full influence, leading to muscle spasticity and seizures (i.e., lockjaw)

G. PRESYNAPTIC INHIBITION: an inhibitory influence can be exerted on the postsynaptic neuron (the activation of inhibitory synapse), which results in hyperpolarization of the postsynaptic cell and depresses all information fed into the cell
   1. Presynaptic inhibition provides the means by which certain inputs to the postsynaptic cell can be selectively altered. Here the postsynaptic cell is not necessarily depressed; rather, its membrane potential is determined by a different group of synaptic inputs and different bits of information
   2. Presynaptic inhibition operates by affecting the transmission of a single excitatory synapse
      a. At excitatory synapses, the amount of chemical transmitter released from the synaptic knob is directly related to the amplitude of the action potential in the axon terminal. The magnitude of the resulting postsynaptic EPSP is directly related to the amount of chemical transmitter released
         i. The excitatory synapse is less effective if the amplitude of the action potential in the presynaptic terminal is depressed (action potentials usually have an amplitude of about 100 mV or −70 to +30 mV)
         ii. This is accomplished in presynaptic inhibition by means of a synaptic junction between 1 neuron and the synaptic knob of a 2nd. When activated, the synapse slightly depolarizes the membrane of the knob of the 2nd neuron, but the depolarization is not great enough to cause an action potential in the 2nd neuron but only brings the membrane potential of the synaptic knob closer to threshold and reduces the amplitude of the action potentials when neuron 2 is fired. The amount of chemical transmitter released from the knob of neuron 2 is decreased, the permeability changes in the postsynaptic cell are smaller, the size of the EPSP is decreased, and the postsynaptic cell is less influenced by inputs from neuron 2

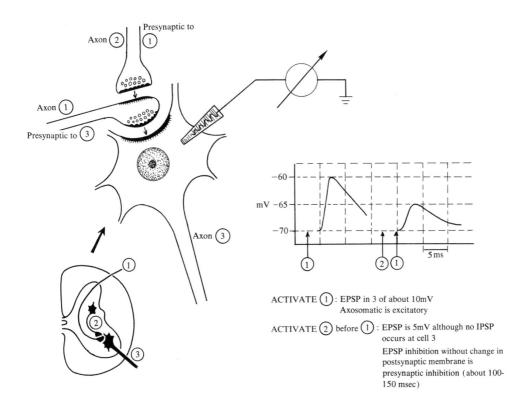

Axon ② Presynaptic to ①

Axon ①

Presynaptic to ③

Axon ③

ACTIVATE ① : EPSP in 3 of about 10mV
Axosomatic is excitatory

ACTIVATE ② before ① : EPSP is 5mV although no IPSP
occurs at cell 3

EPSP inhibition without change in
postsynaptic membrane is
presynaptic inhibition (about 100-
150 msec)

PRESYNAPTIC INHIBITION

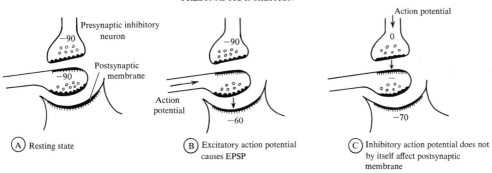

Presynaptic inhibitory
neuron

Postsynaptic
membrane

Action
potential

Action potential

Ⓐ Resting state

Ⓑ Excitatory action potential
causes EPSP

Ⓒ Inhibitory action potential does not
by itself affect postsynaptic
membrane

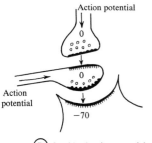

Action potential

Action
potential

Ⓓ Combined action potentials show
blockage of normal EPSP

NEUROTRANSMITTER SUBSTANCES

Proven

1. Acetylcholine
2. Norepinephrine (noradrenaline)
3. Epinephrine (adrenaline)
4. Dopamine (dihydroxyphenylethylamine)

Putative

1. Glycine
2. Aspartic acid
3. Glutamic acid
4. γ-aminobutyric acid (GABA)
5. Serotonin (5-hydroxytryptamine)
6. Histamine
7. Substance D
8. Enkephalin
9. Other peptides

171. PHYSIOLOGIC ACTIVITIES OF NEURONS: NEUROEFFECTORS

I. Neuroeffector communication (effectors)

A. EFFERENT NEURONS innervate muscle cells or glandular cells

    1. The striated skeletal muscles, the smooth muscles of the viscera and blood vessels, and the glands are the "executive organs" or effectors of the nervous system

B. CHEMICAL TRANSMITTERS transmit information from axons to the effector cells

    1. When an action potential reaches the terminal portions of the axon, it causes the release of transmitter, which diffuses to the effector cell and changes its activity

C. THE STRUCTURE OF THE AXON TERMINALS and their anatomic relationships to the effector cells vary, depending on the effector cell type described later

D. THE NEUROEFFECTOR TRANSMITTERS have been well characterized and are either acetylcholine or norepinephrine

    1. The somatic division of the PNS is made up of all the fibers going from the CNS to skeletal muscle cells

        a. The cell bodies of these neurons are found in groups within the brain or spinal cord; their large-diameter, myelinated axons leave the CNS and pass directly to skeletal muscle cells

        b. The transmitter substance released by these neurons is acetylcholine

    2. Fibers of the autonomic division of the PNS innervate cardiac and smooth muscle cells and glands

        a. Anatomic and physiologic differences in the ANS are the basis for further subdivision into sympathetic and parasympathetic components

        b. In both the sympathetic and parasympathetic divisions, the chemical transmitter for the ganglionic synapse between pre- and postganglionic fibers is acetylcholine

        c. The chemical transmitter at the junction between the parasympathetic postganglionic fiber and the effector organ is also acetylcholine

        d. The transmitter between the sympathetic postganglionic fiber and the effector organ is norepinephrine (a member of the chemical family of catecholamines)

E. CHOLINERGIC FIBERS release acetylcholine; those that release norepinephrine are called adrenergic

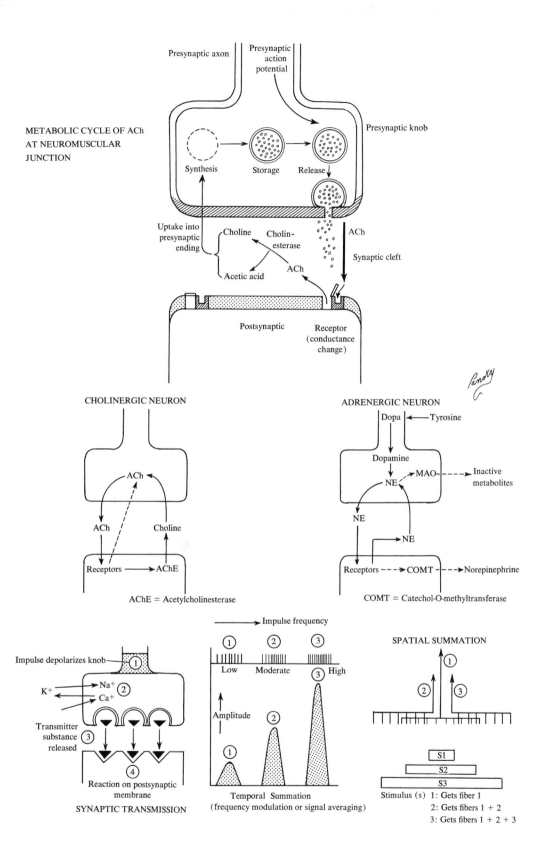

METABOLIC CYCLE OF ACh
AT NEUROMUSCULAR
JUNCTION

Presynaptic axon

Presynaptic action potential

Presynaptic knob

Synthesis   Storage   Release

Uptake into presynaptic ending

Choline   Cholinesterase

Acetic acid   ACh

ACh

Synaptic cleft

Postsynaptic

Receptor (conductance change)

CHOLINERGIC NEURON

ACh

ACh   Choline

Receptors → AChE

AChE = Acetylcholinesterase

ADRENERGIC NEURON

Dopa ← Tyrosine

Dopamine

NE → MAO ---→ Inactive metabolites

NE

NE

Receptors ---→ COMT ---→ Norepinephrine

COMT = Catechol-O-methyltransferase

Impulse frequency

Impulse depolarizes knob ①

K⁺ → Na⁺ ②
      Ca⁺

Transmitter substance released ③

Reaction on postsynaptic membrane ④

SYNAPTIC TRANSMISSION

① ② ③
Low   Moderate   High

Amplitude

① ② ③

Temporal Summation
(frequency modulation or signal averaging)

SPATIAL SUMMATION

① ② ③

S1
S2
S3

Stimulus (s) 1: Gets fiber 1
           2: Gets fibers 1 + 2
           3: Gets fibers 1 + 2 + 3

SENSORY AND MOTOR PATHWAYS, SKELETAL MUSCLE REGULATION, AND SPINAL CORD LESIONS

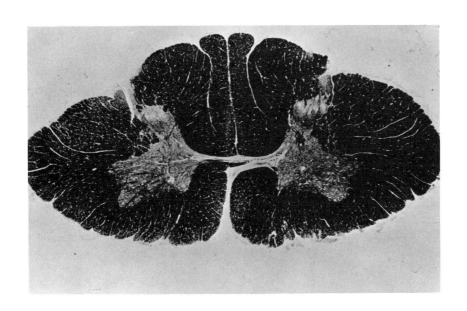

172. SENSATION AND THE SOMATIC SENSORY SYSTEM

I. **Introduction:** the sensations of touch, smell, taste, hearing, vision, awareness of position in space, bodily movements, and pain originate in sensory neurons in response to external or internal environmental stimuli

II. **Somatic receptors** receive specific stimuli and are responsible for 4 subclasses or modalities of sensation
A. Tactile sensations caused by stimulation of mechanoreceptors at the surface of the body
B. Thermal sensations of cold and warmth detected by separate receptors
C. Pain sensations induced by noxious mechanical, thermal, or chemical stimuli
D. Proprioception (position sense) induced by movement of muscles and joints

III. **Stimuli** acting within a receptive field are encoded into signals in sensory neurons. The separate afferent pathways enter the spinal cord or cranial nerves and project to the thalamus and somatic sensory cortex

IV. **A hierarchical arrangement** exists in the somatic sensory pathways. First-order receptor neurons with cell bodies in the dorsal root ganglia converge on second-order neurons in the spinal cord or medulla, which in turn converge on third-order neurons in the thalamus. Thalamic neurons project conscious sensations to the parietal lobe of the cerebral cortex and unconscious signals to interneurons and/or motor neurons to mediate reflex action

V. **Individual receptors** are specific in their response to stimuli. Each of the receptors (nociceptors, thermoreceptors, and mechanoreceptors) responds with a specific sensation when the adequate (natural) stimulus is applied. Often the receptor responds with the same sensation when the receptor or its afferent fiber is electrically stimulated

VI. **Sensory coding**
A. Stimulating a receptor causes impulses to be transmitted along a specific neural pathway or labeled line. An alternative possibility, that a pattern of impulses is transmitted along a neural pathway in a pattern code, is not considered to be important for coding the quality of a sensory stimulation. There is evidence that the frequency of action potentials in the afferent neuron determines the intensity of the perceived sensation

VII. **Sensory information** from each area of the body projects to specific regions of the cerebral cortex and cerebellum, where it is mapped in a pattern that reflects the distribution of sensory receptors in the body. The maps are variable in shape and internal distribution of size, depending on the relative abundance of receptors in the originating organ. For example, the face, fingers, and forelimbs are generally projected to larger areas than those that represent the trunk and back

VIII. **Parallel processing**
A. Two major sensory pathways transmit information in parallel pathways in the spinal cord and brain. These are the dorsal column-medial lemniscus pathway and the anterolateral pathway. They have overlapping functions. If one pathway is damaged, sensory information continues to be distributed in the parallel pathway, although there will be a deficit in the specific aspects of interpretation and integration unique to the damaged pathway

SOMATOSENSORY SYSTEM

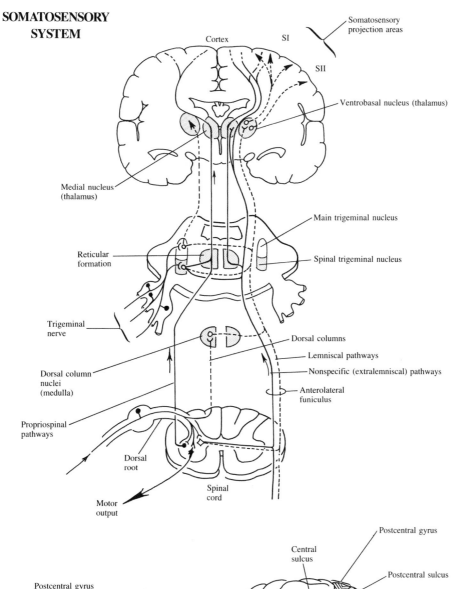

Somatosensory projection areas

Cortex   SI

SII

Ventrobasal nucleus (thalamus)

Medial nucleus (thalamus)

Main trigeminal nucleus

Reticular formation

Spinal trigeminal nucleus

Trigeminal nerve

Dorsal columns

Lemniscal pathways

Nonspecific (extralemniscal) pathways

Dorsal column nuclei (medulla)

Anterolateral funiculus

Propriospinal pathways

Dorsal root

Spinal cord

Motor output

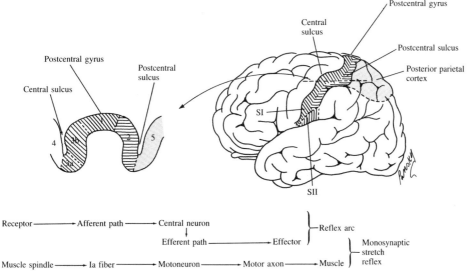

Postcentral gyrus

Central sulcus

Postcentral sulcus

Postcentral gyrus

Postcentral sulcus

Posterior parietal cortex

Central sulcus

SI

4   3b   1   2   5

SII

Receptor ⟶ Afferent path ⟶ Central neuron
                                    ↓   }Reflex arc
                       Efferent path ⟶ Effector

Muscle spindle ⟶ Ia fiber ⟶ Motoneuron ⟶ Motor axon ⟶ Muscle }Monosynaptic stretch reflex

173. GENERAL SOMATIC AFFERENT PATHWAYS: PAIN AND TEMPERATURE—FROM BODY

I. **Introduction:** the general somatic afferent (GSA) system includes those pathways that conduct pain, temperature, touch, pressure, and proprioception. Axons conveying pain and temperature take essentially the same course through the central nervous system and will be considered together

II. **Peripheral receptors for pain (nociceptors)** are naked nerve endings of myelinated (Aδ, fast pain) and unmyelinated (C, slow pain) fibers that spread out among epithelial (epidermal) cells. C fibers are also widely distributed in deeper tissues. Cold and hot spots on the skin correspond to separate zones of free nerve endings of temperature-sensitive receptors (may include Krause end-bulbs and Ruffini's corpuscles)

III. **First-order neurons (neuron I)** are small, unipolar (pseudounipolar) nerve cells
   A. PERIPHERAL PROCESS (DISTAL PART OF AXON) brings sensory information from receptors toward the cell body
   B. CELL BODIES (GANGLION CELLS) of the sensory neurons are located in dorsal root ganglia (no synapses in the ganglia)
   C. PROXIMAL (CENTRAL) BRANCH OF AXON extends to spinal cord in the dorsolateral sulcus at the dorsal root zone
      1. Smallest, finely myelinated fibers (pain and temperature) enter the spinal cord to occupy a lateral position in the entry zone
      2. Largest, most heavily myelinated fibers are medial in position (proprioceptive)
      3. Medium-sized fibers (tactile) are intermediate in position
   D. DORSOLATERAL TRACT (ZONE OF LISSAUER) is where afferent fibers enter the cord
      1. Entering fibers divide into short ascending and descending branches that run longitudinally for 1–3 cord segments
      2. Collaterals are given off to posterior horn along course of dorsolateral tract
      3. Parent fibers and collaterals terminate in the dorsal gray horn (substantia gelatinosa or lamina II)
      4. Short connector neurons may extend from the substantia gelatinosa to more ventral nuclei (laminae) for reflex activity

IV. **Axons of second-order neurons (neuron II):** cell bodies are in substantia gelatinosa
   A. CROSS TO THE OPPOSITE (CONTRALATERAL) SIDE OF THE SPINAL CORD in the anterior white commissure in close proximity to the central canal
   B. AFTER CROSSING, THE AXONS TURN CRANIALLY to enter the lateral spinothalamic tract (and spinotectal tract) in the anterior portion of the lateral funiculus
   C. THIS TRACT EXTENDS WITHOUT INTERRUPTION through the cord to the medulla
   D. SOME FIBERS TERMINATE IN OR SEND COLLATERALS to the brainstem reticular formation in the medulla and pons
   E. IN THE MIDBRAIN, this tract is located at the posterolateral aspect of medial lemniscus
   F. BEFORE THESE SECOND-ORDER NEURONS END in a specific thalamic nucleus, some few fibers enter the reticular system (diencephalon) as an integral part of the reticular activating mechanism of the cerebrum

V. **Third-order neurons (neuron III):** cell bodies are located in the ventral posterolateral nucleus (VPL) of the thalamus
   A. SYNAPSES OCCUR WITH AXON TERMINALS of second-order neurons
   B. AXONS OF THE THIRD-ORDER NEURONS pass through the internal capsule (posterior limb) as a part of the thalamocortical fiber system
   C. THESE AXONS CONTINUE AS THE CORONA RADIATA and finally end in the postcentral gyrus of the parietal lobe (areas 3, 1, 2 of the somesthetic area)

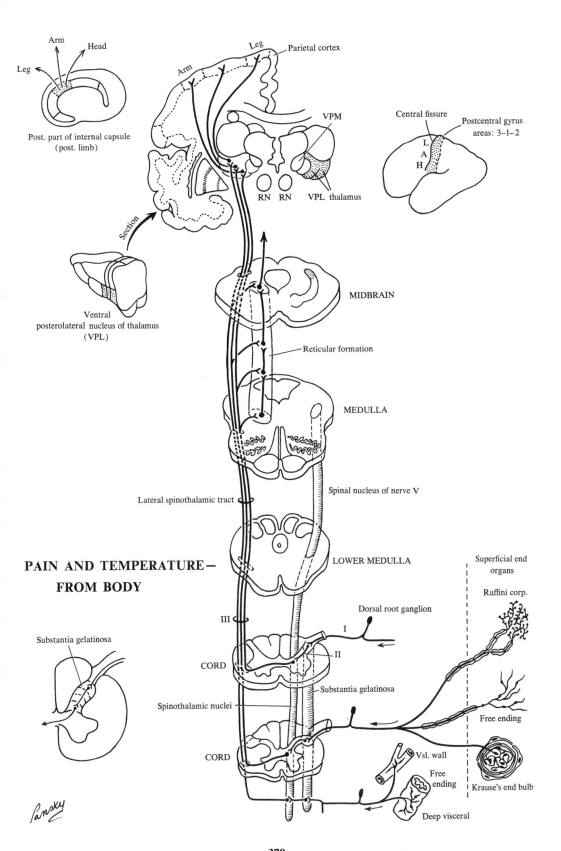

Arm
Head
Leg

Post. part of internal capsule
(post. limb)

Leg
Arm
Parietal cortex

VPM

RN RN
VPL thalamus

Central fissure
Postcentral gyrus
areas: 3–1–2
L
A
H

Section

Ventral
posterolateral nucleus of thalamus
(VPL)

MIDBRAIN

Reticular formation

MEDULLA

Spinal nucleus of nerve V

Lateral spinothalamic tract

LOWER MEDULLA

Superficial end
organs

Ruffini corp.

**PAIN AND TEMPERATURE –
FROM BODY**

Substantia gelatinosa

III

CORD

Dorsal root ganglion

I

II

Substantia gelatinosa

Spinothalamic nuclei

Free ending

CORD

Vsl. wall

Free
ending

Free
ending

Krause's end bulb

Deep visceral

Pansky

174. GENERAL SOMATIC AFFERENT PATHWAYS: PAIN AND TEMPERATURE—FROM FACE

I. Peripheral receptors in the face: same as those of the body

II. First-order neurons (neuron I)

A. PERIPHERAL PROCESSES (DENDRITE IN FUNCTION) are thinly and intermediately myelinated fibers found in

1. Trigeminal (V) nerve: all three divisions
2. Facial (VII) nerve: with receptors for some GSA fibers in the deep face
3. Glossopharyngeal (IX) nerve: with receptors in the middle ear and pharynx
4. Vagus (X) nerve: with receptors in the external auditory meatus

B. CELL BODIES OF FIRST-ORDER NEURONS are small and medium-sized pseudounipolar cells located in

1. Semilunar (Gasserian) ganglion of the trigeminal (V) nerve
2. Geniculate ganglion of the facial (VII) nerve
3. Superior ganglion of the glossopharyngeal (IX) nerve
4. Superior ganglion of the vagus (X) nerve

C. CENTRAL OR PROXIMAL PROCESSES (AXONAL IN FUNCTION)

1. Enter the brainstem at the levels of nerves V, VII, IX, and X
2. Pain and temperature fibers bend inferiorly to form a distinct bundle, the spinal tract of the trigeminal nerve (ipsilaterally)
3. Spinal tract of V descends through the pons and medulla to end in the 3rd or 4th cervical segment, where it merges with the dorsolateral fasciculus of the cord
   a. Pain fibers in the spinal tract (V) are medial to those for temperature
   b. Many collaterals arise from fibers within the spinal tract (V)

D. TERMINATION (SYNAPSE) OF NEURON I: collaterals and axon terminals enter the elongated nucleus of the spinal tract of V (descending nucleus or spinal trigeminal nucleus)

1. This nucleus parallels and adjoins the spinal tract on its medial side
2. This nucleus is continuous with the substantia gelatinosa of the spinal cord

III. Second-order neurons (neuron II)

A. NERVE CELL BODIES OF SECOND-ORDER NEURONS make up the descending nucleus of V

B. AXONS OF SECOND-ORDER NEURONS

1. Cross ventromedially to the opposite side of the brainstem at about the same level of origin in the spinal nucleus of V
2. Approach the lateral spinothalamic tract of the contralateral side
3. Turn rostrally to enter the ventral trigeminal lemniscus or tract (next to the medial lemniscus)
4. Continue through the brainstem in this tract in close association with the medial lemniscus
5. Synapse mainly in the ventral posteromedial (VPM) nucleus of the thalamus
6. Contribute some fibers to the reticular formation (?)

IV. Third-order neurons (neuron III)

A. AXONS ARISE FROM CELL BODIES in the VPM nucleus of the thalamus

B. AXONS PASS THROUGH THE INTERNAL LIMB of the internal capsule and corona radiata

C. TERMINATE IN THE INFERIOR PORTION (FACE AREA) of the postcentral gyrus of the parietal lobe (areas 3, 2, and 1)

V. Pain receptors are activated by strong mechanical pressure (Aδ), extreme heat (especially Aδ fibers), or chemical stimuli (C). The pain response is enhanced by the local presence of prostaglandins. Aspirin may reduce inflammatory pain by inhibiting prostaglandin synthesis

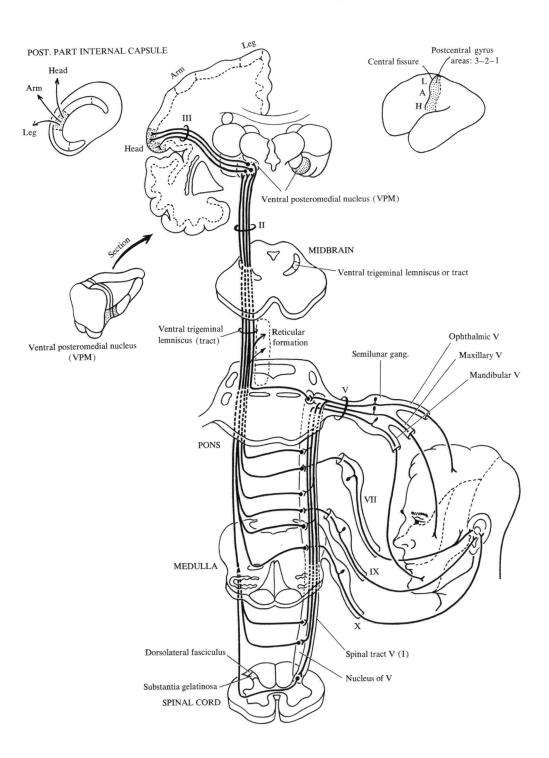

Head

Arm

Leg

Leg

Arm

III

Head

Postcentral gyrus
areas: 3–2–1

Central fissure

L
A
H

Ventral posteromedial nucleus (VPM)

II

Section

Ventral posteromedial nucleus
(VPM)

MIDBRAIN

Ventral trigeminal lemniscus or tract

Ventral trigeminal
lemniscus (tract)

Reticular
formation

Ophthalmic V

Semilunar gang.

Maxillary V

Mandibular V

V

VII

PONS

IX

MEDULLA

X

Dorsolateral fasciculus

Spinal tract V (I)

Substantia gelatinosa

Nucleus of V

SPINAL CORD

PAIN AND TEMPERATURE– FROM FACE

Panosky

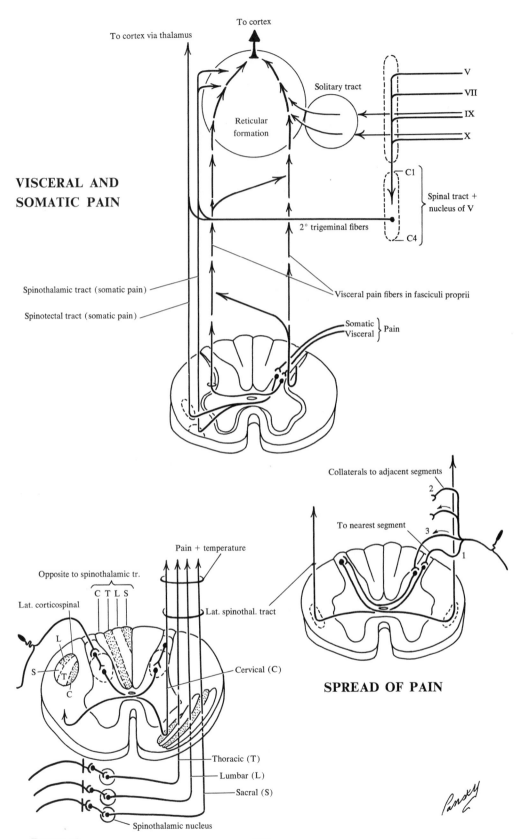

VISCERAL AND SOMATIC PAIN

To cortex

To cortex via thalamus

Solitary tract

Reticular formation

V
VII
IX
X

Spinal tract + nucleus of V

C1

2° trigeminal fibers

C4

Spinothalamic tract (somatic pain)

Spinotectal tract (somatic pain)

Visceral pain fibers in fasciculi proprii

Somatic
Visceral } Pain

Collaterals to adjacent segments

To nearest segment

SPREAD OF PAIN

Pain + temperature

Opposite to spinothalamic tr.

C T L S

Lat. corticospinal

Lat. spinothal. tract

Cervical (C)

L
S
T
C

Thoracic (T)

Lumbar (L)

Sacral (S)

Spinothalamic nucleus

FIGURE 13.

DULL AND SHARP PAIN

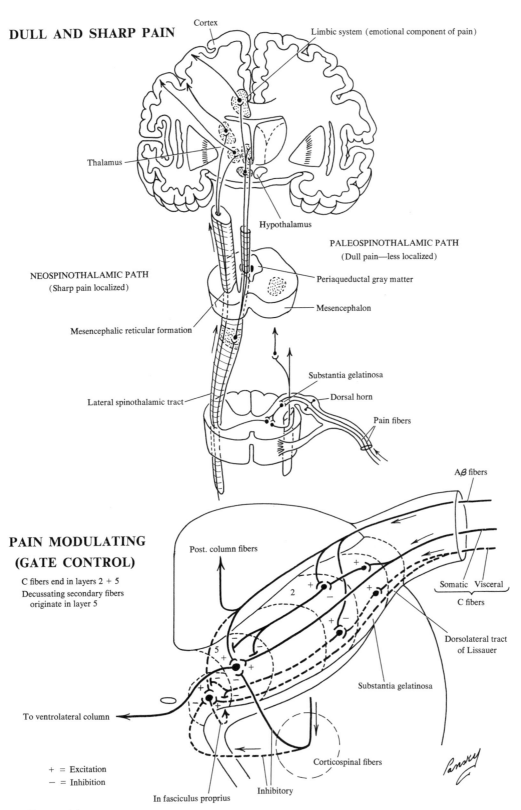

Cortex

Limbic system (emotional component of pain)

Thalamus

Hypothalamus

PALEOSPINOTHALAMIC PATH
(Dull pain—less localized)

NEOSPINOTHALAMIC PATH
(Sharp pain localized)

Periaqueductal gray matter

Mesencephalon

Mesencephalic reticular formation

Substantia gelatinosa

Dorsal horn

Lateral spinothalamic tract

Pain fibers

Aβ fibers

PAIN MODULATING
(GATE CONTROL)

C fibers end in layers 2 + 5
Decussating secondary fibers
originate in layer 5

Post. column fibers

2

Somatic   Visceral

C fibers

5

Dorsolateral tract
of Lissauer

To ventrolateral column

Substantia gelatinosa

+ = Excitation
− = Inhibition

Corticospinal fibers

Pansky

In fasciculus proprius

Inhibitory

FIGURE 14.

175. MODULATION OF PAIN

I. Central pain: Surgical lesions anywhere in the pain pathway, or thalamic lesions induced by stroke, can induce aching, shooting pains, numbness, burning, and cold sensations

II. Gate control hypothesis: collateral fibers from Aβ touch fibers and descending signals from the brain may interact in the substantia gelatinosa (laminae II and III) of the cord, with the pain sensations transmitted in Aδ and C fibers to cells in the substantia gelatinosa (gate cells). Gate cells may regulate dorsal horn cells from which the spinothalamic tract fibers arise

   A. TRANSCUTANEOUS ELECTRICAL STIMULATION of dorsal column sensory nerves can provide long-term diminution of pain (analgesia)

III. Analgesia induced by central stimulation: electrical stimulation of midbrain (serotonin-rich) periaqueductal gray matter or the raphe nucleus of the medulla can relieve chronic pain for many hours. The stimuli appear to selectively inhibit dorsal horn afferent input and trigeminal nerve sensory input

   A. PROFOUND ANALGESIA occurs following injection of morphine into periaqueductal gray matter

   B. BLOCKAGE OF SEROTONIN SYNTHESIS in raphe nuclei results in failure of analgesia by central stimulation or narcotic agents

   C. NALOXONE, a specific morphine antagonist, also blocks morphine analgesia and centrally stimulated analgesia

IV. Opiate receptors: morphine and other opiate narcotics act through specific opiate receptors which are widely distributed in the brain. Endogenous opiates (especially met-enkephalin and leu-enkephalin) formed in the brain, spinal cord, pituitary, and adrenal medulla can react with opiate receptors and produce analgesia when injected into the cerebral ventricles

   A. ENDOGENOUSLY PRODUCED CHEMICAL TRANSMITTER SUBSTANCES, the endorphins (e.g., enkephalin), are related to specific receptor sites. Neurons containing endorphins or having endorphin receptors are found in many sensory relay nuclei in the midbrain, pons, and medulla, and in the periaqueductal gray matter

     1. These sites are involved in analgesia that can be produced by electrical stimulation of brainstem

     2. Endorphin-containing neurons function as a link in the "theoretical" pain control system

     3. The system mediates analgesia produced by the placebo effect, painful stimuli (e.g., acupuncture), opiate drugs, and electrical brain stimulation

V. Enkephalinergic neurons in the substantia gelatinosa may presynaptically inhibit Aδ fibers and C fibers to modulate pain transmission

VI. Stress-induced analgesia: increased stress, such as that occurring in athletes or in soldiers in battle, causes increased circulating levels of β-endorphin, ACTH, and corticosterone. Under the most extreme stress conditions, when survival is paramount, stress-related analgesia may occur

ENDORPHIN SYSTEM

(after Basbaum, A.F. & Fields, H.L., 1978;
Hobson, J.A. & Brazier (eds), 1980)

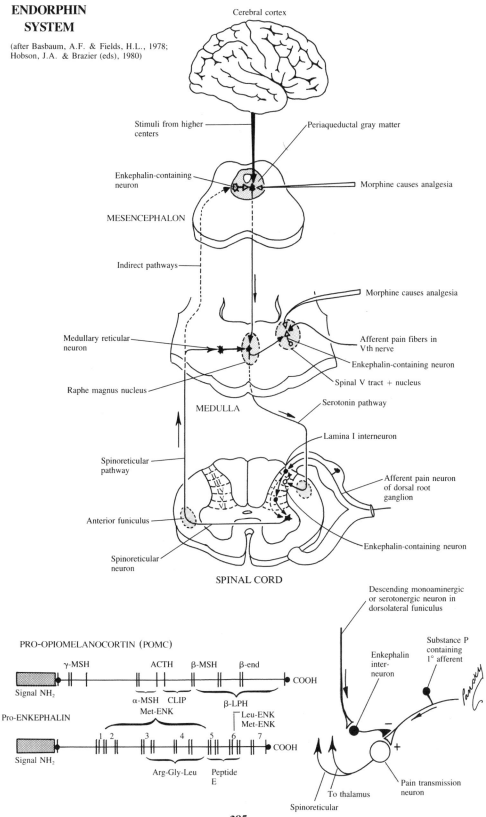

Cerebral cortex

Stimuli from higher centers

Periaqueductal gray matter

Enkephalin-containing neuron

Morphine causes analgesia

MESENCEPHALON

Indirect pathways

Morphine causes analgesia

Medullary reticular neuron

Afferent pain fibers in Vth nerve

Enkephalin-containing neuron

Raphe magnus nucleus

Spinal V tract + nucleus

MEDULLA

Serotonin pathway

Lamina I interneuron

Spinoreticular pathway

Afferent pain neuron of dorsal root ganglion

Anterior funiculus

Enkephalin-containing neuron

Spinoreticular neuron

SPINAL CORD

Descending monoaminergic or serotonergic neuron in dorsolateral funiculus

PRO-OPIOMELANOCORTIN (POMC)

γ-MSH   ACTH   β-MSH   β-end

Signal NH$_2$ α-MSH CLIP   β-LPH   COOH
          Met-ENK   Leu-ENK
                             Met-ENK

Substance P containing 1° afferent

Enkephalin inter-neuron

Pro-ENKEPHALIN

Signal NH$_2$ 1 2   3   4   5 6   7   COOH

Arg-Gly-Leu   Peptide E

To thalamus

Spinoreticular

Pain transmission neuron

-385-

176. GENERAL SOMATIC AFFERENT PATHWAYS: DISCRIMINATIVE (FINE) TOUCH—FROM BODY

I. **Definition:** tactile discrimination is concerned with the sense of deep pressure (pressure touch) spatial localization (2-point discrimination), stereognosis (perception of size, shape, and texture), awareness of movement of body parts, position sense, and vibratory sense. The term "tactile localization" is often used as a collective term to refer to the last 4 senses (above), which also require the involvement of proprioceptive pathways

II. **Receptors for tactile discrimination, conscious proprioception (kinesthesis), stereognosis, etc.**
   A. MEISSNER'S CORPUSCLES (FOR TOUCH) are located in dermal papillae just adjacent to surface epithelium, They are widely dispersed in areas of thin skin, but are close together in palmar and plantar areas
   B. PERITRICHIAL ARBORIZATIONS (FOR TOUCH) are fine nerve endings around hair follicles
   C. TACTILE DISKS OF MERKEL (FOR TOUCH) are platelike expansions of nerve fibers associated with epidermis
   D. FREE NERVE ENDINGS are for touch and conscious proprioception
   E. PACINIAN CORPUSCLES are for pressure vibration and conscious proprioception

III. **Cell bodies of first-order neurons (neuron I)** are large, unipolar (pseudounipolar), and located in the dorsal root (spinal) ganglia
   A. PERIPHERAL PROCESSES (FUNCTION AS DENDRITES) of the unipolar neurons are heavily myelinated
   B. CENTRAL PROCESSES (AXONS) enter the cord through the medial division of the dorsal root

IV. **Course of fibers (axons) of first-order neurons** enter the posterior funiculus and branch into
   A. SHORT BRANCHES, which end in laminae III and IV of the posterior gray horn (then associated with the anterior spinothalamic pathway)
   B. LONG ASCENDING BRANCHES which form the essential part of first-order fibers and continue rostrally in the posterior white columns (fasciculus gracilis and cuneatus) of the same side
      1. Fibers from the lower 6 thoracic, lumbar, and sacral levels form the fasciculus gracilis; from T6 and above, the fasciculus cuneatus
   C. FIBERS IN THE ABOVE 2 FASCICULI enter the lower medulla to synapse in the nucleus gracilis and cuneatus, respectively

V. **Course of axons of second-order neurons (neuron II):** arc anteriorly as the internal arcuate fibers, decussate in the lower medulla as the lemniscal decussation (sensory decussation), and turn rostrally to form the medial lemniscus
   A. IN THE MEDULLA: the medial lemniscus forms a dorsoventrally oriented column just dorsal to the pyramids (cuneate fibers are dorsal to those of the gracilis)
   B. IN THE PONS: the medial lemniscus has its long axis parallel to the horizontal plane with gracilis fibers (from lowest body levels) most lateral in position
   C. IN THE MESENCEPHALON: the lemniscus bends outward to lie along the surface of the brainstem (along with the more superficial spinothalamic tract)
   D. THE SECOND-ORDER FIBERS synapse in the ventral posterolateral (VPL) nucleus of the thalamus on neurons positioned more rostral and dorsal than those receiving simple (crude) touch, pain, and temperature

VI. **Course of axons of third-order neurons (neuron III):** from the VPL nucleus, fibers pass laterally through the posterior limb of the internal capsule and corona radiata to end primarily in the postcentral gyrus and adjacent parietal cortex

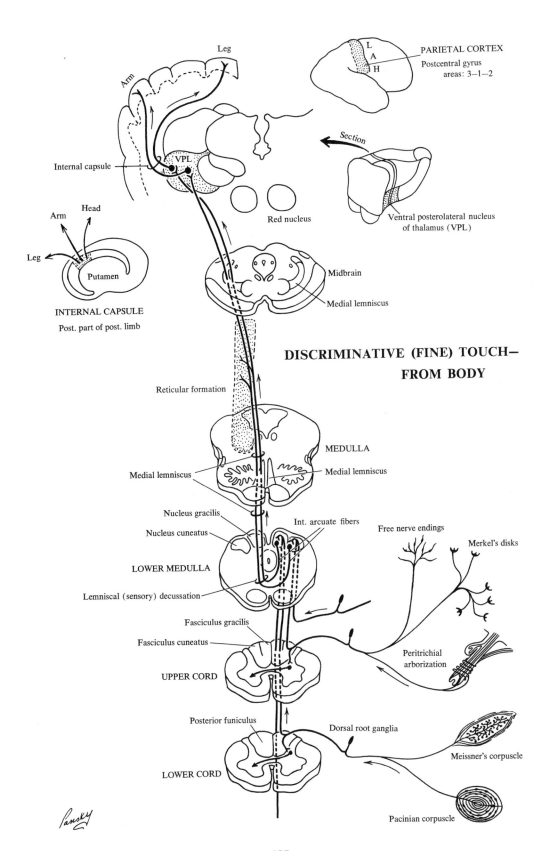

Leg

Arm

PARIETAL CORTEX

Postcentral gyrus
areas: 3—1—2

L
A
H

Internal capsule

VPL

Section

Red nucleus

Ventral posterolateral nucleus
of thalamus (VPL)

Head

Arm

Leg

Putamen

INTERNAL CAPSULE

Post. part of post. limb

Midbrain

Medial lemniscus

**DISCRIMINATIVE (FINE) TOUCH—
FROM BODY**

Reticular formation

MEDULLA

Medial lemniscus

Medial lemniscus

Nucleus gracilis

Int. arcuate fibers

Free nerve endings

Merkel's disks

Nucleus cuneatus

LOWER MEDULLA

Lemniscal (sensory) decussation

Fasciculus gracilis

Fasciculus cuneatus

Peritrichial
arborization

UPPER CORD

Posterior funiculus

Dorsal root ganglia

Meissner's corpuscle

LOWER CORD

Pacinian corpuscle

Pansky

177. GENERAL SOMATIC AFFERENT PATHWAYS: DISCRIMINATIVE (FINE) TOUCH—FROM FACE

I. **Receptors:** for the discriminative general senses from the face and head (anterior to a coronal plane through the ears) are primarily Meissner's corpuscles, which have been found only in primates. They are most abundant in hairless skin (palm of hand and border of lips). In fine or discriminative touch, one can recognize the location of the stimulated point with precision, and one is aware that 2 points are touched simultaneously, even though they are near each other

  A. THE RECEPTORS are numerous and close together at the tip of the tongue, and 2 points only 1.1 mm apart can be distinguished here

   1. The receptors are also numerous on the lips

  B. THE RECEPTORS TRANSMIT THEIR SENSATION via the ophthalmic, maxillary, and mandibular divisions of the trigeminal (V) nerve

II. **Course of first-order neurons (neuron I)**

  A. THE PERIPHERAL PROCESSES follow the ophthalmic, maxillary, and mandibular divisions of the trigeminal (V) nerve

  B. CELL BODIES are located in the semilunar (Gasserian) ganglion

  C. CENTRAL PROCESSES form part of the sensory root of the trigeminal nerve on the lateral aspect of the pons and follow the same course as outlined for crude (light or simple) touch from the head

  D. SYNAPSE OF FIRST-ORDER NEURONS takes place in the main (principal) sensory trigeminal nucleus. This nucleus is often referred to as the cranial equivalent of the nucleus gracilis and nucleus cuneatus

III. **Course of second-order neurons (neuron II)** on their ascent to the thalamus is unclear and disputed. Mostly crossed but some uncrossed

  A. THE CELL BODIES are located in the main sensory nucleus of V

  B. THE AXONS ascend in or near the medial lemniscus

  C. THEY SYNAPSE in the VPL nucleus of the thalamus

IV. **Course of third-order axons (neuron III)** passes through the posterior limb of the internal capsule and corona radiata to end in the postcentral gyrus and adjacent parietal cortex

V. **The sensory pathways** project to specific areas within the postcentral gyrus of the cortex in a pattern (homunculus) that exactly corresponds to an image of the body, but with certain regions (face, hands) more prominent, reflecting the relative number of sensory receptors that they carry

VI. **Somatic sensory neurons** receive stimuli within a defined zone (receptive field). The receptive field is smallest and the number of fields greatest in the most sensitive areas of skin. Fine touch discrimination occurs because each receptive field has an inhibitory surround which is projected to the cortex with the excitatory signal. Firing frequency signals stimulation intensity

VII. **The four Brodmann areas** (3a, 3b, 1, and 2) in the primary sensory cortex (S–I) of the postcentral gyrus each contain a complete representation of body image (homunculus). Cells in area 3a respond mainly to muscle stretch, in 3b to cutaneous receptors, in 2 to deep pressure, and in 1 to rapidly adapting skin receptors. An additional body surface map exists in the secondary somatic cortex (S–II)

  A. COMMUNICATION LINES (labeled lines) transmit each modality from the receptor to a specific cortical location

  B. SOME CORTICAL CELLS in areas 1 and 2 respond to direction of movement (directional sensitivity)

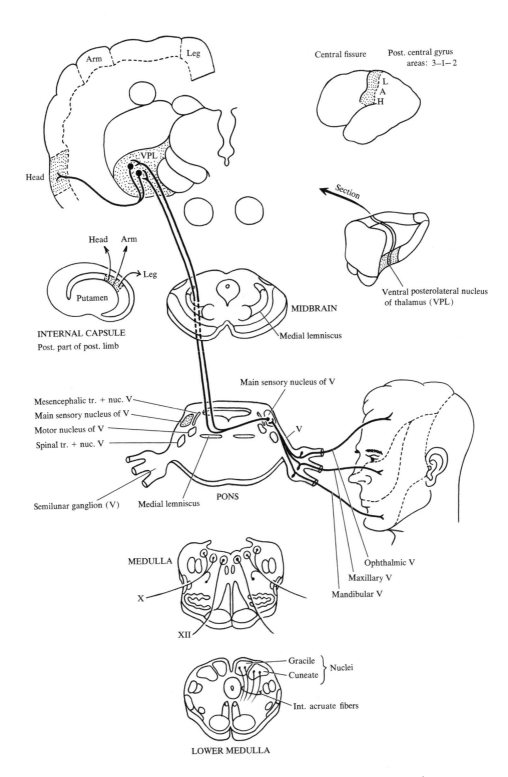

Central fissure

Post. central gyrus
areas: 3–1–2

L
A
H

Arm
Leg

Head

VPL

Section

Head   Arm
Leg

Ventral posterolateral nucleus
of thalamus (VPL)

Putamen

MIDBRAIN

INTERNAL CAPSULE
Post. part of post. limb

Medial lemniscus

Main sensory nucleus of V

Mesencephalic tr. + nuc. V

Main sensory nucleus of V

Motor nucleus of V

V

Spinal tr. + nuc. V

Semilunar ganglion (V)   Medial lemniscus   PONS

Ophthalmic V

MEDULLA

Maxillary V

Mandibular V

X

XII

Gracile
Cuneate

Nuclei

Int. acruate fibers

LOWER MEDULLA

DISCRIMINATIVE (FINE) TOUCH– FROM FACE

Pansky

178. GENERAL SOMATIC AFFERENT PATHWAYS:
LIGHT (CRUDE) TOUCH—FROM BODY

I. **Definition:** light touch is concerned with the sense of simple touch, light pressure, and a very crude sense of tactile localization. This level of tactile sense is stimulated, for example, by the gentle stroking of the skin with cotton

II. **Receptors for light (protopathic) touch: probably include**
   A. FREE NERVE ENDINGS
   B. MERKEL'S (TACTILE) DISKS or platelike expansions of nerve fibers associated with certain epithelial cells in the epidermis
   C. PERITRICHIAL ARBORIZATIONS of fine nerve fibrils around hair follicles which detect any movements of the hair
   D. MEISSNER'S CORPUSCLES in dermal papillae beneath the volar and plantar surfaces
   E. PACINIAN CORPUSCLES which lie deep in the dermis and in tendons and joints (vibration)
   F. GENITAL CORPUSCLES found in the connective tissue of the external genitalia

III. **Cell bodies of first-order neurons (neuron I)** are medium-sized, pseudounipolar, and located in the dorsal root ganglia of spinal nerves
   A. THE PERIPHERAL PROCESSES (FUNCTION AS DENDRITES) are large and well myelinated (intermediate type)
   B. CENTRAL PROCESSES (FUNCTION AS AXONS) group into the medial division of the dorsal root and enter the spinal cord along the dorsolateral sulcus
      1. Entering fibers become part of the posterior funiculus on the ipsilateral side (posterolateral tract of Lissauer)
         a. Fibers from the upper half of the body enter the posterior funiculus laterally (fasciculus cuneatus)
         b. Fibers from the lower half of the body enter the posterior funiculus medially (fasciculus gracilis)
      2. In the appropriate fasciculus, each nerve fiber bifurcates into a long ascending and a short descending branch, both of which give off collaterals
      3. All descending, all collaterals, and some ascending fibers synapse with internuncial neurons of the posterior horn (laminae VI and VII)
      4. Some primary ascending fibers continue superiorly into the lower medulla to synapse (nucleus gracilis and nucleus cuneatus)

IV. **Course of axons of second-order neurons (neuron II)**
   A. AXONS: from cell bodies located in the posterior gray horns (laminae VI and VII) decussate in the anterior white commissure (a few fibers may not cross and ascend on the ipsilateral side)
   B. FIBERS THEN ASCEND in the anterolateral system (in the anterior funiculus) as the anterior spinothalamic tract, which ascends in the spinal cord to the medulla
   C. IN THE MEDULLA AND PONS: the exact course of this tract has not been established
      1. It may become intimately associated with the medial lemniscus
      2. It may join the lateral lemniscus to form a spinothalamic lemniscus
   D. In the diencephalon, this fiber tract ends in the ventral posterolateral nucleus (VPL) of the thalamus

V. **Course of axons of third-order neurons (neuron III):** fibers leave the nerve cell bodies in the VPL nucleus to pass through the posterior limb of the internal capsule and corona radiata, ending in the postcentral gyrus of the parietal lobe of the cerebral cortex (areas 3, 2, and 1)

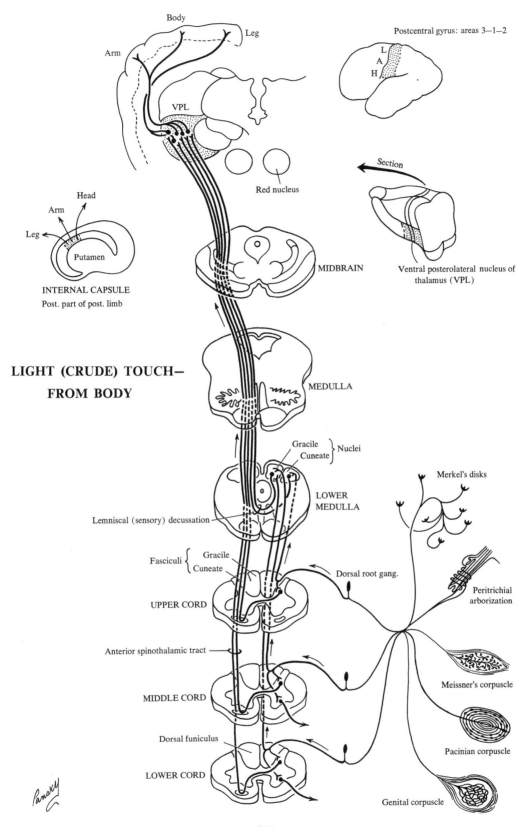

Body

Leg

Arm

VPL

Red nucleus

Postcentral gyrus: areas 3—1—2

L
A
H

Section

Head

Arm

Leg

Putamen

INTERNAL CAPSULE
Post. part of post. limb

MIDBRAIN

Ventral posterolateral nucleus of
thalamus (VPL)

LIGHT (CRUDE) TOUCH—
FROM BODY

MEDULLA

Gracile
Cuneate } Nuclei

LOWER
MEDULLA

Merkel's disks

Lemniscal (sensory) decussation

Gracile
Fasciculi { Cuneate

UPPER CORD

Dorsal root gang.

Peritrichial
arborization

Anterior spinothalamic tract

MIDDLE CORD

Meissner's corpuscle

Dorsal funiculus

Pacinian corpuscle

LOWER CORD

Genital corpuscle

Pansky

-391-

179. GENERAL SOMATIC AFFERENT PATHWAYS: LIGHT (CRUDE) TOUCH—FROM FACE

I. Introduction: the facial area is especially sensitive to light touch

II. Receptors: the same types of receptors mentioned for the body are also found in the skin of the face and in the cornea, conjunctiva, and associated mucous membranes of the nasal and oral cavities

III. Cell bodies of first-order neurons (neuron I) are located in the semilunar (Gasserian) ganglion
A. PERIPHERAL PROCESSES (DENDRITES) convey impulses from the receptors into the cell bodies via the 3 divisions of the trigeminal (V) nerve
B. CENTRAL PROCESSES (AXONS) make up part of the sensory root (portis major) of the trigeminal nerve which enters the lateral aspect of the pons
  1. In the pons, these fibers divide into short ascending and long descending branches
    a. Ascending division synapse in the main (chief) sensory nucleus of the trigeminal nerve
    b. Descending fibers enter the spinal tract of the trigeminal (V) nerve and descend for a short distance to end in the nucleus of V

IV. Cell bodies of second-order neurons (neuron II) are located in
A. SPINAL (DESCENDING) NUCLEUS OF V (ITS ROSTRAL PORTION)
  1. Their axons turn ventromedially toward the midline; from this point, their exact source is disputed
  2. Some may decussate and join the medial lemniscus
  3. Others may join the anterior spinothalamic tract
B. MAIN SENSORY NUCLEUS OF V (LOCATION OF MOST CELL BODIES OF NEURON II)
  1. Some of their axons may decussate in the pontine tegmentum (below the hypoglossal nucleus) and then ascend as the anterior trigeminal (trigeminothalamic) tract, which ends in the ventral posteromedial (VPM) nucleus of the thalamus
  2. Other fibers ascend without crossing as the posterior trigeminal (trigeminothalamic) tract, which also ends in the VPM

V. Course of axons of third-order neurons (neuron III): axons from cell bodies in the VPM nucleus pass laterally through the posterior limb of the internal capsule and corona radiata to end in the postcentral gyrus (facial area)

VI. Primary sensory neurons for simple touch and pressure are also found in the geniculate ganglion of the facial (VII) nerve and the superior ganglia of the glossopharyngeal (IX) and vagus (X) nerves. They include sensation from the external ear; lining of the auditory canal and the tympanic membrane via nerves VII and X; and the back of tongue, pharynx, and larynx via nerves IX and X
A. THE CENTRAL CONNECTIONS OF NEURONS OF THE VIITH, IXTH, AND XTH NERVES correspond to the connections of trigeminal neurons for touch and pressure, most of the afferent fibers forming the trigeminal spinal tract

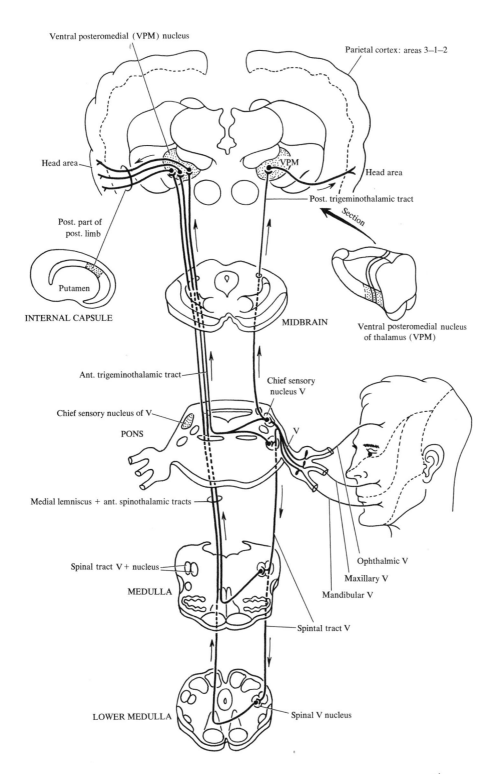

Ventral posteromedial (VPM) nucleus

Parietal cortex: areas 3–1–2

Head area

VPM

Head area

Post. trigeminothalamic tract

Post. part of
post. limb

Section

Putamen

INTERNAL CAPSULE

MIDBRAIN

Ventral posteromedial nucleus
of thalamus (VPM)

Ant. trigeminothalamic tract

Chief sensory
nucleus V

Chief sensory nucleus of V

PONS

V

Medial lemniscus + ant. spinothalamic tracts

Spinal tract V + nucleus

Ophthalmic V

MEDULLA

Maxillary V

Mandibular V

Spintal tract V

LOWER MEDULLA

Spinal V nucleus

LIGHT (CRUDE) TOUCH— FROM FACE

Pansky

180. GENERAL SOMATIC AFFERENT PATHWAYS: CONSCIOUS AND UNCONSCIOUS PROPRIOCEPTION— FROM HEAD AND FACE

I. Receptors at the peripheral ends of the 3 divisions of the trigeminal (V) nerve detect conscious and unconscious proprioception (sense of balance, position, limb movements). They include stationary position (limb position) and kinesthetic sense (limb movement)

A. MUSCLE SPINDLES (p. 416): in muscles of mastication and perhaps in the tongue, extrinsic eye, and facial muscles to detect unconscious proprioception

B. NEUROTENDON (GOLGI) ORGANS found dispersed in dense connective tissue of tendons convey unconscious proprioception

C. FREE NERVE ENDINGS: in tendons, ligaments, and joint capsules are responsible for conscious proprioception

D. PACINIAN CORPUSCLES: located deep in the dermis convey conscious proprioception

II. Cell bodies of first-order neurons (neuron I) are comparatively pseudounipolar and are found in the nucleus of the mesencephalic tract of V (located in the upper pons and mesencephalon)

A. PERIPHERAL PROCESSES (FUNCTION AS DENDRITES) of the first-order neurons are heavily myelinated and take the following course
   1. They enter the 3 divisions of the trigeminal nerve and pass through the semilunar (Gasserian) ganglion of the trigeminal nerve without synapsing
   2. They then enter the lateral aspect of the pons through both motor and sensory roots of the trigeminal nerve and ascend in the mesencephalic tract of V, on the same side, to reach the cell bodies in the mesencephalic nucleus

B. THE CENTRAL PROCESSES of these same neurons function as axons
   1. Pass from the mesencephalic nucleus into the mesencephalic tract of V without crossing
   2. They descend in the mesencephalic tract to synapse in 1 of the following
      a. Main sensory nucleus of V, which is concerned with conscious proprioception
      b. Motor nucleus of V, which deals with motor reflexes (jaw jerk)
      c. Reticular formation
      d. Various other cranial nuclei in the area (facial nucleus)

III. Course of second-order neurons (neuron II)

A. AXONS PASS FROM THE CELL BODIES in the main sensory nucleus of V and ascend through the brainstem in the dorsal trigeminal tract

B. OTHER PATHWAYS are unclear and disputed

C. FIBERS FOR CONSCIOUS PROPRIOCEPTION synapse in the VPM nucleus of the thalamus

IV. Course of third-order neurons (neuron III): the axons pass through the posterior limb of the internal capsule and corona radiata to end in the postcentral gyrus of the parietal cerebral cortex (areas 3, 2, and 1) in the area of the face

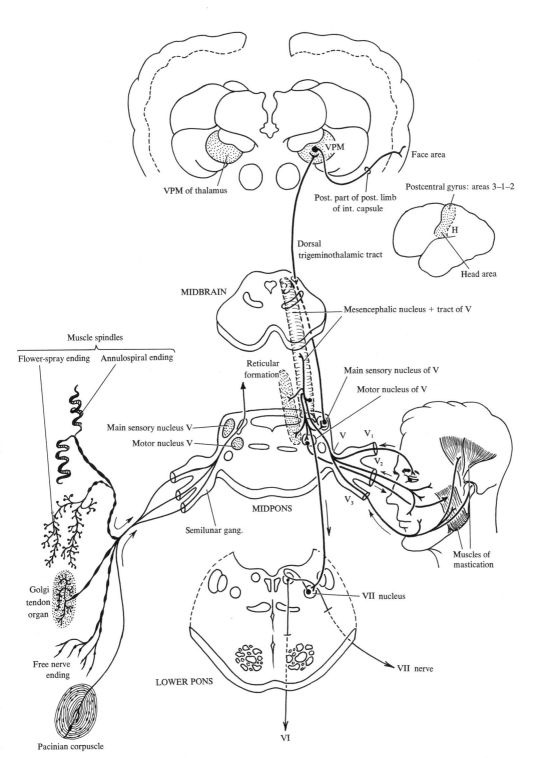

VPM of thalamus

VPM

Face area

Post. part of post. limb of int. capsule

Postcentral gyrus: areas 3–1–2

H

Head area

Dorsal trigeminothalamic tract

MIDBRAIN

Mesencephalic nucleus + tract of V

Muscle spindles

Flower-spray ending   Annulospiral ending

Reticular formation

Main sensory nucleus of V

Motor nucleus of V

Main sensory nucleus V

Motor nucleus V

V   V₁

V₂

V₃

MIDPONS

Semilunar gang.

Muscles of mastication

Golgi tendon organ

Free nerve ending

VII nucleus

VII nerve

Pacinian corpuscle

LOWER PONS

VI

CONSCIOUS AND UNCONSCIOUS PROPRIOCEPTION—FROM HEAD AND FACE

181. AFFERENT CEREBELLAR CONNECTIONS: SPINOCEREBELLAR TRACTS— GENERAL PROPRIOCEPTION

I. **Definition:** spinocerebellar fibers convey unconscious sensory information to the cerebellum about tension, position of muscles, tendons, and joints (proprioception). The cerebellar input provides two somatotypic maps: one in the anterior lobe; the other, reversed, in the posterior lobe

II. **Functions:** anterior and posterior spinocerebellar tracts are primarily concerned with the lower extremities. The posterior spinocerebellar tract conveys information on individual extremity muscles, while the anterior spinocerebellar tract relays information concerning the status of the entire extremity

III. **Receptors for the first-order neurons in muscle, tendons, and joints**
   A. MUSCLE SPINDLES: adapted to register changes in muscle tension
   B. NEUROTENDINOUS (GOLGI TENDON) ORGANS: adapted to register stretch of a tendon due to muscle contraction
   C. PACINIAN CORPUSCLES: located in fascia, intermuscular septa, and joint capsules to record swelling and movements

IV. **Cell bodies of first-order neurons:** found in dorsal root ganglia

V. **Course of axons of first-order neurons (neuron I)**
   A. PROPRIOCEPTIVE AXONS (largest and most heavily myelinated fibers) collect in bundles to make up the medial division of the dorsal root
   B. AXONS ENTER THE POSTERIOR FUNICULUS of the spinal cord and bifurcate into long ascending and short descending branches which give off collaterals
   C. AXONAL TERMINALS AND COLLATERALS SYNAPSE with neurons in the base of the posterior gray horn and in the nucleus dorsalis (of Clarke)
   D. CLARKE'S NUCLEUS is a cell column found only from the 8th cervical through the 3rd lumbar segments. Proprioceptive fibers entering cord segments below L3 ascend in the fasciculus gracilis to reach this nucleus; those entering above C8 ascend in the fasciculus cuneatus to the accessory (lateral) cuneate nucleus in the lower medulla

VI. **Course of axons of second-order neurons (neuron II)**
   A. NERVE CELL BODIES IN THE POSTERIOR FUNICULAR GRAY NUCLEUS (laminae V, VI, and VII) at lumbar and sacral levels send mostly crossed (some uncrossed) fibers through the anterior white commissure to ascend in the anterior or ventral spinocerebellar tract
      1. This is the most peripheral tract in the ventral part of lateral funiculus of cord
      2. In the medulla, the tract is positioned dorsal to the inferior olive and lateral to the lateral spinothalamic tract
      3. In the pons, this tract shifts even more laterally and then turns dorsally in the superior cerebellar peduncle to reach the anterior medullary velum
      4. By way of the anterior medullary velum, it enters the rostral portion of the contralateral (some remain uncrossed) cerebellar vermis. These fibers end as the mossy fibers of the paleocerebellum or anterior lobe
         a. Mossy fibers synapse with granule cells
         b. Granule cells synapse with Purkinje cells
   B. THE NERVE CELL BODIES IN CLARKE'S NUCLEUS (lamina VII, levels C8–L3) send ipsilateral fibers to the periphery of the dorsal portion of the lateral funiculus, where they turn rostrally to form the dorsal or posterior spinocerebellar tract
      1. This tract runs in the superficial dorsolateral part of the lateral funiculus of cord
      2. In the medulla, it lies superficial to the spinal tract of V
      3. Fibers enter the vermis of the cerebellum through the inferior cerebellar peduncle and end as mossy fibers of the paleocerebellum

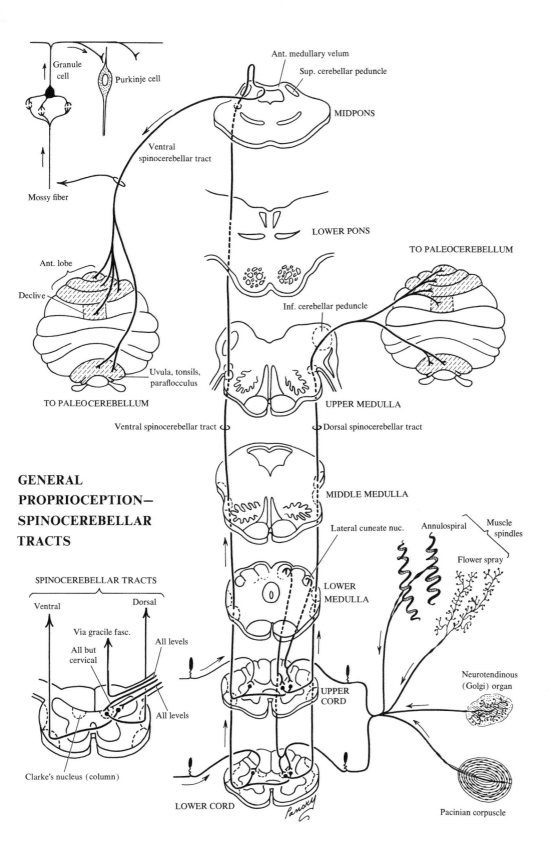

Granule cell

Purkinje cell

Ant. medullary velum

Sup. cerebellar peduncle

MIDPONS

Ventral spinocerebellar tract

Mossy fiber

LOWER PONS

TO PALEOCEREBELLUM

Ant. lobe

Declive

Inf. cerebellar peduncle

Uvula, tonsils, paraflocculus

TO PALEOCEREBELLUM

UPPER MEDULLA

Ventral spinocerebellar tract

Dorsal spinocerebellar tract

GENERAL PROPRIOCEPTION– SPINOCEREBELLAR TRACTS

MIDDLE MEDULLA

Lateral cuneate nuc.

Annulospiral

Muscle spindles

Flower spray

SPINOCEREBELLAR TRACTS

Ventral

Dorsal

Via gracile fasc.

All but cervical

All levels

LOWER MEDULLA

All levels

UPPER CORD

Neurotendinous (Golgi) organ

Clarke's nucleus (column)

LOWER CORD

Pacinian corpuscle

-397-

182. AFFERENT CEREBELLAR CONNECTIONS:
FROM THE BRAINSTEM—PART I

I. Cuneocerebellar tract may be thought of as the upper extremity counterpart of the posterior spinocerebellar tract

A. REPORTS THE INSTANTANEOUS ACTIVITY from muscle spindles associated with the upper limb to the cerebellum, which then modifies muscle group activity for smooth, coordinated movements*

B. NERVE CELL BODIES in the accessory (lateral) cuneate nucleus of the lower dorsolateral portion of the medulla send ipsilateral fibers (dorsal arcuate fibers) to form the cuneocerebellar tract

1. Impulses into the accessory cuneate nucleus are via A alpha afferents
2. This tract enters the cerebellum through the inferior cerebellar peduncle
3. Tract ends primarily in the forelimb region of the anterior lobe and pyramis

II. Vestibulocerebellar tract

A. VESTIBULAR (ARCHICEREBELLAR) FIBERS from the vestibular apparatus (see Vestibular pathways) to the cerebellum follow two routes

1. Direct (primary) ascending branch of the vestibulocochlear nerve courses along the medial side of the inferior peduncle to enter the cerebellum and end as mossy fibers of the homolateral archicerebellum (or flocculonodular lobe)
2. Indirect (secondary) fibers from the inferior and medial vestibular nuclei enter the cerebellum primarily through the juxtarestiform body (medial part of the inferior peduncle) to end bilaterally in the flocculonodular lobe and fastigial nuclei

B. THIS TRACT is essential for monitoring head position and control of equilibrium, balance and coordinated eye movements

III. Trigeminocerebellar tract

A. MESENCEPHALIC NUCLEUS OF V gives rise to secondary proprioceptive fibers which end in the dentate and emboliform nuclei

B. SENSORY NUCLEUS OF V gives rise to secondary fibers which end in the tonsil and anterior quadrangular lobule

C. SPINAL NUCLEUS OF V may give rise to crossed and uncrossed fibers which end in the tonsil and anterior quadrangular lobule

IV. Reticulocerebellar tract

A. FIBERS MAKING UP THIS TRACT originate in the medulla (from 2 regions) and in the pons

1. From the medulla (lateral and paramedian reticular nuclei) fibers enter the inferior cerebellar peduncle to end primarily in the vermis of the anterior lobe (some fibers also pass to the pyramis and uvula of the posterior lobe)
   a. Most transmission within the reticular formation is modality nonspecific
   b. The lateral reticular nucleus, however, may receive tactile impulses from the spinal cord and convey these somatic impulses to numerous cerebellar cortical areas
2. In the pons, fibers from the reticular formation (specifically the reticulotegmental nucleus) enter the cerebellum through the middle peduncle to end in the vermis of the anterior and posterior lobes

B. THIS TRACT MAY SERVE AS AN IMPORTANT FEEDBACK CIRCUIT from the cerebrum to the cerebellum since many of the nuclei also receive input from sensorimotor cortex

---

*The rostral spinocerebellar tract has been described in the cat. It is considered the forelimb equivalent of the anterior spinocerebellar tract.

CEREBELLAR AFFERENTS
FROM THE BRAINSTEM

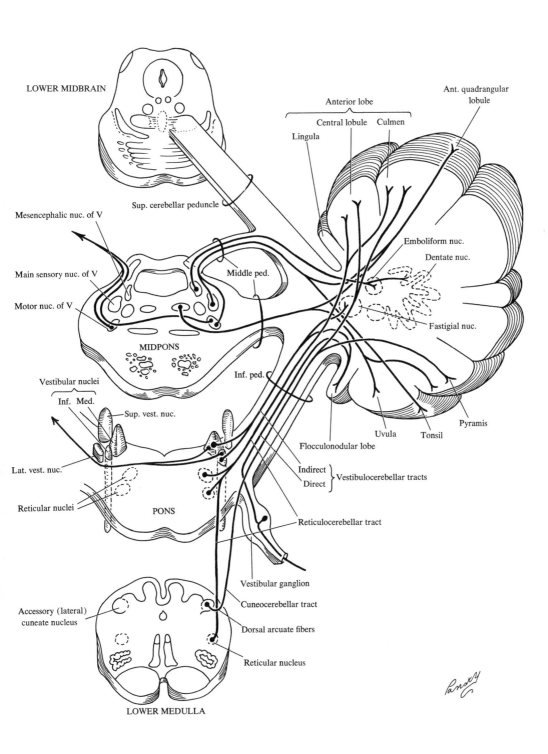

LOWER MIDBRAIN

Anterior lobe

Ant. quadrangular lobule

Central lobule   Culmen

Lingula

Sup. cerebellar peduncle

Mesencephalic nuc. of V

Emboliform nuc.

Dentate nuc.

Main sensory nuc. of V

Middle ped.

Motor nuc. of V

Fastigial nuc.

MIDPONS

Inf. ped.

Vestibular nuclei

Inf. Med.

Sup. vest. nuc.

Uvula   Tonsil

Pyramis

Flocculonodular lobe

Lat. vest. nuc.

Indirect
Direct } Vestibulocerebellar tracts

Reticular nuclei

PONS

Reticulocerebellar tract

Vestibular ganglion

Cuneocerebellar tract

Accessory (lateral) cuneate nucleus

Dorsal arcuate fibers

Reticular nucleus

LOWER MEDULLA

183. AFFERENT CEREBELLAR CONNECTIONS: FROM THE BRAINSTEM—PART II

I. Olivocerebellar tract

A. FIBERS MAKING UP THIS TRACT arise from the inferior olivary complex
   1. Principal (chief) nucleus sends numerous fibers to hemispheres of the anterior and posterior lobes
   2. Medial accessory olivary nucleus sends fibers to the flocculonodular lobe, pyramis, uvula, declive, and fastigial nucleus
   3. Dorsal accessory nucleus sends fibers to the vermis of the anterior lobe and the declive and folium of the posterior lobe

B. COURSE OF THE OLIVOCEREBELLAR TRACT FROM THE NUCLEAR COMPLEX
   1. Fibers cross to the opposite side of the medulla, pass through the contralateral olive, and ascend into the above designated areas of the cerebellar cortex via the inferior peduncle
   2. This tract comprises the bulk of the inferior peduncle

C. MAJOR INCOMING FIBERS OF THE INFERIOR OLIVARY COMPLEX
   1. Fibers ascending in the spino-olivary tract arise from neurons in the dorsal gray horn of the cord on the contralateral side and transmit sensory impulses from the skin, joints, and some muscles
   2. Fibers descend from sensorimotor cortex and the red nucleus

D. INFERIOR OLIVARY COMPLEX serves as a relay for modulation and transmission of information from the cord, sensorimotor cortex, and red nucleus to the cerebellum

II. Tectocerebellar tract arises from the tectum of the mesencephalon (in the colliculi)

A. COURSE OF THE TECTOCEREBELLAR FIBERS
   1. Enter the anterior medullary velum from the inferior aspect of the inferior colliculus, undergo partial decussation in the velum, and end in the anterior lobe, posterior lobe (folium, tuber, pyramis), and adjacent areas of the hemispheres

B. THE TECTUM is the center for both visual and auditory reflexes

C. THE TECTOCEREBELLAR TRACT may be the most important connection between the auditory pathway and the cerebellum

III. Corticopontocerebellar tract

A. FIBERS FROM NERVE CELL BODIES IN THE CEREBRAL CORTEX (especially those concerned with somatic motor function) descend through the internal capsule and mesencephalon as pyramidal, frontopontine, and temporoparieto-occipitopontine tracts

B. IN THE PONS, these fibers terminate on or synapse with pontine nuclei on the same side (in its ventral aspect)

C. AXONS OF THE CELLS OF THE PONTINE NUCLEI form the pontocerebellar tract, which crosses to the opposite side of the pons to enter the cerebellum through the middle cerebellar peduncle, and end as mossy fibers of the neocerebellum (posterior lobe) of the vermis and both hemispheres

IV. Arcuatocerebellar tract arises from the arcuate nuclei located on the ventral surface of the pyramids (medulla)

A. MAJOR ROUTES OF FIBERS LEAVING THE ARCUATE NUCLEI
   1. Ventral external arcuate fibers traverse the ventral and lateral surfaces of the medulla to enter the cerebellum via the inferior peduncle, where most end in cortex with the pontocerebellar fibers
   2. Arcuatocerebellar fibers run dorsally near the midline to the floor of the 4th ventricle, turn laterally as the striae medullares, and enter the cerebellum via the inferior peduncle to end in cortex with pontocerebellar fibers

B. ARCUATE NUCLEI may be displaced pontine nuclei and serve to supplement the corticopontocerebellar tract

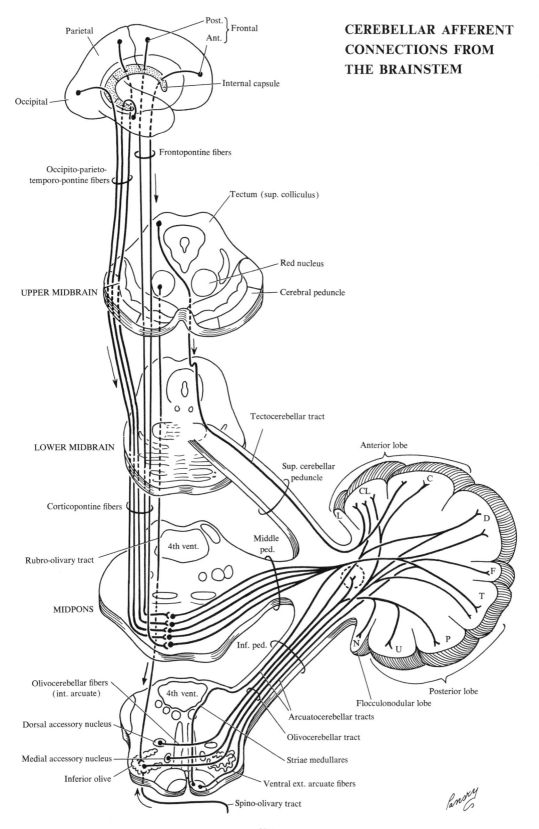

CEREBELLAR AFFERENT
CONNECTIONS FROM
THE BRAINSTEM

Parietal

Post. } Frontal
Ant. }

Internal capsule

Occipital

Frontopontine fibers

Occipito-parieto-
temporo-pontine fibers

Tectum (sup. colliculus)

UPPER MIDBRAIN

Red nucleus

Cerebral peduncle

Tectocerebellar tract

LOWER MIDBRAIN

Anterior lobe

Sup. cerebellar
peduncle

Corticopontine fibers

Middle
ped.

4th vent.

Rubro-olivary tract

MIDPONS

Inf. ped.

Olivocerebellar fibers
(int. arcuate)

4th vent.

Dorsal accessory nucleus

Arcuatocerebellar tracts

Olivocerebellar tract

Medial accessory nucleus

Striae medullares

Inferior olive

Ventral ext. arcuate fibers

Spino-olivary tract

Flocculonodular lobe

Posterior lobe

CL
C
L
D
F
T
N
U
P

Pansky

184. EFFERENT CEREBELLAR CONNECTIONS: CORTICOFUGAL AND NUCLEOFUGAL

I. **Corticofugal pathways** are comprised of fibers that carry impulses away from cell bodies located in the cerebellar cortex

A. DIRECT CORTICOFUGAL FIBERS pass directly from the cerebellar cortex into the brainstem (relatively few in number) and include

    1. Flocculovestibular tract: from flocculus to superior and lateral vestibular nuclei

    2. From anterior lobe (lingula, central lobule, and culmen) to vestibular nuclei

    3. Fibers from the posterior lobe (nodulus and uvula) to vestibular nuclei

    4. Fibers from the nodulus to the reticular formation of the medulla

B. INDIRECT OR CORTICONUCLEAR FIBERS pass from the cerebellar cortex to the intrinsic nuclei of the cerebellum and include

    1. Fibers from the vermis to the fastigial and globose nuclei

    2. Fibers from the medial part of the hemispheres next to the vermis pass to the globose and emboliform nuclei

    3. Fibers from the lateral part of the hemispheres pass to the dentate nucleus

II. **Nucleofugal pathways** are comprised of fibers that carry impulses away from cell bodies located in the intrinsic nuclei out of the cerebellum proper

A. THE FASTIGIAL NUCLEI, through ascending and descending projections, control axial and proximal musculature. The cerebellospinal pathway controls the execution of movement and muscle tone. This includes the smoothing of small oscillations in movement to control "physiological tremor"

    1. Fastigiobulbar tract (or cerebellovestibular tract) arises in the fastigial nucleus and its fibers may cross or remain on the same side

        a. Fibers enter the medial part of the inferior cerebellar peduncle to pass to the brainstem, where they end in the vestibular nuclei

    2. Uncinate bundle (or cerebellospinal tract) arises in the fastigial nucleus

        a. Fibers cross in the inferior cerebellar commissure and ascend a short distance with fibers of the superior cerebellar peduncle

        b. Fibers then curve around (hook = uncinate) the superior peduncle and leave the cerebellum via the inferior peduncle to enter the vestibular nuclei (both crossed and uncrossed fibers; contralateral fibers enter the reticular formation of the pons and medulla)

B. DENTATE FIBERS comprise the majority of efferent pathways from the cerebellum

    1. Dentatorubral tract arises primarily from cell bodies in the dentate nucleus (some fibers from the emboliform and globose nuclei)

        a. Exit the cerebellum via the superior peduncles which form the lateral wall of the 4th ventricle (superior portion)

        b. Enter the pontine tegmentum and pass into the tegmentum of the mesencephalon

        c. Most of fibers cross to the contralateral side in the caudal portion of the mesencephalon; a few uncrossed fibers end in pontine or in reticular nuclei

        d. After crossing, the tract breaks into fasciculi, the majority of which ascend to synapse with both small and large (primarily) cells of the red nucleus (descending fibers end in the reticulotegmental nucleus)

    2. Dentatothalamic tract

        a. Course of fibers is identical to that of above tract except that the fascicles pass around red nucleus without synapsing and continue to ascend into diencephalon, where they end in the ventral lateral (VL) nucleus of the thalamus

        b. Fibers from the VL nucleus ascend to the motor cortex through the internal capsule to influence premotor and motor cortex

    3. Cerebellorubral tract arises mainly in the globose and emboliform nuclei and terminates in the red nucleus

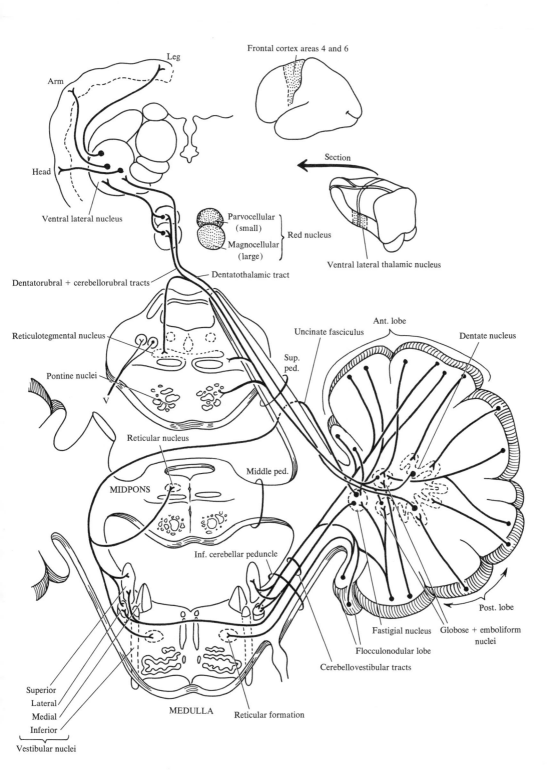

Frontal cortex areas 4 and 6

Leg

Arm

Head

Ventral lateral nucleus

Section

Parvocellular
(small)

Magnocellular
(large)

Red nucleus

Ventral lateral thalamic nucleus

Dentatorubral + cerebellorubral tracts

Dentatothalamic tract

Reticulotegmental nucleus

Uncinate fasciculus

Ant. lobe

Dentate nucleus

Sup. ped.

Pontine nuclei

V

Reticular nucleus

Middle ped.

MIDPONS

Inf. cerebellar peduncle

Post. lobe

Globose + emboliform nuclei

Fastigial nucleus

MEDULLA

Reticular formation

Flocculonodular lobe

Cerebellovestibular tracts

Superior

Lateral

Medial

Inferior

Vestibular nuclei

EFFERENT CEREBELLAR CONNECTIONS

Pandy

185. MOTOR SYSTEMS: PYRAMIDAL (CORTICOSPINAL, TRACT, SOMATIC EFFERENT—PART I

I. Definition: refers to that part of the motor system arising from specific areas of the cerebral cortex and descending through the pyramids to end on alpha motor neurons. The corticospinal tract is a direct link from cerebral cortex to spinal cord

II. Functions: related to precise and skilled movements of the extremities (hand and foot). Deals with the contraction of individual muscles and is the pathway for the selection of the "prime movers" for any muscular activity

III. Cortical representation of muscles (body parts) and movements are somatotopically mapped in the region of the precentral gyrus (motor cortex; area 4)
  A. ELECTRICAL STIMULATION of specific cortical areas elicits individual muscle contractions, but contractions are only fragments of a movement
  B. SYNTHESIS OF FRAGMENTS OF MOVEMENT into a useful movement occurs in the motor cortex (area 4)
  C. COMPLEX MUSCULAR ACTIVITY is controlled through areas of cortex other than the motor strip area
    1. Area 6 of the frontal lobe, mainly the premotor cortex, deals with the development of precise and skilled movements effected in area 4
    2. Areas 3, 2, and 1 of the parietal lobe also contribute efferent fibers
    3. Other cortical areas that also have a somatotopic organization are found in the upper top of the lateral fissure (secondary motor area) and on the medial aspect of the cerebral hemispheres (supplementary motor area)

IV. Upper motor neuron cell bodies are located in cortical gray areas 6, 4, 3, 2, and 1

V. Course of upper motor neuron axons
  A. EFFERENT (CORTICOFUGAL) FIBERS descend by way of corticospinal and corticobulbar pathways to the ventral gray columns of the cord and the cranial motor nuclei of the brainstem, respectively
  B. ALL CORTICOFUGAL FIBERS pass through the corona radiata
    1. Axons forming the corticospinal tract are long (a meter or longer in length)
    2. Pyramidal tract comprises about 1 million fibers; 30% originate in large, medium, and small neurons, including pyramid-shaped Betz cells, in lamina V of the motor cortex (Brodmann's area 4), 30% come from area 6 (premotor cortex), and 40% come from parietal lobe cortex
  C. ALL FIBERS CONVERGE on and pass into the posterior limb of the internal capsule, where they are compactly gathered, but are topographically localized
    1. Corticobulbar fibers are found most anteriorly in the posterior limb, next to the genu
    2. Fibers to the upper extremity are located in the rostral part; those to the lower extremity lie more caudally
    3. Sensory tracts from the thalamus to the parietal lobe lie posterior to the corticospinal fibers
  D. AXONS PASS THROUGH and form the cerebral peduncles (crus cerebri or basis pedunculi) along with other descending cortical fibers
    1. Pyramidal fibers remain in distinct groups as they pass into the peduncles at the rostral border of the midbrain and occupy the middle $\frac{2}{3}$ of the peduncles, with the corticobulbar fibers more medial
    2. Corticobulbar fibers leave the corticospinal pathway at appropriate brainstem levels (some crossed and others uncrossed) to connect with motor nuclei of certain cranial nerves

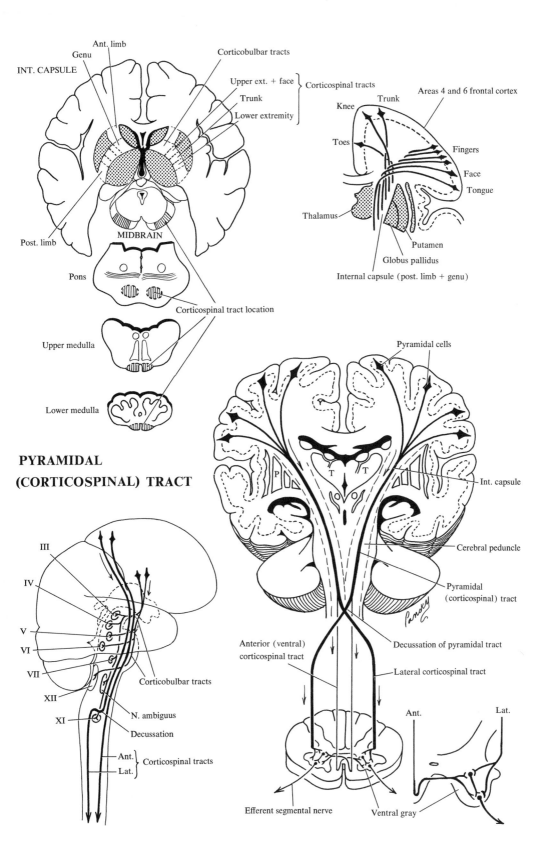

Ant. limb

Genu

INT. CAPSULE

Corticobulbar tracts

Upper ext. + face

Trunk

Corticospinal tracts

Lower extremity

Areas 4 and 6 frontal cortex

Knee

Trunk

Toes

Fingers

Face

Tongue

Thalamus

Post. limb

MIDBRAIN

Pons

Putamen

Globus pallidus

Internal capsule (post. limb + genu)

Corticospinal tract location

Upper medulla

Lower medulla

Pyramidal cells

Int. capsule

**PYRAMIDAL
(CORTICOSPINAL) TRACT**

Cerebral peduncle

Pyramidal
(corticospinal) tract

III

IV

V

VI

VII

XII

XI

Corticobulbar tracts

N. ambiguus

Decussation

Ant.
Lat. } Corticospinal tracts

Anterior (ventral)
corticospinal tract

Decussation of pyramidal tract

Lateral corticospinal tract

Ant.

Lat.

Efferent segmental nerve

Ventral gray

186. MOTOR SYSTEMS: PYRAMIDAL (CORTICOSPINAL) TRACT, SOMATIC EFFERENT—PART II

I. Course of upper motor neuron axons (cont.)

A. IN THE PONS descending pyramidal tract fiber bundles split into smaller irregular bundles (running in the ventral pons) by scattered pontine nuclei; somatotopic localization is observed from medial to lateral as face, upper extremity, and lower extremity

B. IN THE MEDULLA, fiber bundles recombine to form the pyramids on the ventral aspect of the medulla

C. IN THE LOWER MEDULLA, fibers cross the ventral fissure to the opposite side
   1. 75–90% of the fibers decussate (pyramidal decussation) just above the medulla–cord junction
   2. Fibers to the upper extremity cross over higher in the pyramidal decussation than those to the lower extremity

D. IN THE SPINAL CORD, the crossed pyramidal fibers form the lateral corticospinal tract in the lateral white column
   1. This tract gradually becomes smaller as it descends in the spinal cord
      a. 50% of the fibers end in the cervical spinal cord levels
      b. 20% end in the thoracic cord levels
      c. 30% of the fibers continue the full length of the spinal cord
   2. Some (about 10%) of these upper neurons end directly on alpha motor neurons of lamina 9 in the anterior gray horn; the others (as in the brainstem) synapse with connector neurons of laminae 4–7
   3. A few uncrossed fibers may run caudally in the lateral corticospinal tract of the same side (ventrolateral corticospinal tract or Barne's tract)

E. UNCROSSED FIBERS (10–25%) to the trunk muscles descend in the anterior white columns as the anterior corticospinal tracts
   1. These tracts are difficult to trace below the thoracic level, although some few fibers may continue throughout the cord
   2. Before ending, these fibers usually cross through the anterior white commissure to synapse with either anterior horn cells (alpha motor neurons) or connector neurons in laminae 7 and 8

II. Lower motor neurons

A. LOWER MOTOR NEURON CELL BODIES (ALPHA MOTOR NEURONS) are located in nuclei of the medial (for trunk muscles) and lateral (for extremity muscles) divisions of the anterior gray horns of the spinal cord

B. COURSE OF THE LOWER MOTOR NEURON AXONS
   1. Exit spinal cord through the ventral (anterior) roots and enter the spinal nerves at appropriate levels
   2. Are distributed with muscular branches of the dorsal and ventral rami to innervate skeletal muscles of the extremities and trunk

III. The pyramidal tracts contain a small number of fibers other than the motor fibers (as described). These are descending fibers from the sensorimotor strip of cortex surrounding the central sulcus which travel with the pyramidal tract to the spinal trigeminal nucleus and dorsal column nuclei (dorsal gray horn of cord and nuclei cuneatus and gracilis). Cortical activity facilitates or inhibits transmission of data via second-order sensory neurons to the thalamus and somesthetic cortex by this descending pathway

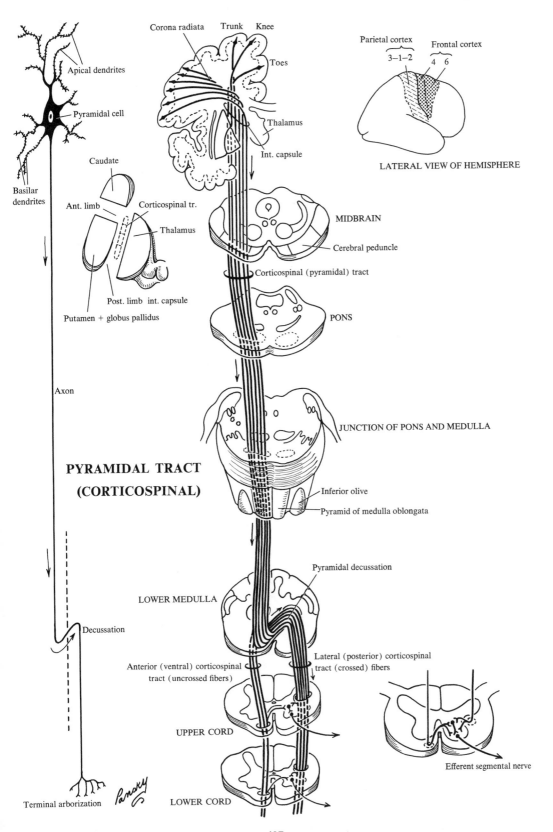

Apical dendrites

Pyramidal cell

Basilar dendrites

Corona radiata   Trunk   Knee

Toes

Thalamus

Int. capsule

Caudate

Ant. limb

Corticospinal tr.

Thalamus

Post. limb int. capsule

Putamen + globus pallidus

Axon

Parietal cortex   Frontal cortex
3–1–2   4   6

LATERAL VIEW OF HEMISPHERE

MIDBRAIN

Cerebral peduncle

Corticospinal (pyramidal) tract

PONS

JUNCTION OF PONS AND MEDULLA

**PYRAMIDAL TRACT
(CORTICOSPINAL)**

Inferior olive

Pyramid of medulla oblongata

Pyramidal decussation

LOWER MEDULLA

Decussation

Lateral (posterior) corticospinal
tract (crossed) fibers

Anterior (ventral) corticospinal
tract (uncrossed fibers)

UPPER CORD

Efferent segmental nerve

Terminal arborization

LOWER CORD

Pansky

-407-

187. MOTOR SYSTEMS: PYRAMIDAL (CORTICOBULBAR) TRACT—SPECIAL VISCERAL EFFERENT

I. Definition and function: the corticobulbar pathways (head and neck) are related to movements of muscles of facial expression and mastication, and tongue musculature

II. Cortical representation and upper motor neuron location: see Corticospinal Tracts

III. Course of upper motor neuron axons
- A. FIBERS CONVERGE as they pass through the corona radiata, with corticobulbar fibers lying in the most anterior portion of the posterior limb of the internal capsule
- B. CORTICOBULBAR FIBERS OCCUPY the medial portion of the middle two thirds of the cerebral peduncles of the mesencephalon
- C. FIBERS PASS INTO THE PONS along with the corticospinal fibers
- D. SOME CORTICOBULBAR FIBERS LEAVE the pyramidal bundle some distance above the nucleus for which they are destined and descend in region of the medial lemniscus
- E. MOST OF THE CORTICOBULBAR FIBERS MAY END in nuclei of the reticular formation (connector neurons) and use the reticular formation, medial longitudinal fasciculus (MLF), and the medial lemniscus to reach cranial motor nuclei
- F. IN THE MIDDLE PONS crossed (and some uncrossed) corticobulbar fibers pass to the motor nucleus of the trigeminal (V) nerve
- G. IN THE LOWER PONS crossed corticobulbar fibers pass to the anterior portion of the nucleus of the facial (VII) nerve, which innervates the lower facial muscles; crossed and uncrossed fibers pass to the posterior portion of this same nucleus, which innervates the upper facial muscles
- H. IN THE MEDULLA both crossed and uncrossed fibers go to the nucleus ambiguus

IV. Lower motor neuron cell bodies are located in
- A. MOTOR NUCLEUS OF THE TRIGEMINAL (V) NERVE
- B. NUCLEUS OF THE FACIAL (VII) NERVE
- C. NUCLEUS OF THE SPINAL ACCESSORY (XI) NERVE located in cervical spinal cord segments 1–5. Uncrossed upper motor neurons synapse with cell bodies of this nucleus
- D. NUCLEUS OF THE HYPOGLOSSAL (XII) NERVE

V. Course of the lower motor neuron axons
- A. DISTRIBUTED WITH THE MANDIBULAR DIVISION OF THE TRIGEMINAL NERVE
- B. DISTRIBUTED WITH THE FACIAL NERVE
- C. NUMEROUS FIBERS ALSO ARE DISTRIBUTED with the glossopharyngeal and vagus nerves and the cranial portion of the spinal accessory nerve
- D. DISTRIBUTED WITH THE SPINAL PORTION OF THE SPINAL ACCESSORY NERVE
- E. DISTRIBUTED WITH THE HYPOGLOSSAL NERVE

VI. Structures innervated by the lower motor neurons
- A. THE MANDIBULAR DIVISION OF THE TRIGEMINAL NERVE: muscles of mastication as well as the tensor veli palatini, tensor tympani, mylohyoid, and anterior belly of the digastric
- B. THE FACIAL NERVE: lower facial muscles (via fibers from anterior part of facial nucleus), upper facial muscles (via fibers from posterior part of facial nucleus), posterior belly of digastric, stapedius, and stylohyoid muscles
- C. THE GLOSSOPHARYNGEAL NERVE: the stylopharyngeus muscle
- D. THE VAGUS NERVE: muscles of the soft palate, pharynx, esophagus, and larynx
- E. SPINAL ACCESSORY NERVE
    1. Spinal portion to trapezius and sternocleidomastoid muscles
    2. Cranial portion to muscles of pharynx and larynx

VII. Subthalamic and peduncular aberrants supply the oculomotor (III), trochlear (IV), and abducens (VI) nuclei to help influence eye movements

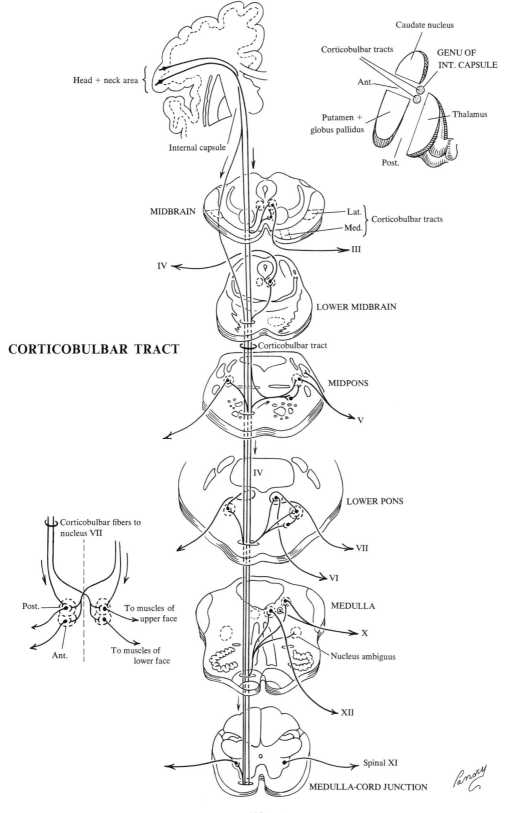

Head + neck area

Internal capsule

Caudate nucleus

Corticobulbar tracts

GENU OF
INT. CAPSULE

Ant.

Putamen +
globus pallidus

Thalamus

Post.

MIDBRAIN

Lat.
Med. } Corticobulbar tracts

III

IV

LOWER MIDBRAIN

CORTICOBULBAR TRACT

Corticobulbar tract

MIDPONS

V

IV

LOWER PONS

Corticobulbar fibers to
nucleus VII

VII

VI

Post.

To muscles of
upper face

MEDULLA

X

Ant.

To muscles of
lower face

Nucleus ambiguus

XII

Spinal XI

MEDULLA-CORD JUNCTION

Pansky

188. MOTOR SYSTEMS: EXTRAPYRAMIDAL PATHWAYS

I. Definition: extrapyramidal system is comprised of all tracts, their nuclei, and feedback circuits that influence somatic motor activity of voluntary muscles except for the pyramidal pathways. It conveys its influence from the cerebral cortex, basal ganglia, and subcortical nuclei indirectly to the spinal cord via multisynaptic connections*

II. Functions: the extrapyramidal tracts are functionally linked to the pyramidal tracts. They initiate or reinforce postural movements of the trunk and limbs, especially refined aspects of finger movement

III. Structures related to and included in the extrapyramidal pathways
A. CEREBRAL CORTEX AND BASAL GANGLIA (CAUDATE, PUTAMEN, AND GLOBUS PALLIDUS)
B. THALAMUS (CENTROMEDIAN NUCLEUS) AND SUBTHALAMIC NUCLEUS
C. RED NUCLEUS AND SUBSTANTIA NIGRA (MESENCEPHALON)
D. NUCLEI IN THE RETICULAR FORMATION (PONS AND MEDULLA)
E. FEEDBACK CIRCUITS, TRACTS, AND PATHWAYS (CORTICORUBROSPINAL, CORTICORETICULOSPINAL, AND VESTIBULOSPINAL)

IV. Main source of afferent fibers into the basal ganglia
A. CEREBRAL CORTEX (ALL CORTICAL AREAS) send nerve fibers into caudate nucleus and putamen (corticostriates), thalamus (ventrolateral), red nucleus and tectum, reticular formation, pons, and substantia nigra
B. CENTROMEDIAN NUCLEUS OF THALAMUS sends fibers into caudate nucleus
C. SUBSTANTIA NIGRA sends fibers to caudate nucleus and putamen (nigrostriates)
D. CAUDATE NUCLEUS AND PUTAMEN send short connector neurons to the globus pallidus to synapse with upper motor neurons

V. Main source of efferent fibers leaving the basal ganglia is via the globus pallidus

VI. Primary upper motor neuron cell bodies of the extrapyramidal system are located in the cerebral cortex and globus pallidus

VII. Course of the primary upper motor neuron axons
A. CEREBRAL CORTICAL FIBERS descend to the level of the superior colliculus of the midbrain and synapse in the red nucleus
B. EFFERENT FIBERS FROM THE GLOBUS PALLIDUS form the ansa lenticularis and the lenticular fasciculus (some fibers may pass to the red nucleus and tegmentum)
C. THE ANSA LENTICULARIS AND LENTICULAR FASCICULUS FIBERS turn medially to join the thalamic fasciculus
   1. Ventral anterior and lateral thalamic nuclei receive fibers from the fasciculus
   2. Efferent fibers from these 2 thalamic nuclei pass to the motor and premotor cortex (areas 4 and 6) to form a feedback circuit for the control of motor activity

VIII. Secondary upper motor neuron cell bodies are located in the red nucleus and reticular formation of the mesencephalon and in other areas of the brainstem

IX. Course of secondary upper motor neuron axons
A. EFFERENT FIBERS FROM THE RED NUCLEUS decussate and form the rubrospinal tract, which descends through brainstem in the lateral funiculus of the spinal cord
B. FIBERS LEAVING THE RUBROSPINAL TRACT synapse in the gray matter of the spinal cord
C. THESE CONNECTOR NEURONS, in turn, synapse either with alpha motor neurons (inhibitory) or with gamma motor neurons in the anterior gray horn

X. Other descending tracts (vestibulospinal, reticulospinal, tectospinal, and MLF), often considered part of the extrapyramidal system, are discussed separately

*The cerebellum and its related pathways are closely associated with but not usually included in the extrapyramidal system.

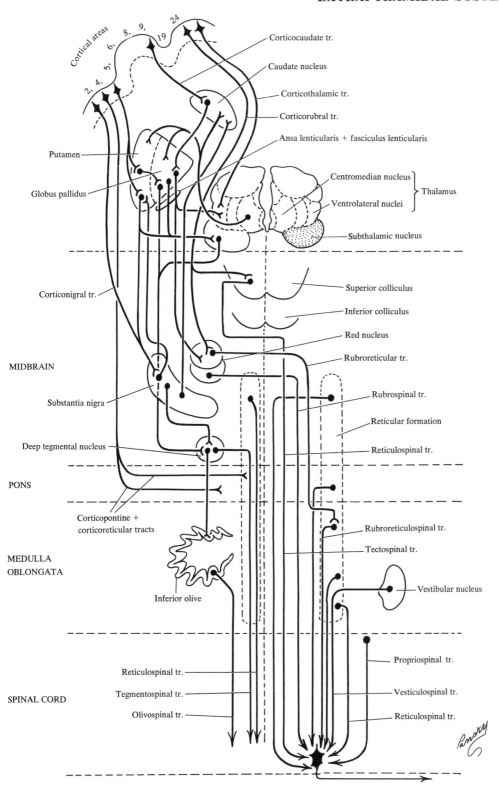

Cortical areas
2, 4, 5, 6, 8, 9, 19, 24

Corticocaudate tr.

Caudate nucleus

Corticothalamic tr.

Corticorubral tr.

Ansa lenticularis + fasciculus lenticularis

Putamen

Centromedian nucleus

Globus pallidus

Ventrolateral nuclei

Thalamus

Subthalamic nucleus

Superior colliculus

Corticonigral tr.

Inferior colliculus

Red nucleus

Rubroreticular tr.

MIDBRAIN

Rubrospinal tr.

Reticular formation

Substantia nigra

Deep tegmental nucleus

Reticulospinal tr.

PONS

Corticopontine +
corticoreticular tracts

Rubroreticulospinal tr.

Tectospinal tr.

MEDULLA
OBLONGATA

Vestibular nucleus

Inferior olive

Propriospinal tr.

Reticulospinal tr.

SPINAL CORD

Tegmentospinal tr.

Vesticulospinal tr.

Olivospinal tr.

Reticulospinal tr.

189. MUSCLE REGULATION

I. **Introduction:** voluntary muscles are regulated through control of the activity of motor units. Each motor unit consists of an individual lower motor neuron and the defined group of muscle fibers innervated by it. A hierarchy of motor centers controls motor unit activity

A. SPINAL CORD MOTOR SYSTEMS (SPINAL REFLEXES): interneurons connect between sensory afferents and lower motor neurons. The pattern of sensory inputs to the interneurons (stimulatory and/or inhibitory) determines motor unit activity
   1. Proprioceptive and nonproprioceptive feedback from muscles, tendons, joint capsules, and skin
   2. Collateral feedback from motor axons
   3. Labyrinthine pathways via the vestibulospinal and reticulospinal tracts

B. HIGHER MOTOR CENTERS: the basal ganglia, cerebellum, and cortical centers initiate and control coordinated postural and directional movements
   1. Brainstem reticular formation pathways originating in or mediated by the rubrospinal and reticulospinal pathways
   2. Indirect cerebellar pathways operating through the reticular formation and cerebral cortex, and reticulospinal and corticospinal tracts
   3. Cerebral cortical areas. Pathways via the pyramidal tracts and corticospinal and reticulospinal tracts

C. CONSCIOUS AND UNCONSCIOUS MOVEMENTS of facial muscles, upper limb muscles during walking or running, and eye muscle movements often accompany other muscular activity

II. **Skeletal muscle functions**

A. INVOLUNTARY (REFLEX) MOVEMENT: in response to environmental stimuli common to all forms of life with contractile tissue. Intrinsic to the segmental organization of the CNS (including the spinal cord and brainstem)

B. MAINTENANCE OF POSITION (POSTURE): including orientation related to gravity and maintenance of stance against gravity, which includes responses related to man's upright position

C. EXECUTION OF VOLITIONAL MOVEMENTS: organized or set into motion by suprasegmental structures and not necessarily dependent on external stimuli
   1. Both B and C require integration, initiation, or both by suprasegmental parts of the brain

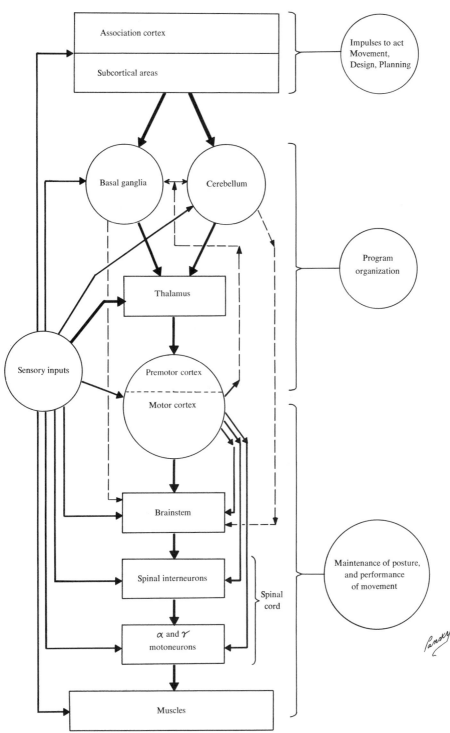

STRUCTURE   ROLE

Association cortex

Subcortical areas

Impulses to act
Movement,
Design, Planning

Basal ganglia   Cerebellum

Program
organization

Thalamus

Sensory inputs

Premotor cortex

Motor cortex

Brainstem

Spinal interneurons

Spinal
cord

Maintenance of posture,
and performance
of movement

α and γ
motoneurons

Muscles

MOTOR SYSTEM FUNCTION: SPINAL + SUPRASPINAL
CENTERS AND CONNECTIONS

190. MUSCLE REGULATION:
AFFERENT AND EFFERENT CONNECTIONS

I. Motor fibers

A. LOWER MOTONEURONS are cells whose axons leave the CNS. The axon branches or telodendria end as neuromuscular junctions or motor end-plates on skeletal muscle fibers, each with a single motor end-plate.

B. THE CELL BODIES are large, polygonal with conspicuous Nissl bodies, and are found in the branchial and somatic brainstem motor nuclei of cranial nerves III, IV, motor V (masticator), VI, VII, ambiguus, accessory (XI), and XII, as well as in the anterior horns of the spinal cord
   1. In the motor nuclei, anatomically indistinct subgroupings correspond to individual muscles, muscle groups, or muscle actions
   2. In the spinal cord, flexor and adductor neurons lie deep to extensor and abductor neurons. Cells for distal muscles lie lateral to those for trunk structures

C. EACH MOTOR AXON, as it passes through the anterior white matter, gives off a recurrent collateral which passes back into the anterior horn to synapse with Renshaw cells (interneurons found in the margins of the anterior horn)

D. MOTOR AXONS are thick, heavily myelinated fibers 10–20 μm in diameter and make up 40% of the large fibers in a muscle nerve. The other 60% are large afferent fibers, mostly from the spindles
   1. When the axon reaches the muscle, it branches extensively, supplying a group of muscle fibers with 1 motor end-plate each
   2. A single axon may supply only 3–10 muscle fibers of small muscles (eye muscles) or it may supply up to 200 fibers of large proximal muscles concerned with postural or coarse voluntary responses
   3. The group of fibers innervated by 1 axon (motor unit) does not correspond to a muscle bundle but extends over adjacent bundles, with much overlapping
   4. The range of tension developed by a single motor unit varies from about 0.1 g (in eye muscle) to about 50 g in a large limb muscle and is the result of fiber size and the number of fibers supplied

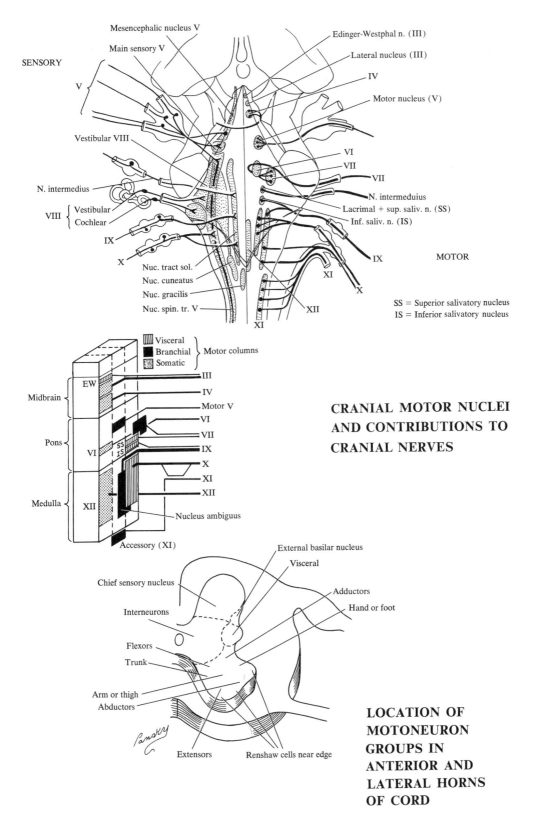

SENSORY

Mesencephalic nucleus V

Main sensory V

V

Vestibular VIII

N. intermedius

VIII { Vestibular
       Cochlear

IX

X

Nuc. tract sol.
Nuc. cuneatus
Nuc. gracilis
Nuc. spin. tr. V

XI

Edinger-Westphal n. (III)

Lateral nucleus (III)

IV

Motor nucleus (V)

VI
VII
VII

N. intermeduius

Lacrimal + sup. saliv. n. (SS)
Inf. saliv. n. (IS)

IX

XI

X

XII

MOTOR

SS = Superior salivatory nucleus
IS = Inferior salivatory nucleus

Visceral
Branchial } Motor columns
Somatic

Midbrain { EW
Pons { VI
Medulla { XII

SS
IS

III
IV
Motor V
VI
VII
IX
X
XI
XII

Nucleus ambiguus

Accessory (XI)

**CRANIAL MOTOR NUCLEI
AND CONTRIBUTIONS TO
CRANIAL NERVES**

Chief sensory nucleus

Interneurons

Flexors

Trunk

Arm or thigh
Abductors

Pansky

Extensors

Renshaw cells near edge

External basilar nucleus

Visceral

Adductors

Hand or foot

**LOCATION OF
MOTONEURON
GROUPS IN
ANTERIOR AND
LATERAL HORNS
OF CORD**

191. MUSCLE REGULATION: SENSORY FEEDBACK FROM MUSCLE AND TENDON—PART I

I. Introduction: naked nerve endings in musculoskeletal tissues convey pain, which may lead to reflex muscle spasm in response to inflammation or ischemia, but are probably not important in regulation of normal muscle activity. Endings dealing with proprioception are found in joint capsules (joint organ), near the attachments of the muscle to the periosteum or tendon (tendon organ), and in the muscle itself (muscle spindle)

A. JOINT ORGAN is associated with a subjective sense of limb position, and its major central connections are through the thalamus to the sensory cortex rather than to the cerebellum

B. TENDON OR GOLGI ORGAN is a mechanoreceptor comprising about 10 extrafusal muscle fibers in a connective tissue capsule. The receptor is stretched either by active contraction or by passive lengthening of the muscle

    1. Each is supplied by 1 to 2 large (12–18 μm) myelinated fibers (Aβ, group IB), which branch within the capsule to form thin, ramifying, unmyelinated branches

    2. Exhibit little or no spontaneous firing without tension, but fire (without adaptation) if tension rises to 100–200 g. The firing rate is proportional to the logarithm of the tension

    3. Arranged in series with extrafusal fibers. Protect against overstretching and rupture of the muscle

C. MUSCLE SPINDLE: a complex, encapsulated, elongated structure in the interstitial connective tissue of the muscle belly. Probably seen in all skeletal muscle, but less easily found in branchial than in somatic muscle

    1. Each contains 2 types of muscle fibers, receives 2 types of motor axons, and originates 2 types of sensory fibers. The muscle fibers are intrafusal, as opposed to the extrafusal fibers that make up the bulk of the muscle

    2. Each contains 1 or 2 nuclear-bag fibers, which are 7–8 mm long and 25 μm wide, centrally dilated, noncontractile "bags" containing several nuclei and show no cross-striations. The rest of the fiber has myofibrils and is striated

        a. Each nuclear-bag fiber has 1–2 medium-sized myelinated gamma (γ) motor axons (2.5–4.5 μm) and 1–6 separate motor end-plates on the cross-striated regions

        b. Each bag is surrounded by spiral sensory fibers, annulospiral endings, from which originate large (10–18 μm) myelinated afferent fibers (Aβ, group IA) at the level of the nuclei

        c. Some spindles also contain small numbers of irregularly branching sensory, flower-spray endings, from which 4–10 μm myelinated afferent fibers originate (Aβ, group II)

    3. The second type of intrafusal fiber is the nuclear-chain fiber, 4–5 mm long and 12 μm in diameter (4–8 per spindle)

        a. Their gamma motor axons are smaller (1–2 μm) and form irregular networks (trail endings) rather than discrete motor end-plates

        b. Both annulospiral (sensory) and flower-spray (sensory) endings are present on chain fibers

    4. Stimulation of the large motor fibers presumably shortens the contractile ends of the bag fibers, increasing tension on the bag. Stimulation of the smaller motor fibers shortens the chain fibers and reduces tension on bag

        a. Differential functions of the annulospiral and flower-spray endings are unclear

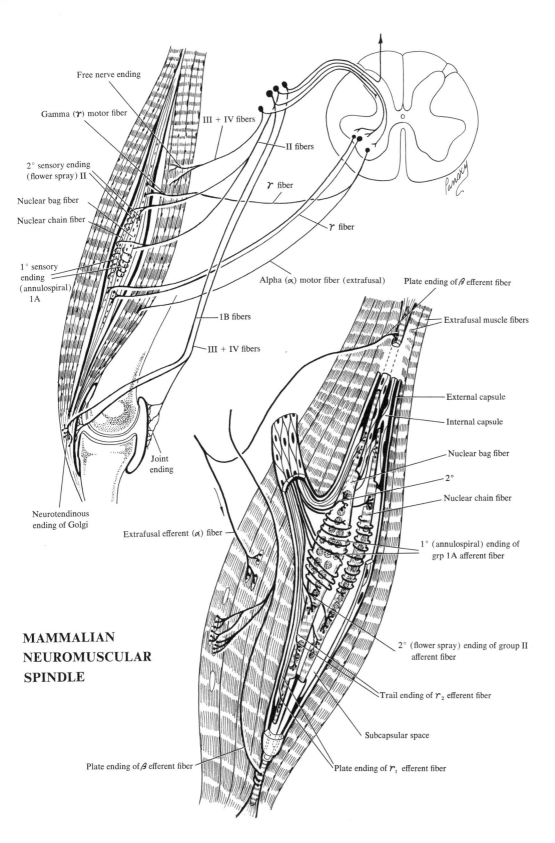

Free nerve ending

Gamma (γ) motor fiber

III + IV fibers

II fibers

2° sensory ending (flower spray) II

γ fiber

Nuclear bag fiber

Nuclear chain fiber

γ fiber

1° sensory ending (annulospiral) 1A

Alpha (α) motor fiber (extrafusal)

Plate ending of β efferent fiber

Extrafusal muscle fibers

1B fibers

III + IV fibers

External capsule

Internal capsule

Nuclear bag fiber

2°

Nuclear chain fiber

Joint ending

1° (annulospiral) ending of grp 1A afferent fiber

Neurotendinous ending of Golgi

Extrafusal efferent (α) fiber

2° (flower spray) ending of group II afferent fiber

Trail ending of γ_2 efferent fiber

MAMMALIAN NEUROMUSCULAR SPINDLE

Subcapsular space

Plate ending of β efferent fiber

Plate ending of γ_1 efferent fiber

192. MUSCLE REGULATION: SENSORY FEEDBACK FROM MUSCLE AND TENDON—PART II

A. MUSCLE SPINDLE (CONT.)
  1. Spindle function
      a. Spindle afferent fibers (IA) are excitatory mainly to motoneurons of the originating (homonymous) muscle; this excitation involves only 1 synapse between the afferent fiber and the motor neuron (the only monosynaptic reflex known in man)
      b. Facilitation of synergists and inhibition of antagonists, acting across the same joint as the originating muscle and confined to the same side, involve interneurons since a single axon cannot be both inhibitory and excitatory
      c. The CNS origin of the spindle and motor fibers is in the same nuclear zone as the motor alpha fibers to the main bulk (extrafusal) muscles. Gamma motoneurons, like alpha motoneurons, are in clusters associated with individual muscles or movements and resemble each other morphologically
          i. One function of gamma neurons is to maintain a constant spindle firing rate by causing the spindle fibers to shorten as the muscle shortens; i.e., alpha and gamma motoneurons fire at the same time
      d. The spindle is stretched when the muscle is stretched (like the tendon organ), but when the muscle actively contracts, the spindle is shortened and the tendon organ is stretched
      e. Both tendon organ and spindle are sensitive to stretching, but the spindle fires at lower levels of tension, 1–2 g
          i. The rate of firing is proportional to the logarithm of tension but is virtually nonadapting
      f. There is a servomechanism to maintain muscle length. If a muscle stretches, the spindle is stretched, the ratio of spindle firing increases, and motoneurons are stimulated and the muscle shortens
          i. Muscle length can be altered directly by stimulating alpha motoneurons (shortening the spindle increases the spindle firing and stimulates alpha motoneurons) or indirectly by stimulating gamma motoneurons
      g. Inhibitory feedback is provided via recurrent motor-axon collaterals that synapse with Renshaw cells which are directly inhibitory to the alpha motoneurons

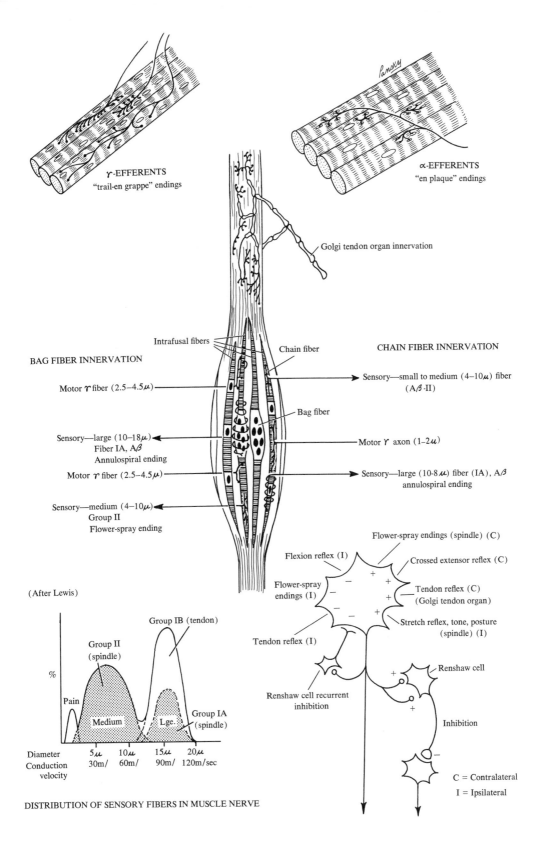

γ-EFFERENTS
"trail-en grappe" endings

α-EFFERENTS
"en plaque" endings

Golgi tendon organ innervation

Intrafusal fibers

Chain fiber

CHAIN FIBER INNERVATION

BAG FIBER INNERVATION

Sensory—small to medium (4–10μ) fiber
(Aβ -II)

Motor γ fiber (2.5–4.5μ)

Bag fiber

Sensory—large (10–18μ)
Fiber IA, Aβ
Annulospiral ending

Motor γ axon (1–2μ)

Motor γ fiber (2.5–4.5μ)

Sensory—large (10-8μ) fiber (IA), Aβ
annulospiral ending

Sensory—medium (4–10μ)
Group II
Flower-spray ending

Flower-spray endings (spindle) (C)

Flexion reflex (I)

Crossed extensor reflex (C)

Flower-spray
endings (I)

Tendon reflex (C)
(Golgi tendon organ)

(After Lewis)

Stretch reflex, tone, posture
(spindle) (I)

Tendon reflex (I)

Renshaw cell

Group IB (tendon)

Renshaw cell recurrent
inhibition

Group II
(spindle)

%

Inhibition

Pain

Renshaw cell

Medium

Lge.

Group IA
(spindle)

C = Contralateral
I = Ipsilateral

Diameter	5μ	10μ	15μ	20μ
Conduction velocity	30m/	60m/	90m/	120m/sec

DISTRIBUTION OF SENSORY FIBERS IN MUSCLE NERVE

193. MUSCLE REGULATION: SEGMENTAL MUSCLE REFLEXES

I. Segmental muscle reflexes: following spinal section, all muscles innervated from segments below the section are flaccid (have no tonus), and movement does not occur in the absence of local sensory input as from muscle spindles, tendon organs, and joint organs [proprioception; skin touch, pressure, temperature, or pain; or deep tissue sensation (pain)]

A. STRETCH EXTENSOR (MYOTATIC, DEEP TENDON) REFLEXES include simple "knee-jerk" or patellar reflexes. Their basis is that the muscle spindle responds to tension by increasing its firing rate, stimulating the alpha motoneurons and increasing the tone of the stretched muscle(s)

1. Characterized as 2 neuron, ipsilateral (restricted to one side), and intrasegmental (each receptor stimulates an afferent neuron, which excites alpha neurons in the same segment)
2. If a flaccid muscle is stretched, it responds by contracting, the latter occurring only in the stretched muscle(s) and physiologically induced by relaxation of the antagonist(s). There is no contralateral response; contraction is rapidly started, concluded, and abolished by dorsal root section
3. The reflex can be elicited without spinal section and is the basis of so-called deep tendon reflexes (T-reflexes) and muscle clonus seen clinically
4. It is monosynaptic and not depressed by drugs that affect interneurons and cause reduced flexor response and reflex spasm
5. Reciprocal innervation. If a flaccid muscle is stretched, it responds by contracting. At the same time, the antagonistic muscle relaxes, due to inhibitory impulses from IA fibers of spindles in the stretched muscle affecting antagonistic lower motor neurons (reciprocal antagonist inhibition). There is no contralateral response; contraction is rapidly started, concluded, and abolished by dorsal root section. It is monosynaptic and excitatory for the stretched muscle and disynaptic and inhibitory for the antagonist muscle

B. CLASP-KNIFE REFLEX: solely dependent on segmental input and cannot be shown in a flaccid spinal prep because it consists of a sudden relaxation of muscle when tension is very high

1. In decerebrate rigidity, if a limb is forcibly bent beyond a certain degree, rigidity abruptly disappears and the limb is freely movable. This is seen also in patients with spasticity
2. The reflex originates in the tendon organs that respond to high tension by firing
   a. Is multisynaptic (involves interneurons) and causes inhibition of the stretched muscle and its synergists and facilitation of the antagonists
   b. Occasionally the corresponding contralateral muscle is also facilitated

C. FLEXOR RESPONSE: painful stimulation to a flaccid limb results in generalized flexion of the limb. Is called withdrawal response and is a protective reflex

1. Accompanied by extension of the contralateral limb; if the cord is intact between hindlimb and forelimb segments, flexion of the diagonally opposite limb also occurs
2. The phenomenon (logical sign) implies topographic representation in segmental sensory and motor gray matter
3. Flexor response with crossed extension is seen in patients with spasticity but may be diffuse in the gray matter and appear as a "mass" reflex including incontinence of bowel and bladder
4. Stimuli reach the cord via Aδ or C fibers and connect with anterior horn cells only via interneurons, accounting for its long duration (compared to stretch reflex) and diffusion into the spinal gray over several segments on same and opposite side
5. All above are multisynaptic and may be intersegmental

STRETCH (MYOTATIC) REFLEX

CLASP-KNIFE REFLEX

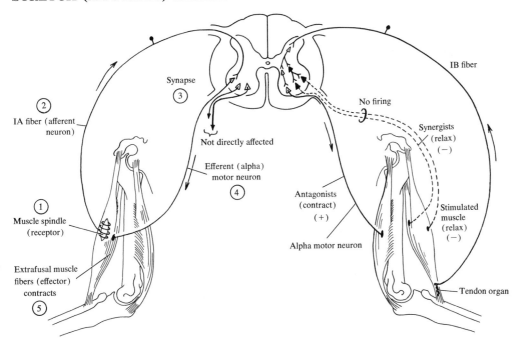

Synapse
③

② IA fiber (afferent neuron)

Not directly affected

Efferent (alpha) motor neuron
④

① Muscle spindle (receptor)

Extrafusal muscle fibers (effector) contracts
⑤

No firing

IB fiber

Synergists (relax) (−)

Antagonists (contract) (+)

Alpha motor neuron

Stimulated muscle (relax) (−)

Tendon organ

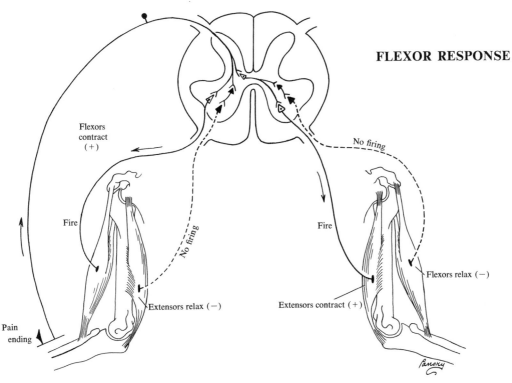

FLEXOR RESPONSE

Flexors contract (+)

No firing

Fire

No firing

Fire

Pain ending

Extensors relax (−)

Flexors relax (−)

Extensors contract (+)

Pansky

FLEXOR RESPONSE

CROSSED EXTENSOR RESPONSE

194. MUSCLE REGULATION:
OTHER REFLEX RESPONSES

I. Gamma reflex loop includes a gamma motoneuron, a neuromuscular spindle, an afferent neuron, an alpha motor neuron, and voluntary muscle

A. CONTRACTION OF A MUSCLE involves shortening of extrafusal muscle fibers in response to asynchronous volleys of discharge in alpha motor neurons and simultaneous activation of the stretch reflex by excitation of gamma motor neurons to muscle spindles (alpha-gamma coactivation). The gamma loop provides servo assistance to maintain sensitivity of the muscle spindle sensor during contraction of a muscle mass and prevent relaxation of the spindle during muscle contraction. Stable activity of the stretch reflex is also maintained during all phases of muscle contraction

B. THE GAMMA CONTROL of spindle sensitivity to stretch by contraction of the intrafusal muscle fibers is important because the gamma neurons influence the quality of sensory input from spindle to spinal cord and play a role in alpha motor neuron output

C. THE GAMMA NEURONS are stimulated by IA fibers from the muscle spindle, from the periphery via cutaneous stimulation by direct connections through interneurons synapsing with the gamma neurons, and actively by upper motor neurons (originating in the brainstem reticular formation)

D. TYPES OF GAMMA EFFERENT FIBERS
   1. Static fusimotor: more numerous; serve to adjust the sensitivity of the annulospiral and flower-spray endings to length alone
   2. Dynamic fusimotor serve to adjust the sensitivity of the annulospiral endings to the velocity of shortening

II. Flexor (superficial) reflex is a 3-neuron, disynaptic, ipsilateral, intersegmental flexor reflex (superficial, nociperceptive, or cutaneous)

A. FLEXOR REFLEX includes skin receptors, afferent neurons, spinal intersegmental interneurons, and alpha motor neurons to flexor muscles of upper extremity

B. REFLEX CAN BE FACILITATED by another 3-neuron reflex arc consisting of the flower-spray ending of the neuromuscular spindle, afferent neuron, spinal interneurons, and alpha motor neurons

III. Polysynaptic extensor reflexes: crossed extensor reflex and extensor thrust reflex

A. CROSSED EXTENSOR REFLEX: stimuli from peripheral receptors that initiate flexion in an extremity are also involved with extension of the contralateral extremity

B. EXTENSOR THRUST REFLEX: with pressure on the sole of the foot, the pressure receptors in the foot project influence that elicit extension of the stimulated extremity

IV. Disynaptic flexor reflex (stretch flexor reflex or "mark time" reflex) is involved in the stepping movements of a spinal animal. The flower-spray receptors are responsible. It is a 3-neuron, disynaptic, intersegmental flexor reflex (see Flexion Reflex)

V. Monosynaptic flexor reflex (jaw jerk or pluck reflex): an antigravity action resulting in teeth clenching

A. REFLEX CAN BE EVOKED by a tap on the jaw, setting off volleys from the annulospiral endings of spindles in flexor muscles of mastication

VI. Polysynaptic flexor reflex (flexor twitch): reflex withdrawal of the hand from a stimulus (cutaneous pain or heat receptors), a primitive protective reflex which is polyneuronal, polysynaptic, ipsilateral, and intersegmental

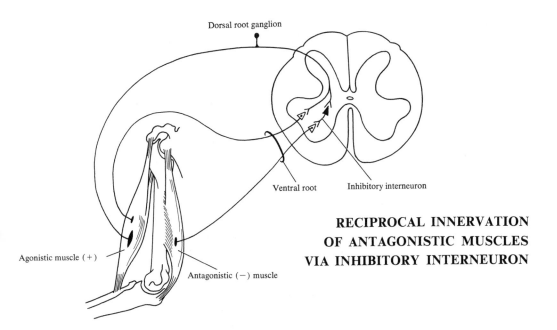

Dorsal root ganglion

Ventral root

Inhibitory interneuron

Agonistic muscle (+)

Antagonistic (−) muscle

**RECIPROCAL INNERVATION
OF ANTAGONISTIC MUSCLES
VIA INHIBITORY INTERNEURON**

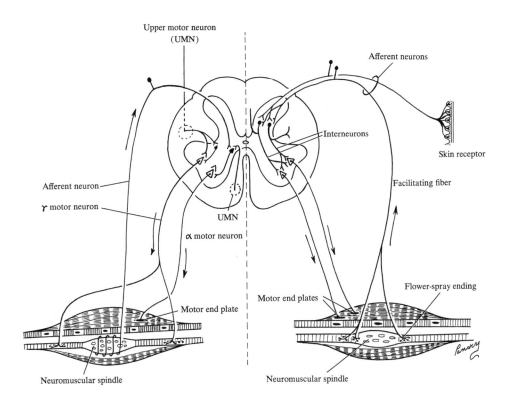

Upper motor neuron
(UMN)

Afferent neurons

Interneurons

Skin receptor

Afferent neuron

Facilitating fiber

γ motor neuron

UMN

α motor neuron

Flower-spray ending

Motor end plates

Motor end plate

Neuromuscular spindle

Neuromuscular spindle

GAMMA REFLEX LOOP

FLEXOR REFLEX LOOP

195. MUSCLE REGULATION: FLEXION REFLEX

I. Flexion reflex is a coordinated functional operation of a reflex arc, but not the entire nervous system which is to respond in neural control systems

A. A PAINFUL STIMULUS to the toe leads to foot withdrawal reflex
  1. Toe receptors are stimulated and transform the energy of the stimulus into the electrochemical energy of action potentials in afferent neurons
  2. Information regarding the intensity of the stimulus is coded both by the frequency of action potentials in single afferent fibers and by the number of different afferent fibers stimulated
  3. The afferent fibers branch after entering the spinal cord, each branch terminating at a synaptic junction with another neuron
  4. In flexion reflex, the 2nd neurons in the pathway are interneurons (between the afferent and efferent limbs of the reflex arc); thus this reflex is one of the large classes of polysynaptic reflexes
  5. The afferent neuron branches have the following functions
     a. Some synapse with interneurons that carry information to higher centers, resulting in a conscious correlate of the stimulus
     b. Other branches synapse with different interneurons that synapse on efferent neurons innervating flexor muscles which, when activated, cause flexion of the ankle and withdrawal of the foot from the stimulus. With an intense stimulus, the afferent discharging zone is large, so that muscles of joints such as the knee and thigh also flex
     c. As the injured leg is flexed away from the stimulus, the opposite leg is extended more strongly to support the added share of the body's weight—the so-called cross-extensor reflex
     d. Other afferents activate interneurons that inhibit motor neurons of antagonistic muscles that oppose the reflex flexion (reciprocal innervation)
B. FLEXION REFLEX NEURAL NETWORKS are simple, yet coordinated and purposeful
  1. Foot flexion away from the stimulus removes the receptors and prevents further damage (negative feedback)
  2. Purposefulness of the reflex response is independent of pain sensation, e.g., withdrawal occurs before the sensation of pain is experienced
  3. The observed response varies with the stimulus strength: low levels of stimulus lead to responses localized to the area of the stimulus, whereas higher levels of stimulation cause more widespread responses

II. Swallowing involves the coordinated interaction of about 20 muscles whose efferent neuron's cell bodies are found in the lower portion of the brain

A. SWALLOWING REQUIRES AN ORDERLY SEQUENCE of excitation of many neurons
  1. A single brief stimulus (10 seconds) triggers off a chain of responses; once initiated, its course is largely unaffected by further sensory input
  2. A more complex network of interacting neurons is built into the swallowing reflex to ensure the timing of the response than is in the flexion reflex, where the stimulus induces a brief and nearly simultaneous response
  3. Swallowing, however, is a less complex reflex than flexion reflex because it elicits a stereotyped all-or-none response whose pattern and magnitude are not altered by the stimulus strength

III. Respiratory movement is mediated by interconnected groups of neurons resulting in a rhythmic firing of phrenic motor neurons, probably not of "pacemaker" neurons

IV. Spontaneously depolarizing neurons, "pacemaker" neurons in some animals, provide a so-called biologic clock for the nervous system and may also function in the sleep-wake cycle, the reproductive cycle, hormone secretion, etc.

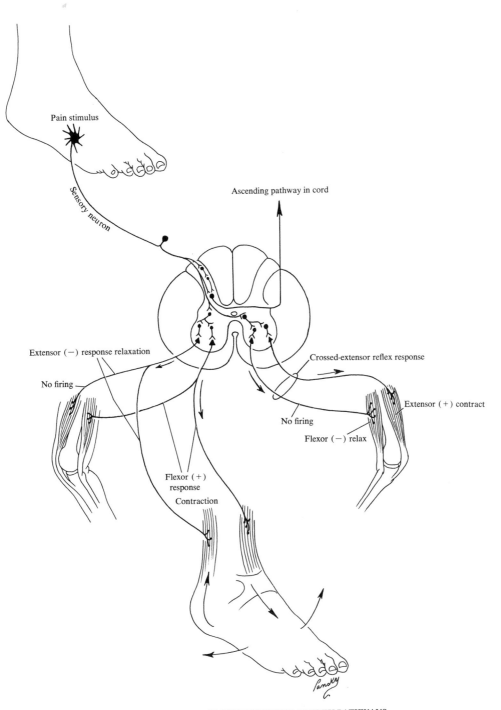

Pain stimulus

Sensory neuron

Ascending pathway in cord

Extensor (−) response relaxation

No firing

Crossed-extensor reflex response

Extensor (+) contract

No firing

Flexor (−) relax

Flexor (+) response Contraction

Pansky

FLEXOR (WITHDRAWAL) + CROSSED-EXTENSOR REFLEX PATHWAYS

196. MUSCLE REGULATION: ORGANIZATION OF SPINAL CORD GRAY MATTER

I. **Stretch, clasp-knife, and flexor reflexes** involve not only contraction of muscle groups but also relaxation of their antagonists

A. ALL REFLEXES AND ALL VOLUNTARY MOVEMENTS (RECIPROCAL INHIBITION) are thus involved, and there appears to be a built-in feature at the segmental level that helps us control and avoid useless concurrent contraction of opposing muscle groups.

B. INFLUENCES ACTING ON MOTONEURONS probably do so via two pools of interneurons, an extensor-positive, flexor-negative pool and an extensor-negative, flexor-positive pool

1. Reciprocal inhibition can be overridden voluntarily, as in pathologic rigidity of basal ganglion disease (mediated via pyramidal fibers)

C. AN EXTENSIVE INTRINSIC ORGANIZATION exists in the spinal gray matter, including some topographic representation in the posterior and anterior horns and interneuronal connections that have implications for posture and progressive movement

1. Most animal progression is associated with rhythmic flexion of the limbs or trunk muscles on one side and then the other, with corresponding contralateral extension

2. In quadripeds, diagonally opposite limbs assume similar positions (in crossed reciprocal innervation) for standing and moving

3. Withdrawal of support of one limb (flexor response) implies transfer of support to another limb, as in the crossed extensor component of response

D. IN THE STEPPING RESPONSE, the flexion response or extensor-thrust response gives way to alternating extension and flexion on the two sides and appears to be independent of suprasegmental influence, implying the existence of successively higher orders of interneurons as well as a switching mechanism. Thus, there may be some circuitry in the spinal gray capable of alternating the expected responses instead of merging or suppressing them

POLYSYNAPTIC REFLEX PATHWAYS

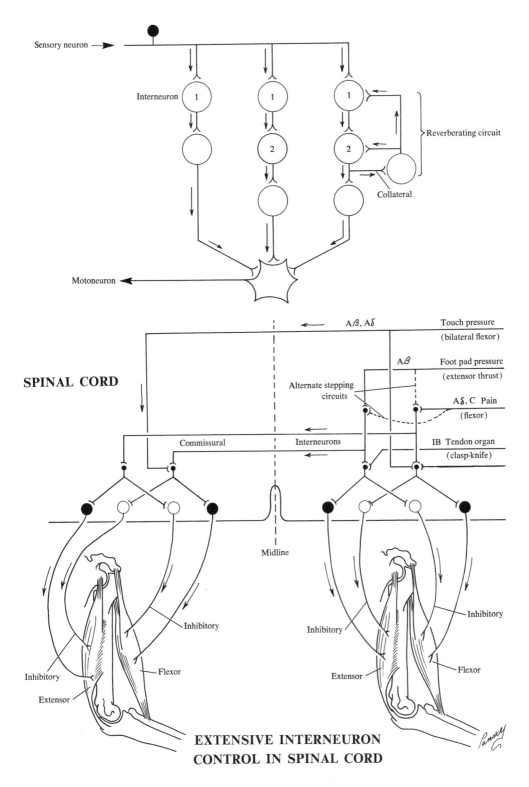

Sensory neuron →

Interneuron

Reverberating circuit

Collateral

Motoneuron ←

SPINAL CORD

Aβ, Aδ   Touch pressure
(bilateral flexor)

Aβ   Foot pad pressure
(extensor thrust)

Alternate stepping circuits

Aδ, C Pain
(flexor)

Commissural   Interneurons

IB Tendon organ
(clasp-knife)

Midline

Inhibitory

Inhibitory   Flexor

Inhibitory   Flexor

Inhibitory

Extensor

Extensor

Flexor

EXTENSIVE INTERNEURON
CONTROL IN SPINAL CORD

-427-

197. MUSCLE REGULATION: POSTURAL TONUS

I. **Origin of postural tonus:** muscles isolated from suprasegmental influences have minimal tonus but respond to segmental sensory input. Both extensor and flexor tonus are low. Animals do not normally collapse under gravity or appear consciously preoccupied with posture maintenance. There is a continuously discharging suprasegmental center responsible for maintaining unbalanced muscle tone
   A. IN QUADRIPEDS: the antigravity or postural muscles are the extensors in all limbs. In primates, postural muscles are extensors of the lower limbs and flexors of upper limbs. Arms are in semiflexed position for dealing with environment
   B. SUGGESTION OF A CENTER(S) IN LOWER BRAINSTEM FOR POSTURAL TONUS whose activity can be modulated by descending influences and volitional movement
      1. If the brainstem is isolated from the cord (spinal animal), all muscles are flaccid; section of midpons or midmesencephalon results in continuous rigidity of postural muscles, and animals are unable to initiate voluntary movement (decerebrate rigidity). Section at higher levels (upper midbrain or above hypothalamus) results in exaggerated postural tone, but these animals can perform purposeful movements (walk, run, chew, etc.) and can override the extensor rigidity
      2. Stimulation of reticular formation in medulla, pons, midbrain, and subthalamus facilitates extensor tonus; stimulation of medial reticular formation of medulla inhibits it. The latter requires activation by more rostral centers; the lateral is autonomous and originates continuous (tonic) discharges. Both centers act on motoneurons via the reticulospinal tract (probably on gamma fibers rather than alpha motoneurons)
         a. Reciprocal inhibition exists. Action is exerted indirectly via interneurons
         b. The lateral division of the vestibular nucleus and vestibulospinal tract also strongly facilitates postural tonus autonomously

II. **Modification of postural tonus for voluntary movement** by impulses in the proprioceptors of the neck muscles or in the labyrinth (reflects head position with respect to gravity and body position with respect to the head). Many pathways with respect to postural reflexes are still unknown

III. **Descending influences on postural tonus**
   A. COEPS: cortically originating fibers of the extrapyramidal system project directly to the medial reticular area of the medulla and are inhibitory to postural tonus
   B. MIDLINE (VERMAL) REGION OF CEREBELLUM (ESPECIALLY ANTERIOR LOBE) project via the fastigiobulbar and other tracts to the postmedullary reticular formation. The vermis is inhibitory to postural tonus. Projections also pass through the lateral ventrolateral thalamic nucleus to cortical areas 4 and 6
   C. INTERMEDIATE AND LATERAL ZONES OF CEREBELLUM project via the red nucleus and thalamic nuclei to areas 4 and 6 of cortex. Appear to facilitate pyramidal (not extrapyramidal) fibers and cause generalized increased muscle tone
      1. There may also be a direct reduction of flexor tonus by decreased stimulation of large-celled red nucleus, and thus a reduction of rubrospinal tract activity
   D. PYRAMIDAL TRACT: probably facilitatory to flexor and extensor stretch reflexes. Lack of this facilitation is more obvious in antigravity (postural) muscles
      1. It is of a monosynaptic nature, with connections to motoneurons, and may have combined or separate facilitatory effects on opposing muscle groups
      2. Unclear whether it is autonomous (intrinsic to cortex) or needs driving by the lateral and intermediate region of cerebellum
   E. POSTURAL TONUS: maintained by the lateral reticular formation and lateral vestibular nucleus, with variable effects originating in lateral and intermediate cerebellum and perhaps pyramidal tracts. It is inhibited by the medial reticular area of the medulla, but this is not autonomous and needs driving by midline cerebellar and COEPS fibers

POSTURAL TONUS

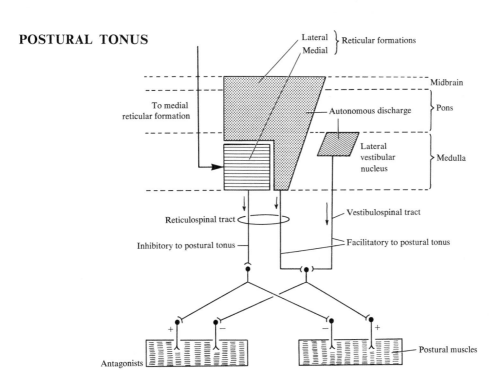

Lateral
Medial } Reticular formations

Midbrain

To medial
reticular formation

Autonomous discharge

Pons

Lateral
vestibular
nucleus

Medulla

Reticulospinal tract

Vestibulospinal tract

Inhibitory to postural tonus

Facilitatory to postural tonus

+   −   −   +

Antagonists

Postural muscles

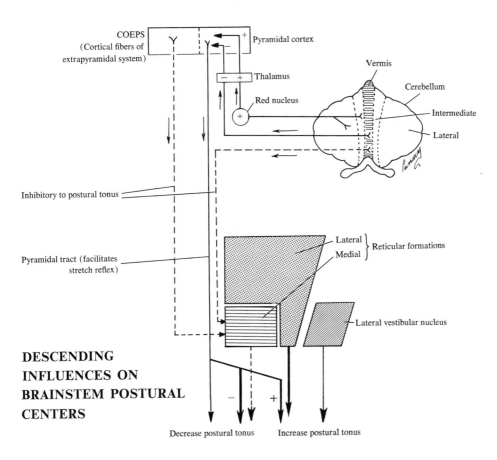

COEPS
(Cortical fibers of
extrapyramidal system)

+ Pyramidal cortex

Vermis

− + Thalamus

Cerebellum

Red nucleus

Intermediate

+

Lateral

Inhibitory to postural tonus

Lateral
Medial } Reticular formations

Pyramidal tract (facilitates
stretch reflex)

Lateral vestibular nucleus

DESCENDING
INFLUENCES ON
BRAINSTEM POSTURAL
CENTERS

−   +

Decrease postural tonus   Increase postural tonus

198. MUSCLE REGULATION: LOCOMOTION

I. Central programming for locomotion resides in a network of neurons in the spinal cord. A rhythmic pattern of neural activity, generated in the spinal cord, produces the rhythmic pattern of alternation between flexion and extension that is characteristic of walking and running behavior

A. DURING NORMAL WALKING or running, rhythmic patterns are generated separately for each limb, but the patterns are coupled together. Thus, when one leg is lifted and moved forward (swing phase), the body and opposite arm usually move forward at the same time and the opposite leg takes the weight of the body (stance phase). As the swinging foot lands on the ground to begin its stance phase, the role of the opposite foot changes so that it moves forward in the swing phase

II. Locomotor centers within the brainstem possess tonically active neurons with connections to the locomotor network of the spinal cord. They transmit signals to rhythmically active spinal cord regions which semiautonomously induce the alternating contractions of extensors and flexors occurring in normal locomotion

A. AXONS FROM NEURONS IN THE RED NUCLEUS cross and descend in the rubrospinal tract. Stimulation causes activation of flexor motor neurons

B. AXONS FROM DEITER'S (LATERAL VESTIBULAR) NUCLEUS descend as uncrossed fibers in the vestibulospinal tract to terminals in the medial anterior spinal cord. When stimulated, they excite extensor motor neurons

C. NUCLEI IN THE PONS AND MEDULLA project crossed and uncrossed fibers down the reticulospinal tract. Both flexor and extensor motor neurons are affected in a reciprocal manner by these fibers

III. Ascending fibers conduct sensory input from muscle spindles, Golgi tendon organs, and joints in the spinocerebellar tracts. These fibers provide information to the cerebellum about muscle activity and central programming in the spinal cord

A. AFFERENT INPUT is essential for fine modulation of the programmed activity of spinal cord locomotor centers
   1. This modulation includes changes required when changing activities during locomotion, including switching from the stance phase to the swing phase
   2. Afferent input is also important in the modification of the intensity of reflex action, as occurs when flexion is suppressed in a leg that is supporting the weight of the body during the stance phase but is permitted when the non-weight-bearing leg is moving in the swing phase

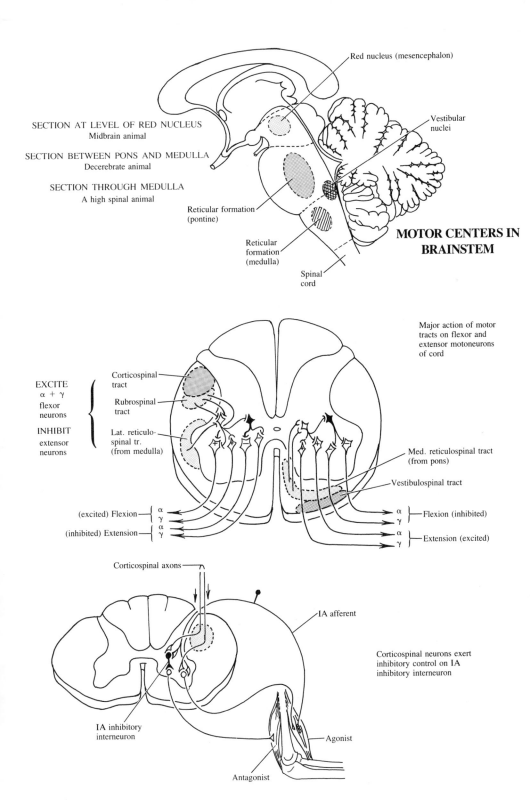

SECTION AT LEVEL OF RED NUCLEUS
Midbrain animal

SECTION BETWEEN PONS AND MEDULLA
Decerebrate animal

SECTION THROUGH MEDULLA
A high spinal animal

Red nucleus (mesencephalon)

Vestibular nuclei

Reticular formation (pontine)

Reticular formation (medulla)

Spinal cord

MOTOR CENTERS IN BRAINSTEM

Major action of motor tracts on flexor and extensor motoneurons of cord

EXCITE
α + γ
flexor
neurons

INHIBIT
extensor
neurons

Corticospinal tract

Rubrospinal tract

Lat. reticulo-spinal tr. (from medulla)

Med. reticulospinal tract (from pons)

Vestibulospinal tract

(excited) Flexion — α γ

(inhibited) Extension — α γ

α γ — Flexion (inhibited)

α γ — Extension (excited)

Corticospinal axons

IA afferent

Corticospinal neurons exert inhibitory control on IA inhibitory interneuron

IA inhibitory interneuron

Agonist

Antagonist

-431-

199. SPINAL CORD LESIONS—PART I

I. Tabes dorsalis (locomotor ataxia)

A. A RELATIVELY UNCOMMON SYNDROME today caused by syphilitic infection of the dorsal root ganglia and dorsal roots with secondary degenerative changes in the dorsal columns, particularly in the fasciculus gracilis bilaterally in the lower segments of the spinal cord

B. WHEN VIEWED GROSSLY the spinal cord, dorsal root ganglia, and dorsal roots commonly appear smaller and atrophic

C. MICROSCOPIC STUDY reveals the following changes
   1. Number of normal nerve fibers in dorsal nerve roots (especially lumbrosacral) is decreased
   2. Cellular infiltration of dorsal root ganglia
   3. Progressive dorsal column degeneration

D. CLINICAL SIGNS AND SYMPTOMS
   1. Paresthesia, hypesthesia, and intermittent attacks of sharp pain, which commonly occur early in neurosyphilis, are caused by irritation of dorsal nerve root fibers
   2. Sensitivity to pain is usually diminished in the course of the disease (particularly about the Achilles tendon and in the testes)
   3. Deep tendon reflexes (knee and ankle jerk) are diminished or absent because of interference with stretch reflex arc in the dorsal roots.
   4. Muscle, joint, and vibratory senses impaired
   5. Positive Romberg sign, hypotonic skeletal muscles, and ataxic gait
   6. Patient walks with the legs apart, head bent forward, eyes looking down (to try to compensate for the loss of kinesthetic sense), knees raised excessively high, and feet slapping down on the ground

II. Syringomyelia

A. SYRINGOMYELIA IS CHARACTERIZED by softening and cavitation around the central canal of the cervical enlargement of the spinal cord. Gliosis and cavitation may extend into other regions as the disease progresses

B. EARLY CLINICAL SIGNS reveal a loss of pain and temperature sensibility with segmental distribution in the upper limbs on both sides. This is due to the interruption of lateral spinothalamic fibers as they decussate ventral to the central canal

C. NO SENSORY IMPAIRMENT in lower limbs and body (spinothalamic tract intact)

D. SIMPLE TOUCH AND PRESSURE SENSES are preserved in the affected dermatomes of the upper limbs (sensory dissociation or dissociate anesthesia)

E. LATE MANIFESTATIONS of this neurologic disorder include
   1. Lower motor neuron disturbances with atrophy of muscles when degeneration extends into the anterior gray horns
   2. Pyramidal tract disturbances in lower limbs (cystic cavity compression)

III. Spinal cord transection (paraplegia)

A. THE FOLLOWING SIGNS AND SYMPTOMS appear immediately following complete transection of the spinal cord
   1. Voluntary muscular activity and somatic and visceral reflex activity in the body below the lesion are absent
   2. Sensibility from the body below the lesion is completely absent
   3. Spinal shock (depressed motor and sensory activity) lasts 2–3 weeks

B. WITH TIME THE DISTAL PORTION OF THE TRANSECTION SPINAL CORD begins to function independently (autonomously) in the following sequence
   1. Minimal reflex activity
   2. Superficial flexor reflex activity noted
   3. Alternation between flexor and extensor spasm activity
   4. Deep extensor reflex activity predominates

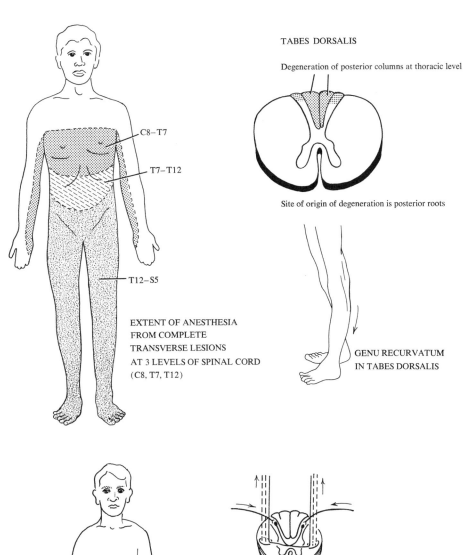

C8–T7

T7–T12

T12–S5

EXTENT OF ANESTHESIA
FROM COMPLETE
TRANSVERSE LESIONS
AT 3 LEVELS OF SPINAL CORD
(C8, T7, T12)

TABES DORSALIS

Degeneration of posterior columns at thoracic level

Site of origin of degeneration is posterior roots

GENU RECURVATUM
IN TABES DORSALIS

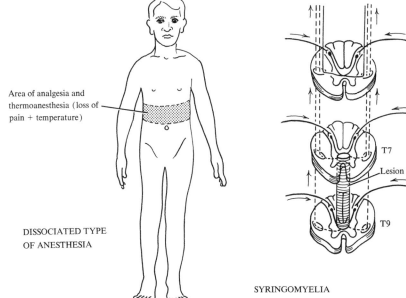

Area of analgesia and
thermoanesthesia (loss of
pain + temperature)

T7

Lesion

T9

DISSOCIATED TYPE
OF ANESTHESIA

SYRINGOMYELIA

-433-

200. SPINAL CORD LESIONS—PART II

I. Lesions of the anterior gray horns and ventral nerve roots (poliomyelitis)

A. ANTERIOR HORN LESIONS interrupt specific alpha and gamma motor neurons. Injury to these neurons results in either lower motor neuron paralysis or paresis (partial paralysis) of the muscle or muscle group innervated

B. LOWER MOTOR NEURON PARALYSIS is diagnosed by recognizing the following
  1. Voluntary muscle movements are completely absent and reflex activity cannot be elicited when all anterior horn cells and/or ventral root fibers to a particular muscle are involved
  2. Paralyzed muscles are flaccid (atonic) and deep tendon reflexes are absent
  3. Muscle fibrillations (single fiber contractions) and fasciculations (muscle twitching) may be recorded or observed respectively
  4. Muscle atrophy becomes apparent and skin becomes dry and cyanotic

C. Acute poliomyelitis or infantile paralysis is caused by a virus that damages anterior horn cells in the cervical and lumbar enlargements
  1. The extent of muscle paralysis depends on the number and distribution of lower motor neurons damaged beyond recovery with time
  2. Muscle atrophy begins several weeks after paralysis

II. Amyotrophic lateral sclerosis, a fatal disease of unknown origin

A. PATHOLOGICALLY, it is characterized by
  1. Degeneration of motor cells in the spinal cord (lower motor neurons), brainstem, and cerebral cortex (upper motor neurons)
  2. Secondary degeneration of axons in both the peripheral nerves and the pyramidal tracts bilaterally

B. CLINICAL SIGNS AND SYMPTOMS associated with this disease include
  1. Characteristics of both upper and lower motor neuron lesions
    a. Most paralyzed muscles show signs of lower motor neuron degeneration which include paralysis, atrophy, fasciculations, and muscle weakness
    b. Some paralyzed muscles show signs of upper motor neuron degeneration which include spasticity, hyperreflexia, and possibly Babinski's sign
  2. This disease is characterized initially by weakness, atrophy, and fasciculations of muscles of the hand and arm, followed by spastic paralysis of legs

C. SENSORY CHANGES are usually not observed with amyotrophic lateral sclerosis

III. Spinal cord hemisection (Brown-Séquard syndrome)

A. UNILATERAL TRANSVERSE LESION (HEMISECTION) of the spinal cord (resulting from a bullet or knife wound, spinal cord tumor, syringomyelia, etc.) will produce the Brown-Séquard syndrome

B. CLINICAL SIGNS AND SYMPTOMS associated with this syndrome include
  1. Upper motor neuron paralysis on the same side below the level of the lesion
  2. Lower motor neuron paralysis on the same side at the level of the lesion
  3. Cutaneous anesthesia on the same side at the segmental level of the lesion
  4. Hyperesthesia on the same side below the level of the lesion and at the segmental level of the lesion on the side opposite the lesion
  5. Loss of 2-point discrimination, proprioception, and vibratory sensibility below the level of the lesion of the same side of the body
  6. Loss of pain and temperature below the level of the lesion on the side opposite the lesion

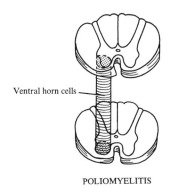

Ventral horn cells

POLIOMYELITIS

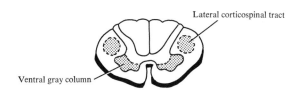

Lateral corticospinal tract

Ventral gray column

AMYOTROPHIC LATERAL SCLEROSIS

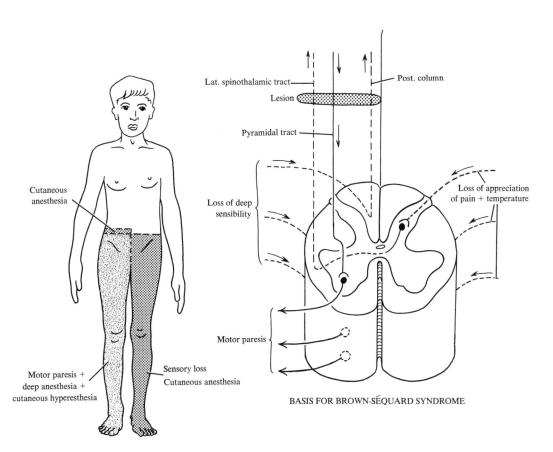

Lat. spinothalamic tract

Post. column

Lesion

Pyramidal tract

Loss of deep sensibility

Loss of appreciation of pain + temperature

Cutaneous anesthesia

Motor paresis

Motor paresis + deep anesthesia + cutaneous hyperesthesia

Sensory loss Cutaneous anesthesia

BASIS FOR BROWN-SÉQUARD SYNDROME

LESION OF RT. HALF OF CORD (ABOUT T10)

Pansky

PART III

Clinical

Neuroscience

Unit Thirteen

INTRACRANIAL VASCULAR ACCIDENTS, TRAUMA, AND EPILEPSY

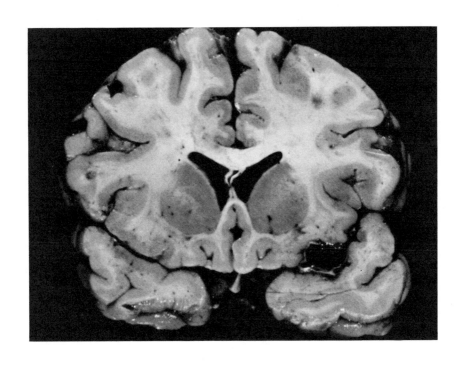

201. INTRACRANIAL VASCULAR ACCIDENTS: GENERAL INFORMATION AND CEREBRAL THROMBOSIS

I. Introduction: "stroke," the best-known disorder of the nervous system, is not primarily a neurologic disease, but one in which nervous tissue is damaged as a result of catastrophe affecting blood vessels carrying oxygen. Under normal body temperature conditions, interruption of oxygen supply for more than 3–4 minutes results in widespread damage and death to nerve cells. With hypothermia or reduced body temperatures, lack of O_2 can be withstood for a considerably longer period

A. STROKE OR CEREBROVASCULAR ACCIDENT refers to the blood supply to the brain being seriously disturbed spontaneously, not due to surgery or trauma. Though some vascular accidents inside the skull do not damage brain tissue, we tend to group all strokes together as intracranial vascular accidents

    1. In general, thrombosis, leading to gradual occlusion, will permit the adaptation of an anastomotic substitution network of supply. By contrast, emboli produce massive and sudden occlusion, after which reirrigation is inadequate and the resulting infarct is usually extensive

II. Characteristics of a cerebrovascular accident: the onset is sudden; the paralysis is maximum at onset; and with survival, the disability tends to improve

III. Etiologic factors

A. ATHEROSCLEROSIS is the chief factor in the production of cerebral infarction

B. CARDIAC EMBOLI are frequent causes of arterial occlusion

C. ARTERITIS is a rare cause of cerebral infarction

D. TRAUMA to the neck or in mouth may lead to internal carotid occlusion

E. VASCULAR MALFORMATIONS, e.g., arterial aneurysms

IV. Types of strokes (clinically)

A. CEREBRAL THROMBOSIS: a clot forms at a diseased point in a cerebral artery

    1. The disease in the artery is an atheroma, although inflammation of the wall may be responsible. An embolism, in most cases, is of cardiac origin

    2. The clot forms, blocks the vessel and cuts off the flow of blood (and O_2) to a particular area of the brain, and produces a zone of damage, an infarct

        a. The clinical result is that the part of the body normally controlled by this zone loses its function, e.g., becomes paralyzed

        b. Not all the cells in the zone are permanently destroyed; after a period of time some cells begin to function again, and other blood vessels open up to take over the job of the blocked artery

        c. The paralysis then begins to recover and may do so partly or completely, according to the degree of the original damage. (*Note:* nerve cells never regrow, but others may take over their function)

    3. Patients tend to be over 50 years old (since atheroma is a major cause)

    4. There is a slight tendency for occlusion to occur during periods of relative immobility, e.g., during sleep or bed rest

    5. The onset is sudden or very rapid, with little headache, and there may not be a loss of consciousness

    6. The form of the stroke varies with the vessel affected, resulting in certain clinical syndromes (e.g., subclavian artery thrombosis may lead to ischemic lesions of the vertebrobasilar artery area after diversion of the arterial flow (subclavian steal syndrome)

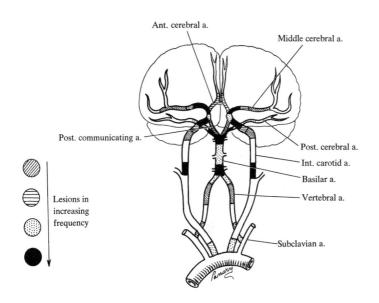

Ant. cerebral a.

Middle cerebral a.

Post. communicating a.

Post. cerebral a.

Int. carotid a.

Basilar a.

Vertebral a.

Subclavian a.

Lesions in increasing frequency

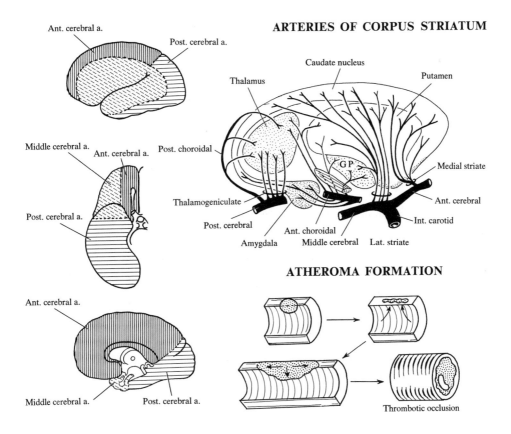

Ant. cerebral a.

Post. cerebral a.

ARTERIES OF CORPUS STRIATUM

Caudate nucleus

Putamen

Thalamus

Post. choroidal

Medial striate

G P

Thalamogeniculate

Ant. cerebral

Middle cerebral a.

Ant. cerebral a.

Post. cerebral a.

Post. cerebral

Int. carotid

Amygdala

Middle cerebral

Lat. striate

Post. choroidal

Ant. choroidal

ATHEROMA FORMATION

Ant. cerebral a.

Middle cerebral a.

Post. cerebral a.

Thrombotic occlusion

-441-

202. INTRACRANIAL VASCULAR ACCIDENTS: CEREBRAL EMBOLI, ARTERIAL INSUFFICIENCY, AND VENOUS THROMBOSIS

I. Cerebral embolism

A. THE CLOT is carried to the cerebral vessel, which may itself be healthy but is too narrow to allow the clot to pass from some other vessel

B. THE SYMPTOMS AND SIGNS are similar to those of thrombosis except that the onset is usually quite abrupt

C. THE EMBOLUS must come from somewhere; the origin may be the heart (cardiac embolism), whose valves have been diseased by rheumatic fever or whose walls have been damaged by a coronary thrombosis

    1. Emboli may also arise from patches of thrombus formed in the large vessels in the neck and chest (carotids, vertebrals, brachiocephalics, and even the aorta). The latter is apparently even more common than primary intracranial thrombosis of such vessels as the middle cerebral artery

II. Cerebral arterial insufficiency

A. SYMPTOMS AND SIGNS of a cerebral thrombosis may develop after a short period of time (only minutes) without any actual thrombosis having occurred—transient ischemic attacks, which may occur as a result of

    1. Minute emboli may break off from a diseased patch in a carotid or vertebral artery, temporarily block a small cerebral or retinal artery, fragment, and pass on into the veins, whereupon flow is restored and symptoms recover

    2. In others, there may be a marked narrowing of a vessel by arteriosclerosis or a sudden loss of blood due to an ulcer which results in a fall of blood pressure, which is then inadequate to force blood through the narrowed vessel

        a. The patient develops signs of insufficient blood in the area supplied by the vessel. When the blood pressure is restored, the symptoms disappear

        b. If one large vessel is blocked, the other may work for both sides

        c. Merely turning the head sharply and kinking the vessel may cause temporary symptoms

        d. If any one of the above is of long duration, infarction may result without an actual thrombosis; this is not an uncommon clinical happening

III. Venous thrombosis: thrombi can form in veins as well as in arteries

A. VENOUS THROMBOSIS is rare but may occur during severe illness, malnutrition, dehydration, or after major surgery or confinement to bed rest

B. PATCH AREAS OF THE CEREBRAL CORTEX are involved, since the veins lie on the surface of the brain

C. IRRITATED CELLS result in recurrent fits, usually followed by weakness of the involved parts of the body

    1. If the thrombus spreads, the fits will also spread to affect more and more of one side of the body, and may even spread to the other side as well

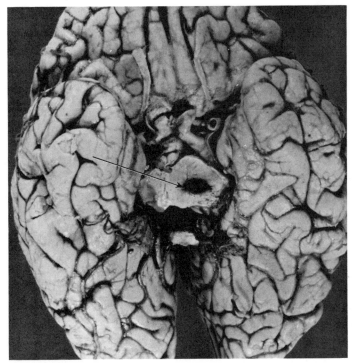

Hemorrhagic infarct in midbrain.

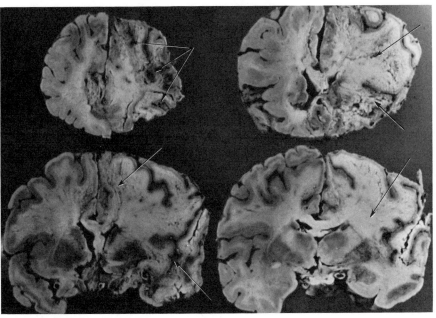

Cerebral infarct following arterial insufficiency.

203. INTRACRANIAL VASCULAR ACCIDENTS: CEREBRAL AND SUBARACHNOID HEMORRHAGE

I. Intracranial hemorrhage: a blood vessel ruptures, allowing blood to rush through cerebral tissue, destroying it, or forming a collection in the brain (hematoma) or permitting it to escape into the subarachnoid space and CSF (a subarachnoid hemorrhage)

A. CEREBRAL HEMORRHAGE: a cerebral vessel ruptures, resulting in blood escaping through brain substance and leading to destruction of much nervous tissue

1. Hemorrhage usually results from a weakness in a vessel wall (i.e., due to an atheroma patch, or an abnormally high vessel pressure as a result of high blood pressure)
2. Occurs in hypertensive patients over age 55 and often during periods of activity
3. Abrupt onset, a bursting headache, vomiting, and rapid loss of consciousness
4. Breathing is stertorous, there is profound paralysis of one side of the body, and the paralyzed cheek flaps in and out with each breath
5. Many patients die rapidly; others remain unconscious for days and then die; still others recover consciousness and may use their limbs, but have disability and speech defects
6. Prognosis is much worse than for a thrombosis or embolism

B. DIFFERENTIATION BETWEEN THROMBOSIS AND HEMORRHAGE: arteriography and other techniques have shown that it is difficult to be certain whether there is a massive cerebral thrombosis (may mimic hemorrhage) or a small hemorrhage

1. Neurologists are reluctant to prescribe anticoagulants without some certainty of what has happened. Patients must be carefully investigated

II. Subarachnoid hemorrhage: bleeding occurs into the CSF between the arachnoid and pia. Often due to a weakness in a blood vessel wall at the base of the brain (in circle of Willis) or to a rupture of abnormally developed vessels lying anywhere. Two major causes are

A. CEREBRAL ANEURYSMS: a weakness in the wall of a pulsating artery at a point where it branches, resulting in a bulge or aneurysm. Called a berry aneurysm

1. The weakness is present at birth and is never due to syphilis
2. May develop due to degenerative changes weakening the vessel wall
3. Referred to by the name of the vessel from which they originate
4. Patient may be young, with normal blood pressure and no other vascular disease
5. Patient rapidly develops intense headaches that spread to back of head and down neck; neck becomes stiff
6. Usually no paralysis occurs, but if force of rupture damages the brain itself, results may resemble an ordinary cerebral hemorrhage
7. Repeated vomiting, loss of consciousness, photophobia, restlessness, and irritability
8. With complications, nearby vessels go into spasm, resulting in cerebral infarction and hemiplegia a few days after the rupture
9. Without complications, the irritant blood is slowly absorbed from the CSF, the headaches diminish in a few days, the neck stiffness decreases, and the patient may recover without any residual trace of disability
10. Hemorrhages do tend to recur, particularly in 1st months, even the 1st fortnight

B. CEREBRAL ANGIOMA (vascular anomalies or arteriovenous aneurysms) consist of a mass of abnormal arteries, capillaries, and veins which have very fragile walls and are liable to rupture and produce subarachnoid hemorrhage

1. Are usually found on or near the cerebral cortex, act as an irritant, and thus produce fits (not seen with berry aneurysms)
2. Blood rushes through these widened vessels, and one may hear a pulsating noise— an intracranial bruit—diagnostic of an angioma in the adult

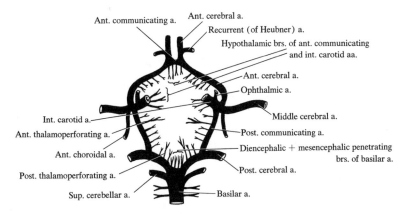

Ant. communicating a.
Ant. cerebral a.
Recurrent (of Heubner) a.
Hypothalamic brs. of ant. communicating and int. carotid aa.
Ant. cerebral a.
Ophthalmic a.
Middle cerebral a.
Int. carotid a.
Ant. thalamoperforating a.
Post. communicating a.
Diencephalic + mesencephalic penetrating brs. of basilar a.
Ant. choroidal a.
Post. cerebral a.
Post. thalamoperforating a.
Sup. cerebellar a.
Basilar a.

CIRCLE OF WILLIS AND ITS PERFORATING BRANCHES

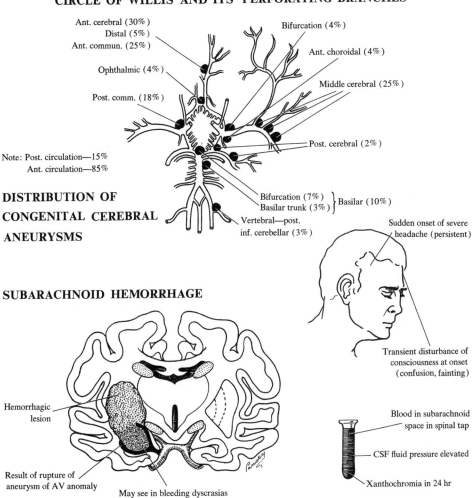

Ant. cerebral (30%)
Distal (5%)
Ant. commun. (25%)
Bifurcation (4%)
Ant. choroidal (4%)
Ophthalmic (4%)
Middle cerebral (25%)
Post. comm. (18%)
Note: Post. circulation—15%
Ant. circulation—85%
Post. cerebral (2%)

DISTRIBUTION OF CONGENITAL CEREBRAL ANEURYSMS

Bifurcation (7%)
Basilar trunk (3%) } Basilar (10%)
Vertebral—post. inf. cerebellar (3%)

SUBARACHNOID HEMORRHAGE

Sudden onset of severe headache (persistent)
Transient disturbance of consciousness at onset (confusion, fainting)

Hemorrhagic lesion
Result of rupture of aneurysm of AV anomaly
May see in bleeding dyscrasias

Blood in subarachnoid space in spinal tap
CSF fluid pressure elevated
Xanthochromia in 24 hr

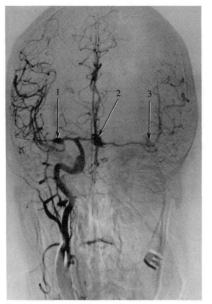

Three aneurysms (subtraction film).

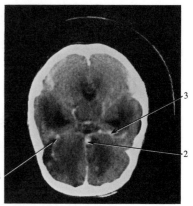

CT scan of same patient showing the three aneurysms:
1. Lt. middle cerebral artery.
2. Ant. communicating art.
3. Rt. middle cerebral art.

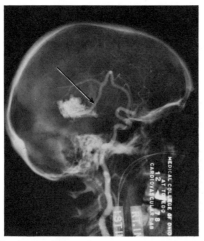

Arteriovenous (AV) malformation (angiogram) feeding vessel indicated by arrow.

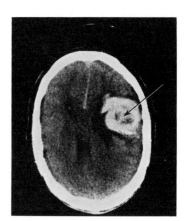

CT scan showing AV malformation.

FIGURE 15. **CT scans: aneurysms and arteriovenous malformations.**

-446-

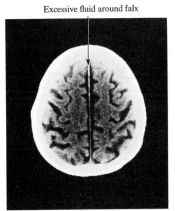

Excessive fluid around falx

Brain atrophy (mild) in 88-yr-old patient.
Note widened sulci.

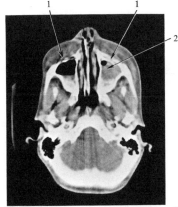

Bilateral frontal lobe contusion. Note fractures(1)
and hematoma (2) with spectrum density.

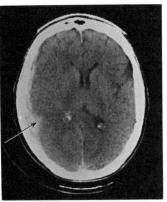

Epidural hematoma (arrow). Note anterior and
medial displacement of rt. choroid plexus.

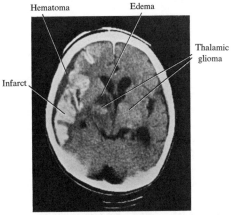

Hematoma   Edema   Thalamic glioma

Infarct

Chronic subdural hematoma (subacute), thalamic glioma,
edema, and displacement of midline structures.

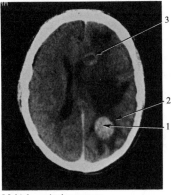

Multiple cerebral metastases.

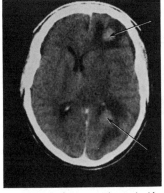

Multiple cerebral metastases (arrows) with
surrounding edema.

FIGURE 16. **CT scans: atrophy, contusion, hematomas, and cerebral metastases.**

204. CEREBRAL ARTERIOSCLEROSIS AND OTHER VESSEL DISEASES

I. **Cerebral arteriosclerosis:** the cerebral arteries narrow with age, reducing the flow of blood through the brain

A. AREAS DEPRIVED OF OXYGEN undergo a slow death of nervous tissue with resultant shrinkage or atrophy of the brain

1. Early symptom is gradual loss of memory, which at first is mainly for recent events, with memory for early years remaining good
2. If the process is severe, the patient also becomes unsteady, tends to take short, shuffling steps, and have attacks of giddiness and sudden falls
3. Epileptic attacks may occur in the early morning hours
4. Patients complain of continuous headaches, become irritable, and may even complain of being persecuted or neglected and are hard to live with
5. May progress to complete loss of mental and physical faculties, leading to arteriosclerotic dementia
6. There is no treatment to prevent or reverse the inevitable deterioration

II. **Oral contraceptives:** cerebral thrombosis is normally rare in young women, but recent high incidence has been related to use of the pill

A. THE CHANCE OF A THROMBOSIS is 3–6 times as great if one of the oral contraceptives contains high doses of estrogen

1. Low-dose estrogen pills are less likely to cause trouble, though they are not totally free from risk

III. **Other diseases of the blood vessels**

A. POLYARTERITIS NODOSA is a rare inflammatory or possibly allergic disease of the arteries, affecting all vessels including those of the brain, spinal cord, and peripheral nerves

1. Repeated vascular catastrophes occur in different parts of the nervous system
2. Small vessels supplying different peripheral nerves become blocked, and the nerves become paralyzed
3. One of the so-called collagen diseases (includes disseminated lupus erythematosus)
4. Blood sedimentation rate is high, blood proteins are abnormal, and diagnosis is confirmed by muscle biopsy

B. TEMPORAL ARTERITIS is an inflammation of the walls of the arteries of the scalp

1. Limited to elderly people ✳
2. Vessels outside of the skull become so inflamed and so intensely painful and tender that the patient cannot bear to touch the head, wear a hat, or even rest head on a pillow
3. Condition subsides eventually, but unfortunately the ophthalmic artery is often affected, resulting in blindness on both sides
4. Treatment: improvement is seen with the use of ACTH and cortisone, which are also used in polyarteritis.

C. SPINAL VASCULAR DISEASE: spinal intramedullary infarcts are rarer than central infarcts

1. Thrombosis or rupture of the spinal arteries is rare
2. If the above do occur, they are usually a complication of another disease
   a. Pressure on the arteries as a result of prolapsed disk or spinal tumor producing thrombosis, or anomalies of the vessels producing hemorrhage
3. A diagnosis of spontaneous spinal thrombosis is almost invariably incorrect

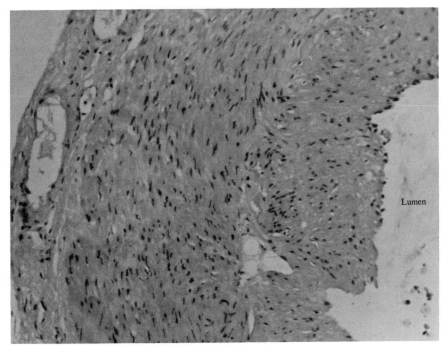

Cerebral arteriosclerosis.

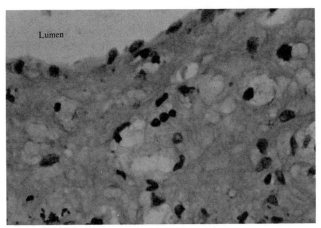

Cerebral arteriosclerosis.

205. TRAUMA: HEAD INJURIES

I. **Introduction:** the skull and vertebral column offer such good protection for the CNS that, although the nervous tissue is delicate, a severe injury is needed for either the brain or spinal cord to be damaged. However, bone strength, thickness, and resilience vary in different people and are influenced by age and disease

II. **Head injuries:** in an open head injury, the brain or meninges are exposed; in a closed head injury, though the skull may be fractured, the coverings of the brain remain intact

A. FRACTURES OF THE SKULL may occur either in the vault or across the base. The fracture itself is of little importance. The danger lies in the damage to the underlying structures such as the meninges, blood vessels, and brain. Thus, the position of the fracture is important. Examples

1. A frontal fracture may pass through the frontal sinuses and breach the dura, creating a pathway for CSF to escape down the nose (CSF rhinorrhea). It also allows air or infection to enter skull, with the danger of recurrent attacks of meningitis

2. Base of the skull fracture may injure the petrous bone, which contains the inner ear, allowing both blood and CSF to escape through the ears (CSF otorrhea). The inner ear is also subject to infection

B. TYPES OF FRACTURES

1. Simple fissure fractures: cracks with no breaks in the skin

2. Compound fractures: when the scalp is breached

3. Comminuted fractures: when the bone is broken into several pieces

4. Depressed fractures are most troublesome, and a piece of bone is driven inward

C. EXTRADURAL HEMATOMA: bleeding between the bone and dura which usually collects and enlarges fairly rapidly

1. Injury to the frontoparietal region may rupture the middle meningeal artery

2. History of injury

a. Patient has head injury and is knocked unconscious or recovers consciousness rapidly. In 1 or 2 hours, there is increased drowsiness, paralysis develops down one side, and there is dilatation of one pupil—all due to pressure of the expanding hematoma on the brain

3. Treatment requires an emergency operation. If left alone, patient will die

D. SUBDURAL HEMATOMA: injury of the fine arteries and veins between arachnoid and dura

1. Collects slowly so that signs and symptoms are like those of extradural bleeding, but may be delayed for days or even weeks and come on more slowly

2. Great variation from hour to hour in the degree of drowsiness. As clot irritates cortex, the patient may have fits and complain of increasing headache

3. In the young, the skull is usually fractured; in the elderly, milder trauma may be the cause of this injury due to the presence of more brittle vessels

4. Treatment: surgical evacuation of the clot

E. TRAUMATIC SUBARACHNOID BLEEDING: tearing of the vessels between the arachnoid and the pia, resulting in bleeding into the subarachnoid space

1. This is not uncommon in head injuries

2. After the patient regains consciousness, he has severe headaches for many days, with marked stiffness of the neck which gradually improves

3. There is no further loss of consciousness and no hemiplegia

4. The patient normally recovers spontaneously

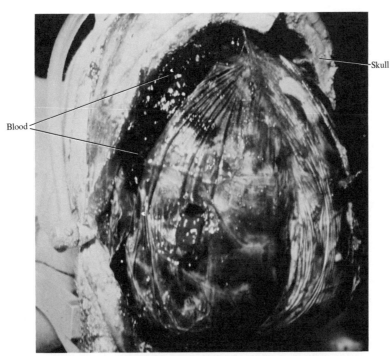

Traumatic injury with subarachnoid space bleeding.

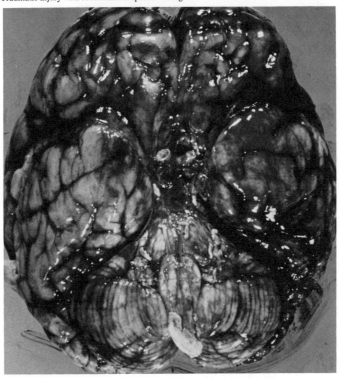

206. TRAUMA AND POSTTRAUMATIC EVENTS: CONCUSSION, CONTUSION, LACERATION, AND CONSCIOUSNESS

I. **Concussion, contusion, and laceration:** the obvious danger from a head injury is damage to the brain. Most head injuries are not penetrating wounds, but the whole brain can be severely "shaken up" without the skull actually being fractured

A. CONCUSSION: a blow or fall on the head, resulting in a temporary loss of brain cell function and loss of consciousness. Need not result in permanent brain damage
   1. Patient is stunned, dazed, or unconscious for minutes or even hours; may be confused for hours after injury. Usually recovers with no abnormal signs

B. CONTUSION: injury is more severe, brain tissue is bruised, and many cells die
   1. Depending on the area damaged, the signs will vary from mild transient weakness of a limb to prolonged unconsciousness and even paralysis

C. LACERATION: brain tissue is actually torn, as in a compound depressed fracture
   1. There is usually bleeding, which results in more severe, prolonged unconsciousness and paralysis, leaving permanent scarring and some permanent disability

II. **Consciousness and the brainstem:** in addition to cranial nerves and fibers in the brainstem, there is a network of cells and fibers called the reticular formation (RF). Impulses from it spread to the cortex and keep the brain in a state of activity and wakefulness. Damage to the RF results in a loss of consciousness

A. HEMORRHAGE, TUMORS, AND HEAD INJURIES can cause this damage
   1. Contusion of the brainstem causes profound loss of consciousness for long periods of time even where there is no sign of severe cerebral damage
   2. Consciousness may return in days or weeks (recovery may be complete)

B. FRACTURES IN THE REGION OF THE BRAINSTEM are more likely to be associated with injury, because the brainstem is so closely applied to the base of the skull

III. **Levels of consciousness:** in head injuries, steadily progressive lowering of consciousness is a serious sign. Failure to regain consciousness lost immediately at the time of an accident may mean cerebral concussion or contusion, but it does not necessarily mean increasing bleeding inside the skull. The following levels of consciousness merge when unconsciousness deepens or lightens

A. DROWSINESS resembles normal sleep, and the patient can be easily awakened

B. STUPOR is a state of unconsciousness in which stimulation produces only restlessness, noisiness, and irritability, but no true cooperation

C. COMA is when no response occurs even to the most painful stimulus

IV. **Aftereffects of head injury:** most patients recover without serious ill effects. A few severe injuries damage the brainstem so that consciousness never returns. Others recover consciousness, but brain damage is so great that they remain mentally abnormal and permanently unable to look after themselves

A. POSTTRAUMATIC EPILEPSY: laceration of brain resulting in a scar which later acts as an irritating focus capable of producing epileptic fits

B. POSTTRAUMATIC VERTIGO: a feeling that the body or outside world is spinning around. Often accompanied by vomiting and unsteadiness, but no loss of consciousness

C. POSTCONCUSSIONAL SYNDROME OR ACCIDENT NEUROSIS rarely follows severe injury but is common after minor injury

V. **Tentorial herniation and dilated pupil:** pressure above tentorium due to expansion inside brain or pressure from the outside forces the uncus of the temporal lobe through the tentorial notch, which squeezes the brainstem, affects the RF, and causes rapid, deep coma. The IIIrd cranial nerve is displaced and stretched. Its autonomic fibers are damaged and pupil dilates on side of the herniation and then on the other side

A. INCREASING COMA AND DILATING PUPILS are also seen with late stages of cerebral tumors, abscesses, and extradural and subdural hematomas

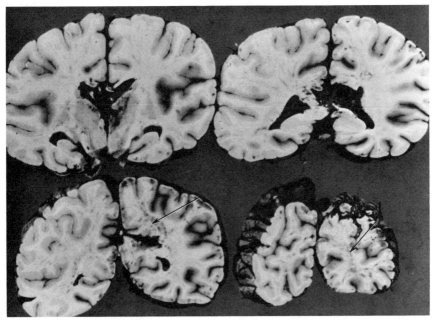

Sections of brain showing extent of hemorrhage in cerebrum and brainstem after traumatic injury.
Note petechial hemorrhages (arrows).

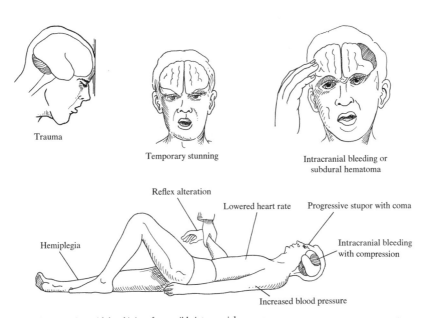

Trauma

Temporary stunning

Intracranial bleeding or
subdural hematoma

Reflex alteration

Lowered heart rate

Progressive stupor with coma

Hemiplegia

Intracranial bleeding
with compression

Increased blood pressure

Observe patient with head injury for possible intracranial pressure.

207. TRAUMA: SPINAL AND PERIPHERAL INJURY

I. **Spinal trauma:** fracture of the bones around the spinal cord is not sufficient for the cord to be damaged; the bones must be displaced as well

A. THE CORD may be crushed, overstretched, or even completely divided

   1. The result is that below the level of the lesion, all neurons carrying motor impulses are cut off from the muscles they innervate, and all neurons carrying sensation from parts below the lesion are unable to transmit to the brain

   2. A severe cord lesion will cause complete paralysis and complete loss of sensation below a clear-cut lesion

      a. A neck injury produces paralysis of the arms, trunk, and legs

      b. A thoracic injury paralyzes the lower part of the trunk and legs

      c. Injury to the lumbar spine may probably miss the spinal cord, which ends at just below L1, but will involve the cauda equina and produce paralysis of and loss of sensation in the legs and the urinary bladder

   3. Immediately after spinal injury at any level, there is usually a flaccid paralysis of the limbs and complete retention of urine as a result of spinal shock, which usually passes off within 3 weeks

      a. Flaccidity then changes to spasticity if the injury is to the cord itself, as a result of damage to the upper motor neurons

      b. If the lesion is in the cauda equina, the legs remain flaccid because, in this case, it is the lower motor neurons that are destroyed

      c. In any case, there is marked loss of skin sensation, the upper level corresponding to the level of the damage

   4. In partial lesions of the cord, the paralysis and sensory loss show a peculiar distribution called the Brown-Séquard syndrome (described under clinical signs of damage to the nervous system in another unit)

II. **Peripheral nerve injuries:** as most peripheral nerves carry both motor and sensory fibers, injury results in weakness of the muscles they supply and loss of all forms of sensation over the particular area of the skin. Vulnerable nerves are

A. ULNAR NERVE INJURY: the nerve lies very near the surface at the elbow and is easily damaged, resulting in wasting of the hand muscles and loss of sensation over the little finger

B. MEDIAN NERVE INJURY is usually involved as it passes through a narrow tunnel at the wrist, resulting in wasting of the thumb muscles and loss of sensation over the thumb and first finger. This has been called carpal tunnel syndrome

C. SCIATIC NERVE may be injured in the buttock by badly placed injections, which results in paralysis of the foot and loss of sensation over the front and back of the lower leg and foot

D. LATERAL POPLITEAL NERVE may be injured just below the knees by the patient's wearing high boots, or by skin traction as a result of disk lesions, or by allowing the patient to lie on the nerve

   1. Injury results in paralysis of dorsiflexion of the foot and loss of sensation between the big toe and 2nd toe

E. OTHER PERIPHERAL NERVE INJURIES are discussed in detail elsewhere

III. **Summary:** in all traumatic lesions, more nervous tissue is "knocked out" of action immediately after the injury than is actually destroyed, and a remarkable degree of recovery is possible. It is not unusual for many weeks, months, or even years to pass before one can even venture to state confidently that no further recovery is possible. However, by the end of the 2nd year, the maximum degree of improvement has usually been reached

GUIDE TO CORD LEVEL DAMAGE

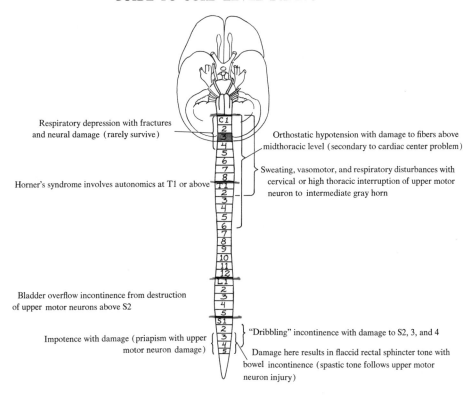

Respiratory depression with fractures and neural damage (rarely survive)

Orthostatic hypotension with damage to fibers above midthoracic level (secondary to cardiac center problem)

Sweating, vasomotor, and respiratory disturbances with cervical or high thoracic interruption of upper motor neuron to intermediate gray horn

Horner's syndrome involves autonomics at T1 or above

Bladder overflow incontinence from destruction of upper motor neurons above S2

Impotence with damage (priapism with upper motor neuron damage)

"Dribbling" incontinence with damage to S2, 3, and 4

Damage here results in flaccid rectal sphincter tone with bowel incontinence (spastic tone follows upper motor neuron injury)

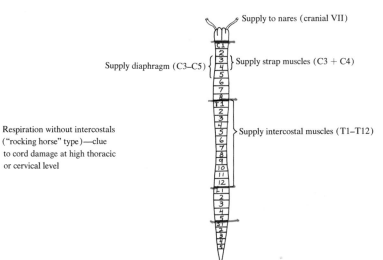

Supply to nares (cranial VII)

Supply diaphragm (C3–C5)

Supply strap muscles (C3 + C4)

Respiration without intercostals ("rocking horse" type)—clue to cord damage at high thoracic or cervical level

Supply intercostal muscles (T1–T12)

RESPIRATION AS CLUE TO CORD LEVEL DAMAGE

Pansky

208. EPILEPSY—PART I

I. Introduction: nerve cells function by discharging small electrical waves. Discharges arising from the brain can be measured by the electroencephalograph (EEG). If the brain cells suddenly produce a burst of larger electrical waves than normal, an epileptic fit may result. Patients whose brains have a tendency to do this repeatedly are said to suffer from epilepsy. This tendency is sometimes weakly hereditary or may develop from an area of damaged tissue. These two types are called idiopathic and symptomatic epilepsy, respectively

A. IDIOPATHIC EPILEPSY
  1. The exact cause of the electrical abnormality in this group is unknown
  2. Microscopic examination of the brain at postmortem shows no abnormality, but there may be some biochemical abnormality in the cells
  3. The tendency seems to be passed on through certain families for many generations. Some members develop fits; most have none
  4. Fits begin in childhood, adolescence (usually around puberty), or early adult life and take the form of generalized convulsions or petit mal attacks
  5. The patients are usually normal between attacks

B. SYMPTOMATIC EPILEPSY: the fits are one of the symptoms of brain damage or disease
  1. The suspicion that epileptic attacks are symptomatic arises when the fits begin later in life (over age 30), the attacks are focal, and there are abnormal signs between attacks or shortly after an attack
  2. There is no family history and patients do not have true petit mal attacks
  3. When other symptoms of brain disease are also present, one is alerted to the possibility of a tumor, because removal may cure the patient

II. Types of epileptic fits: in either of the 2 major types of epilepsy, several types of fits may take place

A. GRAND MAL ATTACKS (major fits)
  1. May start abruptly with loss of consciousness, or may be preceded by a short warning or aura. The type of aura is important since it shows where in the brain the abnormal discharge begins
  2. In a generalized fit, there is often a vague indescribable sensation, followed by a loss of consciousness and rigid contraction of all muscles
    a. The arms are flexed, legs are extended, head is thrown back or turned to one side, teeth are clenched tightly (tongue or cheek may be bitten), and respiratory movements are checked so that the patient becomes cyanosed—this is the tonic phase of the fit
    b. After a few moments, the clonic phase begins; the muscles of the face, limbs, and trunk develop a coarse jerking, which becomes violent. There is often incontinence of urine and occasionally of feces
    c. Breathing becomes stertorous, the face flushes, muscles relax, but the patient remains unconscious, unarousable, with absent corneal and pupillary reflexes and extensor plantars—a postepileptic coma lasting 10 minutes to several hours
    d. This is followed by a period of confusion; the patient is restless and may wander about doing things automatically—postepileptic automatism—knowing nothing of it and possibly endangering himself or others
  3. All grand mal attacks do not show all the features described. There may merely be a sudden loss of consciousness, a loss of consciousness with rigidity, but no clonic movements, or any combination of symptoms
    a. Symptoms may be so mild as to consist only of incontinence of urine at night or so severe that the patient suffocates and dies
  4. Generalized attacks may be idiopathic or symptomatic

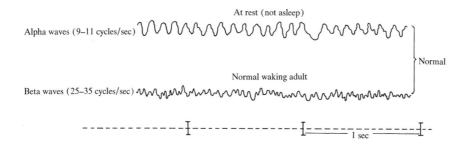

At rest (not asleep)

Alpha waves (9–11 cycles/sec)

Normal

Normal waking adult

Beta waves (25–35 cycles/sec)

1 sec

GRAND MAL

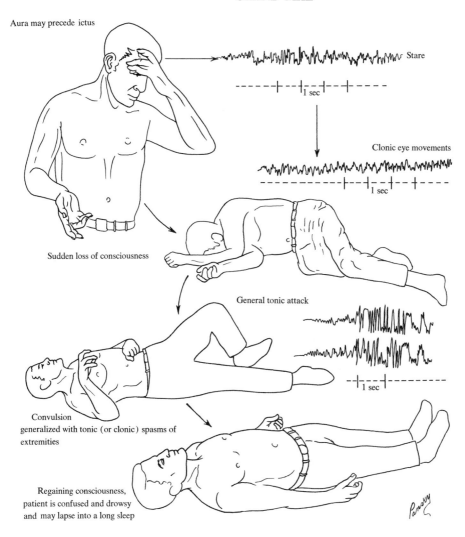

Aura may precede ictus

Stare

1 sec

Clonic eye movements

1 sec

Sudden loss of consciousness

General tonic attack

1 sec

Convulsion
generalized with tonic (or clonic) spasms of
extremities

Regaining consciousness,
patient is confused and drowsy
and may lapse into a long sleep

209. EPILEPSY—PART II

II. Types of epileptic fits (cont.)

B. JACKSONIAN ATTACKS: begin with a twitching of the mouth, thumb, or big toe and spread to other areas of the same side. They then pass to the other side and end in loss of consciousness. The attacks are symptomatic of irritation of the cerebral cortex by a tumor, scar tissue, or some form of inflammation

C. FOCAL EPILEPTIC ATTACKS: affect just one part of the body (many focal attacks show no spread, never become generalized, or cause loss of consciousness).
  1. They are always symptomatic of a localized cerebral lesion, and the form that the attack takes depends on where the lesion lies (accurate account of fits is essential)
     a. A lesion in the motor cortex will cause convulsive twitching of part of the opposite side of the body. This is often followed by a temporary paralysis of the parts involved (Todd's paralysis)
     b. Causes tingling and numbness of the arm, leg, or face if in sensory cortex
     c. In the occipital cortex, flashes of light in the opposite visual field result
     d. In the temporal lobe, vivid scenes, curious sensations of smell or taste, and a feeling that all is very familiar—so-called déjà vu phenomenon—may occur

D. PSYCHOMOTOR EPILEPSY: abnormal discharge results in a sudden disturbance of behavior (rarely dangerous)
  1. Patient may start running around in a circle, or fighting, or just rearranging objects; the behavior stops as suddenly as it began, with knowledge of what has taken place
  2. Originates in temporal lobe and may be preceded by a hallucination of taste, smell, or vision

E. PETIT MAL ATTACKS have often been called minor epilepsy, but these attacks are not minor grand mal attacks, are quite different, and are always idiopathic
  1. Begin in childhood and adolescence; only occasionally persist into adult life
  2. Many attacks occur each day, sometimes even hundreds, consisting of momentary loss of touch with one's surroundings ("absence") so that the patient may pause in conversation; and there is a characteristic EEG

F. MYOCLONIC ATTACKS are a shocklike jerking of the muscles, which may happen in the arms in many epileptics, especially first thing in the morning, and sometimes occur without other signs of epilepsy

G. HYPSARRYTHYMIA is a name given to a gross EEG abnormality seen in infants subject to fits, especially the so-called salaam seizures in which the child throws his arms back and repeatedly bends his head and body forward

H. POSTTRAUMATIC EPILEPSY: follows a serious injury to the brain. All types of attack may be seen (except petit mal)
  1. Fits just after the injury, due to cerebral bruising, are relatively unimportant, but when they develop later, usually within the first 2 years, they are probably due to cerebral scarring and may recur for years

I. STATUS EPILEPTICUS: refers to attacks which usually occur at well-spaced intervals, but one fit may pass into the next so quickly that consciousness is not regained (may last for hours).
  1. The temperature rises to hyperpyrexial levels; unless vigorously treated, the patient may become exhausted and die
  2. An epileptic may develop this condition during an intercurrent illness, but it is a particular danger if treatment is suddenly stopped

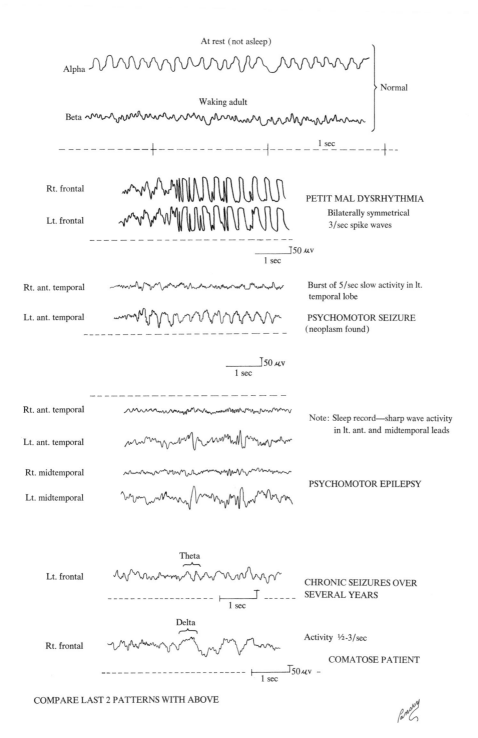

At rest (not asleep)

Alpha

Normal

Waking adult

Beta

1 sec

Rt. frontal

Lt. frontal

PETIT MAL DYSRHYTHMIA

Bilaterally symmetrical
3/sec spike waves

50 μv

1 sec

Rt. ant. temporal

Burst of 5/sec slow activity in lt.
temporal lobe

Lt. ant. temporal

PSYCHOMOTOR SEIZURE
(neoplasm found)

50 μv

1 sec

Rt. ant. temporal

Note: Sleep record—sharp wave activity
in lt. ant. and midtemporal leads

Lt. ant. temporal

Rt. midtemporal

Lt. midtemporal

PSYCHOMOTOR EPILEPSY

Theta

Lt. frontal

CHRONIC SEIZURES OVER
SEVERAL YEARS

1 sec

Delta

Rt. frontal

Activity ½-3/sec

COMATOSE PATIENT

50 μv

1 sec

COMPARE LAST 2 PATTERNS WITH ABOVE

CONGENITAL MALFORMATIONS

AND PERINATAL PATHOLOGY

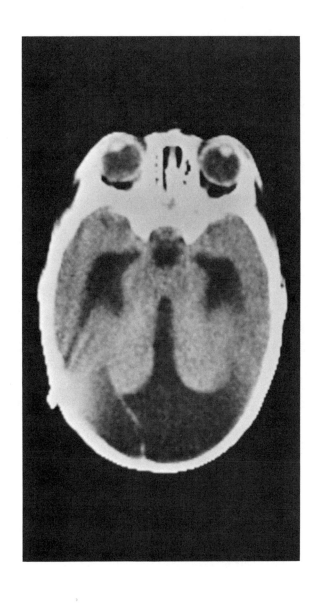

210. PATHOLOGY OF SPINAL CORD AND BRAIN

I. **Introduction:** the cause of abnormal development is not always clear. If the developing fetus is damaged, there is usually some abnormality at birth, but this abnormality depends on the site of damage and stage of development at which it takes place. Malformations of the nervous system may be due to 3 groups of factors

A. EXOGENOUS: such as infections, especially rubella and toxoplasmosis, radiation (x-rays, atomic bomb exposure), and chemical agents (especially various drugs)

B. GENETIC AND CHROMOSOMAL FACTORS: in particular, trisomy 21 or mongolism (Down's syndrome) and trisomy 13–15

C. AN INTERACTION OF THE ABOVE 2: the relatively late completion of brain development in fetal life accounts for the protracted effects of some teratogenic factors

II. **Disturbances as a result of defective closure of the neural groove (dysraphic state)** constitute the most frequent malformations of the nervous system. If there is defective closure of the neural groove in neural tube formation, various malformations will appear, depending on the location and extent of the closure defect. These malformations will be accompanied by developmental defects affecting the subjacent tissue planes (meninges, posterior arch of the vertebrae, dermis), whose development is subject to inductive influences from the neural tube

A. SPINAL CORD: malformations are most often localized to the lumbar and lumbosacral regions and are listed under the generic term of spina bifida

  1. Spina bifida aperta (myelaraphia) is a major form of dysraphia: the open neural groove is exposed to the surface, with its borders in continuity with adjacent epiblast; no leptomeninges are formed; the vertebral arch does not close; and the related skin does not develop

  2. Myelomeningocele is a lesser form of dysraphia: the neural tube has undergone closure but the posterior vertebral arch has not, and the meninges protrude directly under the skin

    a. Various nervous tissue elements, i.e., spinal cord parenchyma and nerve roots, are present in the sac thus formed, which is filled with CSF

    b. Myelomeningoceles are often associated with hydrocephalus due to malformations involving the medulla and cerebellar tonsils, which impede the passage of CSF at the level of the 4th ventricle, resulting in the Arnold-Chiari deformity

  3. Meningocele is essentially similar to (2), but the sac does not contain nervous tissue. Can occur in any part of the spine, but is most common at or just below the occipital region and in the lumbar region

  4. Congenital dermal sinus: the meninges are linked to the epidermal tissues by a narrow aperture across the incompletely closed posterior vertebral arch

  5. Spina bifida occulta is a very frequent malformation affecting one or more spines, shows a closure defect of posterior vertebral arches, and is sometimes accompanied by abnormalities in the overlying skin, i.e., tufts of hair

    a. It does not give rise to any clinical manifestations and is usually found incidentally on radiologic examination of spine ("occulta" means hidden)

    b. Occasionally there is failure of development of underlying nervous tissue, resulting in deformities of feet (pes cavus) or difficulties in controlling urination

B. BRAIN (see also Unit One)

  1. Anencephaly: failure of part of neural tube to close at level of encephalon

    a. There is a lack of development of meninges, cranial vault, skin, and brain

    b. This malformation is fatal within a few hours after birth

  2. Encephalocele: brain protrudes under the skin through bony defect in skull vault

  3. Meningocele: a meningeal sac filled with CSF protrudes under the skin through a bony defect in the skull vault

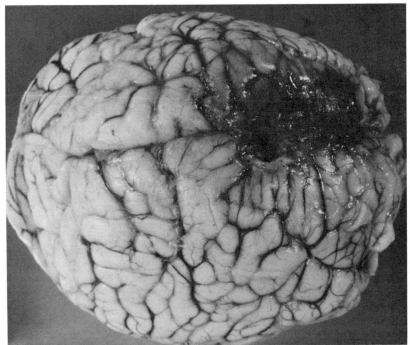

Presence of cysts in the brain cortex known as porencephaly.

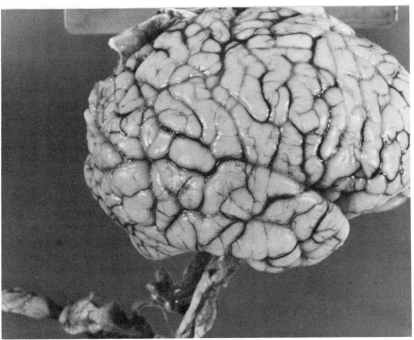

Arnold-Chiari syndrome (cerebellum and medulla oblongata protruded down into spinal cord).

211. AGENESES, DYSGENESES, AND CORTICAL ANOMALIES

I. Ageneses and dysgeneses

A. MICROENCEPHALY describes various destructive processes which occur before birth and result in a decrease in volume of the brain associated with other anomalies

   1. True microencephaly (small brain) is characterized by a decrease in the size of the cranial perimeter and is used only when referring to a brain of small size (less than 900 gm) in the adult, with no degenerative lesions

B. MEGALENCEPHALY may be seen in normal subjects as often as in mentally deficient subjects

   1. The increase in volume of a brain weighing more than 1800 gm may be diffuse or localized

   2. The brain may sometimes be structurally normal, but most often there is either a diffuse glial proliferation or a variable pathologic process

C. CYCLOCEPHALY: rare malformation affecting the tissues derived from the prosencephalon

   1. The telencephalon is not divided, and there is a single central ventricle which communicates with the third ventricle

   2. There is a single median eye (cyclopia), absence of the pituitary, and aplasia of the nasal structures

   3. Due to an induction defect of the prochordal plate as a result of a chromosomal abnormality

D. PORENCEPHALY consists of cerebral cavitation due to localized agenesis of the cortical mantle, leading to the formation of a cavity or a lateral slit through which the lateral ventricle communicates with the convexity

E. HYDRANENCEPHALY is generally the result of destructive processes

   1. It is similar to porencephaly and presents as a large intracerebral cavitation which may at first mimic ventricular dilatation; however, in contrast to hydrocephalus, the cavity is not lined by ependyma but by a glial border, and the thin external wall is formed by leptomeninges and by a thin layer of superficial cortex

   2. Generally involves the anterior part of the cerebral hemispheres

F. AGENESIS AND COMMISSURAL MALFORMATIONS include septal formations (septal agenesis and cysts) and agenesis of the corpus callosum

G. ARHINENCEPHALY may involve only the olfactory bulbs and is associated with other malformations, especially of the cerebellum

   1. Is typical of trisomy 13–15

   2. Indicates a severe disturbance in development of the fore-end of the telencephalon

H. CEREBRAL AND CEREBELLAR AGENESES are rare

II. Cortical anomalies are related to a disturbance in the maturation of the germinative layer around the 2nd month of fetal life

A. AGYRIA (lissencephaly) is the normal state up to 7 months of fetal life but remains with the absence of fissures and convolutions. May be total or involve only a part of the hemisphere

B. PACHYGYRIA: the convolutions are abnormally wide and thick, and the cortex has a 4-layer type of lamination

C. HETEROTOPIAS: seen as isolated phenomena in the shape of masses of gray matter in the centrum ovale near the caudate nucleus, in the cerebellar white matter, in the brainstem, and in the inferior olives. Associated with other malformations

D. MICROGYRIA AND POLYMICROGYRIA are characterized by the grouping of small convolutions that are reduced in size, are irregular, malformed, and with a verrucous appearance. Show large numbers of sulci, tend to involve cerebral and cerebellar cortex, and are frequently bilateral and symmetrical

MICROCEPHALUS

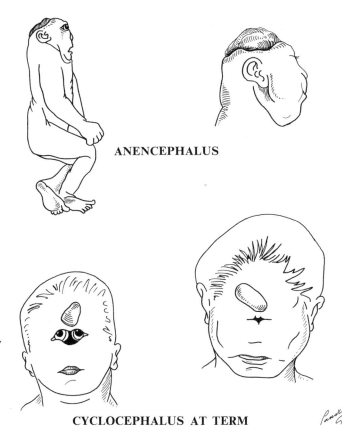

ANENCEPHALUS

Fused globe
Double retina, lens,
 iris, and pupil

CYCLOCEPHALUS AT TERM

212. ARACHNOID CYSTS, HYDROCEPHALUS, AND SYRINGOMYELIA

I. Arachnoid cyst refers to the formation of intracranial extraparenchymatous cysts which are lined internally by arachnoid tissue and contain clear fluid. They are most frequently seen in children and are sometimes associated with cerebral abnormalities

II. Hydrocephalus, strictly speaking, is an increase in cerebral mass due to the excessive amounts of CSF, but in practice, it corresponds to a variable degree of dilatation of the ventricular system. It is often spoken of as "water on the brain" since it is due to the presence of too much CSF at too high a pressure, resulting, in children, in the enlargement of the head since the bones are free to separate at the sutures. In the adult, this cannot happen since the sutures are fixed, resulting in raised intracranial pressure

A. TYPES: in either, the ventricles eventually enlarge
1. Obstructive noncommunicating: if the CSF, after being formed, cannot pass through the ventricular system due to an obstruction in the 3rd ventricle, aqueduct of Sylvius, or 4th ventricle, the lateral ventricles enlarge and "blow" up the brain like a balloon, stretching the cortex. Pressure is highest in the ventricles
2. Communicating: if CSF gets out of the ventricular system, but is blocked in the subarachnoid space and cannot circulate to reach the surface of the brain where it is absorbed, the pressure builds up outside the brain

B. HYDROCEPHALUS may be present at birth, resulting in mechanical difficulties

C. APPEARANCE: the head is large, the fontanelles remain wide open, the sutures can be felt separated, the forehead is bossed, and the eyes may be pushed downward

D. MENTAL RETARDATION is frequent, but high intelligence is sometimes possible

E. THE CONDITION MAY BE ARRESTED if a balance develops between formation of CSF and the defect in its absorption, and the pressure does not increase. Treatment should start before irreparable damage is done by stretching and thinning of the brain

III. Syringomyelia is the result of a cavitation of the spinal cord (syrinx), which extends over several metameric segments, most often involving the cervicothoracic segments. If it forms in the lower brainstem, it is called syringobulbia

A. DESCRIPTION
1. The spinal cord is widened and swollen at its level
2. The syringomyelic cavity, most often single, can be multiple, occupies the center of the cord, and is situated behind the ependyma, involving, in the midline, the crossing fibers of the ascending pain and temperature fiber tracts
3. It may involve the gray matter and fuse laterally with the entering sensory nerve roots or affect, to a variable extent, the anterior horns, resulting in lower motor neuron paralysis with amyotrophy
4. It may extend transversely into the white matter of the lateral and posterior columns, as well as caudally, but seldom involves the lumbar segments
5. It is limited by highly fibrillary glial tissue which often involves blood vessels

B. THEORIES OF DEVELOPMENT
1. The dysraphic theory states that there is a closure defect of the neural tube at the level of the posterior raphe
2. The hydrodynamic theory states that it is due to a disturbance in the outflow of CSF from the 4th ventricle as a result of failure of part of the foramina of Luschka and Magendie to open, resulting in hydrocephalus accompanied by herniation of the medulla and of the cerebellar tonsils (Chiari malformation). This is later followed, as a result of increased CSF pressure, by dilatation of the ependymal canal (hydromyelia). A syringomyelic cavity then might develop as a result of a rupture in the wall of the dilated ependymal canal

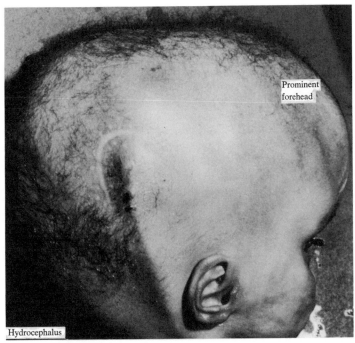

Prominent
forehead

Hydrocephalus

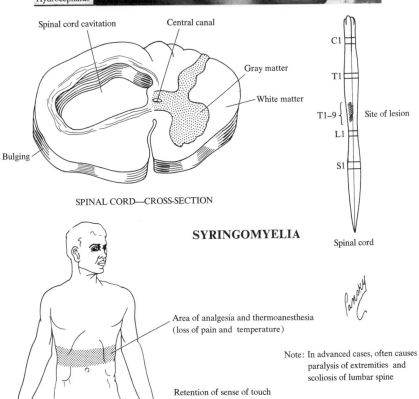

Spinal cord cavitation

Central canal

Gray matter

White matter

Bulging

SPINAL CORD—CROSS-SECTION

C1

T1

T1–9 { Site of lesion

L1

S1

Spinal cord

SYRINGOMYELIA

Area of analgesia and thermoanesthesia
(loss of pain and temperature)

Retention of sense of touch

Note: In advanced cases, often causes
paralysis of extremities and
scoliosis of lumbar spine

– 467 –

213. CRANIOSTENOSIS, BLASTOMATOUS DYSPLASIAS, AND PHAKOMATOSES

I. Craniostenosis: the skull sutures fuse prematurely, the fontanelles close very early, and the skull becomes a rigid box too soon, since the brain is still growing rapidly

A. DESCRIPTION
  1. The brain becomes compressed, pressure rises and is transmitted to the orbits, resulting in the eyes being pushed forward, even to a gross degree
  2. The skull develops a pointed shape (turricephaly or oxycephaly), and the fused sutures can be felt as hard ridges

B. SYMPTOMS
  1. The child may be mentally deficient and epileptic
  2. The child may have other abnormalities (congenital) such as webbed fingers and bronchiectasis

C. TREATMENT
  1. Attempts have been made to split the sutures or break up the skull bones (morcellation) to allow the brain to grow—not too successful
  2. Surgeons have tried placing tantalum foil between the sutures to prevent fusion, but success is limited and variable

II. Blastomatous dysplasias and phakomatoses: in the context of malformations and neuroectodermal dysplastic processes, there are a group of diseases that have in common an association of distinctive malformations of the neuraxis with small tumors (phakomas or lentil-like neoplasia) which involve the skin, the nervous system, or the eyes

A. VON RECKLINGHAUSEN'S NEUROFIBROMATOSIS is a common familial disorder showing
  1. Cutaneous lesions: café-au-lait spots and pigmentary nevi, pendulated tumors and nodular subcutaneous neurofibromas, and so-called royal tumors which may become very extensive
  2. Neural tumors most frequently involve nerve roots and may present as either neurofibromas or as schwannomas (or neurilemomas)
    a. They may involve the cranial nerves, particularly the VIIIth, giving rise to bilateral acoustic schwannomas
    b. Multiple meningiomas and a form of meningiomatosis are also characteristic
  3. Nervous system malformations consist of heterotopias, cortical dysplasias, stenosis of the aqueduct of Sylvius, or syringomyelia
    a. In association with bony malformations of the spine, they indicate the very dysplastic nature of this process

B. TUBEROUS SCLEROSIS (BOURNEVILLE'S DISEASE) is less common and shows an association of the following
  1. Skin lesions, including adenoma sebaceum, involving the skin of the nasolabial fold, angiofibromas, and cutaneous and periungual fibromas
  2. Cortical cerebral malformations: circumscribed nodules with voluminous cells, cerebral tumors of variable size, and visceral tumors (skeletal and cardiac muscle, endocrine, and pulmonary)

C. VON HIPPEL-LINDAU'S DISEASE (retinocerebellar angiomatosis) is familial and characterized by the presence of multiple hemangioblastomas, localized most often in the retina and cerebellum

D. ENCEPHALOTRIGEMINAL ANGIOMATOSIS OR STURGE-WEBER'S DISEASE: a flat, extensive, unilateral cutaneous angioma of the face and a leptomeningeal encephalic angiomatosis, most often in the parieto-occipital region

SCAPHOCEPHALY $\begin{cases} \text{craniosynostosis} \\ (\text{craniostenosis}) \end{cases}$

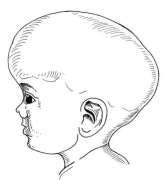

Premature closure of sagittal suture

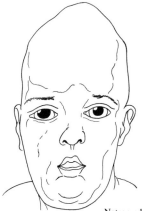

Note: wedge-shaped skull

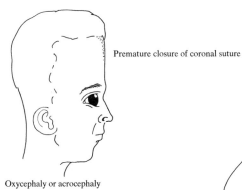

Premature closure of coronal suture

Oxycephaly or acrocephaly
"Tower-like skull"

Note: skull asymmetry

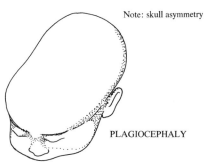

PLAGIOCEPHALY

Premature closure of coronal and lambdoid sutures

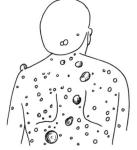

NEUROFIBROMATOSIS

Fibrotic tumors of various sizes: von Recklinghausen's disease

– 469 –

214. PERINATAL AND POSTNATAL PATHOLOGY

I. Perinatal pathology

A. INTRODUCTION: at birth and during the 1st months of life, various etiologic factors may result in definite cerebral lesions and account for various neurologic syndromes. The most common are
   1. Infantile cerebral hemiplegia
   2. Infantile cerebral diplegia or Little's disease
   3. Some forms of congenital epilepsy

B. THE LESIONS ARE TYPICAL of the conditions, but their supervention on the immature brain gives rise to a highly distinct pathologic picture. These lesions are obviously more frequent and more severe in premature infants. One separates the cerebral damage related to the birth process itself (neonatal pathology) from those that are secondary to disturbances in the 1st months of life (postnatal), even though the sequelae are often identical and the lesions of the same type
   1. Neonatal pathology: the pathogenic mechanism consists of
      a. Traumatic mechanism: obstetric injury results in cerebral lesions that are direct or indirect, the effects being usually hemorrhagic as a consequence of blood vessel rupture (often venous)
      b. Circulatory mechanism: either venous and related to increase of intracerebral venous pressure at the time of birth, or arterial and secondary to a fall of blood flow resulting from circulatory arrest
      c. Anoxic and asphyxial mechanisms
   2. The lesions may be acute, leading to death, or cicatrical, resulting in a picture of secondary encephalopathy of neonatal origin
      a. Acute lesions: essentially hemorrhagic, such as subdural hematomas, subarachnoid hemorrhage, and intracerebral hemorrhage
      b. Secondary lesions: responsible for the group of infantile encephalopathies, the sequel of hemorrhage or necrotic lesions
         i. Extensive lobar or hemispheric atrophies suggest vascular occlusion
         ii. Cortical changes: in which ulegyria is the most common appearance, consisting of cortical atrophy involving deeper parts of the cortex of the cerebral gyrus, with relative sparing of its convexity. Also involves the white matter and basal ganglia

II. Postnatal pathology: the cerebral lesions may be secondary to numerous etiologic and pathologic mechanisms

A. EARLY VASCULAR ENCEPHALOPATHIES: some are related to venous lesions, such as thrombosis of the vein of Galen or of the superior longitudinal sinus. Are often secondary to early meningeal infections
   1. Others are related to arterial occlusions whose emboli or thrombic origin is hard to prove. May produce a picture of lobar or hemispheric atrophy

B. POSTEPILEPTIC ENCEPHALOPATHY is characterized by gliosis with atrophy of Ammon's horn, secondary episodes of anoxia due to repeated epileptic fits, or status epilepticus
   1. Believed by some as being of circulatory origin, due to ischemia of the territory of the posterior cerebral artery at the time of birth and as being the cause of the epileptic fits

C. NUCLEAR JAUNDICE: result of hemolytic neonatal anemia due to Rh incompatibility
   1. The lesions are characterized by their topographic distribution, involving the globus pallidus, the subthalamic nuclei, and to a lesser extent, Ammon's horns, the dentate nuclei, and the inferior olives
   2. Their yellow appearance is due to bilirubin

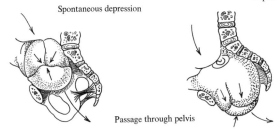

Spontaneous depression

Depressed parietal fracture in newborn

Passage through pelvis

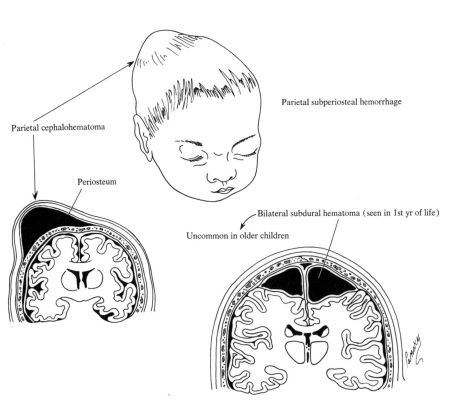

Parietal cephalohematoma

Periosteum

Parietal subperiosteal hemorrhage

Bilateral subdural hematoma (seen in 1st yr of life)

Uncommon in older children

HEAD INJURY SEQUELAE

Scalp: Contusion, hematoma; open wound

Skull: Linear fracture, diastatic fracture, other simple fracture, CSF otorrhea
Depressed fracture, growing fracture, compound fracture, pneumocephalus, fistula

Meninges: Subarachnoid hemorrhage
Epidural hematoma, acute and chronic subdural hematoma, subdural hygroma

Cranial nn.: Paralysis I–XII

Brain: Concussion, edema, contusion, laceration, hematoma, penetrating wound

Unit Fifteen

DEMYELINATING, HEREDITARY,
AND FAMILIAL DISEASES

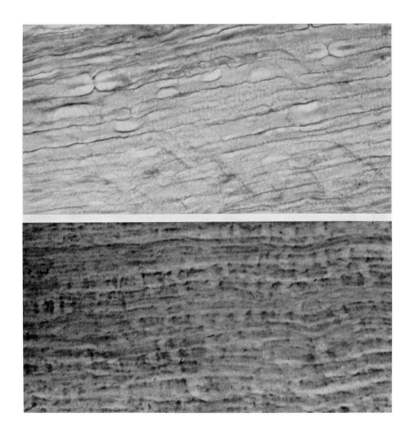

215. DEMYELINATING DISEASES (OF WHITE MATTER) AND MULTIPLE SCLEROSIS

I. Introduction: myelin forms the sheaths that cover the majority of the nerve fibers, whether they lie in the brain and cord or run in the cranial and spinal nerves. When myelinated fibers are collected together, they appear white and form the white matter of the brain and cord. Any damage to nervous tissue can cause the myelin sheaths to degenerate. Nevertheless, there are certain diseases in which this occurs spontaneously without a known cause. Loss of myelin is called demyelination. Without its myelin, the nerve fiber does not conduct impulses properly, and there is a loss of function in that part being supplied. Demyelination is one feature of massive white matter lesions such as cerebral infarcts, hemorrhages, and cerebral tumors, or it may be secondary to degeneration.

If only the myelin sheath is damaged (often the axis cylinder is spared), it can re-form so that function can be restored to the nerve fiber. If the nerve fiber is also destroyed, recovery is less likely. Nerve fibers can grow again, as is often the case in peripheral nerve injuries, but often scar tissue forms and prevents proper growth and recovery. Scar tissue formation in the CNS is formed by the glial cells and is called gliosis. In a healing process, the scar tissue gradually shrinks, but here, too, it may involve nearby healthy tissue resulting in a slowly progressive disability.

Diseases of the white matter result from pathologic processes dissimilar to each other. Some, by virtue of their genetic character, belong to the leukodystrophies and show metabolic changes in the myelin sheaths, while others, like multiple sclerosis, have inflammatory features.

II. Classification of demyelinating disease: 2 types are distinguished in etiologic and histo-pathologic views

A. ETIOLOGIC
1. Multiple sclerosis and Schilder's disease are the result of myeloclastic processes. The etiology is unknown, but an immunopathologic process is suspected
2. The leukodystrophies are familial and are enzymatic disorders of genetic origin (see Genetic and Familial Diseases)

B. HISTOPATHOLOGIC: are distinguished according to whether or not they show inflammatory cellular lesions or not (these may, however, be seen only in the early stages of the disease and disappear in the course of the process)

III. Disseminated or multiple sclerosis: the most common form of demyelinating disease

A. SCLEROSIS means "hardening." At autopsy, hard areas of gliosis are found scattered or disseminated throughout the CNS
1. The cause of the disease remains unknown (probably immunologic or viral)
2. Familial incidence of the disease is not uncommon
3. Starts in young adults, occasionally in adolescence or young children, infrequent in middle life, and rarely begins over the age of 50
4. Women are slightly more often affected than men; country dwellers more than town dwellers
5. It is a disease of temperate climates (exceedingly rare in the tropics)
6. Patches of demyelination, called plaques, develop in some part of the cord, cerebellum, brainstem, cerebral hemispheres, or optic pathways
   a. The patches flare up, take weeks or months to settle down, and after an interval of months or years, flare up again in another site. After several attacks, the healing process of gliosis upsets all attempts at remyelination, and permanent sclerosis and function loss result
7. The peripheral nerves are not affected
8. Inflammatory cellular lesions are seen in recent lesions but are absent in old ones
9. Coexistence of lesions of different ages are seen at various levels of the neuraxis

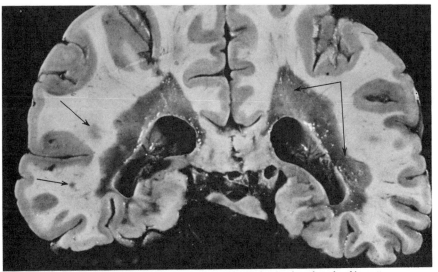

Gross section through brain showing patches and large plaques of demyelination throughout the white matter.

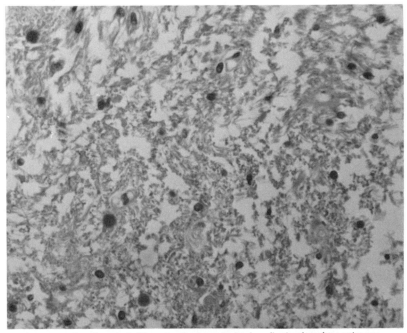

Histologic section through cerebral white matter showing extensive demyelination throughout entire area.

216. SCHILDER'S DISEASE, RETROBULBAR (OPTIC) NEURITIS, AND MYELITIS

I. Schilder's disease, from a histologic point of view, consists of lesions whose structure is like multiple sclerosis

A. DISTINCTION

1. By its more frequent occurrence in children and its continuous evolution
2. Characteristic distribution of lesions
   a. Large areas of bilateral and symmetrical demyelination in the cerebral hemispheres involving the corpus callosum—"butterfly appearance"
   b. Predominantly in the parieto-occipital regions, sparing the gray matter, and involving the optic pathways
   c. In some cases, endocrine disturbances are seen, especially adrenal lesions with cutaneous pigmentation
3. Results in acute onset of fits, rapidly developing blindness, and often spastic paralysis
4. The entire picture is of a progressively slow total disability
5. Several varieties are seen with slight clinical differences

II. Retrobulbar neuritis or optic neuritis: a condition usually due to a patch of demyelination developing in one optic nerve and most commonly in an early stage of disseminated sclerosis. Later, the other nerve may be affected, occasionally both at the same time

A. SYMPTOMS

1. Pain develops in one eye, and it hurts to move it or even press on it
2. Vision very quickly becomes blurred, particularly its central part, and this may progress to a total loss of vision. One may recover after a few days or within weeks or months, and vision may return to normal
3. In some cases, a little area of permanently impaired vision remains at the center of the field, which may interfere greatly with reading, writing, etc.

B. SIGNS

1. If the patch develops in the nerve just behind the globe (retrobulbar), mild swelling of the optic disk (papillitis) can be seen
2. After the acute stage, some nerve fibers atrophy and the disk looks pale or even dead white (optic atrophy)
3. If the patch develops farther back along the nerve, there may be no swelling, but the atrophy can still develop later
4. By charting the visual fields, the blind area is seen to be a central scotoma

III. Transverse myelitis: a plaque of myelination may extend across the spinal cord during one of the relapses of disseminated sclerosis, producing a transverse cord lesion. This may also occur without any evidence of disseminated sclerosis. The lesion is usually found in the thoracic region and its cause is unknown

A. SYMPTOMS

1. Preliminary period of pain around the lower thoracic or abdominal regions, which has been mistaken for pleurisy, cholecystitis, or appendicitis
2. Rapid weakness and loss of sensation in the legs, with urine retention

IV. Bilateral retrobulbar neuritis and transverse myelitis: seen in a condition also called neuromyelitis (Devic's disease), which is a rare condition. It is probably a variety of disseminated sclerosis

A. SIGNS AND SYMPTOMS

1. Patients develop retrobulbar neuritis in both eyes (not necessarily together) followed by a transverse myelitis

TRANSVERSE MYELITIS

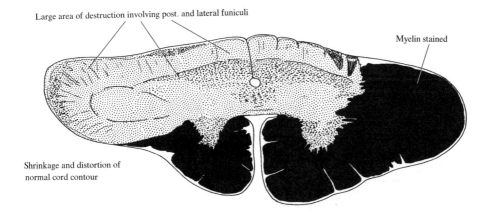

Large area of destruction involving post. and lateral funiculi

Myelin stained

Shrinkage and distortion of
normal cord contour

SPINAL CORD (CROSS-SECTION)

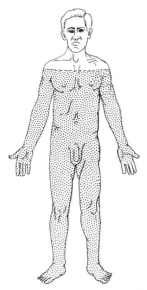

Absence of neurologic function below level
of damaged spinal cord

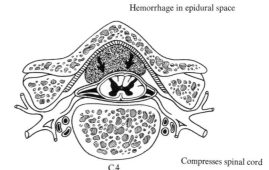

Hemorrhage in epidural space

Compresses spinal cord

C 4

Complete transverse myelitis due to compression
of cord by hemorrhage

217. HEREDOFAMILIAL ATAXIAS

I. **Introduction:** a hereditary disease is passed on from generation to generation. Familial disease affects several members of the same family. The term "heredofamilial" is common; they are diseases of the nervous system which can be traced in several relatives, though the relationship is sometimes a distant one. Most neurologists use the term to refer to the group of uncommon diseases that are degenerative in type and tend to produce abnormalities either of coordination or of the peripheral nerves and thus are classified under the heading of spinal medullary degenerative diseases

II. **Heredofamilial ataxias**
A. GENERAL FEATURES
   1. Group of diseases in which family members develop a progressive ataxia
   2. In any family, the symptoms and signs are much the same, but in different families they vary as to age of onset and the part of the system affected
   3. On the basis of the above, they are given names, usually after the physician who described them, e.g., Marie's ataxia, Charcot-Marie-Tooth disease.
   4. Differences are not of practical importance; certain features are common to all
B. SYMPTOMS
   1. Increasing clumsiness and unsteadiness of the arms and legs
   2. There may be defective vision and difficulty in speech
   3. No headache, vomiting, vertigo, or diplopia, as in posterior fossa tumors
   4. Condition progresses steadily and not by remissions and relapses (as seen in disseminated sclerosis)
C. SIGNS
   1. Ataxia is the major sign, as a result of cerebellar disease
   2. Slurring of speech, nystagmus, and unsteadiness of arms, legs, and gait are often seen
   3. Tendon reflexes are usually absent, but as the pyramidal tracts are affected; in some cases, there may be extensor plantar responses
   4. Pes cavus is common and often found in relatives who are not otherwise affected
   5. Optic atrophy is often seen, but patients rarely complain of visual loss
D. SPECIAL VARIETIES
   1. Hereditary degenerative spinocerebellar disease (Friedreich's ataxia)
      a. The most frequent and best characterized of these diseases
      b. Involves the spinocerebellar tracts, Clarke's column, and posterior columns. Neuronal cell loss from some cranial nerve nuclei (XII, XI, VIII) is frequent and in some cases optic atrophy has been seen
      c. Develops in adolescence
      d. Shows a marked loss of position and vibratory sense, with a postural ataxia due to posterior column degeneration
      e. Weak back muscles and severe scoliosis are common
   2. Olivopontocerebellar degeneration develops in middle life
      a. Dysarthria is a major symptom, as is cerebellar ataxia, and there are spasticity and exaggeration of reflexes
      b. Resembles disseminated sclerosis, but it advances without remissions, is familial, and dysarthria is outstanding
   3. Hereditary spasmotic (spastic) paraplegia (Strumpell-Lorraine disease) is rare
      a. Lesion consists of bilateral atrophy of the pyramidal tracts throughout the length of the cord associated with involvement of the posterior columns and spinocerebellar tracts, but is distinct because there are usually no cerebellar signs and no true ataxia, though reflexes are exaggerated

FRIEDREICH'S ATAXIA

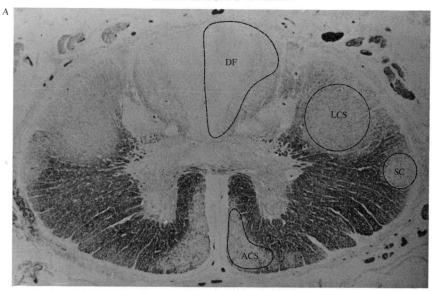

Loss of myelin in dorsal funiculus (*DF*), spinocerebellar (*SC*), lateral corticospinal (*LCS*), and anterior corticospinal (*ACS*) tracts.

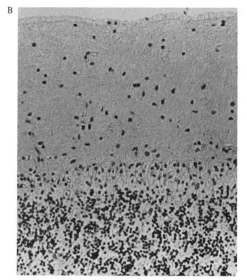

Olivopontocerebellar degeneration as evidenced by the absence of a Purkinje cell layer in the cerebellar cortex. Compare with normal cerebellar cortex shown in *C*.

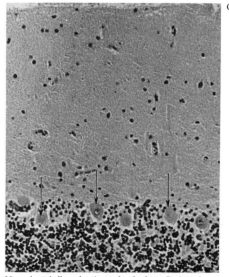

Normal cerebellum showing molecular layer, Purkinje layer (arrows) and granular layer.

218. HEREDITARY NEUROPATHIES

I. **The hereditary neuropathies:** rare diseases due to peripheral nerve degeneration mainly in motor nerves in some cases and in sensory nerves in other cases

A. RADICULOSPINAL DEGENERATIVE DISORDERS

1. Peroneal muscular atrophy (Charcot-Marie-Tooth disease)
   a. Involves both lower motor and peripheral sensory neurons
   b. This is a neuropathy, not purely a degeneration of muscle
   c. It affects either sex and develops in adolescence or early adult life
   d. Symptoms: walking becomes increasingly difficult because the legs have to be lifted high to avoid catching the toes; wasting is first seen in the legs and later in the hands and forearms
   e. Signs
      i. There is a high stepping gait due to bilateral foot drop, each leg being slapped down in turn. There is also pes cavus
      ii. Hands and feet are cold and blue; little glove-and-stocking sensory loss
      iii. All tendon reflexes are absent, and vibration sense is impaired
      iv. Wasting is the most characteristic feature because, though very marked, it stops suddenly halfway up the thighs and forearms—shoulder girdles, upper arms, hip girdles, and upper thighs are normal—giving the extremities the appearance of "inverted bottles"
      v. Condition progresses slowly without incapacity until late in life
2. Hypertrophic polyneuritis has all the features of a polyneuritis but progresses slowly over many years. May resemble (1) but there is great thickening of the peripheral nerves which can be felt at the elbows, below knees, and top of feet
3. Hereditary sensory neuropathy (of Denny-Brown)
   a. Associated with picture of acropathia ulceromutilans, is familial, and involves peripheral sensory neurons. Lesions implicate the spinal ganglia, dorsal nerve roots, and posterior columns
   b. Sensation is severely impaired so that painless sores develop on fingers and toes which may need amputation or may even drop off by themselves
4. Spinal hereditary amyotrophy (Werdnig-Hoffmann disease)
   a. Lesions involve anterior horns of cord and nuclei of certain cranial nerves (VII, XII). Ventral roots are atrophied and demyelinated but posterior columns are not involved
   b. Appears in infancy and one sees a progressive spinal muscle atrophy
   c. May be fatal because of involvement of respiratory muscles. Death usually occurs by the 5th year of life
5. Chronic anterior poliomyelitis produces a degenerative lesion which evolves slowly in the adult, and changes are caused by massive neuronal cell loss and degeneration of ventral nerve roots. May be hard to tell from amyotrophic lateral sclerosis

B. AMYOTROPHIC LATERAL SCLEROSIS (CHARCOT'S DISEASE) produces degenerative lesions of the pyramidal tracts associated with lesions of the lower motor neurons, atrophy of anterior horn cells, and nerves XII, XI, and X

1. Is a chronic, progressive disease of unknown cause in which there is degeneration of both upper and lower motor neurons
2. More common in males (4:1) and appears in 4th and 5th decades
3. There are weakness, wasting, and fine tremors in the thenar, hypothenar, and interossei muscles of the hands; then arms and shoulders are involved. Spastic paraplegia, increased deep reflexes, and (+) Babinski are seen. Death is due to bulbar paralysis within 3 years of onset

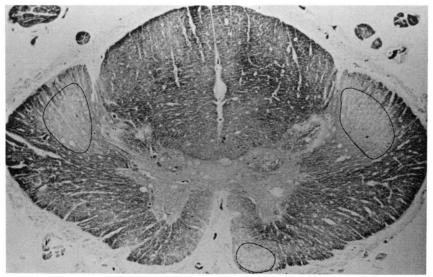

Transverse section through spinal cord showing degeneration of lateral corticospinal tracts on both sides and the anterior corticospinal tract on one side.

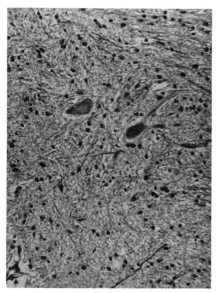

Anterior gray horn from amyotrophic lateral sclerosis patient showing loss of neurons.

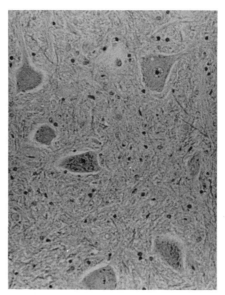

Anterior horn cells from normal anterior gray horn.

AMYOTROPHIC LATERAL SCLEROSIS

219. LEUKODYSTROPHIES

I. **General comments:** it is not uncommon to find examples of more than one of these diseases in the same family and, in fact, there may even be two types in the patient. The diseases in some way may all be linked together. It is very common, on examining relatives, to find unmistakable signs in someone who is entirely symptom-free (called "formes frustes" of the disease). All the diseases progress slowly and no drug therapy is available. Physiotherapy aimed at educating coordination of movements can be helpful but, sooner or later, the disability becomes severe, even though comparatively rarely are the patients bedridden

II. **The leukodystrophies** are diseases affecting children, show a lack of inflammatory cellular lesions and a frequency of genetic factors (familial). Most of them are, apparently, largely the result of a disturbance of myelin synthesis and are often classified under the dysmyelinating diseases
   A. DEGENERATIVE CHANGES take place, particularly in the white matter of the brain, with frequent involvement of the cerebellum
   B. THERE ARE FREQUENT IRREGULAR CHARACTER OF DEMYELINATION and sparing of U fibers
   C. THE PRESENCE OF MACROPHAGIC REACTION whose distribution is diffuse and not perivascular
   D. DISTINCT TYPE OF MYELIN CATABOLISM with formation of nonsudanophilic and metachromatic prelipids
   E. MOST OF THESE DISEASES are the result of a known disturbance of myelin metabolism (see also Metachromatic Leukodystrophy and Krabbe's Disease), and others are of uncertain origin. Thus, the term "dysmyelinating disease" has been used
   F. THE LEUKODYSTROPHIES appear early in childhood and may cause successive children in the family to become increasingly backward, with a loss of intellect, have ataxia, and develop fits and weakness of the limbs which progress to complete paralysis
   G. METACHROMATIC LEUKODYSTROPHY is the best known of the group
      1. Can be diagnosed in life by finding substances called sulphatides in the urine or in biopsies of the kidneys, brain, or peripheral nerves
      2. There is no known method of interfering with the disease process
      3. Others in this group are less common, all are rare, and many are not diagnosed until after death
   H. SUDANOPHILIC LEUKODYSTROPHIES encompass a number of disorders grouped together because of the absence of inflammatory features and the sudanophilic character of the lipid found in the white matter
   I. DIFFUSE OR SIMPLE TYPE ORTHOCHROMATIC LEUKODYSTROPHY: comparable to Schilder's disease. No inflammatory cellular infiltrates but frequent axonal lesions
   J. TIGROID FORMS (E.G., PELIZAEUS-MERZBACHER'S DISEASE): seen very early in life and affects only male children. Has a prolonged clinical course
   K. MASSIVE CONGENITAL FORMS: virtually total absence of central myelin with normal cranial and spinal nerves
   L. SPONGY DEGENERATION OF NEURAXIS (CARAVAN'S DISEASE): spongy aspect of subcortical layers of cerebral and cerebellar white matter

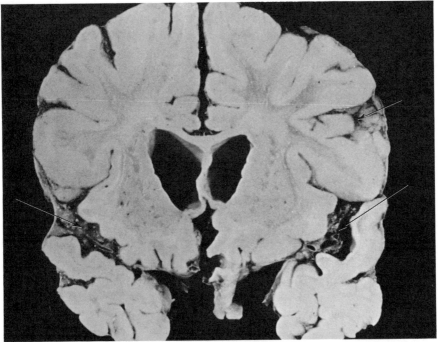

Coronal section through brain presenting with enlarged ventricles and sulci, narrow gyri, and corpus callosum and grayish-white matter.

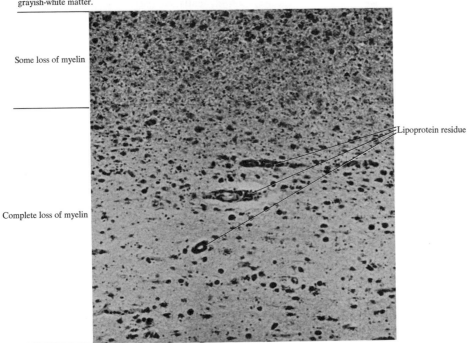

Some loss of myelin

Lipoprotein residue

Complete loss of myelin

Section through grayish-white matter shown in gross brain.

LEUKODYSTROPHY

NEUROLOGIC DISTURBANCES WITH SKELETAL ABNORMALITIES; DEGENERATIVE DISEASES

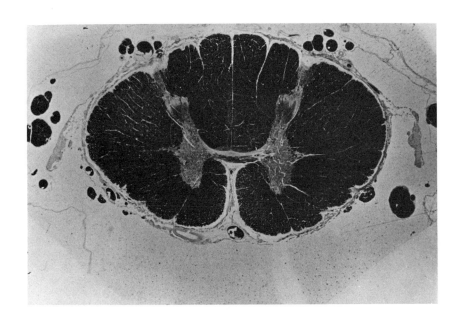

220. INTERVERTEBRAL DISK DISEASE

I. Introduction: since the skull and spine closely encase the brain and spinal cord, many peripheral nerves run very close to bones, joints, and ligaments. Thus, skeletal abnormalities may involve nerve tissue. Certain commonly recurring conditions are seen in both neurologic and orthopedic departments

II. Intervertebral disk disease: the disks are ''cushions'' of cartilage separating the vertebral bodies and are held in place by a fibrous ring, the anulus fibrosus, which joins the bodies of the vertebrae together

A. ACUTE DISK PROLAPSE: if the anulus ruptures, the disk will bulge through the gap (prolapse)

   1. In the cervical region, will compress the cord, causing an acute quadriplegia
     a. This is uncommon and the protrusion more likely will be to the side and compress a nerve root causing severe pain down the arm
     b. Requires urgent surgery

   2. Thoracic disk prolapse: since there is little room here, it usually compresses the cord and results in paraplegia
     a. Requires urgent surgery

   3. Lumbar disk prolapse is very common
     a. It misses the lower end of the cord usually and presses on the lumbar and sacral nerves, resulting in sciatica
       i. Results in pain in the back, passing down the back of the thigh and either over the front of the lower leg toward the big toe (L4, L5 lesions) or down the calf to the little toe (L5, S1 lesions)
       ii. Muscle spasms make the back rigid, scoliosis develops, and the straight-leg-raising test (Lasègue's sign) is limited to the affected side
     b. 95% of lumbar disk lesions settle down after complete bed rest for about 4–5 weeks. With severe symptoms, about 5% require removal
     c. Massive prolapse can compress the entire cauda equina resulting in flaccid paralysis with urinary retention and requires urgent surgery

B. CERVICAL SPONDYLOSIS

   1. X-rays of the necks of middle-aged and older patients have shown degenerative changes in the disks with overgrowth of bone (osteophytes) forming hard bars
   2. The bars compress nerve roots, thus these spondylitic changes are responsible for most cases of "fibrositis," "rheumatism," or "neuritis" in the arms and shoulders
     a. The bars may press on the blood vessels supplying the anterior horn cells and pyramidal tracts in the cord which results in a wasting and numbness in the arms with progressive spastic paraparesis

   3. Treatment
     a. Immobilization of the neck in a well-fitting collar for many months
     b. Neck manipulation is highly dangerous
     c. Few are suitable for surgery
     d. Immobilization can be made permanent with a bone graft

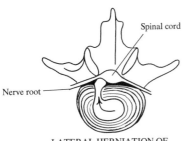

LATERAL HERNIATION OF
INTERVERTEBRAL DISK

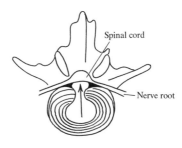

CENTRAL HERNIATION OF
INTERVERTEBRAL DISK

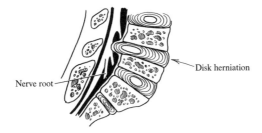

INTERVERTEBRAL DISK
PROLAPSE

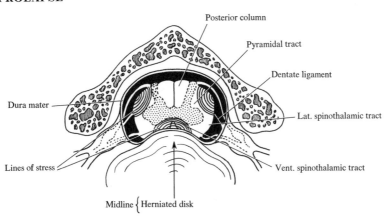

Note: Strain greatest ventral (not clinically tested); secondary
stress on pyramidal tracts—hands usually spared since
leg area is most lateral

221. CARPAL TUNNEL SYNDROME AND PRESSURE NEUROPATHIES

I. The carpal tunnel syndrome

A. As THE MEDIAN NERVE enters the wrist to supply the muscles of the thumb and give sensation to the thumb and first two fingers, it passes through the carpal tunnel where soft tissue and fat may become swollen and thickened and compress the nerve
   1. Carpal tunnel compression is common in middle-aged women resulting in severe pain, burning, and tingling of the hand, especially at night
   2. Muscles may waste after a long time
B. Treatment
   1. Splinting the wrist during the night offers some relief
   2. Injections of ACTH or local infiltration of hydrocortisone may help
   3. To cure, the tunnel must be opened surgically and the pressure relieved

II. The thoracic outlet syndrome: as the nerves pass through the root of the neck to reach the arms, they are compressed by tight bands of muscle in the region. These patients, predominantly women, after carrying heavy weights or after an increase in body weight with a drooping of the shoulders, develop tingling in the entire arm

A. TREATMENT: physiotherapy to correct posture and division of the muscle bands

III. Cervical ribs are congenital anomalies consisting of small ribs attached to the 7th cervical vertebrae and, as a result of their position, they raise and stretch the nerves that run over the 1st rib

A. THE CONDITION results in wasting of the small muscles of the hand and loss of sensations over the little and ring fingers and part of the forearm
B. TREATMENT: if symptoms are severe, the small ribs can be removed surgically

IV. Pressure neuropathies: certain nerves, especially the ulnar at the elbow and the popliteal outside the knee, lie superficial and can be compressed by resting on them for periods of time (result of bedrest, application of a poorly padded cast, or even use of adhesives for skin traction)

A. THE HAND CAN WASTE AND SENSATION IS LOST over the ring and little fingers, if the ulnar nerve is involved
B. IN THE FOOT, if the popliteal nerve is involved, it can lead to footdrop
C. TREATMENT: adequate protection for the nerves involved

V. Bony abnormalities at the base of the skull

A. THE SKULL BASE AND POSTERIOR FOSSA may be abnormally flat (platybasia)
   1. Associated with basilar invagination or basilar impression, resulting in the first cervical vertebrae being almost inside the posterior fossa so that parts of the cerebellum (cerebellar tonsils) are squeezed down through the foramen magnum
   2. These patients usually have short necks and low hairlines and become increasingly ataxic
B. THE BONY ABNORMALITY may be congenital or as a result of bone softening due to Paget's disease
C. BASILAR IMPRESSIONS are also associated with Arnold-Chiari deformity and syringomyelia

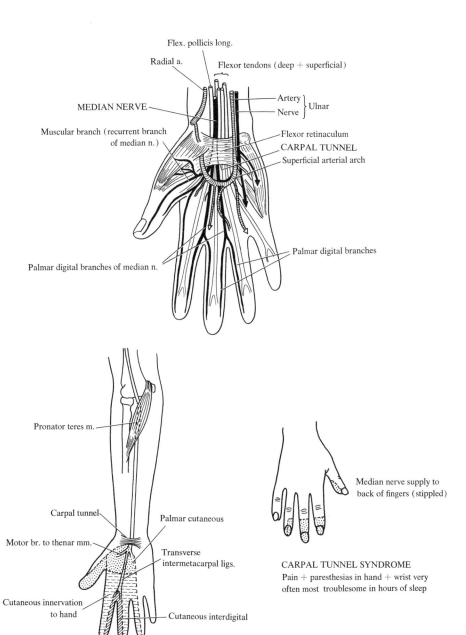

Flex. pollicis long.

Radial a.

Flexor tendons (deep + superficial)

MEDIAN NERVE

Artery
Nerve } Ulnar

Muscular branch (recurrent branch of median n.)

Flexor retinaculum

CARPAL TUNNEL

Superficial arterial arch

Palmar digital branches of median n.

Palmar digital branches

Pronator teres m.

Carpal tunnel

Palmar cutaneous

Motor br. to thenar mm.

Transverse intermetacarpal ligs.

Cutaneous innervation to hand

Cutaneous interdigital

SITES OF MEDIAN NERVE COMPRESSION

Median nerve supply to back of fingers (stippled)

CARPAL TUNNEL SYNDROME

Pain + paresthesias in hand + wrist very often most troublesome in hours of sleep

222. DEGENERATIVE CORTICAL DISEASE

I. Introduction: when the cells and fibers cease to function, shrink, and gradually disappear, the process is vaguely called degeneration. The degenerative diseases of the CNS form a heterogeneous group of disorders in which changes are mainly neuronal and occur sporadically, independent of any known inflammatory, toxic, or metabolic factor. These changes can occur as a result of ischemia from the narrowing of blood vessels in arteriosclerosis, when all other systems are comparatively normal; yet certain portions of the brain and cord die. Genetic factors determine a large number of these conditions, which frequently have a familial character, but the pathologic mechanisms are unknown. The degenerative process may be diffuse or it may select particular parts of the nervous system, e.g., cerebral cortex (giving rise to demential disorders), basal ganglia (causing disturbances in which the symptomatology is extrapyramidal), or a number of spinocerebellar systems (seen in the group of hereditary degenerative diseases). The symptoms and signs, when pooled, form a syndrome that is recognizable, e.g., parkinsonism, motor neuron disease, or the cerebral atrophies. The diseases generally occur late in adult life and often in the presenile age group. Memory is most noticeably disturbed, at first for recent events, but gradually all memory is lost

II. Degenerative cortical disease: is the most frequent neuropathologic cause of organic dementia. Senile dementia and Alzheimer's disease account for about 80% of cases in this group

A. CEREBRAL (CORTICAL) ATROPHY: shrinkage of brain tissue may be result of injury, inflammation, or interference with blood supply, but cells and fibers can atrophy for reasons unknown. The gyri thin out, sulci widen, and ventricles enlarge

1. Senile dementia causes discrete cerebral atrophy, symmetric thinning of frontal lobe cortex, and a moderate degree of ventricular dilatation. Lesions include
   a. Senile plaques (seen mainly in the superficial cortical layers)
   b. Neuronal cell changes: neurofibrillary degeneration and tangles, granulovacuolar degeneration, excess lipopigments, and absence of Nissl bodies
   c. Vascular changes: thickening of the capillaries and terminal arteriolar vessel walls, deep arteriolar hyalinosis, and atherosclerosis of arteries

2. Alzheimer's disease: causes a striking cerebral atrophy and diffusely affects all the cerebral convolutions. Parieto-occipital gross atrophy may account for early symptoms of aphasia, apraxia, and agnosia with dementia. Lesions are those of senile dementia, including senile plaques, and neurofibrillary and granulovacuolar degeneration, but no vascular changes are noted.
   a. Occurs earlier than senile dementia; even seen in young adults and often on a familial basis. Neuronal cell loss is considerable. Alzheimer's and Pick's diseases have been called presenile dementias
   b. Investigators have recently demonstrated an abnormal concentration of a microtubule-associated protein (tau) in the neuron cell bodies of Alzheimer's patients. Normally, this protein is synthesized in the body and is transported to the axon in high concentrations

3. Pick's disease: is frequently hereditary and presents with circumscribed frontotemporal cerebral atrophy which spares the posterior $\frac{1}{3}$ of the superior temporal gyrus. This accounts for the rarity of aphasic phenomena. Parietal cortex is seldom involved; occipital cortex never
   a. There are neuronal losses associated with dense astrocytic gliosis, white matter lesions, and basal ganglia changes of the caudate nucleus head.

ALZHEIMER'S DISEASE

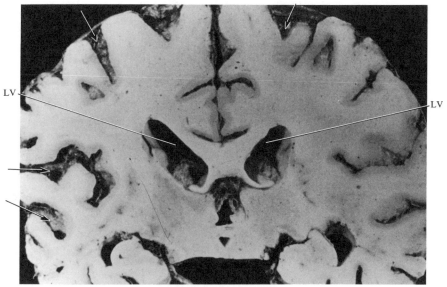

Coronal section showing narrowing of gyri and widening of sulci (arrows).
Note also the enlarged lateral ventricles (LV).

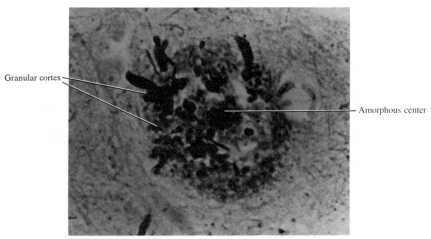

A senile plaque found between neurons in a patient with senile dementia—
Alzheimer's disease complex.

223. PARKINSONISM AND HUNTINGTON'S CHOREA

I. Subcortical degenerative diseases involve the basal ganglia to a variable degree and account for many of the extrapyramidal syndromes

A. PARKINSONISM: In addition to Parkinson's disease proper, which is of degenerative origin, a number of processes of different etiology involve identical structures, resulting in a series of related parkinsonian syndromes

  1. Parkinson's disease, once called paralysis agitans or shaking palsy, is the most frequent of the subcortical degenerative disorders
     a. Lesions involve pigmented nuclei of the brainstem, especially the substantia nigra (origin of dopaminergic pathway) and nuclei of the upper brainstem. Sympathetic ganglia are often involved
  2. Nondegenerative parkinsonian syndrome: may show lesions in similar structures
     a. Postencephalitic parkinsonian syndrome develops years after an encephalitis
        i. Is diffuse, affects younger people, mainly unilateral, likely to produce mental changes, and chiefly involves the substantia nigra
        ii. May cause attacks called oculogyric crises: eyes (sometimes head) are forced to turn upward for minutes at a time; no loss of consciousness
        iii. History of encephalitis or a serious illness is prevalent
     b. Arteriosclerotic parkinsonism: due to defective blood supply to basal nuclei
     c. Syphilitic parkinsonism: is a rare complication of syphilitic infection and has seldom been systemically investigated from a pathologic viewpoint
     d. Posttraumatic parkinsonism
     e. Vascular origin of parkinsonism is only exceptionally pure
     f. Parkinsonism following large doses of certain tranquilizer drugs
  3. Symptoms: three major forms are tremor, muscle rigidity, and akinesia (inability to initiate movements or perform them quickly enough). Any one may dominate

B. HUNTINGTON'S CHOREA OR CHRONIC CHOREA is a familial disease of genetic origin, chiefly involving the corpus striatum and, to some extent, the cerebral cortex (frontal and parietal regions show neuronal loss). Atrophy of the putamen and caudate nucleus preferentially selects the head of the caudate and is accompanied by ventricular dilatation of the frontal horns

  1. Involuntary movements
     a. Athetosis: consists of slow writhing movements of the limbs, face, and tongue, seen mostly after brain damage at birth, or after rhesus incompatibility
     b. In chorea, the movements are very rapid, never repeated, and accompanied by flapping of the tip of the tongue and grimacing
     c. When both (a) and (b) are present, we speak of choreoathetosis
     d. Chorea may follow rheumatic fever and is called Sydenham's chorea
        i. It may occur in pregnancy
        ii. Seen as a rare complication of the use of oral contraceptives
        iii. May occur as a familial disturbance in middle life
        iv. Usually accompanied by mental deterioration (called Huntington's chorea)
     e. Hemiballism is a wild violent flinging movement of a limb, usually as a result of a vascular accident in the subthalamic nucleus

PARKINSON'S DISEASE

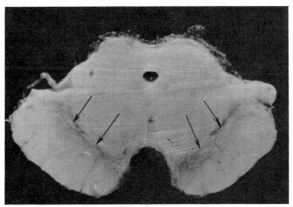

Section through midbrain showing decreased neuromelanin pigment in the substantia nigra.

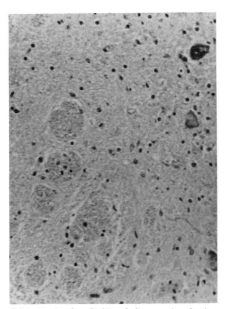

Substantia nigra from Parkinson's disease patient showing reduction in number of pigmented neurons.

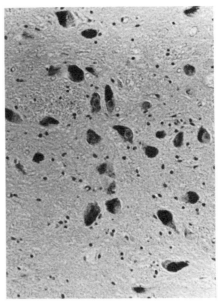

Normal substantia nigra to show neuromelanin pigment in neuron.

224. RARE SUBCORTICAL, MOTOR NEURON, AND CEREBELLAR DEGENERATIVE DISEASES

I. Rarer forms of subcortical degeneration

A. STRIATONIGRAL DEGENERATION: rare, with clinical features of extrapyramidal rigidity involving the substantia nigra and atrophy of corpus striatum

B. PURE PALLIDAL ATROPHY: often familial, rare, and involves subthalamic nuclei

C. ISOLATED ATROPHY OF SUBTHALAMIC NUCLEI: often associated with other pallidonigral degeneration

D. HALLERVORDEN-SPATZ DISEASE: neuronal cell loss with deposition of iron pigment

E. SHY-DRAGER DISEASE: extrapyramidal syndrome associated with orthostatic hypotension

F. PROGRESSIVE SUPRANUCLEAR PALSY (STEELE-RICHARDSON-OLSZEWSKI DISEASE): clinical picture of hypertonia, supranuclear ocular palsies, and disturbance of wakefulness

II. Motor neuron disease: degenerative changes here affect the motor system only, namely, the anterior horn cells, the nuclei of some cranial nerves, and the fibers of the pyramidal tracts. Three syndromes develop

A. PROGRESSIVE MUSCULAR ATROPHY

B. PROGRESSIVE BULBAR PALSY

C. AMYOTROPHIC LATERAL SCLEROSIS

D. ALL OF THE ABOVE ARE PART OF THE SAME DISEASE, the difference in their signs depends on the part of the motor system affected, thus, they are grouped together
   1. The cause is unknown

E. SYMPTOMS
   1. The disease most frequently develops between the ages of 40 to 60
   2. Weakness may first be noticed in the hands, shoulders, or neck muscles
   3. Occasionally the legs are first affected, usually as a painless footdrop
   4. The hand and shoulder girdle muscles begin to waste, and gradually the weakness and wasting spread throughout the body until the trunk muscles no longer give support. Weakness of the tongue and throat muscles makes speech difficult and swallowing impossible. When the latter symptoms begin, we speak of the disease being progressive
   5. In amyotrophic lateral sclerosis, pyramidal tract degeneration causes a spasticity that may predominate over wasting and weakness at first, but not for long

F. SIGNS
   1. Muscle wasting is usually first seen in the shoulder girdles and hands, where the weakness appears greater than the wasting suggests. Later, the wasting becomes widespread and the tongue becomes small, wrinkled, and paralyzed
   2. Bundles of muscle show a constant jumping and flickering (spoken of as fasciculation) which are clearly seen in the tongue
   3. The tendon reflexes are greatly exaggerated, despite muscle wasting, and are combined with fasciculation and normal sensation

G. Treatment: there is no effective treatment, and the disease progresses inevitably to death, usually from respiratory infection in from 6 months to 2 years

III. Cerebellar degenerative diseases: the complex group of cerebellar atrophies comprises all the degenerative disorders that involve cerebellar cortex and its efferent and afferent pathways. They may be focal or diffuse

A. FOCAL CEREBELLAR ATROPHY

B. DIFFUSE CEREBELLAR ATROPHY

C. OLIVOPONTOCEREBELLAR ATROPHY

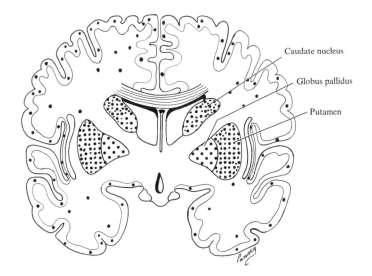

Caudate nucleus

Globus pallidus

Putamen

DISTRIBUTION OF LESIONS

HUNTINGTON'S DISEASE

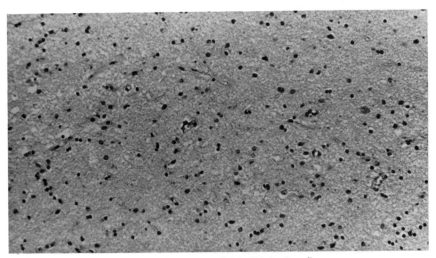

Histologic preparation showing loss of neurons and increased glial cells in basal ganglia.

Unit Seventeen

DEFICIENCY, METABOLIC, AND TOXIC DISEASES

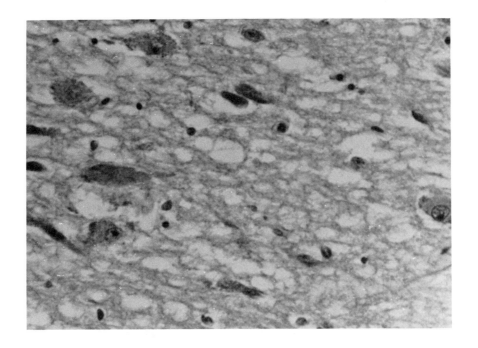

225. POLYNEURITIS AND POLYNEUROPATHY

I. **Introduction:** the diseases discussed here are caused by a poison or toxin acting on the nervous system, a shortage of some substance essential for nervous tissue to function properly, or both. The poison may be produced by abnormalities inside the body (e.g., uremia) or it may be taken in from the outside (e.g., drug addiction). A vital substance may become deficient as a result of not enough being taken into the body (starvation), or because it is not absorbed properly (as in pernicious anemia) or because the body fails to use it properly once it is absorbed. The peripheral nerves and the posterior columns of the cord are particularly vulnerable to chronic toxic and deficiency states, thus, the initial discussion deals with polyneuritis and polyneuropathy before the more common diseases are discussed

II. **Polyneuritis and polyneuropathy**
A. DEFINITIONS: neuritis means inflammation of a nerve; polyneuritis refers to inflammation of many nerves. However, nerves do not often become inflamed in the true sense of the word, and the term "polyneuropathy" is preferred by many. "-Opathy" means something wrong but not necessarily an inflammation, and exactly what is wrong may not be certain
  1. In polyneuritis or polyneuropathy, the symptoms usually begin in the fingers and hands, toes, and feet on both sides (thus, the alternative terminology of peripheral neuritis)
B. SYMPTOMS consist of numbness, pins and needles, or burning; or they can be motor such as weakness and wasting (discussed later)
  1. Symptoms spread upward to involve the calves, thigh and hip muscles, and the arms and shoulder girdles. In severe cases, even the trunk, facial, and respiratory muscles are involved
  2. Reflexes are characteristically absent, and there is a loss of sensation in the hands and forearms as well as feet and lower legs (glove-and-stocking anesthesia)
  3. The condition may range from a mild tingling in all 4 limbs to total paralysis of all muscles, requiring a tracheotomy and artificial respirator
C. THE ABSENCE OF MENINGEAL IRRITATION SIGNS, the slow onset, the symmetry of the paralysis, and the presence of sensory loss distinguish polyneuritis from poliomyelitis
D. THE EXACT CAUSE of many cases has never been determined, but fortunately recovery is quite common
E. ACUTE "INFECTIVE" POLYNEURITIS is a well-known syndrome but may not be of toxic or metabolic origin
  1. Develops rapidly, often after a few weeks of vague pyrexial illness and sometimes as a complication of known infection
  2. Symptoms are of polyneuritis, but the motor paralysis may be severe and bilateral facial paralysis is very characteristic
  3. Diagnosis is confirmed by a very high protein in the CSF (over 250 mg%) but no cell rise
  4. After 2 or 3 weeks of great disability, the patient slowly recovers and usually makes a complete recovery
  5. It may represent one of the autoimmune diseases where the patient's own body defenses react to damage his own tissue
  6. Early use of ACTH may help in the acute but not in the chronic stage
  7. Condition has been called the Guillain-Barré syndrome (after 2 French neurologists who described it during World War I)

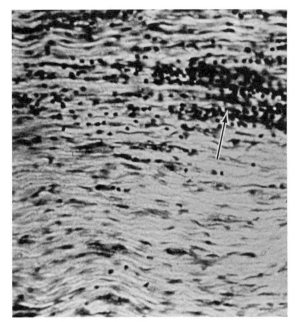

Guillain-Barré syndrome: polyneuritis.

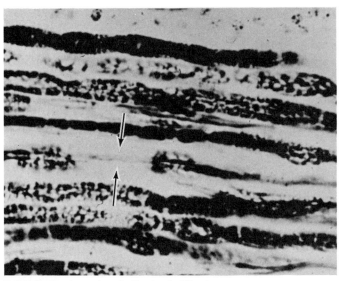

Guillain-Barré syndrome: demyelinating at node of Ranvier.

226. DEFICIENCY DISEASES—PART I

I. Deficiency diseases

A. SUBACUTE COMBINED DEGENERATION is usually associated with pernicious anemia, the underlying cause of which is atrophy of the mucous membrane of the stomach, resulting in the lack of absorption of factors for the development of RBCs and for the nutrition of nervous tissue

    1. The vital substance of deficiency is vitamin B_{12} (isolated from liver)

        a. In its absence, the number of RBCs that develop is small, yet they are larger than normal and carry more hemoglobin than usual, resulting in macrocytic anemia

    2. The deficiency also results in a gradual degeneration of the posterior columns and pyramidal tracts in the spinal cord, together with a polyneuropathy, thus accounting for its name of "subacute combined" degeneration

    3. Symptoms: patient is usually middle-aged, white-haired, and often a woman who complains of numbness and tingling in the fingers and toes, soreness of the calves (due to peripheral neuritis), and unsteadiness in walking, particularly in the dark (result of posterior column degeneration). Patient also admits to soreness of the tongue

B. OTHER VITAMIN B_{12}–DEFICIENCY SYNDROMES: absorption of vitamin B_{12} may be defective in a variety of intestinal diseases and conditions

    1. Steatorrhea: where the absorption of fat is deficient

    2. Surgical removal of the stomach

    3. Crohn's disease (regional enteritis)

    4. Chronic diarrhea

    5. In the above conditions, there are deficiencies of other vitamins as well, such as vitamin B_1, pyridoxine, nicotinic acid, and folic acid

    6. The symptoms and signs may be suggestive of polyneuritis or of subacute combined degeneration, but they are not associated with a histamine-fast achlorhydria or the typical marrow of pernicious anemia

C. VITAMIN B_1 (THIAMINE) DEFICIENCY: leads to beriberi

    1. In conditions of starvation, inadequate diets, or severe gastric disturbances, not enough vitamin B_1 may be absorbed, and a polyneuritis may develop

    2. In severe cases, the heart enlarges and fails, and edema of the limbs develops

    3. If changes take place in the upper brainstem resulting in paralysis of eye movement, ataxia, confusion, and coma, this is called Wernicke's encephalopathy

D. NIACIN (NICOTINIC ACID) DEFICIENCY leads to pellagra (B-group deficiency)

    1. Symptoms may be those of polyneuropathy, or a syndrome similar to subacute combined degeneration occurs. However, in addition, there is a scaly brown rash on those parts of the skin exposed to light

E. FOLIC ACID DEFICIENCY: since folic acid is concerned with blood formation, deficiencies cause anemias and polyneuropathies. Severe mental changes also occur

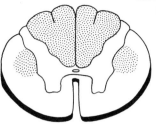

Gradual degeneration of posterior columns
and pyramidal tracts

SUBACUTE COMBINED DEGENERATION $\left\{\begin{array}{l}\text{Posterior column + pyramidal tract deg.}\\\text{Polyneuropathy}\end{array}\right.$

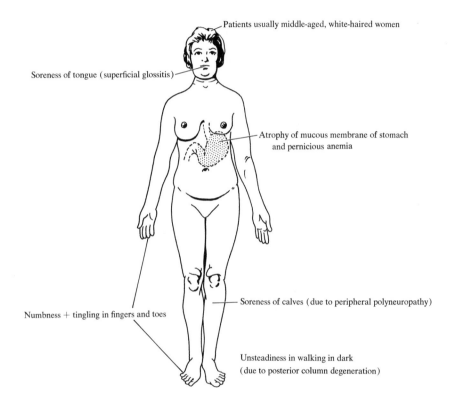

Patients usually middle-aged, white-haired women

Soreness of tongue (superficial glossitis)

Atrophy of mucous membrane of stomach
and pernicious anemia

Soreness of calves (due to peripheral polyneuropathy)

Numbness + tingling in fingers and toes

Unsteadiness in walking in dark
(due to posterior column degeneration)

VITAL DEFICIENCY OF VITAMIN B₁₂

227. DEFICIENCY DISEASES—PART II

II. Neurologic complications of certain metabolic diseases: used clinically to describe a number of cerebral disturbances secondary to general metabolic disorders. Lesions are nonspecific and related to visceral lesions or anoxic disturbances

A. DIABETES MELLITUS
1. Polyneuritis: a common complication of poorly controlled diabetes. Particularly sensory, with severe burning sensation in the hands and feet, numbness of extremities, and increased difficulty in walking
2. Diabetic coma: a danger present in poorly controlled patients
   a. General malaise for some hours or days, and increased thirst, followed by vomiting. Dehydration and drowsiness occur; patient passes into coma if untreated
3. An additional infection (e.g., a carbuncle) may lead to coma even in a well-controlled patient since control breaks down, urine is loaded with sugar and acetone, and blood sugar is high

B. HYPOGLYCEMIC COMA is the result of too low blood sugar and may lead to coma
1. Patient begins to sweat profusely, becomes ataxic, confused, and may have convulsions and lose consciousness. Prolonged hypoglycemia may cause permanent brain damage

C. SPONTANEOUS HYPOGLYCEMIA: very low blood sugar in patients not on insulin
1. Symptoms vary from transient disturbances of behavior, and episodes of ataxia and drowsiness, to fits and loss of consciousness with profuse sweating
2. Most common cause is a tumor of cells in the pancreas, which secretes insulin

D. POTASSIUM DEFICIENCY: blood potassium levels fall very low and patient develops a weakness of the limbs which may lead to total flaccid paralysis
1. Can occur in patients receiving cortisone and in some forms of nephritis
2. A similar paralysis may occur spontaneously (in familial periodic paralysis)
   a. Occurs while patient is asleep, especially after heavy meals
   b. Several members of the family are usually affected

E. CARCINOMATOUS NEUROPATHY: cause unknown
1. A slow degenerative process of the sensory fibers in the peripheral nerves and the posterior columns that can occur in patients who have carcinoma of the bronchus
2. Patients develop progressive numbness, weakness, and wasting of the legs, with increasing ataxia due to loss of position sense
3. A similar degeneration of the cerebellum produces cerebellar ataxia and occurs in patients with unsuspected carcinoma. The primary site is commonly the bronchus

F. RESPIRATORY ENCEPHALOPATHIES: secondary to chronic bronchopulmonary disease and attributable to hypoxia and hypercapnia

G. HEPATIC INSUFFICIENCY: as a result of terminal coma in hepatic cirrhosis and severe hepatitis, postcaval encephalopathy, and Wilson's disease
1. Lesions predominate in the globus pallidus, but may be seen in the dentate nucleus and cerebral cortex

H. DISORDERS OF IRON METABOLISM: seen in primary and secondary hemochromatosis. The blood-brain barrier provides efficient protection against diffusion of iron pigments into the CNS, and consequently cerebral lesions are very constricted. Hemosiderin is seen in the choroid plexus, area postrema of the medulla, pineal gland, and other vestigial remnants such as the paraphysis and subfornical organ

I. DISORDERS OF CALCIUM METABOLISM
1. Hypoparathyroidism may lead to Fahr's disease, with a massive perivascular deposit of pseudocalcium in the basal ganglia, especially the caudate nucleus
2. When the serum calcium levels approach the critical level, increased neuromuscular excitability develops, leading to muscle twitching and occasionally to a convulsive disorder resembling epilepsy, although there is no loss of consciousness

HYPOGLYCEMIC STATE — MECHANISMS AND SYMPTOMS

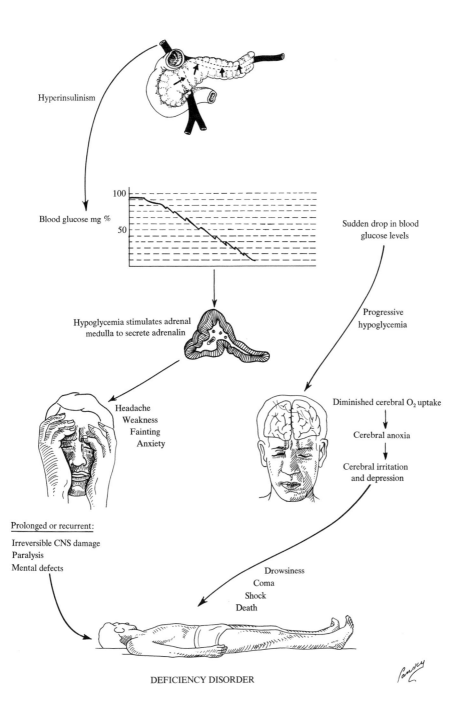

Hyperinsulinism

Blood glucose mg %

Sudden drop in blood
glucose levels

Hypoglycemia stimulates adrenal
medulla to secrete adrenalin

Progressive
hypoglycemia

Headache
Weakness
Fainting
Anxiety

Diminished cerebral O$_2$ uptake

Cerebral anoxia

Cerebral irritation
and depression

Prolonged or recurrent:

Irreversible CNS damage
Paralysis
Mental defects

Drowsiness
Coma
Shock
Death

DEFICIENCY DISORDER

228. INBORN ERRORS OF METABOLISM

I. **Inborn errors of metabolism** (genetic metabolic diseases due to an enzyme defect): some people are born with a tendency for one of the vital processes of the body to go wrong. As a result, some substance, normally broken down and excreted, builds up in the body to toxic levels. There are many such diseases. Fortunately all are rare and few affect the nervous system

A. DISORDERS OF METAL METABOLISM
1. Wilson's disease (hepatolenticular degeneration): due to a deficiency of a substance called ceruloplasmin, which normally carries copper in the blood
   a. Copper cannot be maintained at its normal blood level, excessive amounts are excreted in the urine and, more important, are deposited in various organs, including the liver and basal ganglia (lenticular nucleus)
   b. Copper is also deposited in the cornea of the eye, producing a diagnostic golden brown ring, the Kayser-Fleischer ring
   c. Several members of a family are affected
   d. Symptoms
      i. Severe involuntary movements develop in youth with progressive disabling and mental deterioration
      ii. Liver failure later develops
   e. Treatment: Penicillamine is used which extracts copper from the CNS and increases kidney secretion of copper. Striking improvement is seen in the clinical condition, which appears to be maintained for many years in cases that are treated early
2. Kinky hair disease (Menkes' disease): defect of intestinal copper absorption. Abnormalities in hair and neuropsychiatric manifestations

B. DISORDER OF PIGMENT METABOLISM: porphyria (hematoporphyrinuria)
1. Porphyrins are substances concerned in the formation of blood pigments. If metabolism is disturbed, large amounts are excreted and detected in the urine. In some cases, the urine may turn to a port wine color after exposure to light and in other cases special tests are needed
2. Symptoms
   a. The skin of the patient blisters easily on exposure to sunlight
   b. Some patients have abdominal pain attacks and mental disturbances lasting several weeks, during which time they develop a polyneuritis. The attacks often vanish spontaneously and may follow administration of barbiturates

C. DISORDERS OF AMINO ACID METABOLISM: cause of many syndromes of mental retardation in childhood which may be associated with various neurologic manifestations
1. Phenylketonuria: due to the absence of phenylalanine hydroxylase
   a. The neurologic findings are variable and poorly defined
   b. Can be detected by adding a few drops of ferric chloride to the urine
2. Hartnup's disease: due to a disorder of tryptophan absorption (resembles pellagra)
3. Homocystinuria: due to a deficit of cystathione synthetase
   a. Disease causes alterations in blood vessel walls, with fibrosis of the intima and degeneration of elastic fibers. Also see foci of cerebral necrosis
4. Leucinosis or maple sugar urine disease: identified by presence of alphahydroxybutyric acid in urine with characteristic odor. Due to decrease in decarboxylase activity
5. Tyranosis: related to deficiency in oxidation of parahydroxyphenylpyruvic acid

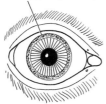
Kayser-Fleischer ring (pathognomonic)

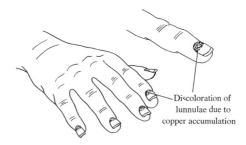
Discoloration of lunnulae due to copper accumulation

WILSON'S DISEASE (HEPATOLENTICULAR DEGENERATION)

Golden or green-brown Kayser-Fleischer rings at periphery of cornea

Serum ceruloplasm decreased; serum + urinary copper increases

Choreoathetosis and staggering possible

Blue-brown discoloration of lunulae

Tremors, incoordination, speech difficulty, and deteriorating handwriting

Personality changes—episodes of acute schizophrenia

Dysphagia, rigidity, transient hemiparesis

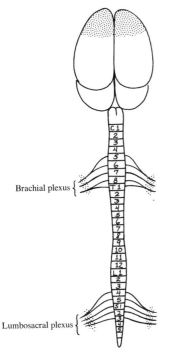

Brachial plexus {

Lumbosacral plexus {

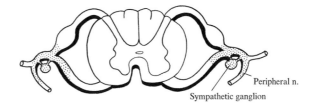
Peripheral n.
Sympathetic ganglion

PORPHYRIA

Weakness and atrophy of muscles of all extremities

Tenderness of peripheral nerve trunks

Impairment of judgment and lack of emotional control

Large amounts of porphobilinogen in urine

Acute episodes of encephalopathy, neuropathy, abdominal pain, jaundice

Increased coproporphyrin and protoporphyrin in feces

METABOLIC BRAIN DISEASES

229. DISORDERS OF CARBOHYDRATE, MUCOPOLYSACCHARIDE, AND LIPID METABOLISM

I. Disorders of carbohydrate metabolism

A. GLYCOGENOSES: some forms of glycogen storage disease (MacArdle's and Forbes' diseases) involve the skeletal musculature with an exceptional CNS involvement. Pompe's disease, caused by acid maltase deficiency, does implicate the CNS

B. GALACTOSEMIA: deficit of galactose-1-phosphate-N-uridyltransferase (CNS lesions)

C. LAFORA'S DISEASE: exact metabolic disturbance uncertain. May be due to excessive storage of amylopectin. Picture of progressive familial myoclonic epilepsy

D. SUBACUTE NECROTIZING ENCEPHALOPATHY (LEIGH-FEIGIN DISEASE) involves basal ganglia, tegmentum of brainstem, corpora quadrigemina and inferior olives. Neurons are spared and there is gliosis. Metabolic disorder poorly understood

II. Disorders of mucopolysaccharide metabolism: a systemic disturbance of acid mucopolysaccharides or glycosaminoglycans (excreted in urine) is accompanied by a lipid neuronal storage disorder that essentially implicates the gangliosides

A. HURLER'S DISEASE (TYPE I MUCOPOLYSACCHARIDOSIS): with excessive chondroitin sulfate B and heparin sulfate

B. HUNTER'S DISEASE (TYPE II MUCOPOLYSACCHARIDOSIS) differs from Hurler's by its sex-linked recessive character and absence of corneal opacities

C. SANFILIPPO'S DISEASE (TYPE III MUCOPOLYSACCHARIDOSIS): only heparin sulfate accumulates in excessive amounts

III. Disorders of lipid metabolism: an abnormal accumulation, mostly in neurons of the CNS, of a specific lipid metabolite. May or may not be associated with PNS lesions and visceral lesions. Classifications refer to the lipids involved and enzyme(s) defect(s) rather than the clinical features. The chief lesions are of intense neuronal swelling with distention of the cell bodies. Cerebellar lesions are common

A. THE SPHINGOLIPIDOSES: are the most important group among the neurolipidoses and are characterized by an excess of sphingolipids

B. THE GANGLIOSIDOSES (AMAUROTIC IDIOCIES) are characterized by accumulation of gangliosides
   1. GM2 gangliosidoses: classified according to their specific enzyme deficiency
      a. Tay-Sachs disease: due to accumulation of GM2 gangliosides and ceramide trihexoside as a result of deficiency of hexosamidase A
      b. Sandhoff disease: due to deficiency of both hexosamidases A and B
   2. GD3 gangliodisosis: congenital amaurotic idiocy
   3. GM1 gangliosidosis: late amaurotic infantile idiocy

C. THE CEREBROSIDOSES
   1. Gaucher's disease: deficiency of glucocerebroside-B-glucosidase. In acute stage one sees macrophages (Gaucher cells) in the CNS
   2. Krabbe's disease: deficiency seems to be a galactoside-B-galactosidase. Characterized by conspicuous involvement of myelin of the white matter

D. THE SULFATIDOSES: characterized by a deficit of arylsulfatase
   1. Metachromatic leukodystrophy: deficit of arylsulfatase A
   2. Austin's disease: deficit of arylsulfatases A, B, and C

E. NIEMANN-PICK'S DISEASE displays several forms both clinically and genetically
   1. All show an accumulation of sphingomyelin with a sphingomyelinase deficit
   2. CNS involvement is accompanied by an infiltration of foamy cells and changes in the endothelial and perithelial elements of blood vessel walls

F. REFSUM'S DISEASE: excessive storage of phytanic acid due to a deficit of phytanic oxidase. The PNS is affected by a demyelinating neuropathy

G. THE LIPOPROTEINOSES: acanthocytosis with a deficit of β-lipoprotein and Tangier's disease with a deficit of α-lipoprotein

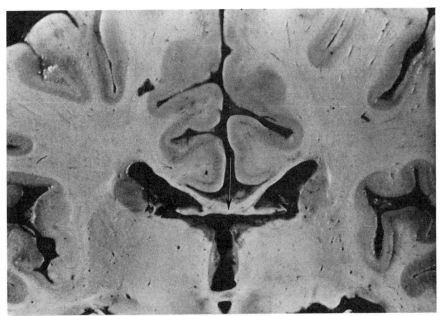

Gross section of the brain showing widened sulci, narrowed gyri, a thinned corpus callosum, and enlarged ventricles typical of brain atrophy observed in Tay-Sachs disease and other ganglioside storage diseases.

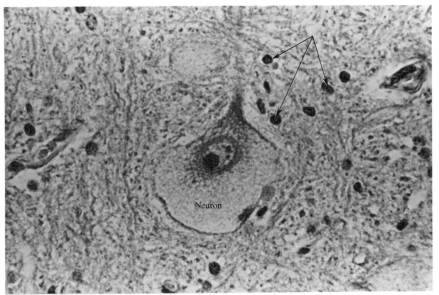

Histologic section showing swollen neuron and increased number of glial cells (arrows) characteristic of Tay-Sachs disease.

TAY-SACHS DISEASE

230. TOXIC DISEASES: ALCOHOLISM; CONDITIONS CAUSED BY BARBITURATES, ANTIEPILEPTIC DRUGS, AND OTHER AGENTS

I. Toxic diseases

A. ALCOHOLISM: patient develops a disturbance of coordination, resulting in double vision, slurring of speech (dysarthria), unsteadiness of the arms and legs, and ataxia of gait (these develop during a drinking bout, but usually clear up some hours afterward, usually with a hangover)

  1. Excessive alcohol intake over a long period of time, however, has other side effects since the toxin not only acts on the brain and liver but also causes damage to the stomach, preventing absorption of vitamins, especially B_1

     a. The direct effects are gross tremulousness of the entire body followed by wild confusion and agitation, with terrifying visual and auditory hallucinations and absolute wakefulness—delirium tremens (DTs)

  2. After years of heavy drinking (chronic alcoholism), the deficiency of vitamin B_1 causes severe damage to the brainstem, leading to mental confusion and paralysis of eye movements, double vision, ataxia of the arms and legs, and a polyneuropathy—Wernicke's encephalopathy

     a. Korsakoff's syndrome: in alcoholics is regarded as a late chronic stage of Wernicke's encephalopathy. Predominant involvement of the mamillary bodies accounts for memory disturbances with fixation amnesia, confabulation, and temporospinal disorientation with polyneuritis

B. BARBITURATES AND ANTIEPILEPTIC DRUGS

  1. If a large dose of barbiturates is suddenly ingested, coma develops rapidly, with complete flaccidity of the limbs, absent reflexes, and often collapse of the circulation

  2. Excessive amounts of some drugs taken over a long time produce slurring of speech, ataxia of the limbs, and recurrent drowsiness

  3. Many poorly controlled epileptics develop toxic effects from a combination of too many medications. They also develop drowsiness, gross ataxia, nystagmus, and dysarthria, but often with sufficient headache and vomiting to arouse suspicion of a cerebellar tumor. Folic acid deficiency also occurs

C. OTHER TOXIC CONDITIONS AND DISEASES

  1. Tranquilizers: may cause slowing reactions, even in moderate doses, and defective judgment and slight unsteadiness; even these signs may be more marked if the drugs are mixed with alcohol

     a. Some cause troublesome polyneuropathy

     b. Large doses over a long time produce a syndrome almost indistinguishable from parkinsonism

  2. Amphetamines (to a lesser extent, belladonna): cause excitement, tremor, rapid pulse, collapse, and delirium

  3. Lead: produces a polyneuropathy, especially wrist drop

     a. The types of lead used in paints and gasolines can cause a cerebral disturbance called lead encephalopathy with headaches, mental changes, fits, and coma

  4. Mercury: is rarely a problem, but it causes severe tremors of the arms and legs (hatter's shakes). Certain plant sprays contain a form of mercury which causes degeneration of the posterior columns, with severe postural ataxia

  5. Arsenic: when taken by mouth over a long time, causes polyneuropathy

  6. Oral contraceptives: may cause cerebral thrombosis, papilledema, chorea, and migraine headaches

D. GENERAL COMMENTS

  1. Most toxic effects of any substance can be overcome by complete withdrawal

  2. If intoxication is chronic, permanent damage may have occurred in the nervous system and cure cannot be complete (includes drugs and alcohol addiction)

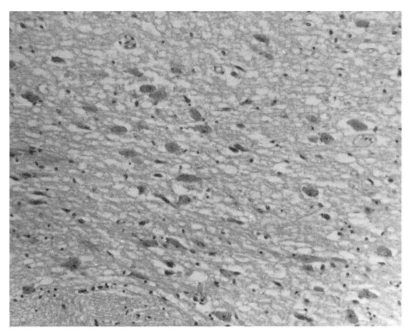

Wernicke's encephalopathy—thiamine deficiency.

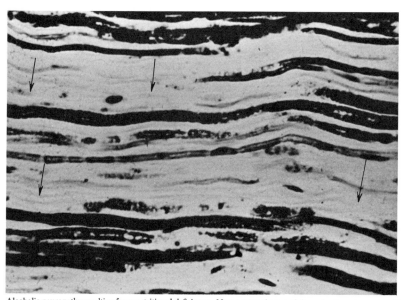

Alcoholic neuropathy resulting from nutritional deficiency. Note scattered areas of demyelination along nerve fibers (arrows).

Unit Eighteen

DISEASES OF MUSCLE

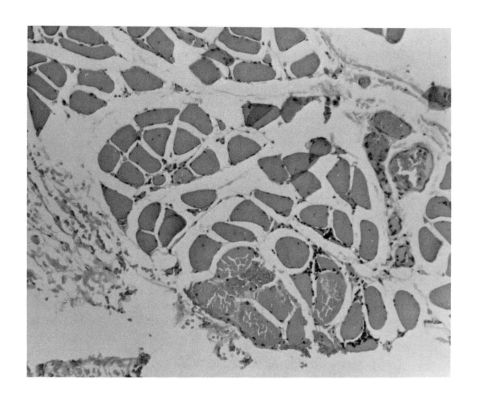

231. THE MYOPATHIES OR PROGRESSIVE MUSCULAR DYSTROPHIES

I. **Introduction:** diseases of the muscles have always come under the province of the neurologists since muscles are a vital portion of the motor system, even though the nervous system is normal. It is not always easy to tell whether or not muscle weakness and wasting are due to a lesion of the motor nerves or to disease limited to the muscle itself. Muscle diseases may be degenerative and are called myopathies or progressive muscular dystrophies, or they may be inflammatory and called myositis or polymyositis. In addition, there is a condition called myasthenia gravis in which impulses cannot ''get across'' from the nerve into the muscle to be supplied

II. **The myopathies or progressive muscular dystrophies** are heredofamilial diseases in which the muscle fibers slowly degenerate and waste, eventually producing severe disability. Since the muscles are often replaced by fat, the patient superficially may not appear wasted. Several types exist, varying in the age groups and particular muscles affected as well as in the gravity of prognosis

A. PSEUDOHYPERTROPHIC MUSCULAR DYSTROPHY (DUCHENNE TYPE) is the most severe and serious type, so named because the muscles are unusually large or hypertrophied, though very weak
   1. Almost, though not completely, confined to boys
   2. Progression is such that very few reach adult life, and its rapid course is usually fatal by the age of 20
   3. Mode of inheritance: though mainly affecting boys, the disease is passed on by apparently unaffected female carriers (recessive, sex-linked). It can be detected by a combination of chemical, electrical, and biopsy examination of muscles. There is a 50% chance of any son of a carrier to be affected, and a daughter may become a carrier
   4. Symptoms: child walks badly and has difficulty in picking himself up after falling and in climbing stairs. Early onset in muscles of pelvic girdle

B. FACIOSCAPULOHUMERAL DYSTROPHY (OF LANDOUZY-DÉJERINE)
   1. Wasting and weakness affect the face, shoulder girdles, and upper arms (often one side more than the other)
   2. Of slow evolution, compatible with a relatively long life. Onset in middle childhood or sometimes in the 2nd decade of life and affects both sexes
   3. The least progressive of the dystrophies
   4. The muscles do not hypertrophy
   5. Inherited from an affected individual, not via a carrier, and is transmitted as an autosomal dominant trait

C. LIMB GIRDLE MYOPATHIES OR DYSTROPHIES involve muscles of the hip and shoulder girdles or both and affect older patients
   1. Weakness and wasting progress slowly, and disability is great in 10 years
   2. Usually no history in preceding generation, and transmission is recessive but sex-linked

D. OCULAR MYOPATHIES
   1. Degenerative changes develop in the muscles of the lids and eyes in late middle life (occasionally earlier)
   2. Progressive ptosis and difficulty in moving eyes; finally, fixed in midline
   3. The lesion is now known to lie in the muscles
   4. There is no treatment, but plastic surgery may be needed to raise eyelids

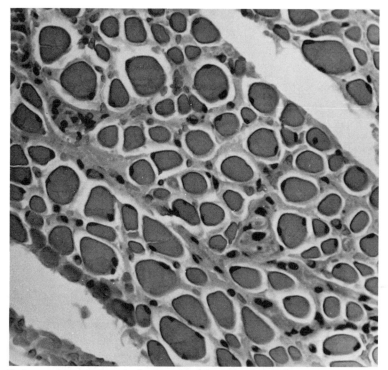

Duchenne-type muscular dystrophy showing great variation in size of muscle fibers.

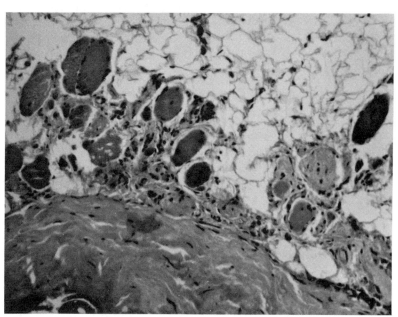

Progressive muscular dystrophy showing atrophy and wasting of muscle fibers.

232. MYOTONIA, MYOSITIS, AND MYASTHENIA GRAVIS

I. Myotonia is a name given to a condition in which muscles continue to contract for some time after the need has passed. Seen in 2 major diseases: dystrophia myotonia and myotonia congenita

A. DYSTROPHIA MYOTONIA (MYOTONIA ATROPHICA): a heredofamilial disorder which affects men more than women and appears to begin in young adult life
   1. Patients complain of increasing weakness of the hands and arms, and occasionally of muscular wasting
   2. Occasionally have difficulty in letting go after picking up an object or when shaking hands and may realize that their vision is failing

B. MYOTONIA CONGENITA is a hereditary condition present from infancy
   1. Dystrophic features (muscle wasting, baldness, cataracts, testicular atrophy, or ovarian atrophy) are usually absent
   2. Patient complains of muscle stiffness rather than muscle weakness

II. Myositis and myasthenia gravis

A. MYOSITIS: acute myositis (viral or infectious) is rare; parasitic myositis is less exceptional and is common in the tropics and subtropics
   1. Muscles can be infected from dirty wounds or careless injections. They become swollen and tender as inflammatory cells enter from the bloodstream
   2. Whether allergic or a toxic state, it can occur with inflammatory lesions of the skin (dermatomyositis) or of the blood vessels (polyarteritis nodosa)
   3. When many muscles are involved, we speak of polymyositis
   4. Patients (of any age) feel generally unwell and complain of increasing weakness of the limbs, with swelling and tenderness of the muscles

B. MYASTHENIA GRAVIS: "myasthenia" means weakness of muscles. The term usually is used for a weakness that worsens with use and rapidly recovers when muscle is at rest
   1. Mainly affects young women, but may occur in either sex at any age
   2. Caused by an abnormality which prevents impulses from branches of motor nerves from passing across the nerve endings in the muscle fibers at the neuromuscular junctions
   3. The patient's eyes are often affected: eyelids droop, eyes squint, and there is double vision (symptoms vary from hour to hour)
   4. Patients may be normal on awakening but get worse as day wears on: voice gets weak and hoarse near the end of each sentence; chewing and swallowing become more difficult as meal progresses; arm and back muscles become weak after use (varies greatly); and recovery is complete on resting
   5. Months may pass without any symptoms
   6. Any muscle may be affected, so signs will vary greatly. Most commonly seen are ptosis, squinting, hoarseness, and a nasal voice, all varying from time to time during the examination

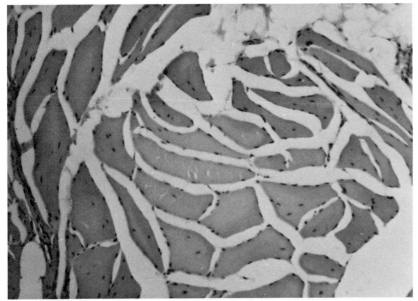

Myotonic dystrophy: low-power cross-section of muscle fibers showing their variation in size.

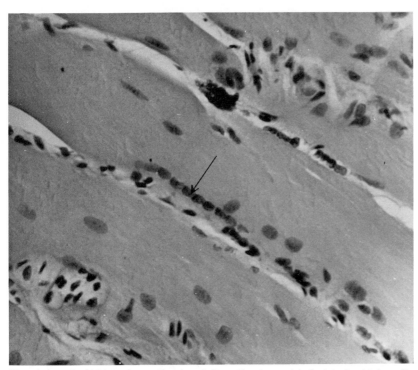

Myotonic dystrophy: high-power long. section of muscle fibers. Note characteristically chained nuclei along cell membrane.

Unit Nineteen

TUMORS

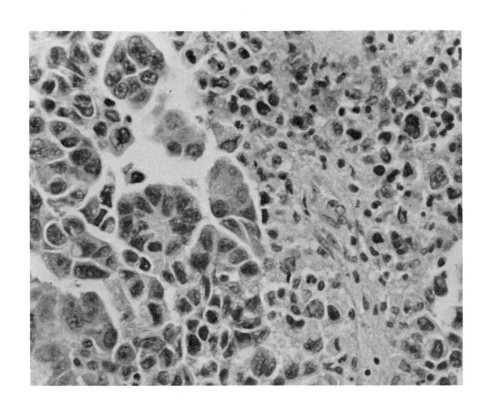

233. TUMORS OF THE CENTRAL NERVOUS SYSTEM

I. **Introduction and classification:** tumors or neoplasms that affect the nervous system are said to be primary if they grow from cells of the nervous system itself or secondary (metastases) if they are carried to the nervous system by the blood from a primary tumor elsewhere (e.g., the bronchus, breast). Nervous system tumors are classified into 2 groups: those of the CNS and those of the PNS. Tumors of the PNS include tumors of the peripheral nerve trunks, sympathetic ganglia, adrenal medulla, and chemoreceptors. Tumors of nerve cells and fibers are very rare. Neoplasms grow instead from supporting tissue: gliomas from glia, meningiomas from meninges, ependymomas from ventricular lining, neurofibromas from nerve sheaths, etc. Tumors are also classified topographically and histologically

    CNS tumors can be divided into 2 groups: intracranial (erroneously called cerebral tumors) and extrinsic (tumors causing spinal root and cord compression). Most extrinsic tumors are benign, grow slowly, and compress nervous tissue while remaining separate from it. They do not spread elsewhere. Most intracranial (intrinsic) tumors are malignant, grow rapidly, and destroy nervous tissue by spreading through it. Metastases to other parts of the body from even malignant cerebral tumors are rare. Some intrinsic tumors might be called relatively benign since they grow slowly and do not produce much destruction for years. Some extrinsic tumors may become malignant. All metastases are malignant

II. **Cerebral tumors**
A. GENERAL FEATURES: intracranial tumors produce symptoms in 3 ways
   1. Space-occupying lesion: a new mass fills up any remaining space and causes pressure inside the skull to rise
   2. Tumors may compress the ventricles or parts of the CSF pathways and dam up flow, and pressure is built up
   3. The tumor presses on or destroys part of the brain directly, producing symptoms and signs that gradually become worse and more extensive
B. SYMPTOMS AND SIGNS OF INCREASED INTRACRANIAL PRESSURE: headaches, vomiting (usually at height of headache), diplopia, blurring of vision on movement of head, drowsiness (increasing and progressive and leading to coma), dilatation of pupil with failure to react to light, and papilledema (swelling of optic disk)
C. DIRECT EFFECTS OF TUMOR ON BRAIN: weakness of the opposite side and fits. The latter may take the form of convulsions affecting all parts of the body, or may be focal fits affecting, e.g., one forearm alone
   1. The closer to the surface the tumor lies, the more likely it is to produce focal fits, but as it grows, the amount of the limb affected becomes greater. After the twitching stops, part of the body remains paralyzed for some time (Todd's paralysis)
   2. Temporal lobe tumors cause sudden hallucinations of taste, smell, or vision
   3. Occipital lobe fits may be also seen as blobs and flashes of light
   4. Special features develop according to the lobe affected
     a. If in frontal lobe: there are also loss of memory and personality changes
     b. If in temporal or occipital lobe: optic radiation damage can lead to visual loss in the half of each visual field opposite the lesion
     c. Tumors in the dominant hemisphere cause progressively increasing speech disturbances (dysphagia)
     d. Tumors in the brainstem are common in children and cause increasing paralysis of eye movements and of other cranial nerves, loss of facial sensation, and difficulty in swallowing
     e. Tumors in the cerebellum cause unsteadiness (ataxia) and early signs of elevated intracranial pressure

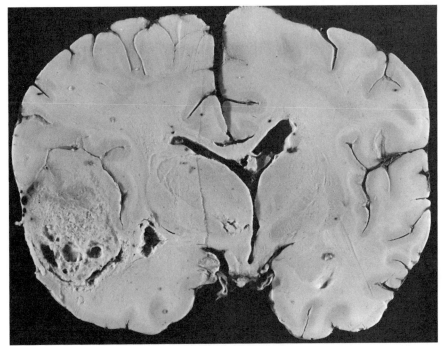
Intracranial tumors: a glioblastoma shown in gross brain.

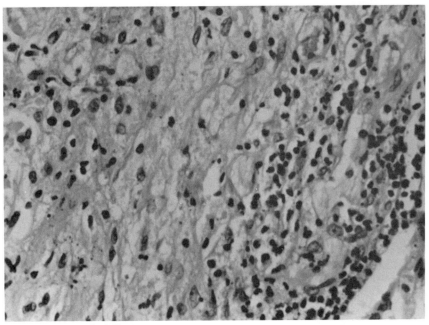
Histologic preparation of a glioblastoma.

234. PRIMARY NEOPLASMS: CEREBRAL TUMORS— PART I

I. Primary neoplasms

A. CEREBRAL TUMORS

1. Meningiomas: benign extrinsic tumors seen between 20 and 60 years, with a maximal incidence at 45 years, accounting for 13–18% of primary intracranial tumors and about 25% of intraspinal tumors. Predominantly seen in females
   a. Grow from arachnoid cells to irritate and compress the brain surface
   b. Patients have a long history of headaches, focal fits, and focal paralysis
2. Gliomas: most cerebral tumors, unfortunately, fall into this group
   a. Develop from glial cells such as astrocytes, oligodendrocytes, and microglia
   b. Are always intrinsic: produce signs by affecting nerve cells and fibers and compressing surrounding tissue they have not destroyed
   c. Astrocytomas: histologically benign, but are susceptible to malignant changes
      i. Account for 20–30% of tumors in the glioma group
      ii. Are of varying malignancy, some being relatively benign and growing slowly, and others growing rapidly in a few weeks or months (glioblastomas). Rapidly fatal astrocytic neoplasms account for about 50% of tumors in the glioma group
      iii. Occur at any age and in any region of the CNS
   d. Oligodendrogliomas or oligodendrocytomas account for 5% of the gliomas
      i. Are seen between ages of 30 and 50 years, and involve the cortex and white matter of the cerebral hemispheres. Are slow-growing
   e. Microgliomas: are very rare
   f. Ependymoma: like a medulloblastoma, but usually seen in childhood and adolescence. Accounts for about 6% of the gliomas
   g. Choroid plexus papilloma: 0.5% of intracranial tumors and 2% of glioma group
      i. Seen in 1st decade of life in the ventricular system
   h. Colloid cysts of 3rd ventricle, near foramina of Monro, account for 2% of intracranial gliomas and are seen in young adults, rarely in children
3. Hemangioblastomas: tumors of immature blood vessels, invariably in the cerebellum. Produce signs of raised intracranial pressure and marked ataxia
   a. Tend to form a cyst, with tumor itself remaining as a nodule in the cyst wall
4. Medulloblastoma: a malignant neoplasm accounting for $\frac{1}{3}$ of the posterior fossa tumors. Is always confined to children under 12 years of age
   a. Develops from primitive cells in roof of the 4th ventricle and spreads into cerebellum (vermis) and brainstem. Blocks CSF pathways, causing intracranial pressure
5. Ganglioneuromas and gangliogliomas are rare, small neoplasms seen in children and young adults in the floor of the 3rd ventricle or temporal lobe
6. Pineal tumors are a variety arising in the pineal region
7. Schwannomas (neurilemomas or neurinomas) are benign tumors arising from Schwann cells. Found on cranial nerves, spinal nerve roots, peripheral nerve trunks, and at nerve endings
   a. Most frequent type is the acoustic neuroma, which grows from the VIIIth cranial nerve sheath and lies in the cerebellopontine angle
      i. Compresses the VIIIth nerve and produces deafness on same side; VIIth nerve involvement produces weakness of facial muscles; and Vth nerve involvement produces loss of sensation over the eye and forehead
      ii. Compresses the cerebellum and produces ataxia
8. Neurofibromas form multiple tumors of nerve roots and peripheral nerves as part of the picture of von Recklinghausen's neurofibromatosis
9. Melanomas are a wide variety of disorders ranging from a simple increase in normal leptomeningeal pigmentation to highly malignant melanomas

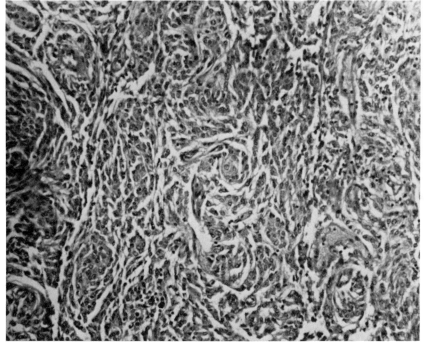
Cerebral tumor: meningioma.

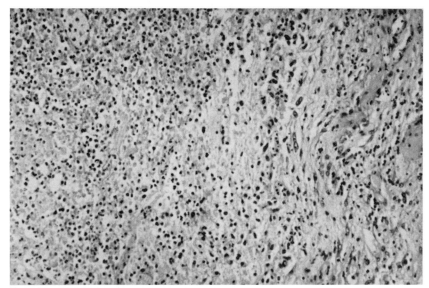
Cerebral tumor: glioblastoma.

235. PRIMARY NEOPLASMS: CEREBRAL TUMORS—PART II

I. Primary neoplasms (cont.)

A. CEREBRAL TUMORS (CONT.)

10. Sarcomas: intracranial; are rare and account for only 1–3% of tumors
    a. Fibrosarcoma: derived from fibroblasts, are well circumscribed but nonencapsulated, and are firm tumors. Seen in dura, leptomeninges, perivascular spaces, tela choroidea, or stroma of choroid plexuses
    b. Reticulum cell sarcomas: of controversial origin; seen between ages 50 and 70 years; more invasive and less circumscribed than fibrosarcomas. Most often involve the cerebral hemispheres

✱ 11. Pituitary adenomas: form a special group. Gland in sella turcica contains 3 major types of cells: basophils stain blue, eosinophils stain red, and chromophobes do not stain. Each cell type can produce a tumor

don't need for test

   a. Basophil tumors: small tumors that are probably the result of a disturbance of the close relationship between the pituitary and adrenal glands; found in association with tumors or hypertrophy of the adrenals
      i. Are not space-occupying tumors and result in <u>Cushing's syndrome</u>
   b. Eosinophil adenomas: cause an increased output of growth hormone
      i. They develop before adult life and produce a very feeble giant
      ii. If bone growth has stopped, they cause acromegaly
      iii. Tumors can increase in size until they burst out of the sella and press on the optic chiasma, causing bitemporal hemianopia
   c. Chromophobe adenomas: are quite common and grow to considerable size
      i. Their cells have no special function, but because of their size, they compress the pituitary gland, causing a loss of hormones and resulting in hypopituitarism
      ii. The tumors expand upward and compress the optic chiasma, resulting in bitemporal hemianopia; if not treated, blindness follows

✱ 12. Craniopharyngiomas are benign and account for 3% of all intracranial tumors
    a. Are encapsulated, solid, and/or cystic; are related to the pituitary gland and stalk; and often arise from Rathke's pouch
    b. They are seen in children and adolescents, and are most common of the supratentorial tumors seen in childhood

13. Cholesteatomas (''pearly tumors''): are rare cystic tumors resulting from the inclusion of epiblastic elements in areas from which they are normally absent
    a. Histologically, two groups are distinguished: epidermoid and dermoid cysts

14. Lipomas: rare, benign growths that favor the corpus callosum, suprasellar and pineal regions. In the spinal cord, they are intradural and extramedullary and occur at the thoracic cord level. They are associated with other congenital anomalies

15. Teratomas: very rare within the CNS (0.1% of primary intracranial growths). Found mostly in 1st decade of life and tend to favor the midline
    a. Are composed of various derivatives of the 3 primitive germ cell layers: i.e., epidermal, dermal, glandular, muscular, vascular, cartilaginous elements

16. Hemangioblastomas: account for 1–2.5% of all intracranial tumors and are seen at any age, but most frequently in young and middle-aged people
    a. Most often seen in cerebellum (7% of primary tumors of posterior fossa)
    b. Are most often solitary, but frequently multiple, and in the latter case fall into the category of Hippel-Lindau's disease (familial; retinal and cerebellar hemangioblastoma with visceral lesions)

17. Chordomas: originate from interosseous notochordal remnants. Found at level of sella turcica and clivus in cranial cavity; in spine at sacrococcygeal region

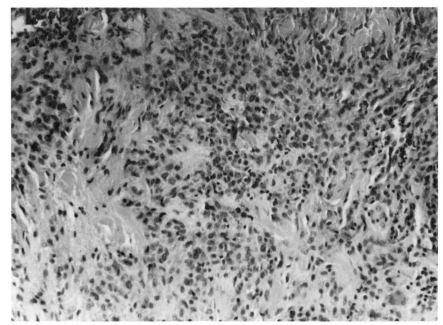

Schwannoma.

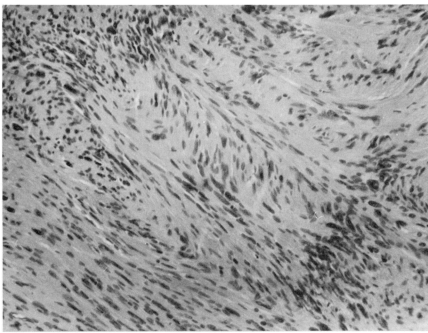

Neurofibroma.

236. SECONDARY AND SPINAL TUMORS

I. **Secondary tumors:** metastatic neoplasms from primary visceral cancer are some of the most frequent histologic types of intracranial and intraspinal tumors. They are found in any region of the cranial cavity or spinal canal, and may involve the cerebral hemispheres, the cerebellum, the brainstem, or less often, the spinal cord, spinal or cranial nerve roots, and meningeal coverings

A. METASTASES may be solitary, but most often are multiple; their size varies from a very small size to that of a large egg. They are generally well circumscribed, either firm or soft, and may show necrosis or cystic degeneration

B. THE TWO CHIEF PRIMARY SITES are bronchopulmonary carcinoma in the male and mammary carcinoma in the female

   1. Other sites may include malignant melanoma, renal carcinoma, and GI carcinoma

C. HISTOLOGICALLY they resemble their primary sources, but atypical features are not rare

II. **Spinal tumors:** most are benign, whereas most brain tumors are malignant

A. FIVE MAJOR TYPES are seen

   1. Neurofibromas grow on the nerve roots
   2. Meningiomas grow from the arachnoid
   3. Hemangioblastomas grow from immature blood vessels
   (*Note:* types 1–3 are all benign, extrinsic, and compress the cord. They can be removed with complete cure)
   4. Astrocytomas*
   5. Ependymomas*

B. ALL SPINAL TUMORS produce similar symptoms which depend on the level of the spine at which they develop

   1. Spinal tumors are suspected with slowly progressive paralysis of the lower body and a clear-cut level on the skin, below which sensation is lost
      a. Neurofibroma may cause pain in the nerve root distribution for months before it is large enough to compress the cord. It then causes a Brown-Séquard syndrome and later, as it grows larger, a total transverse lesion
      b. Meningiomas and hemangioblastomas behave like neurofibromas but with more cord signs and less primary root pain

C. THE INTRINSIC TUMORS OF THE CORD are common in the cervical region

   1. They first damage those sensory fibers that are crossing each other in the center of the cord and thus produce a large patch of loss of sensation to pain and temperature over the arm and upper part of body (like syringomyelia)
   2. They later spread to involve the motor fibers in the pyramidal tracts and cause paralysis of the lower limbs

D. TUMORS OFTEN DEVELOP BELOW THE CORD itself among the fibers of the cauda equina. Due to their position, they cause a progressive flaccid paralysis of the legs with loss of sensation over the buttocks and the back of the legs, and paralysis of the bladder and rectum (if the tumor is benign and not left too long before removal, all these signs are capable of reversing)

E. THE DIAGNOSIS OF SPINAL CORD COMPRESSION is confirmed by finding a block on the Queckenstedt test on lumbar puncture and usually a high CSF protein level. Surgical exploration is necessary to determine the exact position of the tumor following a myelogram—necessary because an examination alone finds that a curable extrinsic tumor may masquerade as an inoperable intrinsic one

---

*Types 4 and 5 grow in the cord substance, are intrinsic, and cannot be removed, but their growth rate is slow. Secondary deposits grow in spaces about the cord and develop rapidly; removal is unsuccessful.

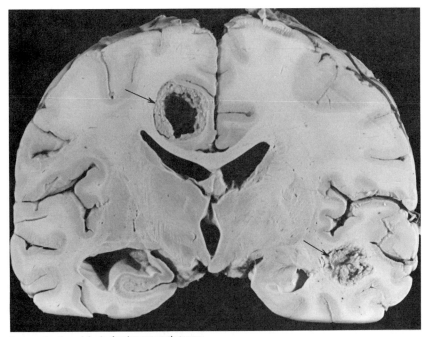

Gross section through brain showing metastatic tumors.

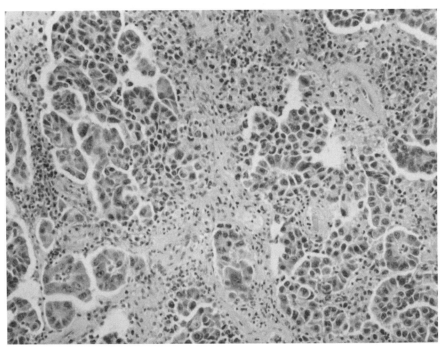

Histologic section of a metastatic carcinoma.

237. OTHER SPINAL TUMORS

I. Peripheral nerve tumors rarely cause much trouble and consist entirely of neurofibromas

A. THESE TUMORS MAY CAUSE PAIN AND SENSORY LOSS in the skin area that the nerve is supplying, but they are usually symptomless

B. THESE TUMORS CAN OFTEN BE REMOVED WITHOUT NERVE SACRIFICE, but it is rarely necessary

II. von Recklinghausen's disease

A. THE OCCURRENCE OF LARGE NUMBERS OF NEUROFIBROMAS throughout the body, involving cranial, spinal, and peripheral nerves

B. THE SKIN IS DARK IN COLOR; there are brown so-called café-au-lait patches and blue nevi, together with numerous little wartlike tags called mollusca fibrosa

C. THE SKIN FEATURES may be present by themselves, and the disease is often familial

III. Congenital tumors may occur anywhere in the nervous system (also see Primary Tumors) and consist of 3 major types

A. EPIDERMOIDS contain cheesy keratin

B. DERMOIDS contain hair or teeth

C. TERATOMAS may have any type of human tissue

D. ALL THESE TUMORS GROW SLOWLY and, if conveniently situated, may be completely removed

E. OTHER SPACE-OCCUPYING LESIONS due to aberrations in development include
   1. Chordomas grow at the base of skull, compressing the brainstem and cranial nerves
   2. Colloid cysts, usually of the 3rd ventricle, cause blockage of CSF flow due to their position; symptoms of high pressure buildup develop, at first intermittently on change of position and then constantly
      a. There are often mental changes and curious episodes in which the legs suddenly give way (drop attacks)
   3. These tumors are very rare

IV. Bone tumors involve the skull and/or the spine

A. BENIGN: osteomas, chondromas, aneurysmal bone cysts, cholesteatomas, etc.

B. MALIGNANT
   1. Primary: osteosarcoma, myeloma, etc.
   2. Secondary: metastases from mammary gland, prostate, kidney, lungs, or thyroid gland

V. Structural changes from expanding intracranial space-occupying lesions

A. FOCAL CHANGES
   1. Changes in the tumor mass with an increase in volume, with local extension, hemorrhage, or cyst formation
   2. Changes involving the adjacent neural parenchyma, resulting in compression and displacement of the parenchyma and/or destruction and invasion of the neural parenchyma

B. REGIONAL CHANGES
   1. Cerebral edema as a result of circulatory disturbances
   2. Disturbances of CSF circulation
   3. Cerebral herniations

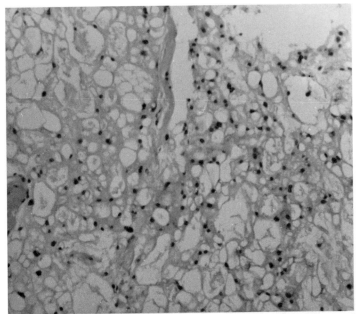

Chordoma: a malignant tumor arising from the embryonic remains of the notochord (low power).

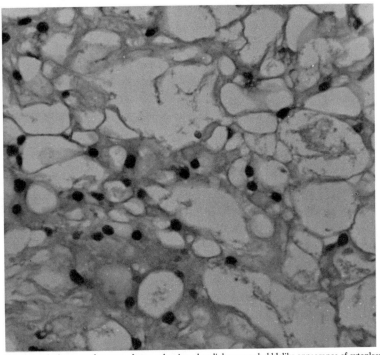

Chordoma: high power of an area of tumor showing physaliphorous or bubblelike appearance of cytoplasm.

INFECTION AND
INFLAMMATION

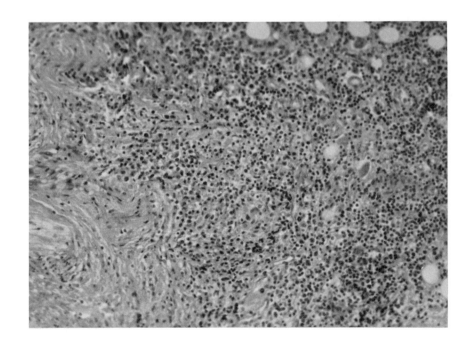

238. BACTERIAL AND PYOGENIC INFECTIONS

I. **Introduction:** the same variety of pathogenic organisms (bacterial, fungal, parasitic, rickettsial or viral) that affect the organs of the body can be involved in the nervous system and its coverings. Some are selective, but not all. Illnesses resulting may be acute (poliomyelitis), subacute (some forms of encephalitis), or drawn out and chronic (neurosyphilis). Infection first leads to an inflammation of the tissues. Inflammation of the meninges (most commonly pia and arachnoid) is called meningitis; of the brain, encephalitis; and of the spinal cord, myelitis. The terms "meningism" or "meningismus" are often used to indicate signs of irritation of the meninges without necessarily meaning they are directly infected.

Inflammatory lesions consist of either disseminated or localized perivascular cuffings and inflammatory infiltrates of variable cell types depending on etiology. Their presence does not necessarily indicate an infectious etiology, since an inflammatory infiltrate of reactive or symptomatic character is well known to occur in other pathologic processes (polymorphonuclear leukocytes in early stages of cerebral infarction or lymphocytic infiltrations are seen in some demyelinating diseases)

II. **Bacterial infections**

A. PYOGENIC INFECTIONS: caused by pus-forming bacteria, the commonest being meningococci, pneumococci, and influenza bacillus. Less often, the agent is staphylococcus, *Listeria monocytogenes,* enterococcus, streptococcus, or others. Some pyogenic infections may be localized in the epidural space (epidural abscess) or in the subdural space (subdural abscess or empyema), but these are exceptional. Most often they involve the leptomeninges or the cerebral parenchyma

1. The organism reaches the meninges directly after open skull fractures or after injury with fracture of base of skull, and spreads from an adjacent focus of suppuration (otitis, sinusitis, mastoiditis) or as the result of blood-borne dissemination from an infective focus (lung, GI tract, etc.)

2. Meningitis

   a. Meningococcal meningitis is spread by nasopharyngeal infection

   b. Pneumococcal meningitis often follows pneumonia or occurs when a fracture of the skull involves the paranasal sinuses

   c. Influenzal meningitis appears especially in infants

3. Pyogenic encephalitis, like meningitis, is caused by pyogenic bacteria or viruses. If the meninges are involved, it is called meningoencephalitis

   a. General features of encephalitis: involvement of neuronal cell bodies, resulting in their destruction and engulfment by macrophages; predominantly perivascular infiltrates; microglial proliferation with nodules

     i. Symptoms consist of moderate headaches, vomiting, confusion, delirium, increasing drowsiness (which passes into coma), and epileptic fits

     ii. Unless meninges are involved, there is little neck stiffness, Kernig's sign is negative, and the CSF may be under normal pressure

     iii. The seriousness of the disease depends on the organisms involved: if due to pyogenic bacteria, it will kill the patient unless treated; if viral, the patient may recover without serious side effects. In addition, some patients show personality changes with a tendency toward outbursts of temper, fits, and difficulty in learning, which may alter their entire lives

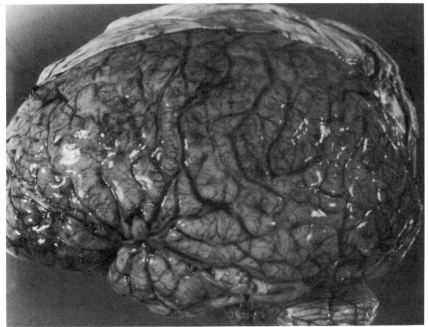

Hemophilus meningitis.

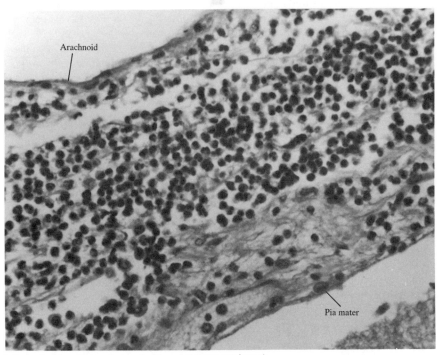

Inflammation of meninges and subarachnoid space (between membranes).

MENINGITIS

239. TUBERCULOSIS AND OTHER INFECTIONS AND INFESTATIONS

II. Bacterial infections (cont.):

B. TUBERCULOSIS is caused by the tubercle bacillus, is difficult to diagnose, and consists of tumorlike masses of cells on the center of which are the bacilli. May behave like a tumor or may rupture into the meninges to cause a meningitis. The masses are called granulomas

1. Epidural tuberculous abscess: usually a complication of tuberculosis of the spine (Pott's disease) involving either the vertebral bodies or the disks

2. Subdural tuberculous abscess: less frequent and usually discovered incidentally after surgery or at necropsy

3. Tuberculous meningitis differs from purulent meningitis in that the meninges over the skull base are predominantly involved; the exudate is essentially made up of lymphocytes, mononuclear cells, and tubercles with areas of caseous necrosis; arterial lesions are constant and responsible for foci of ischemic necrosis

   a. Onset is much less acute, and patients may have been vaguely unwell for weeks, with gradually increasing headaches and listlessness

   b. Eventually the same symptoms and signs as in other cases of meningitis develop, but the CSF shows fewer cells; polys and lymphocytes are seen in equal numbers; and both sugar and chlorides are very low

   c. Once a fatal disease, but now treated by prolonged course of streptomycin injections, together with isonicotinic acid hydrazide (INAH) given orally. Majority respond to streptomycin given intramuscularly; others may need injection directly into CSF. Treatment takes about 3 months

4. Cerebral tuberculoma; a spherical or multiloculated lesion which may be single or multiple, with sites of predilection being the cerebellum, pontine tegmentum, and paracentral lobule. May spontaneously become cystic, fibrous, and calcified, but the danger lies in spilling into the meninges

III. Mycoses are fungal infections of the CNS that are seen in special circumstances, such as following prolonged antibiotic treatment, in the course of hematologic or lymphoreticular disease, or as a complication of immunosuppressive therapy

A. THE MOST FREQUENT FORMS are candidiasis and torulosis (or cryptococcosis)

B. RARE FORMS: aspergillosis, mucormycosis (or phycomycosis), and coccidioidomycosis

IV. Rickettsial infections: forms of infections (murine or endemic typhus, exanthematic or epidemic typhus, or Rocky Mountain spotted fever) may cause nervous system lesions involving the gray matter of the cerebral hemispheres and brainstem

V. Parasitic infections: parasites may affect the nervous system but vary greatly in different geographic areas. The chief causative parasites are protozoa and metazoa

A. TAPEWORM *(Taenia solium)* can at times develop its larval stage not, as normally, in the pig, but in human muscle and brain; the condition is called cysticercosis

1. Muscle cysts are harmless, but cysts of the brain can cause epilepsy and mental changes

B. TRICHINELLA SPIRALIS is a parasite that affects man if eaten in uncooked meat

C. HYDATID DISEASE: cysts formed by the parasite *Echinococcus*. Is rare and may develop in the brain, skull, or meninges (behaves like a brain tumor)

D. LEPTOSPIROSIS: caused by an organism *(Leptospira canicola)* carried by rats and dogs. Causes canicola fever, meningitis, and inflammation of the eyes

BACTERIAL CAUSES OF FEVER IN CHILDREN

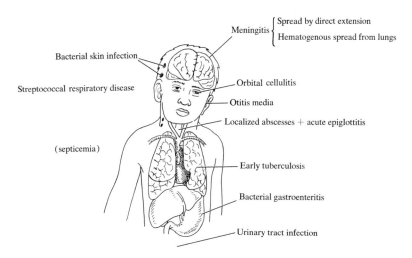

Meningitis {
Spread by direct extension
Hematogenous spread from lungs

Bacterial skin infection

Streptococcal respiratory disease

Orbital cellulitis

Otitis media

Localized abscesses + acute epiglottitis

(septicemia)

Early tuberculosis

Bacterial gastroenteritis

Urinary tract infection

Convulsions not uncommon and justify lumbar puncture in child with febrile illness

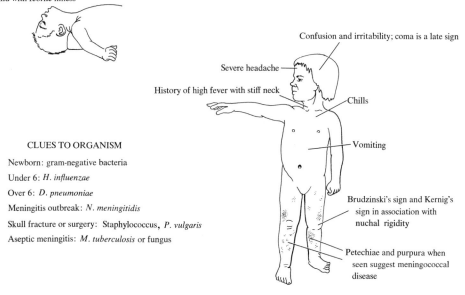

Confusion and irritability; coma is a late sign

Severe headache

History of high fever with stiff neck

Chills

Vomiting

CLUES TO ORGANISM

Newborn: gram-negative bacteria

Under 6: *H. influenzae*

Over 6: *D. pneumoniae*

Meningitis outbreak: *N. meningitidis*

Skull fracture or surgery: Staphylococcus, *P. vulgaris*

Aseptic meningitis: *M. tuberculosis* or fungus

Brudzinski's sign and Kernig's sign in association with nuchal rigidity

Petechiae and purpura when seen suggest meningococcal disease

CLINICAL PICTURE OF
BACTERIAL MENINGITIS
OVER AGE 2

240. SYPHILIS, SARCOIDOSIS, LEPROSY, AND BRUCELLOSIS

I. Syphilis: CNS involvement is a sequel of primary luetic disease that has escaped notice or has been inadequately treated and appears years after the original infection. The organism is a spirochete, *Treponema pallidum,* and causes a chronic inflammation of the nervous tissue and its coverings

A. SUBACUTE SECONDARY SYPHILITIC MENINGITIS: leptomeningeal invasion resulting in meningitis with lymphocytes, plasma cells, and perivascular infiltrates

B. TERTIARY SYPHILIS: varied clinical picture since many parts of CNS are involved

    1. Chronic meningitis leads to fibrous organization and ultimate occlusion of the CSF pathways or resorption with concomitant infiltration of arterial walls

        a. Episodes of thrombosis occur, resulting in sudden paralysis of 1 of the cranial nerves or of a limb

    2. General paresis (general paralysis of the insane): inflammatory meningovascular lesion associated with parenchymatous changes of encephalitic type

        a. Characterized by much cortical atrophy, neuronal depopulation, and proliferation of rod-shaped microglia as scattered foci of different ages

        b. Symptoms are those of slow mental deterioration with grandiose delusions

        c. There are often fits and signs of upper motor neuron disease such as spasticity, exaggerated reflexes, and extensor plantars

        d. Argyll Robertson pupils are common

    3. Tabes dorsalis consists of degeneration of the posterior columns and spinal roots, with involvement of dorsal root ganglia (''tabes'' = wasting)

        a. Spinal cord involvement presents with demyelination and gliosis of the posterior columns secondary to radiculoganglionic lesions

        b. There is an inflammatory process and the treponema are absent

        c. Root irritation causes ''lightning'' pains like a sudden outburst of needlelike pins in the shins, calves, and arms, as well as ''girdle pains'' which resemble a constricting band around the waist

        d. Lesions result in a loss of position sense, so patient becomes ataxic and walks with legs wide apart and a stamping gait—helpless in the dark

        e. Sensory fiber degeneration results in an upset reflex arc so that tendon reflexes are absent and bladder sensations are disturbed, leaving no desire to pass urine, resulting in overfilling or overflow incontinence

        f. Gross destructive changes occur in knees and other joints, but, due to sensory loss, are painless—the disorganized joint of Charcot

        g. Most characteristic sign of neurosyphilis is in the eyes—the so-called Argyll Robertson pupil (small, irregular pupil which does not react to light but does constrict on looking at near objects)

        h. In active disease, the CSF shows raised lymphocyte count and a tabetic (luetic) Lange curve

II. Sarcoidosis: origin is uncertain; granulomas develop that are similar to tuberculomas but do not contain bacilli. Found in brain, meninges, about peripheral nerves, and in glands of body

III. Leprosy: leprosy bacillus causes peripheral nerve thickening. Sensation is lost in parts supplied by the nerves and trophic changes take place: areas of skin and bone slough off, resulting in mutilation of hands, feet, and face

IV. Brucellosis (Malta fever): leptomeningeal involvement common in the septicemic phase of this disease. Different forms of neurobrucellosis correspond to a variety of clinicopathologic pictures which frequently involve the VIIIth (acoustic) nerve

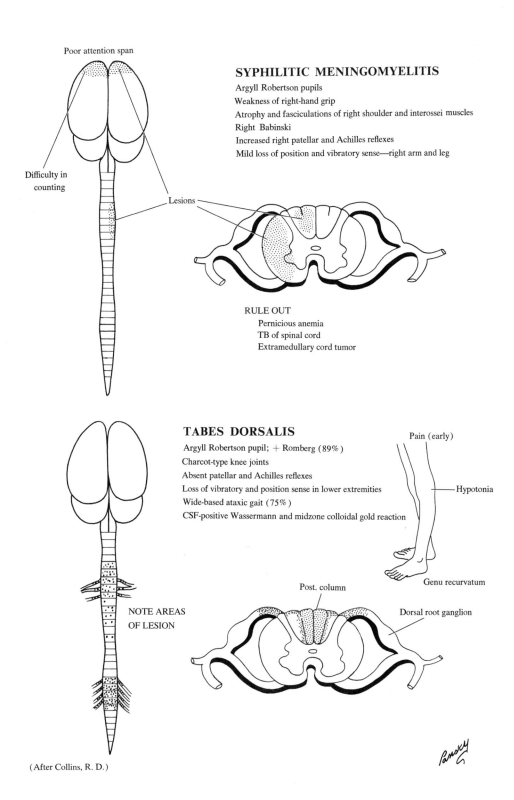

SYPHILITIC MENINGOMYELITIS

Argyll Robertson pupils

Weakness of right-hand grip

Atrophy and fasciculations of right shoulder and interossei muscles

Right Babinski

Increased right patellar and Achilles reflexes

Mild loss of position and vibratory sense—right arm and leg

Poor attention span

Difficulty in counting

Lesions

RULE OUT

Pernicious anemia

TB of spinal cord

Extramedullary cord tumor

TABES DORSALIS

Argyll Robertson pupil; + Romberg (89%)

Charcot-type knee joints

Absent patellar and Achilles reflexes

Loss of vibratory and position sense in lower extremities

Wide-based ataxic gait (75%)

CSF-positive Wassermann and midzone colloidal gold reaction

Pain (early)

Hypotonia

Genu recurvatum

NOTE AREAS OF LESION

Post. column

Dorsal root ganglion

(After Collins, R. D.)

241. VIRAL DISEASES AND ENCEPHALITIDES OF RNA VIRUSES—PART I

I. Viral diseases: lesions show variable forms. Some have an acute course (in viral encephalitides); others show an immunologic phenomenon and develop latent infections (herpes zoster)

A. SOME LESIONS are nonspecific, related to immunoallergic reactions, and are secondary to viral infection, involving the meninges and white matter; other lesions are directly related to penetration of the CNS by the virus and involve mainly the gray matter (polioencephalitis)

B. ONE GROUP shows animal transmissibility and resembles a comparative neuropathology, suggesting a viral mechanism related to the ''slow-virus diseases''

C. VIRAL INFECTIONS WITH NONSPECIFIC NERVOUS SYSTEM MANIFESTATIONS
1. Acute viral lymphocytic meningitis: response common to infections caused by a number of different viruses
2. Postinfectious perivenous encephalitis: may complicate a variety of viral diseases, particularly exanthemas (measles, chickenpox, rubella, or smallpox)
    a. Produces a clinical picture of an acute disseminated encephalomyelitis that may show up during convalescence from the viral infection
    b. The basic mechanism appears to be immunoallergic, but this is questionable given the finding of viral antigen in glial cells of the lesion

D. VIRAL ENCEPHALITIS is directly related to viral penetration of the CNS as a result of direct spread along the peripheral nerves (particularly the olfactory) or, more commonly, in the course of a viremia. The inflammatory process often involves the meninges (meningoencephalitis), the cord (encephalomyelitis), or both (meningoencephalomyelitis), as well as the nerve roots. Diagnosis and treatment of these conditions are often very difficult because a virus is rarely isolated and the diagnosis is usually presumptive. Much more insight is obtained by showing changes in the antibodies in the blood than by attempting to find the organism
1. Encephalitides of RNA viruses: due to enteroviruses (Coxsackie group and echoviruses) present as a lymphocytic meningitis
    a. Poliomyelitis: most frequent form of the disease; does not correspond to a typical encephalitis, but rather to an acute anterior poliomyelitis
        i. The lesions selectively involve the motor neurons of the anterior horns
        ii. Also involve other regions: frontal gyri, hypothalamus, reticular formation, and posterior horns (polio encephalitis)
        iii. The inflammatory infiltrates, edema, vasodilatation, microglial and macrophage proliferation can be severe
        iv. Following resolution, the residual lesions consist of atrophy of the anterior horns, with neuronal loss and gliosis
        v. About 50% of the patients are adults
        vi. Onset is acute, with pyrexia, headache, vomiting, neck stiffness, and pain in the back and limbs; may last 2–4 days and may subside completely; called nonparalytic poliomyelitis
        vii. In a few cases, after the nonparalytic period, there is a 48-hour interval and then the paralytic stage begins, with pain and extreme tenderness of various muscles. The muscles rapidly become weak or totally paralyzed. Danger lies in involvement of respiratory or throat muscles, preventing swallowing, coughing, or proper breathing
        viii. When the acute stage subsides, many muscles may recover completely, but others may waste, at times almost completely
    b. An identical or fairly similar picture is seen with other enteroviruses, in particular with Coxsackie virus, possibly as a result of polio immunization

VIRAL SYNDROMES IN A FEBRILE CHILD

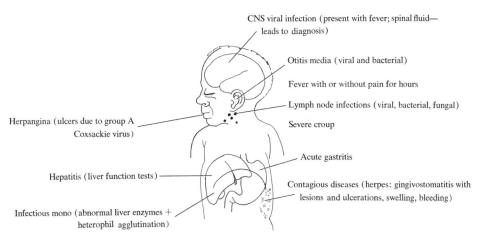

CNS viral infection (present with fever; spinal fluid— leads to diagnosis)

Otitis media (viral and bacterial)

Fever with or without pain for hours

Lymph node infections (viral, bacterial, fungal)

Severe croup

Herpangina (ulcers due to group A Coxsackie virus)

Acute gastritis

Hepatitis (liver function tests)

Contagious diseases (herpes: gingivostomatitis with lesions and ulcerations, swelling, bleeding)

Infectious mono (abnormal liver enzymes + heterophil agglutination)

CLINICAL CONSIDERATIONS

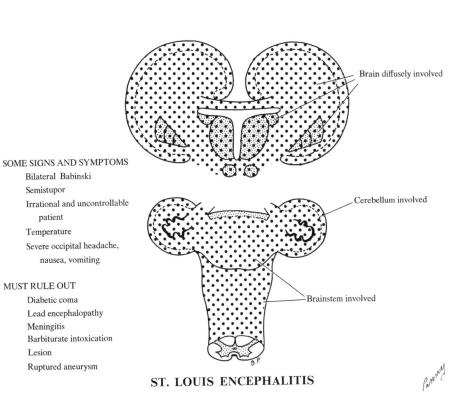

Brain diffusely involved

Cerebellum involved

Brainstem involved

SOME SIGNS AND SYMPTOMS
- Bilateral Babinski
- Semistupor
- Irrational and uncontrollable patient
- Temperature
- Severe occipital headache, nausea, vomiting

MUST RULE OUT
- Diabetic coma
- Lead encephalopathy
- Meningitis
- Barbiturate intoxication
- Lesion
- Ruptured aneurysm

ST. LOUIS ENCEPHALITIS

242. ENCEPHALITIDES OF RNA VIRUSES—PART II

I. Viral diseases (cont.)

D. VIRAL ENCEPHALITIS (CONT.)

2. Encephalitides due to arboviruses (arthropod-borne) have a distinct geographic distribution, as often indicated by their names. The best known of these are

   a. Mosquito-borne encephalitides: St. Louis encephalitis, eastern and western equine encephalitis, Japanese B encephalitis seen in the Far East, and Murray Valley encephalitis (Australia)

   b. Tick-borne encephalitis: Russian spring-summer and Central European encephalitis

   c. Human rabies is due to a rhabdovirus and is transmitted to man by the bite of an infected animal (cat, dog, bat, wild animal, etc.) and is always fatal once it has occurred

   i. Identified by the presence of characteristic cytoplasmic inclusions in the neurons, e.g., Negri bodies accompanied by inflammatory cellular infiltrates

   ii. Polyneuritis and encephalomyelitis have been reported after rabies vaccination with either the attenuated or the inactivated virus and are probably related to inadequate inactivation of virus or to immunopathologic process

   d. Encephalitides due to paramyxoviruses: in addition to influenza and mumps, which may be associated with different forms of encephalitis, measles infection can cause 2 types of encephalitis: acute postinfectious encephalitis and subacute sclerosing panencephalitis

   i. Mumps encephalitis tends to occur during the worst stage of the acute illness and is often accompanied by signs of meningitis as well

   (a) Lasts 2–3 days and is not usually followed by serious side effects

   ii. Measles encephalitis, in general, usually occurs as the child is getting over the worst phase and is beginning to recover, when signs of cerebral disturbances develop and the child becomes very ill again. There is a danger of serious sequelae

   iii. Subacute sclerosing panencephalitis is seen in children several years after a known episode of measles

   (a) The lesions involve the gray matter (cortex and basal ganglia), where inclusion bodies are seen in neuronal and glial nuclei as well as in the cytoplasm of neurons (inclusion body encephalitis)

   (b) Lesions may involve the white matter, where one sees a diffuse demyelination with marked astrocytic proliferation

   (c) One sees perivascular inflammatory infiltrates in both white and gray matter

   (d) Although the measles virus has been implicated, the precise mechanism of this prolonged viral infection is not understood

   (e) Is a rare condition and usually begins as a slowly progressive mental deterioration followed by fits, tremulousness, ataxia, and finally extreme spasticity and rigidity, accompanied by attacks of drooping of the head and limbs followed by tremor

   (f) EEG shows an almost diagnostic abnormality

   (g) No treatment exists to check its fatality and long illness

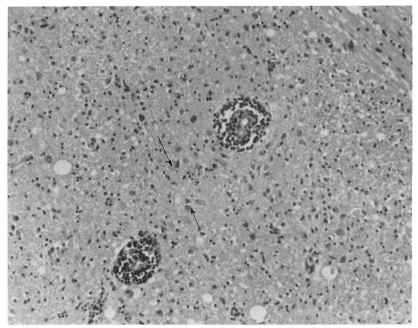

Note hypertrophied astrocytes (arrows) and perivascular cuffing in white matter.

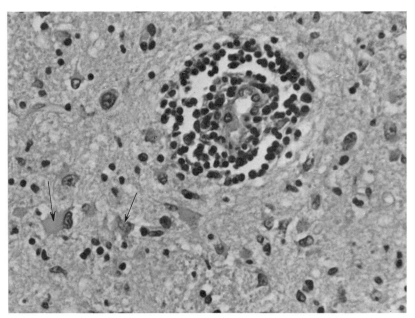

High power of one perivascular cuff and reactive hypertrophied astrocytes (arrows).

SUBACUTE SCLEROSING PANENCEPHALITIS

243. ENCEPHALITIDES OF DNA, NONIDENTIFIED, AND SLOW VIRUSES

I. Viral diseases (cont.)

D. VIRAL ENCEPHALITIDES (CONT.)

3. Encephalitides of DNA viruses: due to herpesvirus group
   a. Herpes simplex encephalitis: due to a subgroup A virus; topographically involves predominantly the temporal lobes and limbic areas
      i. Severe inflammatory infiltrates which may result in hemorrhagic necrosis with marked cerebral edema
      ii. Intranuclear inclusions may be seen in the neurons and in some glial cells
      iii. Clinical picture develops rapidly over 3–4 days and may be accompanied by focal fits and paralysis of one part of the body. Speech disturbances and visual field loss may also be seen
   b. Cytomegalovirus encephalitis: caused by a herpesvirus of subgroup B. Seen as part of a multisystem viral infection or as an isolated CNS infection associated with chorioretinitis. Usually caused by early fetal infection and may lead to microcephaly, hydrocephalus, and intracranial calcifications with subependymal and subcortical lesions
   c. Herpes zoster varicella: the virus of shingles (also causes varicella); selectively involves neurons of 1 or more spinal root ganglia and the posterior horn of the corresponding metameric segment. Trigeminal nerve ganglia are also involved
      i. Illness begins with severe pain in distribution of nerve root, often around the trunk, or above the eyes in the case of ophthalmic herpes, and red patches appear on the skin. The latter change to vesicles which run together, crust over, and heal, leaving pale depressed scars
      ii. In geniculate herpes, the vesicles develop in the ears, and often facial paralysis and deafness result
4. Encephalitides due to papoviruses: progressive multifocal leukoencephalopathy seen in complications of some immunodeficiency diseases, e.g., chronic lymphatic leukemia, Hodgkin's disease.
   a. White matter of occipital lobe is site of limited foci of demyelination
5. Due to nonidentified viruses
   a. Encephalitis lethargica: known as epidemic encephalitis of von Economo or "sleeping sickness." Is rare and is characterized by preferential involvement of midbrain and basal ganglia
      i. Brainstem involvement leads to paralysis of eye movements; patients are drowsy by day and wakeful by night
      ii. Face becomes rigid, and there is considerable personality disturbance
      iii. Recovery usually occurs but with an aftereffect, namely, a type of parkinsonism called postencephalitic parkinsonian syndrome
   b. Uveomeningoencephalitides: inflammatory encephalitic, meningitic, and uveal (choroid, ciliary body, and iris) lesions of unknown etiology (seen in Behçet's disease)

E. DUE TO TRANSMISSIBLE "SLOW VIRUS": characterized by animal transmissibility after injection of cerebral biopsy fragments

1. Creutzfeldt-Jacob disease: characterized by dementia associated with various neurologic manifestations. Fatal in a few months
   a. See neuronal cell depopulation with spongiosis and dense astrocytic gliosis of the cerebral cortex, especially occipital lobes
   b. Resembles animal virus diseases (scrapie in sheep and mink encephalopathy)
2. Kuru: a human disease of New Guinea. A progressive, fatal encephalopathy

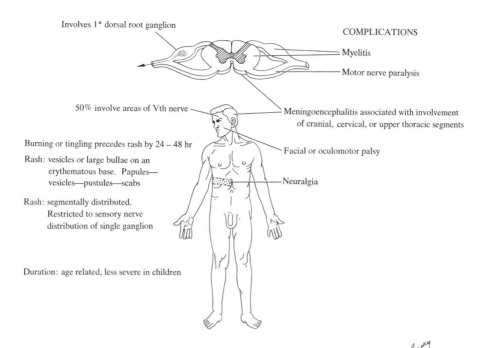

Involves 1° dorsal root ganglion

COMPLICATIONS

Myelitis

Motor nerve paralysis

50% involve areas of Vth nerve

Meningoencephalitis associated with involvement of cranial, cervical, or upper thoracic segments

Burning or tingling precedes rash by 24 – 48 hr

Rash: vesicles or large bullae on an erythematous base. Papules—vesicles—pustules—scabs

Facial or oculomotor palsy

Neuralgia

Rash: segmentally distributed. Restricted to sensory nerve distribution of single ganglion

Duration: age related, less severe in children

HERPES ZOSTER: AN ACUTE INFECTION BY A FILTERABLE VIRUS

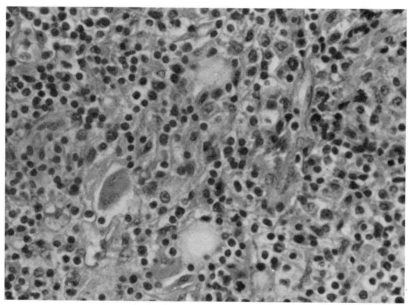

Histologic preparation showing cellular infiltration.

APPENDIXES

APPENDIX I. ATLAS OF THE BRAINSTEM

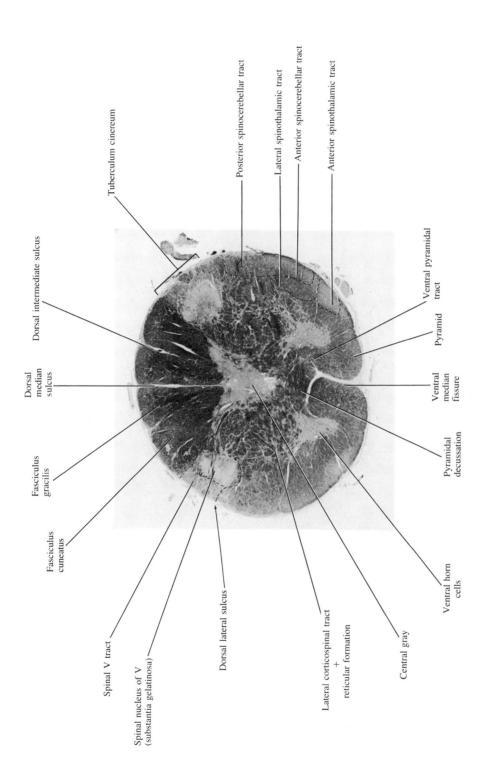

Tuberculum cinereum

Dorsal intermediate sulcus

Dorsal median sulcus

Fasciculus gracilis

Fasciculus cuneatus

Spinal V tract

Spinal nucleus of V (substantia gelatinosa)

Dorsal lateral sulcus

Lateral corticospinal tract + reticular formation

Central gray

Ventral horn cells

Pyramidal decussation

Ventral median fissure

Pyramid

Ventral pyramidal tract

Posterior spinocerebellar tract

Lateral spinothalamic tract

Anterior spinocerebellar tract

Anterior spinothalamic tract

1. CORD-MEDULLA JUNCTION

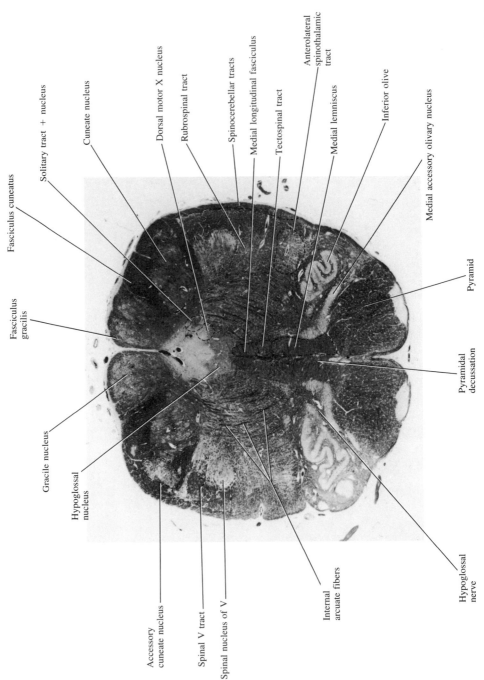

Anterolateral
spinothalamic
tract

Dorsal motor X nucleus

Rubrospinal tract

Spinocerebellar tracts

Medial longitudinal fasciculus

Tectospinal tract

Medial lemniscus

Inferior olive

Medial accessory olivary nucleus

Cuneate nucleus

Solitary tract + nucleus

Fasciculus cuneatus

Fasciculus
gracilis

Gracile nucleus

Hypoglossal
nucleus

Accessory
cuneate nucleus

Spinal V tract

Spinal nucleus of V

Internal
arcuate fibers

Hypoglossal
nerve

Pyramidal
decussation

Pyramid

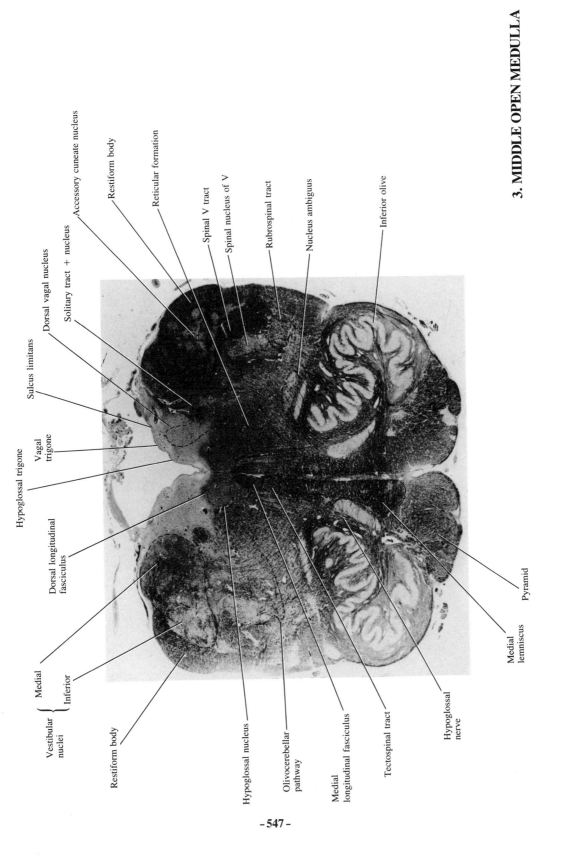

Vestibular nuclei { Medial
Inferior

Restiform body

Hypoglossal nucleus

Olivocerebellar pathway

Medial longitudinal fasciculus

Tectospinal tract

Hypoglossal nerve

Medial lemniscus

Pyramid

Inferior olive

Nucleus ambiguus

Rubrospinal tract

Spinal nucleus of V

Spinal V tract

Reticular formation

Restiform body

Accessory cuneate nucleus

Solitary tract + nucleus

Dorsal vagal nucleus

Sulcus limitans

Vagal trigone

Hypoglossal trigone

Dorsal longitudinal fasciculus

3. MIDDLE OPEN MEDULLA

-547-

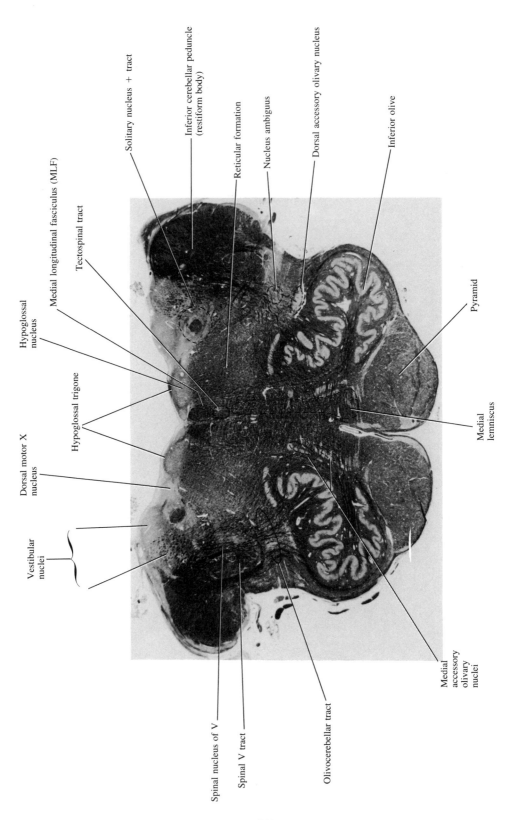

Vestibular nuclei

Dorsal motor X nucleus

Hypoglossal trigone

Hypoglossal nucleus

Tectospinal tract

Medial longitudinal fasciculus (MLF)

Solitary nucleus + tract

Inferior cerebellar peduncle (restiform body)

Reticular formation

Nucleus ambiguus

Dorsal accessory olivary nucleus

Inferior olive

Pyramid

Medial lemniscus

Medial accessory olivary nuclei

Olivocerebellar tract

Spinal V tract

Spinal nucleus of V

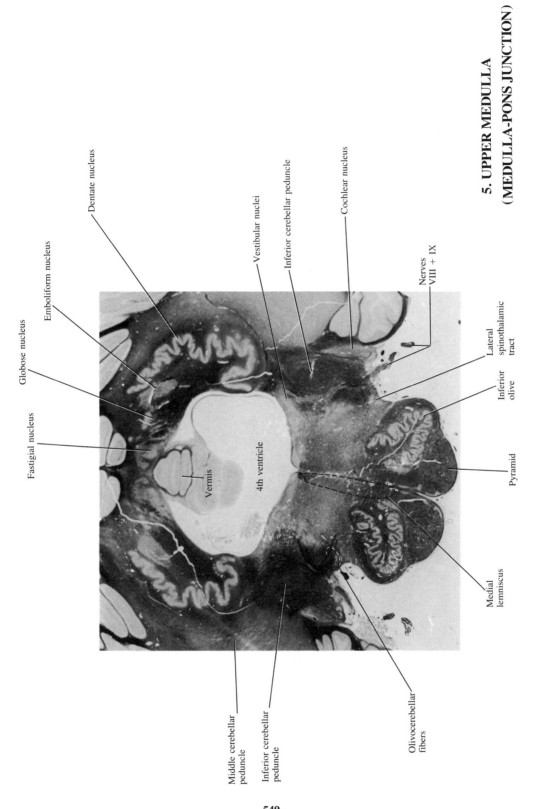

5. UPPER MEDULLA (MEDULLA-PONS JUNCTION)

Dentate nucleus

Emboliform nucleus

Globose nucleus

Fastigial nucleus

Vermis

4th ventricle

Vestibular nuclei

Inferior cerebellar peduncle

Cochlear nucleus

Nerves VIII + IX

Lateral spinothalamic tract

Inferior olive

Pyramid

Medial lemniscus

Olivocerebellar fibers

Middle cerebellar peduncle

Inferior cerebellar peduncle

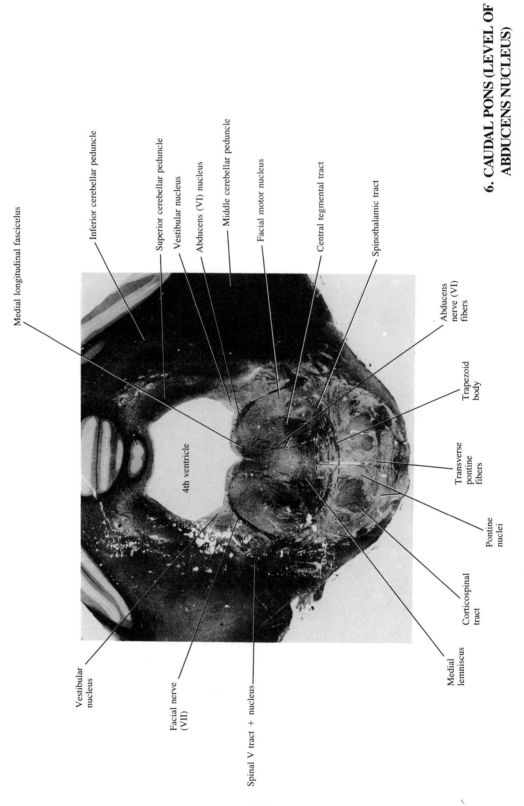

Medial longitudinal fasciculus

Inferior cerebellar peduncle

Superior cerebellar peduncle

Vestibular nucleus

Abducens (VI) nucleus

Middle cerebellar peduncle

Facial motor nucleus

Central tegmental tract

Spinothalamic tract

Abducens nerve (VI) fibers

Trapezoid body

Transverse pontine fibers

Pontine nuclei

Corticospinal tract

Medial lemniscus

Spinal V tract + nucleus

Facial nerve (VII)

Vestibular nucleus

4th ventricle

6. CAUDAL PONS (LEVEL OF ABDUCENS NUCLEUS)

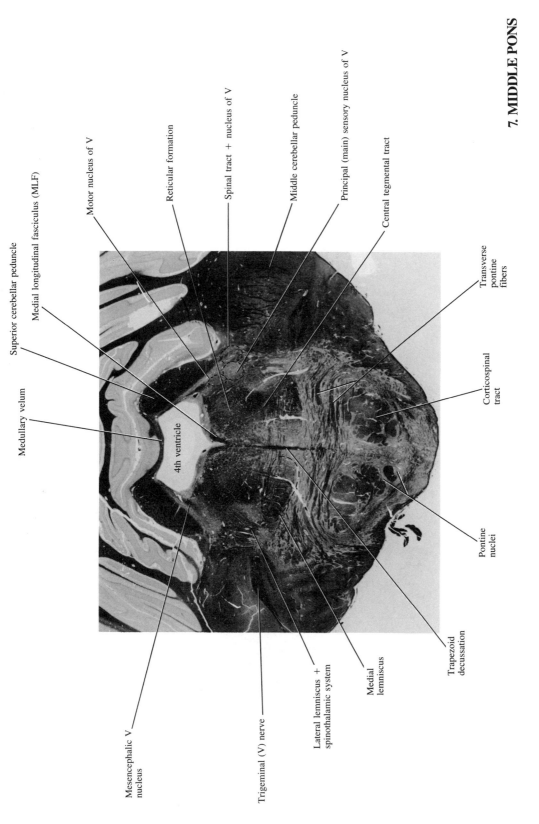

Superior cerebellar peduncle

Medial longitudinal fasciculus (MLF)

Motor nucleus of V

Reticular formation

Spinal tract + nucleus of V

Middle cerebellar peduncle

Principal (main) sensory nucleus of V

Central tegmental tract

Transverse pontine fibers

Corticospinal tract

Pontine nuclei

Medullary velum

4th ventricle

Mesencephalic V nucleus

Trigeminal (V) nerve

Lateral lemniscus + spinothalamic system

Medial lemniscus

Trapezoid decussation

7. MIDDLE PONS

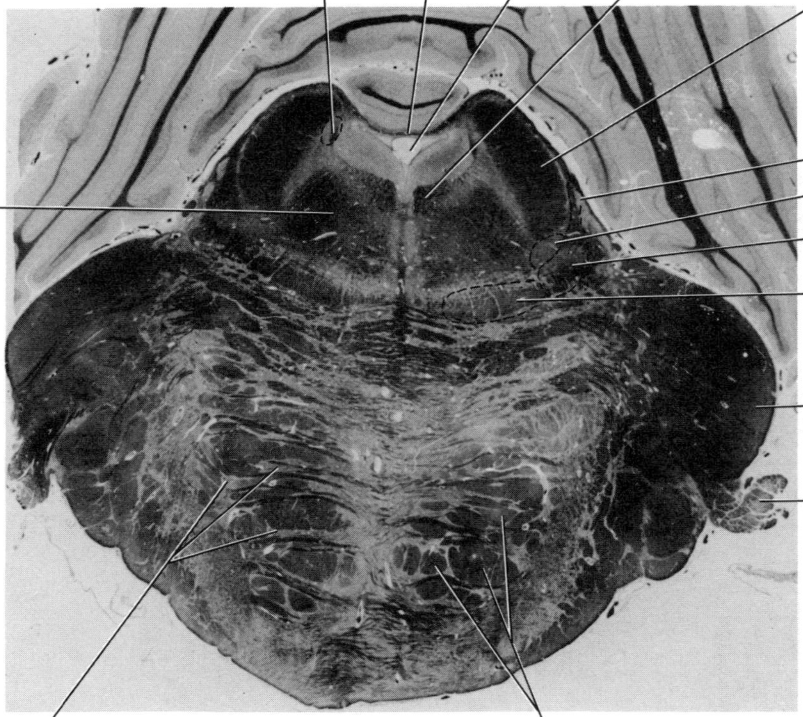
Mesencephalic nucleus
+ tract of V
Medullary velum
Rostral IVth ventricle
Medial longitudinal fasciculus
Superior cerebellar peduncle
Lateral lemniscus
Nucleus of lateral lemniscus
Spinothalamic tract
Medial lemniscus
Middle cerebellar peduncle
Nerve V
Reticular formation
Pontine nuclei
Corticospinal tract

8. JUNCTION OF PONS AND MESENCEPHALON

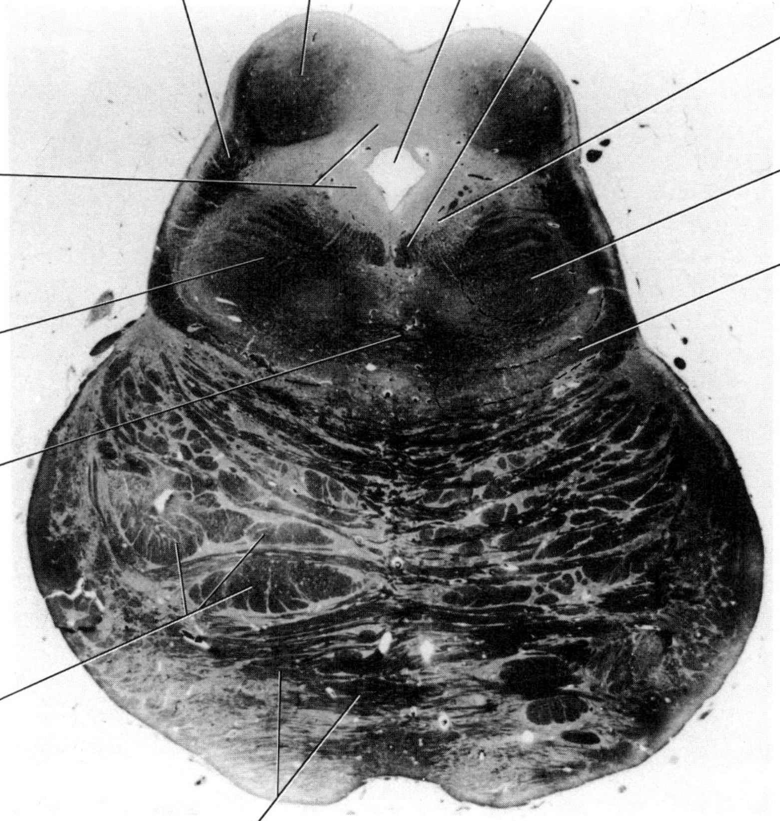

Cerebral aqueduct

Medial longitudinal fasciculus

Trochlear (IV) nucleus

Superior cerebellar peduncle

Medial lemniscus

Inferior colliculus

Lateral lemniscus

Periaqueductal gray

Central tegmental tract

Decussation of superior cerebellar peduncle

Corticospinal, corticobulbar, corticopontine fibers

Transverse pontine fibers

9. MESENCEPHALON (LEVEL OF INFERIOR COLLICULUS)

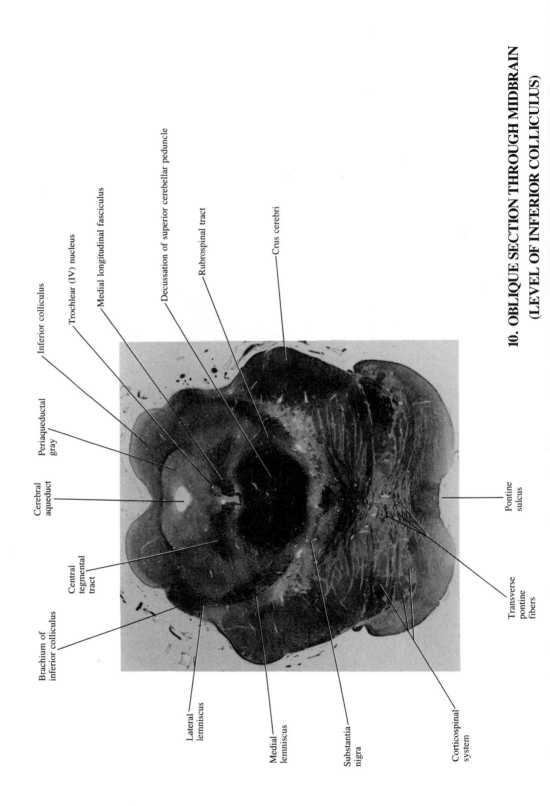

Inferior colliculus

Trochlear (IV) nucleus

Medial longitudinal fasciculus

Decussation of superior cerebellar peduncle

Rubrospinal tract

Crus cerebri

Periaqueductal gray

Cerebral aqueduct

Central tegmental tract

Brachium of inferior colliculus

Pontine sulcus

Transverse pontine fibers

Lateral lemniscus

Medial lemniscus

Substantia nigra

Corticospinal system

10. OBLIQUE SECTION THROUGH MIDBRAIN (LEVEL OF INFERIOR COLLICULUS)

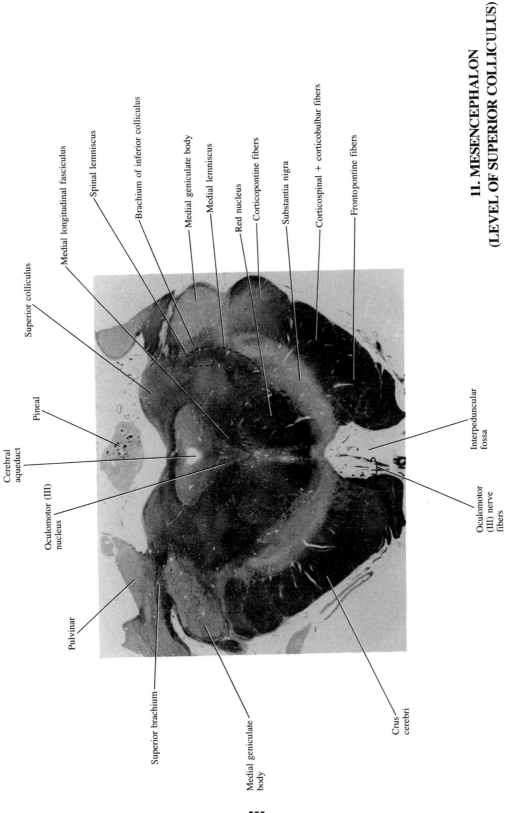

Cerebral aqueduct

Pineal

Oculomotor (III) nucleus

Pulvinar

Superior brachium

Medial geniculate body

Crus cerebri

Superior colliculus

Medial longitudinal fasciculus

Spinal lemniscus

Brachium of inferior colliculus

Medial geniculate body

Medial lemniscus

Red nucleus

Corticopontine fibers

Substantia nigra

Corticospinal + corticobulbar fibers

Frontopontine fibers

Interpeduncular fossa

Oculomotor (III) nerve fibers

**11. MESENCEPHALON
(LEVEL OF SUPERIOR COLLICULUS)**

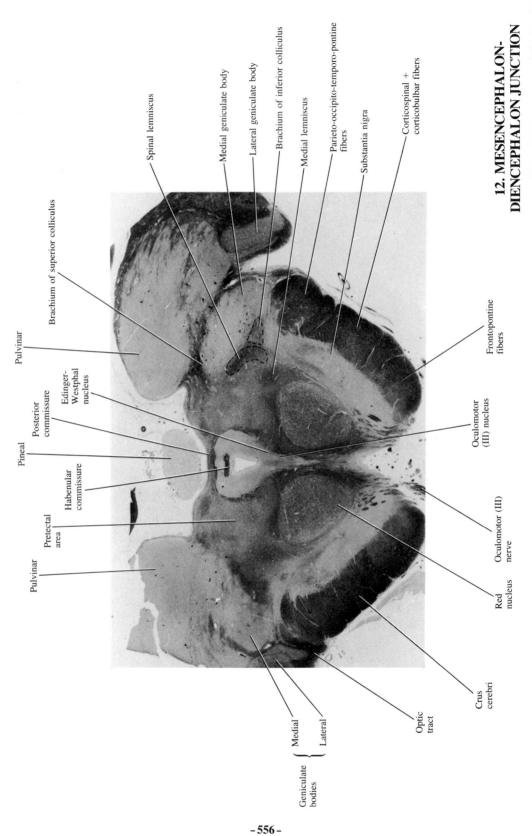

Spinal lemniscus

Medial geniculate body

Lateral geniculate body

Brachium of inferior colliculus

Medial lemniscus

Parieto-occipito-temporo-pontine fibers

Substantia nigra

Corticospinal + corticobulbar fibers

Brachium of superior colliculus

Pulvinar

Edinger-Westphal nucleus

Posterior commissure

Pineal

Habenular commissure

Pretectal area

Pulvinar

Geniculate bodies { Medial / Lateral

Optic tract

Crus cerebri

Red nucleus

Oculomotor (III) nerve

Oculomotor (III) nucleus

Frontopontine fibers

12. MESENCEPHALON-DIENCEPHALON JUNCTION

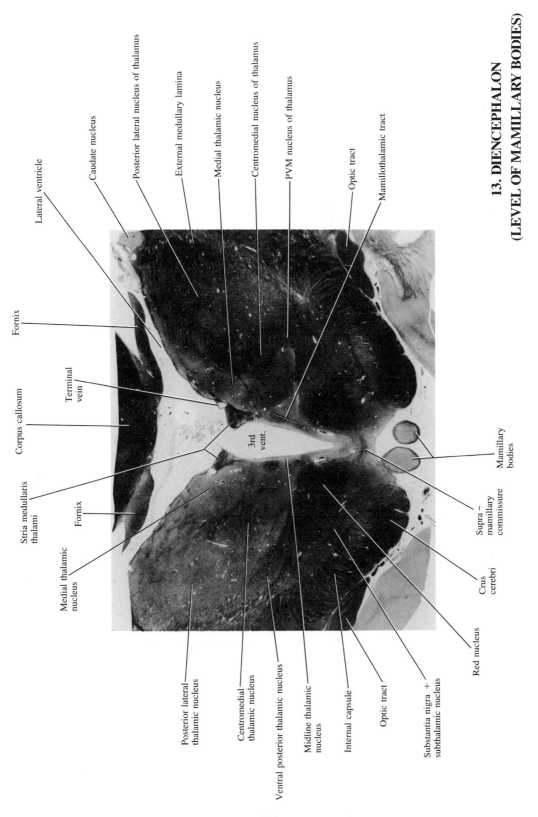

Lateral ventricle

Fornix

Corpus callosum

Terminal vein

Stria medullaris thalami

Medial thalamic nucleus

Fornix

Caudate nucleus

Posterior lateral nucleus of thalamus

External medullary lamina

Medial thalamic nucleus

Centromedial nucleus of thalamus

PVM nucleus of thalamus

Optic tract

Mamillothalamic tract

3rd vent.

Mamillary bodies

Supra – mamillary commissure

Crus cerebri

Red nucleus

Substantia nigra + subthalamic nucleus

Optic tract

Internal capsule

Midline thalamic nucleus

Ventral posterior thalamic nucleus

Centromedial thalamic nucleus

Posterior lateral thalamic nucleus

13. DIENCEPHALON (LEVEL OF MAMILLARY BODIES)

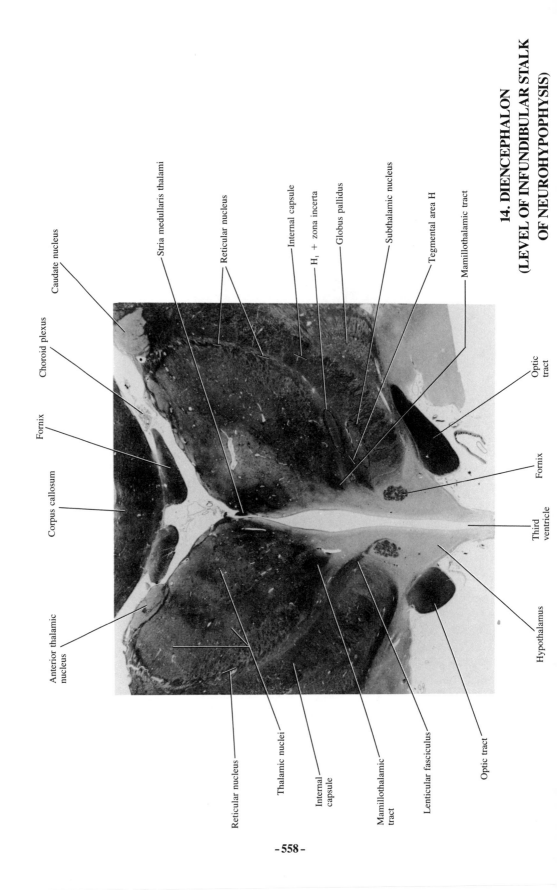

14. DIENCEPHALON
(LEVEL OF INFUNDIBULAR STALK
OF NEUROHYPOPHYSIS)

Caudate nucleus

Stria medullaris thalami

Reticular nucleus

Internal capsule

H₁ + zona incerta

Globus pallidus

Subthalamic nucleus

Tegmental area H

Mamillothalamic tract

Choroid plexus

Fornix

Corpus callosum

Anterior thalamic nucleus

Optic tract

Fornix

Third ventricle

Hypothalamus

Reticular nucleus

Thalamic nuclei

Internal capsule

Mamillothalamic tract

Lenticular fasciculus

Optic tract

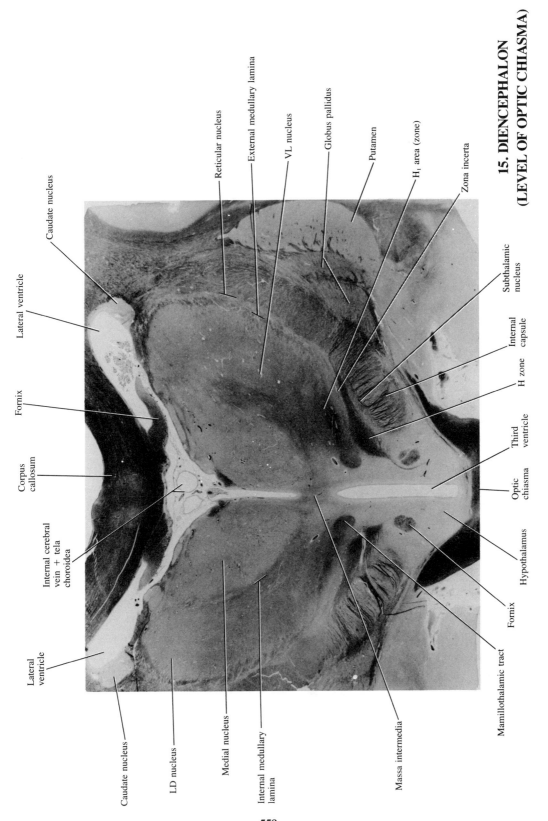

Caudate nucleus

Lateral ventricle

Fornix

Corpus callosum

Internal cerebral vein + tela choroidea

Lateral ventricle

Caudate nucleus

LD nucleus

Medial nucleus

Internal medullary lamina

Massa intermedia

Mamillothalamic tract

Fornix

Hypothalamus

Optic chiasma

Third ventricle

H zone

Internal capsule

Subthalamic nucleus

Reticular nucleus

External medullary lamina

VL nucleus

Globus pallidus

Putamen

H₁ area (zone)

Zona incerta

15. DIENCEPHALON (LEVEL OF OPTIC CHIASMA)

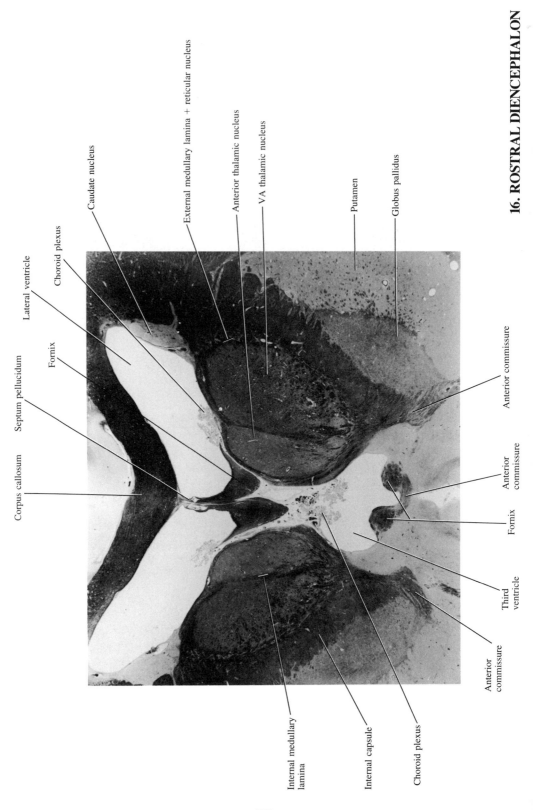

Corpus callosum

Septum pellucidum

Fornix

Lateral ventricle

Choroid plexus

Caudate nucleus

External medullary lamina + reticular nucleus

Anterior thalamic nucleus

VA thalamic nucleus

Putamen

Globus pallidus

Anterior commissure

Anterior commissure

Fornix

Third ventricle

Anterior commissure

Choroid plexus

Internal capsule

Internal medullary lamina

16. ROSTRAL DIENCEPHALON

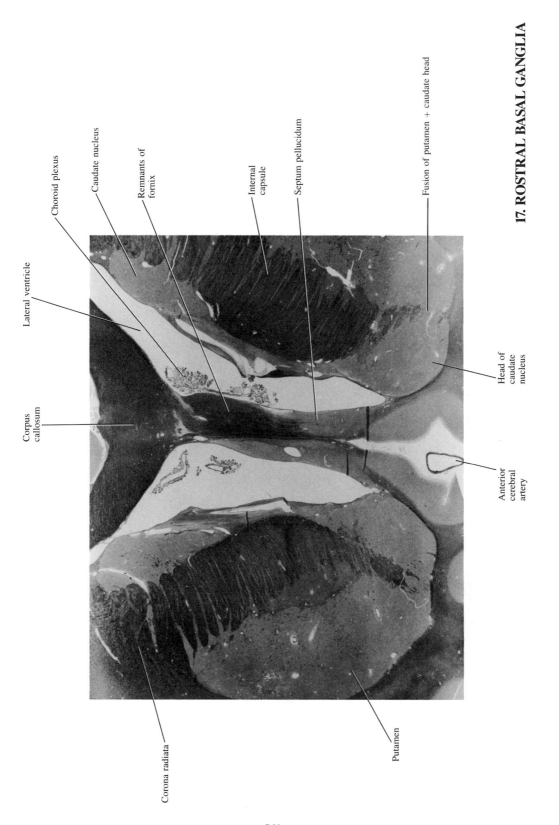

Lateral ventricle

Choroid plexus

Caudate nucleus

Remnants of fornix

Internal capsule

Septum pellucidum

Fusion of putamen + caudate head

Corpus callosum

Head of caudate nucleus

Anterior cerebral artery

Corona radiata

Putamen

17. ROSTRAL BASAL GANGLIA

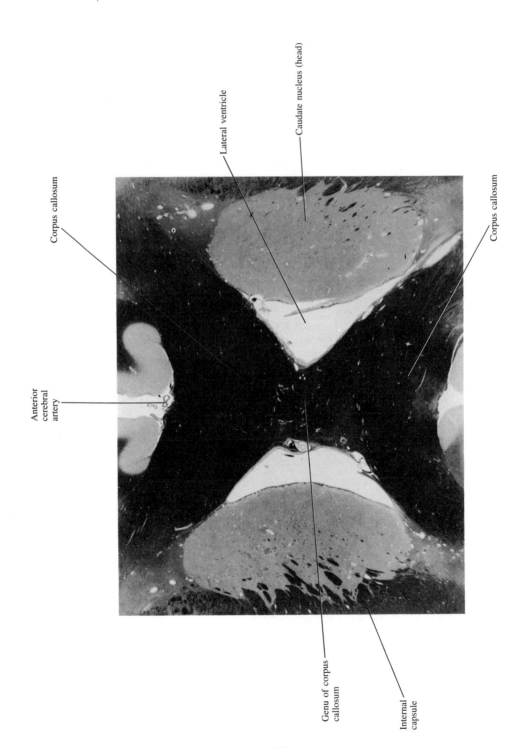

Lateral ventricle

Caudate nucleus (head)

Corpus callosum

Corpus callosum

Anterior cerebral artery

Genu of corpus callosum

Internal capsule

APPENDIX II. GLOSSARY

Abdominal reflexes. Abdominal muscle contractions on stroking the skin

Action potential (spike). An all-or-none depolarization occurring in nerve or muscle, resulting when the membrane potential reaches a threshold value

Adiadochokinesia. Inability to perform rapidly alternating movements. Also called dysdiadochokinesia

Agnosia. Lack of ability to recognize significance of sensory stimuli (auditory, visual, tactile, etc.)

Agraphia. Inability to express thoughts in writing, due to a central lesion

Air encephalography (pneumoencephalography). X-raying the ventricles and subarachnoid spaces after introduction of air by lumbar or cisternal puncture

Akinesia. Inability to start a movement; slowness in movement

Ala cinerea. Vagal triangle in floor of 4th ventricle

Alexia. Loss of power to grasp the meaning of written or printed words and sentences

Allocortex. Phylogenetic older cerebral cortex, consisting of 3 layers. Includes paleocortex and archicortex

Alpha adrenergic. Subgroup of receptors, synaptic terminals, axons, and neurons that are part of the sympathetic system and localized mostly in vascular smooth muscle and iris

Alveus. Thin layer of white matter covering ventricular surface of hippocampus

Amacrine. A long fiber (amacrine nerve cells of retina)

Ammon's horn. The hippocampus has the configuration of a ram's horn (also called cornu ammonis)

Amygdala. Like an almond. Amygdaloid nucleus or amygdala in temporal lobe of cerebral hemisphere

Analgesia. Inability to feel pain

Anarthria. Inability to pronounce words

Anesthesia. Inability to feel touch

Aneurysm. A bulge in a weak arterial wall

Angioma. A collection of abnormal arteries, capillaries, and veins

Anopsia. A defect of vision

Anosmia. Loss of sense of smell

Anterior horns. Part of the spinal gray matter containing motor nerve cells

Anticonvulsants. Drugs used to control epileptic fits

Aphasia. A defect in the power of expression by speech or of comprehending spoken or written language

Aphonia. Inability to make sounds

Apoplexy, cerebral. Old term for massive cerebral hemorrhage

Apraxia. Inability to carry out purposeful movements in the absence of paralysis

Arachnoid mater. The weblike middle layer of the meninges forming the outer boundary of the subarachnoid space

Archicerebellum. Phylogenetically old part of cerebellum

Archicortex. Three-layered cortex included in limbic system; located mostly in hippocampus and dentate gyrus of temporal lobe

Argyll Robertson pupil. An irregular pupil reacting to convergence but not to light

Arteriography (-gram). X-raying blood vessels after injecting an artery with opaque dye

Association areas. Areas whose neurons integrate sensory and other information. In parietal lobe, they integrate sensory data from somesthetic, auditory, visual, and taste areas. In temporal lobe, posterior part, they integrate visual and auditory data; in anterior part, concerned with past experiences. In frontal lobe, they deal with judgment, reasoning, abstraction, planning, behavior patterns, etc.

Association fibers. Nerve fibers connecting different areas of the cerebral cortex in the same hemisphere.

Astereognosis. Inability to recognize objects by shape and size, by touching or feeling them

Astrocyte. A neuroglial cell of ectodermal origin

Astrocytoma. A tumor arising from astrocytes

Asynergy. Disturbance of proper association in contraction of muscles that assumes that the different components of an act follow in correct sequence, at the right moment, and correct degree for accurate execution of act

Ataxia. Unsteadiness. A loss of power of muscle coordination with irregularity of muscle action

Athetosis. Slow, writhing, involuntary, bizarre movements, especially of fingers and toes

Auditory area. Located in superior temporal gyrus of temporal lobe

Aura. Warning symptoms (of an attack of epilepsy or migraine)

Autonomic. Autonomic system. Efferent or motor innervation of smooth muscle, glands, and heart muscle

Axolemma. The plasma membrane of an axon

Axon. The longest process of a nerve cell. Efferent process conducting impulses to other neurons or muscle fibers and gland cells.

Axon hillock. Region of nerve cell body from which axon arises. Contains no Nissl material

Axon reaction. Changes in cell body of a neuron after damage to its axon

Axoplasm. Cytoplasm of the axon

Babinski reflex (sign). Extension of the great toe on scratching the sole

Baroreceptor. A sensory nerve terminal that is stimulated by changes in pressure

Basal ganglia. Islands of gray matter in each cerebral hemisphere. They are associated with the postural aspect of voluntary activity and habitual activity.

Beta adrenergic. Subgroup of receptors, synaptic terminals, axons, and neurons that are part of the sympathetic system and localized in the heart, lungs, other internal organs, and in the CNS

Betz cells. Large pyramidal cells of the precentral gyrus associated with motor activity

Bitemporal hemianopia. Loss of the outer halves of both visual fields

Brachium. Denotes a large bundle of fibers connecting one part with another (in CNS)

Bradykinesia. Abnormal slowness of movements

Brainstem. Portion of brain comprising midbrain, pons, and medulla. Contains vital centers, ascending and descending tracts, nuclei of cranial nerves, and reticular formation

Broca's area. Area for speech in frontal lobe

Brown-Séquard syndrome. Signs produced by damage to one half of the cord

Bruit (intracranial). A sound in time with the heart beat

Bulbar. Concerning the medulla

Burr hole. A hole drilled in the skull

Café-au-lait patches. Brown skin marks in neurofibromatosis

Calamus scriptorius. An area in the caudal part of the floor of the 4th ventricle shaped like a pen point.

Calcar. Any spur-shaped structure. Calcar avis is an elevation on the medial aspect of the lateral ventricles, at the junction of the posterior and inferior horns. Calcarine sulcus is responsible for calcar avis

Capsule. Fibrous barrier around a tumor or abscess

Carpal tunnel. Channel in the wrist through which the median nerve passes

Cauda equina. Lumbar and sacral spinal nerve roots in lower part of spinal canal below spinal cord

Caudate nucleus. Part of corpus striatum named because of its long extension or tail

Central nervous system (CNS). Part of nervous system consisting of the spinal cord and brain (cerebrum, cerebellum, midbrain, pons, and medulla)

Cephalgia. Headache

Cerebellum. Second largest part of brain, located in postcranial fossa, below tentorium

Cerebral. Pertaining to the brain

Cerebral aqueduct (aqueduct of Sylvius). Connection between 3rd and 4th ventricles

Cerebral hemorrhage, massive. Massive bleeding into an area of brain substance, with resultant loss of function of that part of the brain

Cerebrovascular accident (CVA). Same as massive cerebral hemorrhage

Charcot's joints. Painless disorganized joints in tabes

Chiasma (-al). The X-like crossing of the two optic nerves

Chordotomy. Division (cutting) of lateral spinothalamic tract for intractable pain

Chorea. Very rapid, irregular, spasmotic involuntary movements of limbs or facial muscles attributed to degenerative changes in neostriatum

Choroid. Forming a delicate membrane. Choroid or vascular eye coat; choroid plexuses in brain ventricles

Chromatolysis. Dissolution. Dispersal of Nissl material of neurons after axon section or in viral infections of nervous system

Cinerea. Gray matter. Tuber cinereum in basal portion of hypothalamus; tuberculum cinereum in elevation on medulla due to spinal V tract and nucleus; and ala cinerea in vagal triangle in floor of 4th ventricle

Cingulum. Bundle of association fibers in white matter of cingulate gyrus on medial surface of cerebral hemisphere

Circle of Willis. The arteries at the base of the brain

Cisterns. Collection of CSF (e.g., cisterna magna)

Claustrum. Thin sheet of gray matter (unknown function) between lenticular nucleus and insula

Clava. Slight swelling on dorsal surface of medulla due to gracile nucleus

Clonus (-ic). Rhythmic contractions of a spastic muscle on stretching

Coma. Deep loss of consciousness

Commissural fibers. Nerve fibers traveling from one side of the brain to the other

Concussion (of brain). Temporary impairment of brain function due to trauma; often characterized by loss of memory of the traumatic incident, unconsciousness, temporary visual disturbances, and loss of equilibrium

Coordination. Smooth, efficient movement

Corneal reflex. Normal blinking on touching of the cornea

Corona. A crown. Corona radiata are fibers radiating from internal capsule to cerebral cortex

Corpus callosum. Large mass of commissural fibers located at the bottom of the longitudinal fissure and connecting one hemisphere to the other

Corpus striatum. Furrowed or striped. A mass of gray matter, or motor function, at base of cerebral hemisphere

Cortex. The surface layer (gray matter) of the cerebral and cerebellar hemispheres

Cortical areas. Areas of the cerebral cortex that have specific functions

Cranial nerves. Twelve pairs arising from or running to the brain and brainstem

CSF. The cerebrospinal fluid bathing the brain and spinal cord

Cuneus. A wedge-shaped gyrus on medial surface of cerebral hemisphere

Deficiency disease. One due to lack of a normal food substance

Degeneration. Death of tissue, often of unknown origin or cause

Déjà-vu. A sense of familiarity

Dementia. Deterioration of intellect

Demyelinative. Due to loss of the myelin sheath

Dendrite. A nerve cell process on which axons of other neurons terminate. Also peripheral process of a primary sensory neuron

Dentate. A nucleus of the cerebellum or a gyrus in the temporal lobe

Diabetes insipidus. Passage of large amounts of sugar-free urine

Diencephalon. Part of cerebrum consisting of epithalamus, thalamus, subthalamus, and hypothalamus. Formed from embryonic prosencephalon

Diplopia. Seeing double

Discriminative sense. Ability to detect difference in size, shape, texture, and number of stimuli

Disk (optic). The optic nerve leaving the eye, seen through an ophthalmoscope

Disorientation. Confusion as to time, place, and person

Dissociated sensory loss. Loss of pain sense with light touch preserved

Dominant hemisphere. The cerebral hemisphere controlling speech

Dorsiflexion. Bending backward (usually wrist or ankle)

Drop attacks. Sudden falls without loss of consciousness

Dura mater. The outer, toughest sheet of the meninges

Dural sinuses. Spaces in layers of the dura mater that contain venous blood

Dys-. Difficulty in; e.g., dysphasia, dysarthria (difficulty in expressing thoughts), dysmetria: disturbance of power to control range of movement in muscle action

Dystrophy. Degeneration with loss of function

Electrocorticography. Recording electrical activity direct from the cortex

Electroencephalography (EEG). Recording the electrical activity of the brain

Electromyography (EMG). Recording the electrical activity of muscle

Emboliform. In form of a plug. Nucleus of cerebellum

Encephalitis. Inflammation of the brain

Encephalopathy. A disorder of brain function

Endoneurium. Delicate connective tissue sheath around an individual nerve fiber of a peripheral nerve

Engram. A latent memory picture (psychology)

Entorhinal. The anterior part of the parahippocampal gyrus of the temporal lobe. Functions as olfactory association cortex

Ependyma. Lining epithelium of ventricles of brain and central canal of spinal cord

Euphoria. A casual cheerfulness

Epidural (extradural) hematoma. A hematoma is a somewhat localized mass of blood that has escaped from a blood vessel; the blood is usually partly clotted and discolored. An epidural hemorrhage becomes an epidural hematoma

Epilepsy. Chronic disorder of brain function, often associated with seizures and loss of consciousness

Epineurium. Connective tissue sheath around a peripheral nerve

Epithalamus. Region of diencephalon above thalamus (includes pineal gland)

Extensor plantar response. The big toe goes up when the sole is scratched

Exteroceptor. Sensory receptor serving to acquaint one with the environment

Extracranial. Outside the skull

Extradural. Between the dura and the skull

Extrapyramidal. Motor fibers arising from cells other than the pyramidal cells (often basal ganglia)

Extrinsic. Outside and separate from nervous tissue

Falx cerebri. Sheet of dura mater separating the 2 cerebral hemispheres

Fasciculation. Flickering contractions of muscle fibers

Fasciculus. A bundle of nerve fibers

Fastigial. Top of a gabled roof. Nucleus of cerebellum

Fimbria. Band of nerve fibers along the medial edge of the hippocampus, continuing as the fornix. A fringe

Flaccid. Limp, floppy, loss of tone

Flexor plantar response. The normal downward movement of the big toe when the sole is scratched

Flexor spasm. Painful contraction of muscles in spastic limbs

Focal. Arising from, or limited to, one part

Focal epilepsy. A fit affecting one part of the body, arising from one point in the brain (an attack which begins with specific muscles and then spreads to adjacent ones)

Foramen. An opening

Foramen magnum. Opening in the base of the skull (occipital bone) where the brain becomes the spinal cord

Foramina of Luschka. There are 2: they connect the 4th ventricle and the subarachnoid space

Foramen of Magendie. One of the 3 connections between the 4th ventricle and the subarachnoid space

Forceps. U-shaped configuration of fibers constituting anterior and posterior portions of corpus callosum (minor and major)

Fornix. Efferent tract of hippocampus arching over thalamus and ending in mamillary body

Fortification spectra. Zigzag patterns seen in migraine

Fossa. A compartment of the skull holding a part of the brain

Fovea. Pit or depression. Fovea centralis is depression in center of macula lutea of retina

Functional. Due to a disturbance in the working, not the structure, of some part

Funiculus. Area of white matter of cord containing several functionally different fasciculi

Ganglia. Collections of nerve cells outside the central nervous system

Generalized fits. Convulsions affecting all parts of the body

Genu: Knee. Anterior end of the corpus callosum: genu of the facial nerve

Girdle pains. Constricting pains around the trunk

Glia. The supporting cells and fibers (neuroglia) of the nervous system

Glioblastoma. The most malignant glioma

Glioma. A tumor arising from one of the neuroglial cells, usually from astrocytes. Tumors growing from the supporting cells

Gliosis. Overgrowth of glial fibers

Globus pallidus. A pale ball. Medial part of lenticular nucleus of corpus striatum

Glove-and-stocking anesthesia. Impaired sensation over the periphery of all four limbs

Gracilis. Slender. Fasciculus gracilis of cord and medulla; nucleus of medulla

Grand mal. Epilepsy characterized by major fits (an epileptic attack characterized by loss of consciousness and generalized seizures)

Granule. Grain. Small neurons of cerebellar cortex and stellate cells of cerebral cortex

Gray matter. Nervous tissue containing nerve cells (neuron cell bodies, fibers, and neuroglial cells)

Gyri. The folds of the cerebral cortex (elevated areas of cerebral cortex between the sulci)

Habenula. Small swelling in epithalamus near the posterior end of the roof of the 3rd ventricle

Hallucinations. Sensory impressions of something that is not there

Hemianopia. Loss of half of the visual field

Hemiballismus. Half-jumping. Violent form of motor restlessness

Hemiplegia (-paresis). Paralysis (weakness) of one half of the body

Heredofamilial. Passed both from generation to generation and to several members of one family

Herpes simplex. The virus of "cold sores"

Herpes zoster. Shingles

Hiatus. A large gap

Hippocampus. "Seahorse." A gyrus that is an important part of the limbic system. Produces elevation on floor of inferior horn of lateral ventricle

Homeostasis. Tendency toward stability in internal environment of organism

Homonymous. The same on both sides

Hydrocephalus. "Water head." Excessive accumulation of CSF in ventricles

Hyper-. Increased

Hypertrophy. Enlargement

Hyperventilation. Excessively deep and fast breathing

Hypo-. Decreased

Hypoglycemia. Abnormally low blood sugar

Hypopituitarism. Loss of pituitary function

Hypothalamus. A region of the diencephalon below the thalamus serving as the main controlling center of the autonomic nervous system

Idiopathic. Of unknown cause

Impulses. Electrical waves traveling along nerves

Inborn. Part of an individual's makeup

Indusium. To put on. Indusium griseum is a thin layer of gray matter on the upper surface of the corpus callosum (gray tunic)

Infarct. A zone of tissue deprived of blood supply

Infundibulum. Funnel-shaped stalk connecting the hypothalamus with the neural lobe of the pituitary gland

Insula. An island. Cerebral cortex concealed from the surface and lying at the bottom of the lateral fissure. Island of Reil

Interceptor. One of sensory end organs in viscera

Intracranial. Inside the skull

Intracranial hypertension. High pressure inside the skull (not high blood pressure)

Intrinsic. Inside the substance of the nervous system

Involuntary movement. Muscular activity not under the patient's control

Ischemia. Shortage of blood supply

Isocortex. Cerebral cortex having six layers (neocortex)

Jacksonian fits. Convulsions starting at one point and spreading to involve wider areas

Kernig's sign. Inability to straighten the knee with the hip flexed in meningeal irritation

Kinesthesia. Sense of perception of movement

Konicortex. Areas of cerebral cortex containing large numbers of small neurons (typical of sensory areas)

Labyrinth. The semicircular canals of the inner ear

Lange curve. A test carried out by adding CSF to colloidal gold solution

Lasègue's sign. Limitation of straight leg raising

Lemniscus. A ribbon or band. A bundle of nerve fibers in CNS (medial and lateral lemniscus)

Lenticular. Shape of lentil or lens. A nucleus of the corpus striatum

Leptomeninges. Arachnoid and pia mater

Leukodystrophy. Degeneration of white matter

Lightning pains. Needlelike pain in the limbs in tabes

Limbus. C-shaped configuration of cortex on medial surface of cerebral hemisphere consisting of cingulate and parahippocampal gyri. Limbic system: limbic lobe, hippocampal formation, and parts of diencephalon

Limen. Threshold. Limen insulae is basal part of insula

Lobe. A major division of the cerebral hemispheres or cerebellum based on surface marking

Local, localize. One point affected; to determine the exact point affected

Locus ceruleus. Small dark spots on either side of the floor of the 4th ventricle, marking the position of nerve cells that have melanin pigment

Longitudinal fissure. Cleft that separates the cerebrum into right and left cerebral hemispheres

Lower motor neuron. Cells and fibers of motor cranial and spinal nerves

Macroglia. Larger types of neuroglial cells (astrocytes, oligodendrocytes, and ependymal cells) of ectodermal origin

Macrosmatic. Sense of smell strongly or acutely developed

Macula. A spot. Macula lutea is a spot of posterior pole of eye having a yellow color when seen with red-free light. Maculae sacculi and utriculi are sensory areas in vestibular part of membranous labyrinth

Mamillary. Breast-shaped like nipple. Mamillary bodies are small swellings on basal surface of hypothalamus

Massa intermedia. Bridge of gray matter connecting thalami of two sides across 3rd ventricle (in 70% of brains)

Medulla. The lowest portion of the brainstem

Meninges. The 3 membranes clothing the brain and cord and lining the skull and vertebral canal

Meningioma. A benign tumor growing from the arachnoid

Meningism. Signs of irritation of the meninges, not due to infection

Meningitis. Inflammation of the meninges (usually bacterial in origin)

Meningocele. A bulge of the meninges through a breach in the bony coverings

Meningoencephalitis. Inflammation of both the brain and meninges

Meningomyelocele. A meningocele containing spinal cord tissue

Mesencephalon. The midbrain. Second of 3 primary brain vesicles

Mesocortex. Cerebral cortex intermediate between 3-layered and 6-layered cortex (in cingulate gyrus)

Mesoglia. Microglial cells of mesodermal origin

Metathalamus. Medial and lateral geniculate bodies (nuclei)

Metencephalon. Pons and cerebellum. Anterior division of rhombencephalon or posterior brain vesicle

Microglia. Neurologic cells of mesodermal origin belonging to reticuloendothelial system of cells of body

Microsmatic. A relatively poorly developed sense of smell

Mitral. Bishop's miter. Mitral cells of the olfactory bulb

Mixed nerve. One containing both motor and sensory fibers

Molecular. Denotes tissue containing large numbers of fine nerve fibers. Has a punctate appearance in silver-stained sections. Layers of cerebral and cerebellar cortices

Monoplegic (-paresis). Paralysis (weakness) of one limb

Motor. Concerned with movement

Motor areas. Located in precentral gyrus of frontal lobe; control specific skeletal muscles of opposite side of body

Multiple sclerosis. Disseminated sclerosis

Myasthenia. Weakness of muscle

Myelencephalon. Medulla oblongata. Posterior of 2 divisions of rhombencephalon or posterior primary brain vesicle

Myelin. The white sheath of nerve fibers

Myelitis. Inflammation of the spinal cord

Myelography (myelogram). X-raying the vertebral canal by introducing opaque fluid into the CSF

Myoclonus. Shocklike muscle contractions

Myopathy. Degenerative disease of muscle (muscular dystrophy)

Myositis. Inflammation of muscle

Myotonia. Contraction of muscle persisting after the need for it is passed

Neck stiffness, rigidity, retraction. Signs of irritation or infection of the meninges

Neocerebellum. Newest part of cerebellum seen in mammals (especially well developed in man). Ensures smooth muscle action in finer voluntary movements

Neocortex. Six-layered cortex, characteristic of mammals. Most of cerebral cortex in man

Neostriatum. Striped or grooved. Newer part of corpus striatum, consisting of caudate nucleus and putamen; the striatum

Nervous system. All nerve tissue of the body and associated supporting tissue. Divided into a central nervous system (CNS) and a peripheral nervous system (PNS)

Neuralgia. Pain in the distribution of a nerve

Neuraxes. A short form referring to brain and spinal cord. May refer to an axon

Neurofibril. Small nerve. Delicate filaments in cytoplasm of neurons

Neuroglia. Accessory or interstitial cells of CNS include astrocytes, oligodendrocytes, microglial cells, and ependymal cells

Neurolemma. Delicate sheath around a peripheral nerve fiber consisting of a series of neurolemma cells or Schwann cells

Neuromuscular junction. A point where a nerve fiber ends in a muscle

Neuron. The nerve cell, its fibers, and all its branches

Neuropil. Complex net of nerve cell processes occupying the intervals between cell bodies in gray matter

Neurosyphilis. Involvement of the nervous system in the later stages of syphilis

Neurotransmitter. Substance synthesized in a neuron, transported to its synaptic endings, released as a result of an action potential, and binding to receptors on the postsynaptic membrane (muscle or nerve) to produce an electrical change in the postsynaptic cell

Nociceptive. Responsive to injurious stimuli

Nucleus (-ei). A collection(s) of nerve cell bodies

Nystagmus. Rhythmic involuntary oscillation of the eyes

Obex. Barrier. Small transverse fold overhanging the opening of the 4th ventricle into central canal of closed position of medulla

Oculomotor. Concerned with eye movement

Olfactory. Concerned with the sense of smell

Oligodendrocyte. A neuroglial cell of ectodermal origin. Forms myelin sheath in CNS

Oligophrenia. Mental retardation

Operculum. Cover or lid. Frontal, parietal, and temporal opercula bound the lateral fissure of the cerebral hemisphere. Conceal insula

Ophthalmoplegia (-ic). Paralysis of eye movement

Opisthotonos. Backward arching of the whole body

Optic. Concerned with the eyes or visual pathways

Organic. Due to structural disease

Otorrhea. Running from the ear

Overflow incontinence. Constant dribbling due to overdistention of an insensitive bladder

Pachymeninx. Thick membrane. The dura mater

Paleocerebellum. Phylogenetically old part of the cerebellum functioning in postural changes and locomotion

Paleocortex. Ancient, 3-layered olfactory cortex

Paleostriatum. Ancient, striped, or grooved phylogenetically older and efferent part of corpus striatum. Globus pallidus or pallidum

Pallidum. Pale. Globus pallidus of corpus striatum. Medial part of lenticular nucleus.

Pallium. Cloak. Cerebral cortex with subjacent white matter. Sometimes used synonymously with cortex

Panencephalitis. Inflammation of the whole brain

Papilledema. Swelling of the optic nerve seen with an ophthalmoscope

Para-. Alongside (e.g., paraventricular)

Paraplegia (paraparesis). Paralysis (weakness) of both legs and lower part of trunk

Parkinsonism. The tremor and rigidity of Parkinson's disease

Perikaryon. Cytoplasm around the nucleus. Sometimes refers to cell body of a neuron

Perineurium. Connective tissue sheath surrounding a bundle of nerve fibers in a peripheral nerve

Peripheral nervous system (PNS). Nerves, ganglia, and special receptors

Pes cavus. Very high arches of the feet

Petit mal. Minor epilepsy; frequent episodes of detachment from one's surroundings (an epileptic attack characterized by brief loss of consciousness and sometimes mild facial twitchings)

Photophobia. A dislike of light

Pia mater. Thin innermost layer of meninges, attached to surface of brain and spinal cord. Forms inner boundary of subarachnoid space

Pineal gland. A structure lying in the center of the skull, frequently calcified and visible on x-ray. A part of epithalamus of diencephalon

Plantar reflex. Toe movement when the sole is scratched

Plaques. Patches; usually applied to areas of demyelination in disseminated sclerosis

Pneumoencephalography. *See* Air encephalography

Polioencephalitis. Brainstem infection of poliomyelitis virus

Poly-. Many (e.g., polyneuritis, inflammation of many nerves)

Pons. Bridge. Part of the brainstem lying between the medulla and midbrain. A bridge between right and left halves of cerebellum

Position (postural) sense. Knowledge of where each part of the body is without looking at it

Posterior columns. Spinal cord tracts carrying position, vibration, and discriminative sensations

Posterior horns. Part of the spinal gray matter receiving sensory roots

Premotor areas. Areas in frontal lobe just anterior to motor areas; control general muscular activity of opposite side of body

Presenile dementia. Cerebral atrophy in middle age

Pressure cone. The forcing of brain tissue through a foramen as a result of high pressure above, e.g., tentorial or cerebellar pressure cones

Projection fibers. Nerve fibers traveling to or from lower parts of the brain and spinal cord

Proprioceptor. One of the sensory endings in muscles, tendons, and joints. Gives information about movement and position of body parts

Prosencephalon. Forebrain consisting of telencephalon (cerebral hemispheres) and diencephalon. Anterior primary brain vesicle

Pseudohypertrophy. Apparent, but not true, enlargement

Psychomotor epilepsy. Disturbance of behavior due to epileptic discharges

Psychoneurotic. Of psychologic, not disease, origin

Ptosis. Drooping of the eyelid

Pulvinar. Posterior projection of thalamus beneath which the medial and lateral geniculate bodies are found

Putamen. Shell. The larger and lateral part of the lenticular nucleus of the corpus striatum

Quadriplegia. Paralysis of all 4 limbs

Queckenstedt test. Rise in lumbar CSF pressure when the jugular vein is compressed

Reflex. An automatic response to a stimulus

Relapse. A further attack of a disease

Remission. A period of recovery from a disease

Restiform. Rope form. Restiform body: main mass of inferior cerebellar peduncle containing afferent fibers to cerebellum

Retrobulbar. Behind the eye

Rhinal. Related to the nose

Rhinencephalon. In man, refers to components of the olfactory system

Rhinorrhea. Running from the nose

Rhombencephalon. The pons and cerebellum (metencephalon) and medulla (myelencephalon). Posterior primary brain vesicle

Rigidity. Stiffness due to equal resistance in all muscles

Roots. Nerve fibers as they leave or enter the stem or cord

Rostrum. Recurved part of corpus callosum passing backward from the genu to the lamina terminalis

Saltatory conduction. Electrical conduction along myelinated nerves in which the nerve impulse ''jumps'' from one node of Ranvier to the next

Satellite. Satellite cells: flattened cells of ectodermal origin, forming a capsule for nerve cell bodies in dorsal root ganglia and sympathetic ganglia. Satellite oligodendrocytes adjacent to nerve cell bodies in CNS

Scanning (brain-). Detecting the distribution of a radioactive substance in the brain

Schwann cells. Derived from neural crest; responsible for ensheathing and myelinating peripheral axons

Sclerosis. Hardening

Scotoma. A patch of blindness

Sella turcica. A saddle-shaped cavity at the base of the skull containing the pituitary gland

Sensory. Concerned with feeling

Sensory level. The point where sensation changes from abnormal to normal

Septal area. Cortex and underlying septal nuclei, the latter extending into the septum pellucidum beneath genu and rostrum of the corpus callosum on the medial side of the frontal lobe. Also called the medial olfactory area

Septum pellucidum. Triangular double membrane separating the anterior horns of the lateral ventricles. Fills interval between the corpus callosum and forms in the median plane

Smell areas. Located in the gyrus on the medial surface of the temporal lobe

Somatic. Denotes the body, exclusive of viscera

Somesthetic. The consciousness of having a body. Somesthetic senses are general senses of pain, temperature, touch, pressure, position and movement, and vibration

Space-occupying lesion. A tumor or other growing lesion

Spastic ataxia. A combination of spasticity and unsteadiness

Spinal cord. Part of the CNS in the spinal canal. Structure: gray matter on inside, white matter on outside

Spinal cord segments. There are 8 cervical, 12 thoracic, 5 lumbar, 5 sacral, and 1 coccygeal (31 in all). Each segment gives rise to a pair of spinal nerves

Splenium. Thickened posterior extremity of the corpus callosum

Spondylosis. Degenerative changes in bones and the disk of the spine

Status epilepticus. Fits following each other in rapid succession

Stenosis. A narrowing

Stereotaxis. Accurately placing small lesions in the depths of the brain

Strabismus. A squint

Stretch reflex. Contraction of muscle following its sudden stretching

Stria terminalis. Furrow limit. A slender strand of fibers running along the medial side of the tail of the caudate nucleus, originating in amygdaloid nucleus and mostly ending in the septal area and hypothalamus

Striatum. Furrowed. More recent part of the corpus striatum (neostriatum) consisting of the caudate nucleus and putamen

Stroke, cerebral. Same as massive cerebral hemorrhage

Stupor. Unconsciousness, but rousable

Subarachnoid space. Between arachnoid and pia mater containing CSF

Subdural. Between dura and arachnoid

Subdural hemorrhage. Bleeding, usually due to trauma (blow to the head), in which veins entering sinuses are broken, e.g., cerebral veins entering the sagittal sinus. Subdural hemorrhage is usually venous in origin

Subiculum. A layer. Transitional cortex between the parahippocampal gyrus and the hippocampus

Substantia gelatinosa. Column of small neurons of the apex of the dorsal gray horn throughout the spinal cord

Substantia nigra. A large nucleus, with motor functions, in the midbrain. Many cells contain melanin

Subthalamus. Region of the diencephalon beneath the thalamus, containing fiber tracts and subthalamic nucleus

Sulci. Furrows on the surface of the brain

Superficial reflex. Various muscle contractions following stimuli to the skin surface

Symptomatic. Representing a disease process

Synapse. Site of contact between neurons, at which site one neuron is excited or inhibited by another neuron

Syndrome. A collection of signs and symptoms recurring frequently enough to be recognizable

Syringomyelia. A condition characterized by central curvature of the spinal cord and gliosis around the cavity

Syrinx. A cavity in the brainstem or spinal cord

Tapetum. A carpet of fibers of the corpus callosum sweeping over the lateral ventricle and forming the lateral wall of its posterior and inferior horns

Taste area. Located on the ventral aspect of the postcentral gyrus of the parietal lobe

Tectum. Roof of the midbrain consisting of paired superior and inferior colliculi

Tegmentum. To cover. The dorsal portion of pons. Also the major part of the cerebral peduncle of midbrain between the substantia nigra and tectum of the midbrain

Teichopsia. Flashes of light in migraine

Tela choroidea. Like a membrane. The vascular connective tissue, continuous with that of pia mater, which continues into core of choroid plexuses

Telencephalon. Cerebral hemispheres. Anterior of the 2 divisions of the prosencephalon or anterior primary brain vesicle

Telodendrea. Terminal branches of axons

Tentorial herniation. Forcing part of the brain through the tentorial hiatus

Tentorium (cerebelli). Tentlike sheet of dura separating the cerebellum from the cerebral hemispheres (occipital lobes)

Thalamus. Large mass of gray matter of diencephalon in each cerebral hemisphere, bordering on the 3rd ventricle with sensory and integrative functions

Theca. The space formed by the meninges containing CSF

Tic. A recurrent spasm (e.g., tic douloureux)

Tinnitus. Ringing in the ears

Todd's paralysis. Temporary paralysis of a part following a fit

Tone. The tension present in a muscle at rest

Torcular. To twist. Confluence of dural venous sinuses at internal occipital protuberance

Tracts. Collection of nerve fibers in the CNS having similar functions

Trapezoid body. Transverse fibers of the auditory pathway found at the junction of the dorsal and basal portions of the pons

Trophic. Changes occurring in tissues that have lost their nerve supply

Uncinate. Hook-shaped. Uncinate fasciculus: association fibers connecting cortex of the basal surface of the frontal lobe with the temporal pole; a bundle of fastigiobulbar fibers (of Russell) that curves over the superior cerebellar peduncle on the way to the inferior cerebellar peduncle

Uncus. A hook. Portion of the rostral end of the parahippocampal gyrus of the temporal lobe. A landmark for lateral olfactory area

Upper motor neuron. The motor cells and fibers running from the cerebral cortex to cranial nerve nuclei and anterior horn cells

Uvula. Portion of inferior vermis of cerebellum

Vallecula. Midline depression on the inferior aspect of the cerebellum

Vagus. Tenth (X) cranial nerve. Once called the pneumogastric nerve.

Ventral horn. Ventral part of spinal cord gray matter, in which motor neurons are found

Ventricles. Cavities in the brain containing CSF [fluid-filled (CSF) interconnected spaces in the brain, 4th, 3rd, and 2 lateral ventricles]

Ventriculography. X-raying the ventricles by injecting air directly into them

Vermis, worm. Midline portion of cerebellum

Vertigo. A sense of rotation

Vestibular. Concerned with the inner ear labyrinth and its connections

Vibration sense. Ability to detect tuning fork vibration

Visual field. That part of the visual space ''seen'' by each eye

White matter. The part of the brain and spinal cord that has a light appearance and contains myelinated fibers and neuroglial cells

Xanthochromic. Yellow colored

Zona incerta. Gray matter in the subthalamus, representing the rostral extension of the reticular formation of the brainstem

APPENDIX III. REFERENCES

Affifi, A. K., and Bergman, R. A.: *Basic Neuroscience: A Structural and Functional Approach,* 3rd ed. Urban and Schwartzenberg, Baltimore, 1986.

Allen, D. J.; Allen, J. S.; DiDio, L. J. A.; and McGrath, J. A.: Scanning Electron Microscopy and X-ray Microanalysis of the Human Pineal Body with Emphasis on Calcareous Concretions. *J. Submicrosc. Cytol.,* **13:**675–95, 1981.

Barr, M. L., and Kiernan, J. A.: *The Human Nervous System: An Anatomical Viewpoint,* 4th ed. Harper and Row, Publishers, Inc., New York, 1983.

Bullock, T. H.; Orkand, R.; and Grinnell, A.: *Introduction to the Nervous System.* W. H. Freeman and Company, San Francisco, 1977.

Carpenter, M. B.: *Human Neuroanatomy,* 8th ed. The Williams and Wilkins Company, Baltimore, 1983.

Carpenter, M. B.: *Core Text of Neuroanatomy,* 3rd ed. The Williams and Wilkins Company, Baltimore, 1985.

Chusid, J. G.: *Correlative Neuroanatomy and Functional Neurology,* 19th ed. Lange Medical Publications, Los Altos, California, 1985.

Cooper, J. R.; Bloom, F. E.; and Roth, R. T.: *The Biochemical Basis of Neuropharmacology,* 3rd ed. Oxford University Press, New York, 1978.

Daube, J. R.; Sandock, B. A.; Reagan, T. L.; and Wesmoreland, F. F.: *Medical Neurosciences,* 2nd ed. Little Brown & Company, Boston, 1986.

Dunkerley, G. B.: *A Basic Atlas of the Human Nervous System.* F. A. Davis Company, Philadelphia, 1975.

Escourocle, R., and Poirier, J.: *Manual of Basic Neuropathology,* 2nd ed. W. B. Saunders Company, Philadelphia, 1978.

Fitzgerald, M. J. T.: *Neuroanatomy, Basic and Applied.* Bailliere Tindall Company, Philadelphia, 1985.

Gilman, S., and Newman, S. W.: *Manter and Gatz's Essentials of Clinical Neuranatomy and Neurophysiology,* 7th ed. F. A. Davis Company, Philadelphia, 1987.

Guyton, A. C.: *Textbook of Medical Physiology,* 7th ed. W. B. Saunders Company, Philadelphia, 1986.

Guyton, A. C.: *Basic Neuroscience,* W. B. Saunders Company, Philadelphia, 1987.

Heimer, L.: *The Human Brain and Spinal Cord, Functional Neuroanatomy and Dissection Guide.* Springer-Verlag, New York, Heidelberg, Berlin, 1983.

House, E. L.; Pansky, B.; and Siegel, A.: *A Systematic Approach to Neuroscience,* 3rd ed. McGraw-Hill Book Company, New York, 1979.

Jensen, D.: *The Principles of Physiology,* 2nd ed. Appleton-Century-Crofts, New York, 1980.

Kandel, E. R., and Schwartz, J. H.: *Principles of Neural Science,* 2nd ed. Elsevier/North-Holland, New York, 1985.

Kiernan, J. A.: *Introduction to Human Neuroscience.* J. B. Lippincott Company, Philadelphia, 1987.

Kuffler, S. W., and Nicholls, J. G.: *A Cellular Approach to the Function of the Nervous System.* Sinauer Associates, Inc., Sunderland, Mass., 1976.

Nauta, W. J. H., and Feirtag, M.: *Fundamentals of Neuroanatomy.* W. H. Freeman and Company, New York, 1986.

Nieuwenhuys, R.; Voogd, J.; and van Huijzen, C.: *The Human Central Nervous System: A Synopsis and Atlas.* Springer-Verlag, New York, Heidelberg, Berlin, 1978.

Nobak, C. R., and Demarest, R. J.: *The Human Nervous System: Basic Principles of Neurobiology,* 3rd ed. McGraw-Hill Book Company, New York, 1986.

Nolte, J.: *The Human Brain: An Introduction to Its Functional Anatomy.* C. V. Mosby Company, St. Louis, 1981.

Otsuka, M., and Hall, Z. W. (eds.): *Neurobiology of Chemical Transmission*. John Wiley & Sons, Inc., New York, 1979.

Peele, T. L.: *The Neuroanatomic Basis for Clinical Neurology,* 3rd ed. McGraw-Hill Book Company, New York, 1977

Robbins, S. L., and Kumar, V.: *Basic Pathology,* 4th ed. W. B. Saunders Company, Philadelphia, 1987.

Romero-Sierra, C.: *Neuroanatomy: A Conceptual Approach*. Churchill Livingstone, New York, 1986.

Schade, J. P., and Ford, D. H.: *Basic Neurology,* 2nd ed. Elsevier Scientific Publishing Company, New York, 1973.

Schmidt, R. F. (ed.): *Fundamentals of Neurophysiology,* 3rd ed. Springer-Verlag, New York, Heidelberg, Berlin, 1985.

Scientific American: The Brain, Vol. 241, No. 3, September, 1979.

Snell, R. S.: *Clinical Neuroanatomy for Medical Students,* 2nd ed. Little Brown & Company, Boston, 1987.

Eleventh International Anatomical Nomenclature Committee in Mexico City: *Nomina Anatomica,* 5th ed. Excerpta Medica Foundation, Amsterdam, 1980.

Thorek, P.: *Anatomy in Surgery*. J. B. Lippincott Company, Philadelphia, 1951.

Vander, A. J.; Sherman, J. H.; and Luciano, D. S.: *Human Physiology: Mechanisms of Body Function,* 4th ed. McGraw-Hill Book Company, New York, 1985.

Walton, J. N.: *Essentials of Neurology,* 4th ed. J. B. Lippincott Company, Philadelphia, 1975.

Warwick, R., and Williams, P. L.: *Gray's Anatomy,* 36th British ed. W. B. Saunders Company, Philadelphia, 1980.

Watson, C.: *Human Neuroanatomy: An Introductory Atlas,* 3rd ed. Little Brown & Company, Boston, 1985.

Wilkinson, J. L.: *Neuroanatomy for Medical Students*. PSG Publishing Company, Littleton, Mass., 1986.

Williams, P. L., and Warwick, R.: *Functional Neuroanatomy of Man*. W. B. Saunders Company, Philadelphia, 1975.

Willis, W. D., Jr., and Grossman, R. G.: *Medical Neurobiology,* 3rd ed. C. V. Mosby Company, St. Louis, 1981.

INDEX

Figures in **boldface** type refer to pages on which illustrations appear.

thoracic disk prolapse, 486, **487**
Disk disease, 486, **487**
Disorientation, 566
Dissociated sensory loss, in hydrocephalus, 566
Dizziness, 308
DNA viruses, 540
Dominant hemisphere, **213**, 566
Dorsal longitudinal fasciculus, 180, **181**
Dorsal nucleus, 68, **69**
Down's syndrome, 462
Drowsiness, 452
Duchenne, muscular dystrophy, 512, **513**
Ductus reuniens, **47, 291**
Dura mater, 100, **101, 109,** 566
  cerebrospinal fluid circulation, 108, **109**
  innervation, 100, **101**
  layers, 100, **101, 109**
  spinal, 64, **65**
Dural sinuses, venous, 100, **101, 109,** 566
Dysarthria, 474, 478, 508
Dysdiadochokinesis, 254
Dysgenesis, 464
Dyslexia, 212, 324
Dysmetria, 254
Dysraphic states, 462
Dysraphic theory, in syringomyelia, 466
Dystonia, 194, **195**
Dystrophy, **513,** 566

Ear, anatomy, 290, **291**
  congenital malformations, 52, **53**
  development, 46–**51**
    external, 48, **49**
    internal, 46–**49**
    middle, 50, **51**
  inner, 290, **291, 297**
    receptors, 296, **297**
    vestibular system, 46–**49**
  middle, 290, **291, 295**
  outer, 290, **291, 295**
Eardrum (tympanic membrane), 48, **49,** 290,
  291, 294, **295**
Echinococcus, 534
Economo's cortical map, **203,** 204
Edinger-Westphal nucleus, 84, **85**
Effectors, 56
Efferent groups, general visceral, lateral, 14–**19**
  somatic, medial, 14–**19**
  special visceral, intermediate, 14–**19**
Efferent pathways, 402–**411**
Electric charges, membrane potentials, 338–**341**
Electric field, 340, **341**
Electrical force, 338, **339**
Electrical phenomena, 338, **339**
Electroencephalogram (-graphy), 456, **457,** 566
Electromagnetic radiation, 310
Electromyography, 566
Emboli, 440, 442, 566
Emboliform nucleus, **240, 243,** 248, **249,** 566
Embryology, of CNS. *See* Development
Embryonic disk, 4, **5**

Emmetropic, 312
Encephalitidies, arboviruses, 536
  mosquito-borne, 536
  tickborne, 536
  of DNA viruses, 540
  enteroviruses, 536
  of RNA viruses, 536, 538
  of slow viruses, 540
  viral, 536
Encephalitis, 530, 556, 566
  lethargica, 540
  viral, 536, 538, 540
Encephalocele, 462
Encephalomyelitis, 536
Encephalon, 90, **91**
Encephalopathy, 508, 566
  early vascular, 470
  postepileptic, 470
Encephalotrigeminal angiomatosis, 468
End brain, 90, **91**
End-bulbs of Krause, **237,** 378, **379**
  alpha efferents, **295**
Endings, annulospiral, 416, **417**
  "en plaque," **419**
  flower spray, 416, **417, 419**
  gamma efferents, **419**
  joint, **417**
  "trail-en grappe," **419**
Endolymph, 46, **47, 51,** 290, **291,** 300, **301**
Endolymphatic duct, 290, **291,** 300, **301**
Endoneurium, 566
Energy, 338, **339**
Engram, 216, 566
  dynamic, 216
  structural, 216
Entorhinal cortex, **267, 269,** 566
Ependymal cells, 8, **9,** 566
  layer, 8, **9**
Ependymoma, 520, 524
Epidemic encephalitis of von Economo, 558
Epidermoids, 526
Epidural, hematoma, 450, 566
  space, 100
Epiglottis, 284, **285**
Epilepsy, 456–**459,** 566
  fits (seizures), 456
  idiopathic, 456
  posttraumatic, 452
  symptomatic, 456
Epileptic seizures (fits), 456, 458
  focal epileptic, 458
  grand mal, 456
    subphases, 456
  jacksonian, 458
  myoclonic, 458
  petit mal, 458
  psychomotor, 458
  status epilepticus, 458
Epineurium, 566
Epithalamus, 80, **81,** 566
  pineal body, 80, **81,** 170, **171**

Sodium-potassium pump, 346, **347**, 350, **351**
Soma, 56, 356, **357**
Somatic afferent group, 142, **143**, 378–**395**
Somatic afferent system, general. *See* General
 Somatic afferent (GSA) system
Somatic efferent system, 404, **405**, 406, **407**
Somesthetic cortex, 208, **209**
Somites, 4, **5**, 10, **11**
Somnambulism, 220
Sound, decibel ratings, **293**
 frequency and pitch, **293**
 transmission, **293**, 294
Sound waves, 292, **293**, 296, **297**
Space-occupying lesions, 518, 572
Spasmotic dystonia, **195**
Spastic paraparesis, 486
Spastic paraplegia, 480
Spatial facilitation, **369**
Spatial localization, 386
Special somatic afferent fibers, 136, 152
Special visceral afferent fibers, 136
Special visceral efferent fibers, 408, **409**
Speech areas, **211**
Sphingolipidoses, 506
Spikes. *See* Action potentials
Spina bifida, 12, **13**
Spina bifida aperta, 462
Spina bifida occulta, 12, **13**, 462
Spinal centers and connections, **413**
Spinal cord, 62–**79**, 573
 alar plates, 6, 7. *See also* Alar plates,
 arteries, 78, **79**, 114, **117**, 118
 astrocytes, 58, 60
  fibrillar, 58
  protoplasmic, 58
 basal plates, 6, 7. *See also* Basal plates
 blood supply, 78, **79**
 brachial plexus, 64, **65**, 74, **75**. *See also*
  Brachial plexus
 central canal, 66, **67**, **107**, **109**
 cervical plexus, 72, **73**. *See also* Cervical
  plexus
 congenital malformation, 12, **13**
 conus medullaris, **63**, 64, **65**
 coverings, 64, **65**
 development of, 6–**13**
 floor plate, 6, **7**
 glial cell differentiation, 58, 60
 gray matter and white matter, 66, **67**
 hemisection, **63**
 horns, 66–**69**. *See also* Horn(s)
  intermediate, lateral, 66–**69**
  motor, anterior, 66–**69**
  sensory, posterior, 66–**69**
 interneuron control, 426, **427**
 lesions, 432–**435**
  amyotrophic lateral sclerosis, 434, **435**
  anterior gray horns (paralysis), 434, **435**
  hemisection (Brown-Sequard syndrome),
   434, **435**
  locomotor ataxia, 432, **433**

syringomyelia, 432, **433**
 tabes dorsalis, 432, **433**
 transection (paraplegia), 432
 ventral nerve roots (poliomyelitis), 434, **435**
location and length, 12, **13**, 62, **63**
lumbosacral plexus, 64, **65**. *See also*
 Lumbosacral plexus
malformations, 462
mantle layer, 6, **7**, 8, **9**
marginal layer, 6, **7**, 8, **9**
mesenchyme cells, 6, **7**, 8, **9**
morphology, 64, **65**, 66, **67**
 cauda equina (tail of a horse), **63–65**
 cervical enlargement (brachial plexus), 64,
  65
 cervical (8) segments, 64, **65**
 coccygeal (1) segments, 64, **65**
 lumbar enlargement (lumbosacral plexus),
  64, **65**
 lumbar (5) segments, 64, **65**
 sacral (12) segments, 64, **65**
 segments, 62, **63**, 64, **65**
 thoracic (12) segments, 64, **65**
motor plates, 6, **7**, 14, **15**, 18, **19**, 22, **23**. *See
 also* Basal plates
myelination, 10, **11**
nerve. *See* Nerve(s)
nerve cell differentiation, 8, **9**
neural crest cells, 10, **11**, 34–**37**
neural tube, 6, **7**, 10, **11**
neuroblasts, 6, **7**, 8, **9**
nuclei, 68, **69**
oligodendroglia cells, 60
plates, alar (sensory), 6, **7**, 16–**19**, 22, **23**
 basal (motor), 6, **7**, 14, **15**, 18, **19**, 22, **23**
 floor, 6, **7**
 roof, 6, **7**, 16, **17**, 24, **25**
segments, 66, **67**, 573
sensory plates. *See* Alar plates
spinal nerves, 72, **73**
sulcus limitans, 6, **7**, **9**
tracts, 70, **71**
tumors, 524
venous drainage, 78, **79**
Spinal fluid. *See* Cerebrospinal fluid
Spinal hereditary amyotrophy, 480
Spinal medullary degenerative diseases, 478
Spinal nerves. *See* Nerve(s)
Spinal nuclei, 68, **69**
Spinal sensory fibers, 416, **417**
Spinal trauma, 454
Spinal vascular disease, 448
Spindle, of muscle, 416, **417**
Spine. *See* Spinal cord
Spinocerebellar tracts, 70, **71**, 252, 396, **397**,
 545
Spinoreticular, 262, **263**
Spinotectal tract, **71**
Spinothalamic tract, **67**, **71**, 88, **89**, 378, **379**,
 390, **391**, **545**
Spinous processes, 62, **63**

Spitz-Holter valve, 466, 490
Splanchnic nerves, 228, **229**
Splenium, of corpus callosum, **81**, 94, **95**, 573
Spondylosis, 486, 573
Spongy degeneration, leukodystrophies, 482
Staining techniques, del Rio-Hortega's silver
  method, 60
  gold sublimate method of Cajal, 58
  hematoxylin and eosin (H & E), 58
Stapedius, 50, **295**
Stapes, 290, **291, 295**
  development, 50, **51**
Status epilepticus, 573
Steatorrhea, 500
Steele-Richardson-Olszewski disease, 494
Stellate cells, 200, **201**, 246, **247**
Stepping response, 426
Stereocilia, 300, **301**
Stereopsis (solid vision), 324
Stimulus, adequate, 376
Stomach, surgically removed, 500
Stratum, calcarium, 98, **99**
  lacunosum, **267**
  moleculare, **267**
  oriens, **267**
  radiatum, **267**
Stretch extensor reflexes, 420, **421**, 426
Stria(e), acoustic, **299**
  longitudinal, **81**, 94, **95, 99**, 266, **271**
  medullaris, **87, 89, 557**
  medullaris thalami, 80, **81, 169, 265, 275**
  olfactory, 278–**281**
  terminalis, **187, 265, 275**, 573
Striate (visual) cortex, 210, **211**
Striatonigral degeneration, 494
Striatum, 186, **187**, 573
Stroke, 440, 573
Strumpell-Lorraine disease, 478
Stupor, 452, 573
Sturge-Weber's disease, 468
Subacute combined degeneration, **501**
Subacute necrotizing encephalopathy, 506
Subarachnoid cisternae, 110, **111**. *See also*
  Cisterns
Subarachnoid hemorrhage, 444, **445**, 470
  treatment, 444
Subarachnoid pillars, **103**
Subarachnoid space, of brain, 102, **103**, 573
  cerebrospinal fluid circulation, 108, **109**
  of spinal cord, 64, **65**
Subcallosal gyrus, **81**, 94, **95**
Subclavian artery thrombosis, 440
Subclavian steal syndrome, 440
Subcortical cerebellar nuclei, 248, **249**
Subcortical degeneration, 492, 494
Subdural space, of brain, 100, **101, 109**
  of spinal cord, 64, **65**
Subicular complex, 266, **267**
Subiculum, 266, **267**, 573
Sublingual gland, 148, **149**, 226, **227**, 231–235
Submandibular ganglion, 166

Submandibular gland, 148, **149**, 226, **227**, 231–
  235
Substantia gelatinosa, 68, **69**, 378, **379**, 573
Substantia nigra, **81**, 84, **85, 411**, 190, **191**, 492,
  554, 556, 573
Subthalamus, 80, **81**, 176, **177**, 573
  subthalamic fibers, **177**
  subthalamic nucleus, **81**, 176, **177**
  atrophy, 494
Subthreshold stimulus, 352, **353**
Sudanophilic leukodystrophies, 482
Sulcus(i), calcarine, 97, **267**
  central, **91, 93**
  cingulate (cinguli), 94, **95**
  circular, 92, **93**
  collateral, **95–97**
  of corpus callosum, **95**
  cruciate, **83**
  fimbriodentate, **267**
  hippocampal, **95, 267**
  hypothalamic, **169**
  inferior temporal, 92, **93**
  interparietal, 92, **93**
  lateral, **91, 93, 97**
  limitans, 6, **7**, 86, **87, 547**
  median, 86, **87**
  occipital, 92, **93**
  olfactory, 96, **97**
  orbital, 96, **97**
  parietal, 92, **93**
  parieto-occipital, **91, 95**
  parolfactory, **95**
  postcentral, 92, **93**
  posterior intermediate, of cord, **67, 545**
  posterior lateral, of cord, **67**
  posterior median, of cord, 66, **67, 545**
  posterolateral, 96, **97, 545**
  preolivary, 96, **97**
  rhinal, 96, **97**
  subparietal, **95**
Sulcus limitans, 6, **7, 9, 87, 89**
Sulfatidoses, 506
Sulphatides, 482
Sulphonamides, 530
Summation, spatial, 368, **369, 373**
  temporal, 368, **369, 373**
Superior cerebellar peduncle, 86, **87**, 240, **241**
Superior colliculus. *See* Colliculus, anterior
Superior fovea (locus coeruleus), **87, 89**
Superior salivatory nucleus. *See* Salivatory
  nucleus
Suprarenal gland, cortex, 38, **39**
  congenital, 38, **39**
  development, 38, **39**
  hyperplasia of, 38, **39**
  medulla, 38, **39**
Supraspinal centers and connections, 413
Suprathreshold stimulus, 352, **353**
Sustentacular cells, 278, **279**
Swallowing, 424
Sydenham's chorea, 194, 492